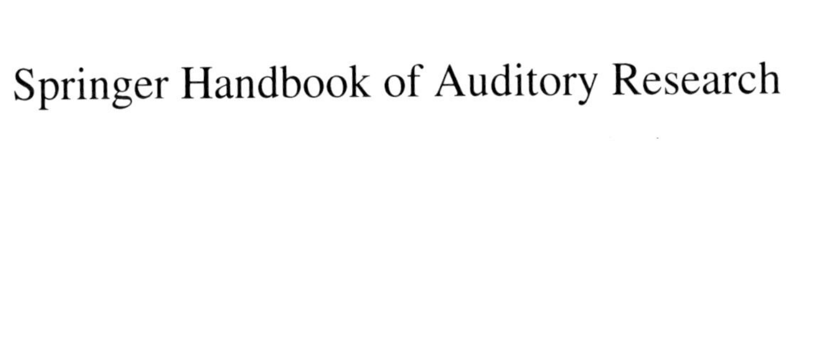

Springer Handbook of Auditory Research

For further volumes:
http://www.springer.com/series/2506

Colleen G. Le Prell · Donald Henderson
Richard R. Fay · Arthur N. Popper
Editors

Noise-Induced Hearing Loss

Scientific Advances

Editors
Colleen G. Le Prell
Department of Speech, Language,
and Hearing Sciences
University of Florida
Gainesville, FL 32610, USA
colleeng@phhp.ufl.edu

Richard R. Fay
Marine Biological Laboratory
Woods Hole, MA 02543, USA
rfay@luc.edu

Donald Henderson
Department of Communication
Disorders and Science
Center for Hearing Research
College of Arts and Sciences
State University of New York at Buffalo
Buffalo, NY 14214, USA
donaldhe@buffalo.edu

Arthur N. Popper
Department of Biology
University of Maryland
College Park, MD 20742, USA
apopper@umd.edu

ISSN 0947-2657
ISBN 978-1-4419-9522-3 e-ISBN 978-1-4419-9523-0
DOI 10.1007/978-1-4419-9523-0
Springer New York Dordrecht Heidelberg London

Library of Congress Control Number: 2011934032

Cover image: Figure 2, page 63, shows an example of expansion pattern of a hair cell lesion.

Printed on acid-free paper

Springer is part of Springer Science+Business Media (www.springer.com)

Series Preface

The Springer Handbook of Auditory Research presents a series of comprehensive and synthetic reviews of the fundamental topics in modern auditory research. The volumes are aimed at all individuals with interests in hearing research including advanced graduate students, post-doctoral researchers, and clinical investigators. The volumes are intended to introduce new investigators to important aspects of hearing science and to help established investigators to better understand the fundamental theories and data in fields of hearing that they may not normally follow closely.

Each volume presents a particular topic comprehensively, and each serves as a synthetic overview and guide to the literature. As such, the chapters present neither exhaustive data reviews nor original research that has not yet appeared in peer-reviewed journals. The volumes focus on topics that have developed a solid data and conceptual foundation rather than on those for which a literature is only beginning to develop. New research areas will be covered on a timely basis in the series as they begin to mature.

Each volume in the series consists of a few substantial chapters on a particular topic. In some cases, the topics will be ones of traditional interest for which there is a substantial body of data and theory, such as auditory neuroanatomy (Vol. 1) and neurophysiology (Vol. 2). Other volumes in the series deal with topics that have begun to mature more recently, such as development, plasticity, and computational models of neural processing. In many cases, the series editors are joined by a co-editor having special expertise in the topic of the volume.

Richard R. Fay, Falmouth, MA
Arthur N. Popper, College Park, MD

Volume Preface

Hearing impairment is the third-most-prevalent chronic disability in the United States, with a major portion of the loss related to exposure to noise in the environment. Hearing loss attributed to noise is called noise-induced hearing loss (NIHL). This volume describes the effect of environmental noise on hearing, provides important background on the subject, and also explores the broader issues currently arising on effects of noise on nonhuman vertebrates.

In Chapter 1, Le Prell and Henderson provide an introduction that outlines the problem. The first section of the book (Chapters 2 and 3) more carefully defines the scope of the problem. Rabinowitz, in Chapter 2, describes the prevalence and significance of NIHL and the public health importance of this health issue. More specific concerns about exposure to sound in the military are discussed in Chapter 3 by Grantham.

The second section of the book (Chapters 4–8) details the relationships among noise exposure and anatomical trauma, physiological changes, and perceptual deficits. In Chapter 4, Henderson and Hamernik review the classic measures of sound, such as sound pressure level, frequency, and duration, and their relationship to NIHL. Importantly, this chapter extends this discussion to the special hazards of impact/impulse noise. Hu, in Chapter 5, describes noise-induced pathological changes in the cochlea, including both apoptotic and necrotic pathways of cell death and the relationship between pathology and hearing loss. In Chapter 6, Young describes the impact of noise on the auditory nerve, including implications for loudness, pitch, and temporal coding.

In Chapter 7, Shrivastav reviews known psychophysical changes associated with NIHL in detail, specifically including processing of speech in noise, and contrasts the changes that occur subsequent to noise exposure with those that occur as a consequence of age-related hearing loss (ARHL). In the last chapter in this section, Kaltenbach and Manz (Chapter 8) carefully describe the effects of noise on the central nervous system (CNS), with special emphasis on neural plasticity and development of tinnitus as a consequence of neural changes that develop in the days and months post-noise exposure.

The third section of the book (Chapters 9–11) focuses on factors influencing susceptibility to NIHL. First, Gong and Lomax (Chapter 9) carefully review the genetics of NIHL. The links between NIHL and ARHL emerge in further detail in Chapter 10 by Bielefeld. Finally, in Chapter 11, Morata and Johnson address interactions between noise and a variety of chemicals.

The final section of this book (Chapters 12–14) addresses issues of protection and repair. In Chapter 12, Casali reviews the specific characteristics of different personal protection devices (PPD). In Chapter 13, Le Prell and Bao expand on the notion of protection, although the topic is intervention using novel pharmaceuticals currently under development and other potential therapeutic agents. Finally, in Chapter 14, Yamasoba, Miller, Ulfendahl, and Altschuler explore the next new frontiers in hearing science.

Issues of noise and its effects on hearing are emphasized in this volume, but many earlier volumes in the Springer Handbook of Auditory Research have themes and chapters germane to the issues in this volume. For example, otoacoustic emissions are broadly considered in *Active Processes and Otoacoustic Emissions* (Volume 30, edited by Manley, Fay, and Popper), while the effects of noise on the ear, and repair of such damage, are discussed in detail in Volume 31 on *Auditory Trauma, Protection, and Repair* (edited by Schacht, Popper, and Fay) and Volume 33 on *Hair Cell Regeneration, Repair, and Protection* (edited by Salvi, Popper, and Fay). Of course, another source of hearing loss is aging, and this is considered in depth in *The Aging Auditory System* (Volume 34, edited by Gordon-Salant, Frisina, Popper, and Fay). Finally, issues of loudness are also associated with hearing loss, and this topic is considered in *Loudness* (Volume 37, edited by Florentine, Popper, and Fay).

Colleen G. Le Prell, Gainesville, FL
Donald Henderson, Buffalo, NY
Richard R. Fay, Falmouth, MA
Arthur N. Popper, College Park, MD

Contents

Contributors

Richard A. Altschuler Department of Otolaryngology and Department of Cell & Developmental Biology, Kresge Hearing Research Institute, 1150 West Medical Center Drive, Ann Arbor, MI 48109-5616, USA
shuler@umich.edu

Jianxin Bao Department of Otolaryngology, Washington University, 4560 Clayton Avenue, St. Louis, MO 63110, USA
jbao@wustl.edu

Eric C. Bielefeld Department of Speech and Hearing Science, The Ohio State University, 1070 Carmack Road, Columbus, OH 43220, USA
bielefeld.6@osu.edu

John G. Casali Department of Industrial and Systems Engineering, Auditory Systems Laboratory, Virginia Tech, Blacksburg, VA 24061, USA
jcasali@vt.edu

Tzy-Wen Gong Department of Otolaryngology-Head/Neck Surgery, Kresge Hearing Research Institute, The University of Michigan, 1150 W. Medical Center Drive, Ann Arbor, MI 48109-5648, USA
tzywen@umich.edu

Marjorie A. M. Grantham US Army Public Health Command (Provisional), 5158 Blackhawk Road, Aberdeen Proving Ground, MD 21010-5403, USA
marjorie.grantham@us.army.mil

Roger P. Hamernik Auditory Research Laboratory, State University of New York at Plattsburgh, Plattsburgh, NY 12901, USA
hamernrp@plattsburgh.edu

Donald Henderson Department of Communication Disorders and Science, Center for Hearing Research, College of Arts and Sciences, State University of New York at Buffalo, 3435 Main Street, Buffalo, NY 14214, USA
donaldhe@buffalo.edu

Bohua Hu Department of Communicative Disorders and Sciences, Center for Hearing and Deafness, State University of New York at Buffalo, 3435 Main Street, Buffalo, NY 14214, USA
bhu@buffalo.edu

Ann-Christin Johnson Karolinska Institutet, Alfred Nobels allé 10, 5tr, Stockholm 141 83, Sweden
ann-christin.johnson@ki.se

James A. Kaltenbach Department of Neurosciences, NE-63, The Cleveland Clinic, 9500 Euclid Avenue, Cleveland, OH 44195, USA
kaltenj@ccf.org

Colleen G. Le Prell Department of Speech, Language, and Hearing Sciences, University of Florida, 101 S. Newell Road, Gainesville, FL 32610, USA
colleeng@phhp.ufl.edu

Margaret I. Lomax Department of Otolaryngology-Head/Neck Surgery, Kresge Hearing Research Institute, The University of Michigan, 1150 W. Medical Center Drive, Ann Arbor, MI 48109-5648, USA
mlomax@umich.edu

Ryan Manz Head and Neck Institute, The Cleveland Clinic, 9500 Euclid Avenue, Cleveland, OH 44195, USA
rmmanz@gmail.com

Josef M. Miller Department of Otolaryngology and Department of Cell & Developmental Biology, Kresge Hearing Research Institute, 1150 West Medical Center Drive, Ann Arbor, MI 48109-5616, USA
josef@umich.edu

Thais C. Morata National Institute for Occupational Safety and Health, 4676 Columbia Parkway, Cincinnati, OH 45226-1998, USA
tmorata@cdc.gov

Peter M. Rabinowitz Yale Occupational and Environmental Medicine Program, Yale University School of Medicine, 135 College Street, New Haven, CT 06510, USA
peter.rabinowitz@yale.edu

Mini N. Shrivastav Department of Speech, Language, and Hearing Sciences, University of Florida, 101 S. Newell Road, Gainesville, FL 32611, USA
mnshriv@ufl.edu

Mats Ulfendahl Center for Hearing and Communication Research, Karolinska University Hospital Solna, Stockholm SE-171 76, Sweden
mats.ulfendahl@ki.se

Tatsuya Yamasoba Department of Otolaryngology and Head and Neck Surgery, University of Tokyo, Hongo 7-3-1, Bunkyo-ku, Tokyo 113-8655, Japan
tyamasoba-tky@umin.ac.jp

Eric D. Young Department of Biomedical Engineering, Johns Hopkins School of Medicine, 720 Rutland Ave, Baltimore, MD 21205, USA
eyoung@jhu.edu

Chapter 1
Perspectives on Noise-Induced Hearing Loss

Colleen G. Le Prell and Donald Henderson

1 Introduction

Hearing impairment is the third most prevalent chronic disability in the United States, and hearing loss in the speech frequency region (pure-tone average threshold at 0.5, 1, 2, and 4 kHz ≥ 25 dB) is currently estimated to affect 29 million Americans ages 20–69 years based on 2003–2004 data (16% of population; Agrawal et al. 2008). When the higher frequencies are considered (pure-tone average at 3, 4, and 6 kHz ≥ 25 HL), the number affected doubles (Agrawal et al. 2008). Consistent with this, the National Institutes of Health has estimated that some 15% of Americans between the ages of 20 and 69 have hearing loss at higher test frequencies, suggesting the hearing loss may have been caused by exposure to loud sound (National Institute on Deafness and Other Communication Disorders 2002).

Age is clearly one major risk factor for hearing loss (Cruickshanks et al. 1998); others include nutritional status (Spankovich et al. 2011) and cardiovascular risk factors including diabetes, hypertension, and obesity, as well as smoking (Agrawal et al. 2009). Hearing loss can also be caused by use of drugs that are harmful to the auditory system, such as aminoglycoside antibiotics or the chemotherapeutic cisplatin (for review, see Campbell and Le Prell 2011). For many people, however, hearing loss is caused not by aging, drugs, or existing disease conditions; it is an injury induced by exposure to loud sound that can come from a variety of sources including machinery (producing occupational noise exposure), loud music (concerts or personal music players), stadium sporting events, power tools, lawn care equipment, firearms, household appliances, and other sources too numerous to list. Indeed, noise insult appears to be the main cause of preventable acquired hearing loss.

C.G. Le Prell (✉)
Department of Speech, Language, and Hearing Sciences, University of Florida, 101 S. Newell Road, Gainesville, FL 32610, USA
e-mail: colleeng@phhp.ufl.edu

C.G. Le Prell et al. (eds.), *Noise-Induced Hearing Loss: Scientific Advances*, Springer Handbook of Auditory Research 40, DOI 10.1007/978-1-4419-9523-0_1,

Hearing loss is typically attributed to noise exposure if the configuration of the patient's audiogram is "notched" and the patient also reports a positive history of noise exposure. However, not all individuals identified as having an audiometric notch report a history of noise exposure, and not all individuals reporting a history of noise have an audiometric notch (Hong 2005; Nondahl et al. 2009; Osei-Lah and Yeoh 2010). The definition of a notch of course affects the measured prevalence of audiometric notches (Nondahl et al. 2009), just as the definition of a hearing loss importantly influences the measured prevalence of hearing loss (Le Prell et al. 2011). Despite the lack of a precise relationship between presence of an audiometric notch and history of noise exposure, the "notched" audiogram in combination with the noise history is the most used clinical metric for assessing potential NIHL. The problem of NIHL impacts a number of disciplines, and the purpose of this volume is to provide a current review of the state of the science across disciplines. This book has been authored by leading scientists and clinicians, and is divided into four key sections, as described in the following sections.

2 Scope of the Problem

The first section of the book (Chaps. 2 and 3) outlines the scope of the problem. Here, we note that according to the National Institute for Occupational Safety and Health (NIOSH), more than 22 million American workers are exposed to hazardous levels of occupational noise, with 75% of workers within the mining industry and more than 33% of workers within the manufacturing industry exposed to loud sound (Murphy and Tak 2009). In Chap. 2, Rabinowitz describes the prevalence and the public health importance of the NIHL health issue. In addition to discussing the scope of NIHL, the complications of NIHL (i.e., socioacoustics, financial burden) are discussed. Finally, the shifting impact of NIHL from the United States to developing countries is discussed.

Noise trauma presents a staggering problem for the U.S. military, and the problem is growing given the current engagements in Afghanistan and Iraq. Single-year disability costs related to hearing loss for the military exceeded $900 million for 2006, with increasing costs largely related to the noise insults experienced by soldiers serving in current military operations [Operation Enduring Freedom (OEF), Operation Iraqi Freedom (OIF)] (United States Army Center for Health Promotion and Preventive Medicine (USACHPPM) 2007). In Chap. 3, Grantham reviews not only the financial cost of hearing loss to the United States Veterans Administration (VA), but also the impact on the United States Department of Defense (DOD), and their response to the problem.

Although the volume does not include a chapter specifically dedicated to the problem of hearing loss in children, this is an important topic. Recent reports have suggested an *increasing* prevalence of NIHL in children. The Third National Health and Nutrition Examination Survey (NHANES III) evaluated a sample of 6,166 children ages 6–19 years from 1988 to 1994. Using a criterion of low-frequency pure-tone-average (LFPTA) threshold≥16 dB HL at 0.5, 1, and 2 kHz, 7.1% were reported to have hearing loss in one or both ears (Niskar et al. 1998). That number increased to 12.7% using a criterion of high-frequency pure-tone-average (HFPTA) threshold≥16 dB HL at 3, 4, and 6 kHz, and 14.9% when either LFPTA or HFPTAs were considered (Niskar et al. 1998). More recently, in the 2005–2006 NHANES data set, collected from 1,771 participants ages 12–19, there was a 16.4% prevalence rate for any high-frequency hearing loss (Shargorodsky et al. 2010). This represents a 31% increase in prevalence of any hearing loss (defined as unilateral or bilateral LFPTA or HFPTA>15 dB), with an increase from 14.9% in NHANES III to 19.5% in NHANES 2005–2006 (Shargorodsky et al. 2010).

Niskar et al. (2001) extended the analysis of the NHANES III data by evaluating the prevalence of audiometric notches, with a notch defined as (1) thresholds ≤15 dB HL at 0.5 and 1.0 kHz; (2) 3, 4, or 6 kHz threshold at least 15 dB worse than thresholds at 0.5 and 1 kHz; and (3) 3, 4, or 6 kHz threshold at least 10 dB worse than 8-kHz threshold. Using that criterion, they detected a 15.5% prevalence of audiometric notches within the NHANES III data set (Niskar et al. 2001). The popular press has recently highlighted the potential risk that personal music player use could cause NIHL over time, particularly in adolescent and young adult populations, given the growing popularity of music players incorporating MPEG audio layer 3 (MP3) technology (for examples, see de Vries 2005; Castle 2008; Cunningham 2009; Kean 2010). It has been widely suggested that modern digital audio players are potentially more dangerous than the personal stereos of previous generations because of their smaller size and convenience, larger storage capacity, and longer battery life. It is clear that music players can produce sounds sufficiently intense to damage the inner ear (Katz et al. 1982; Fligor and Cox 2004; Hodgetts et al. 2007). Further, some survey studies suggest a subset of users/listeners engage in potentially risky listening behaviors, such as extended listening duration and/or listening at a high volume (see Vogel et al. 2008; Danhauer et al. 2009; Shah et al. 2009; Vogel et al. 2009). Some researchers suggested the potential for small (2–3 dB) but statistically significant threshold shifts in users of personal music players (Meyer-Bisch 1996; Kim et al. 2009; Le Prell et al. 2011). However, other studies found no relationship between music player use and threshold elevation (Shah et al. 2009). Thus, it remains unclear the extent to which music player use contributes to permanent noise-induced cochlear dysfunction or hearing loss in adolescent and young adult populations (for discussion, see Morata 2007; Hodgetts et al. 2009; Maria et al. 2009; Williams 2009).

3 Dose–Response Relationships

The second section of the book (Chaps. 4–8) details the relationships among noise exposure and anatomical trauma, physiological changes, and perceptual deficits. There is a clear dose–response relationship in which increasing levels of noise result in increasing threshold shifts (Wang et al. 2002); this relationship holds relatively constant across species (for review, see Yoshida et al. 2000).

In Chap. 4, Henderson and Hamernik review the current noise standards and present experimental and epidemiological data that is inconsistent with the current concept of a safe exposure. Of particular interest is complex noise exposure consisting of background and higher level impulses, impacts, or noise bursts. Henderson and Hamernik clearly explain the concept of kurtosis, a measure of dynamic level change during an acoustic signal, and they review data supporting kurtosis as an additional metric of noise risk. U.S. standards for permissible noise exposure, as regulated by the U.S. Occupational Safety and Health Administration (OSHA), were originally published in 1968 and remain largely unchanged today. Indeed, permissible noise exposure for U.S. personnel is significantly greater than that allowed in numerous other countries, including, for example, Canada, China, Brazil, Mexico, and the European Union based on key differences in action level (85 dB A vs. 90 dB A) and exchange rate (3 dB vs. 5 dB) (for excellent summary, see Suter 2007).

In Chap. 5, Hu describes noise-induced pathological changes in the cochlea. In the last 10 years there have been major changes in the understanding of how the ear is damaged by high-level noise. For example, Hu and others have reported that high-level noise increases free radical activity in the cochlea and the cells of the cochlea (especially outer hair cells) are damaged or destroyed. It has also been reported that cochlear cells die by both apoptosis and necrosis. In the current chapter, Hu describes the biochemical/mechanical process involved in sensory cell death. The discoveries reported by Hu are fundamental to the growing field of therapeutic auditory pharmacology (see Le Prell and Bao, Chap. 13).

In Chap. 6, Young describes the effects of noise and NIHL on the auditory nerve, response to sound, and neural codes. The auditory nerve provides all the nerve data for auditory perception. Young's chapter describes the systematic changes to the neural code when the ear is damaged by noise. Not only are the thresholds for the eighth nerve neurons elevated, but their tuning is also compromised. The eighth nerve changes lead to the audiological symptoms of recruitment and poor speech perception in noise. Also, knowledge of the neural code from a noise-damaged ear provides the targets for future clinical intervention with hearing aids.

Although the audiometric notch configuration is generally well documented in patients with NIHL, there is less information on the nature of suprathreshold auditory processing deficits in this population. Kujawa and Liberman (2006, 2009) have suggested difficulty processing speech in noise is one likely correlate of noise exposure, even when the noise insult produces only temporary changes in hearing, and they have provided data that such deficits may increase when combined with normal aging. In Chap. 7, Shrivastav reviews known psychophysical changes associated with NIHL in detail, specifically including processing of speech in noise, and contrasts the changes that occur subsequent to noise exposure with those that occur as a consequence of age-related hearing loss (ARHL). Differences between NIHL and ARHL have significant implications for amplification strategy given that deficits extend beyond simple loss of audibility.

In the last chapter in this section, Kaltenbach and Manz (Chap. 8) describe the effects of noise on the central nervous system (CNS), with special emphasis on neural plasticity and development of tinnitus as a consequence of neural changes. Importantly, tinnitus was the most prevalent service-connected disability among veterans receiving compensation at the end of fiscal year 2009 (U.S. Department of Veterans Affairs 2010); tinnitus can be debilitating for a subset of those affected. New understanding of tinnitus, based on CNS plasticity, has come from a combination of anatomical, physiological, and pharmacological studies in animals and from human studies using modern brain imaging techniques. Data generated using each of these approaches are reviewed and integrated in Chap. 8.

Finally, it is important to recognize the relationship between hyperacusis and NIHL. Hyperacusis is an increased sensitivity to sound, such that sound levels that would not trouble most individuals are bothersome or even painful, inducing discomfort, pain, annoyance, dislike, fear, or other negative emotions (Katzenell and Segal 2001; Baguley 2003; Jastreboff and Jastreboff 2003). Reports of increased sound sensitivity are prevalent in musicians (Schmuziger et al. 2006; Jansen et al. 2009). Hyperacusis is distinct from loudness recruitment, in which rate of loudness growth is abnormally steep, but with no emotional fear or distress component. Noise exposure induces changes in the rate of loudness growth, with both neural amplitude (Sendowski et al. 2004) and reaction time (May et al. 2009) used as metrics for equal loudness. Hyperacusis and tinnitus are well linked, with one report of up to 80% of human patients seen for tinnitus treatment reporting hyperacusis when surveyed (Dauman and Bouscau-Faure 2005). Hyperacusis often has no known medical etiology, but can occur after facial nerve dysfunction (with loss of the stapedial reflex), and central hyperexcitability has been suggested (Baguley 2003). The association of hyperacusis and NIHL in some patients suggests that noise-induced damage to the cochlea is another possible cause of hyperacusis (Katzenell and Segal 2001). One possibility is that central changes after NIHL, such as loss of tonic inhibition or new neural connections and hyperstimulation, amplify a reduced neural input after hearing loss, thus producing both tinnitus and hyperacusis (Nelson and Chen 2004). Changes in the CNS, occurring as a consequence of plastic changes in the brain with loss of hair cells and neurons, have been detected in the cochlear nucleus and auditory cortex, and these changes might also induce abnormal perceptions of loudness (Cai et al. 2009). Like tinnitus, hyperacusis is challenging to measure in animal models. An increase in startle response amplitude may provide evidence for hyperacusis in animal subjects (see also Ison et al. 2007; Turner and Parrish 2008). To date, there are no drug therapies that effectively reduce hyperacusis. Given the lack of understanding and minimal research in this area, hyperacusis is not discussed further in this volume.

4 Variability in Vulnerability

The third section of this book (Chaps. 9–11) focuses on factors influencing susceptibility to NIHL. First, Gong and Lomax (Chap. 9) review the genetics of NIHL. Specifically, they discuss candidate genes including oxidative stress genes, connexin and cadherin mutations, and potassium recycling genes. Advances in this area have been made possible by advances in technology that for the first time allow rapid screening and identification of multiple genes. The interaction of environmental factors with genetic profile and application of these high-throughput human genome scanning techniques are reviewed. Another gene, not reviewed in this chapter as it has not yet been well identified in human analogue, is vezatin, an ubiquitous integral protein in the hair cell adherens junction that may play a role in both NIHL and ARHL (Avraham 2009). In the absence of vezatin, after genetic "knockout," vulnerability to noise is increased, and spontaneous, progressive, hearing loss is common (Bahloul et al. 2009).

The links between NIHL and ARHL emerge in further detail in Chap. 10 by Bielefeld. This chapter expands on some of the ARHL themes raised by Shrivastav (Chap. 7), as well as some of the public health issues discussed by Rabinowitz (Chap. 2). Specifically, Bielefeld reviews human and animal data in the vectors of ARHL and NIHL, compares and contrasts putative causes of ARHL and NIHL, and addresses in detail the influence of early (noise-induced) hearing loss on later ARHL.

The last chapter in this section (Chap. 11, by Morata and Johnson) also addresses interactions among insults, although the focus is noise plus chemicals, rather than noise plus aging. A number of industrial chemicals have been shown to contribute to NIHL. Noise that does not by itself induce a hearing loss, when combined with chemicals that by themselves do not induce a hearing loss, can result in significant permanent hearing loss. This synergistic toxicity is alarming, given the number of workers exposed to both noise and industrial chemicals, and the lack of any regulatory framework for evaluating combined risks. Morata and Johnson provide an in-depth discussion of the challenges in addressing increased hearing hazards and potential issues to consider in developing novel strategies for evaluation of risk, which are clearly, urgently, needed.

5 Protection and Repair

The final section of this book (Chaps. 12–14) addresses issues of protection and repair. In Chap. 12, Casali reviews the specific characteristics of different personal protection devices (PPDs). The use of PPDs appears to be a relatively simple solution to preventing NIHL, but Casali explains the problems and challenges associated with PPDs as a protective approach to preventing NIHL. Real-life effectiveness is contrasted with theoretical, best possible, attenuation of noise insults. The chapter describes the difference in performance with PPDs and explains advances in PPD (i.e., active protection and musician's plug).

In Chap. 13, Le Prell and Bao expand on the notion of protection, using novel pharmaceuticals currently under development. The development of "drugs" that protect the inner ear has been driven by advances in our understanding of noise-induced cell death. As reviewed in detail elsewhere (Henderson et al. 2006; Le Prell et al. 2007), noise is no longer considered strictly a mechanical insult. Acoustic overstimulation *can* produce a mechanical lesion, but there is clearly a role for oxidative stress in which free radical production and accumulation directly initiate apoptotic cell death events that drive cell death over the days and weeks immediately following a loud noise insult. Other cell death pathways are also activated subsequent to noise, such as the c-Jun N-terminal kinase (JNK) group of mitogen-activated protein (MAP) kinases that phosphorylate the transcription factor c-*Jun*, and there are indeed novel JNK blockers that are now being evaluated as novel drug agents that may protect the inner ear against noise-induced trauma. It is possible to envision a future in which ever better mechanical devices (see Casali, Chap. 12) are accompanied by an increasing array of "drug" options that can supplement the level of protection provided by such devices, and, moreover, potentially provide some level of protection to users who choose not to use such devices, such as individuals who listen to personal music players at potentially harmful levels. The face of hearing conservation may be different indeed should one or more "drug" strategies emerge as clearly beneficial in controlled human clinical tests.

The last chapter in the book (Chap. 14, by Yamasoba, Miller, Ulfendahl, and Altschuler) boldly explores the next new frontiers in hearing science. The successes of the cochlear implant to return hearing to those who have lost, or never had, this sense are first highlighted, and new directions suggested. The possibilities for reengineering the cochlea using viral vectors, or stem cells, to induce new tissue growth and cellular differentiation in the cochlea are then reviewed, ending with a "status report" and suggestions for the steps needed before clinical translation can occur. As detailed in this chapter, many advances in the understanding of the mechanisms associated with NIHL have now illuminated a path forward, toward prevention and perhaps treatment for those who have already lost their hearing, but more basic research is still needed. Moreover, the difficult-to-fund parametric dose–response measurements of efficacy and safety must be performed in animals and then in humans. Indeed, translational research is now demanded such that clinical trials can ultimately be initiated. Together, the authors provide a compelling vision for the future of hearing research and the prevention of NIHL.

6 Summary

Together, the chapters in this book provide a comprehensive overview of the problem of NIHL. From the populations most affected, to the most common deficits and comorbidity issues, to the current best practices, to the next generation of prevention and repair strategies, all aspects of NIHL are reviewed. Great progress has been made, and this book celebrates the accomplishments and scientific advancements to date. Indeed, there have been a number of discoveries and advances that have increased our understanding of the mechanisms of NIHL. These advances have the potential to impact how NIHL can be prevented and how our noise standards can be made more appropriate. However, as the authors of the chapters in this book make clear, great challenges remain. We hope this book provides a useful guide to the literature and serves as an interdisciplinary roadmap for researchers and clinicians seeking to advance the field ever further. The combination of different methodological and experimental approaches, the diverse range of aspects of human auditory perception, and integration of noise risk with other insults (age, chemical exposure) will inspire novel insights and advance future research.

References

Agrawal, Y., Platz, E. A., & Niparko, J. K. (2008). Prevalence of hearing loss and differences by demographic characteristics among U.S. adults: Data from the National Health and Nutrition Examination Survey, 1999 to 2004. *Archives of Internal Medicine*, 168(14), 1522–1530.

Agrawal, Y., Platz, E. A., & Niparko, J. K. (2009). Risk factors for hearing loss in U.S. adults: Data from the National Health and Nutrition Examination Survey, 1999 to 2002. *Otology & Neurotology*, 30(2), 139–145.

Avraham, K. (2009). Noise stresses the junctions to deaf. *EMBO Molecular Medicine*, 1, 85–87.

Baguley, D. M. (2003). Hyperacusis. *Journal of the Royal Society of Medicine*, 96(12), 582–585.

Bahloul, A., Simmler, M.-C., Michel, V., Leibovici, M., Perfettini, I., Roux, I., Weil, D., Nouaille, S., Zuo, J., Zadro, C., Licastro, D., Gasparini, P., Avan, P., Hardelin, J.-P., & Petit, C. (2009). Vezatin, an integral membrane protein of adherens junctions, is required for the sound resilience of cochlear hair cells. *EMBO Molecular Medicine*, 1, 125–138.

Cai, S., Ma, W. L., & Young, E. D. (2009). Encoding intensity in ventral cochlear nucleus following acoustic trauma: Implications for loudness recruitment. *Journal of the Association for Research in Otolaryngology*, 10(1), 5–22.

Campbell, K. C., & Le Prell, C. G. (2011). Potential therapeutic agents. *Seminars in Hearing*, *in press*.

Castle, S. (2008, October 12). Did you hear? MP3 players threaten hearing loss. Retrieved from http://www.nytimes.com/2008/10/12/technology/12iht-noise.4.16883369.html?_r=1.

Cruickshanks, K. J., Wiley, T. L., Tweed, T. S., Klein, B. E., Klein, R., Mares-Perlman, J. A., & Nondahl, D. M. (1998). Prevalence of hearing loss in older adults in Beaver Dam, Wisconsin. The Epidemiology of Hearing Loss Study. *American Journal of Epidemiology*, 148(9), 879–886.

Cunningham, K. (2009, September 3). The iPod effect—Excessive MP3 player use can lead to permanent hearing loss. Retrieved December 17, 2010 from http://www.associatedcontent.com/article/2124553/the_ipod_effect_excessive_mp3_player.html.

Danhauer, J. L., Johnson, C. E., Byrd, A., DeGood, L., Meuel, C., Pecile, A., & Koch, L. L. (2009). Survey of college students on iPod use and hearing health. *Journal of the American Academy of Audiology*, 20(1), 5–27;quiz 12–19.

Dauman, R., & Bouscau-Faure, F. (2005). Assessment and amelioration of hyperacusis in tinnitus patients. *Acta Oto-Laryngologica*, 125(5), 503–509.

de Vries, L. (2005, August 25). MP3s may threaten hearing loss. *CBS News*. Retrieved from http://www.cbsnews.com/stories/2005/08/25/health/webmd/main796088.shtml.

Fligor, B. J., & Cox, L. C. (2004). Output levels of commercially available portable compact disc players and the potential risk to hearing. *Ear and Hearing*, 25(6), 513–527.

Henderson, D., Bielefeld, E. C., Harris, K. C., & Hu, B. H. (2006). The role of oxidative stress in noise-induced hearing loss. *Ear and Hearing*, 27(1), 1–19.

Hodgetts, W., Szarko, R., & Rieger, J. (2009). What is the influence of background noise and exercise on the listening levels of iPod users? *International Journal of Audiology*, 48(12), 825–832.

Hodgetts, W. E., Rieger, J. M., & Szarko, R. A. (2007). The effects of listening environment and earphone style on preferred listening levels of normal hearing adults using an MP3 player. *Ear and Hearing*, 28(3), 290–297.

Hong, O. (2005). Hearing loss among operating engineers in American construction industry. *International Archives of Occupational and Environmental Health*, 78(7), 565–574.

Ison, J. R., Allen, P. D., & O'Neill, W. E. (2007). Age-related hearing loss in C57BL/6 J mice has both frequency-specific and non-frequency-specific components that produce a hyperacusis-like exaggeration of the acoustic startle reflex. *Journal of the Association for Research in Otolaryngology*, 8(4), 539–550.

Jansen, E. J., Helleman, H. W., Dreschler, W. A., & de Laat, J. A. (2009). Noise induced hearing loss and other hearing complaints among musicians of symphony orchestras. *International Archives of Occupational and Environmental Health*, 82(2), 153–164.

Jastreboff, P. J., & Jastreboff, M. M. (2003). Tinnitus retraining therapy for patients with tinnitus and decreased sound tolerance. *Otolaryngologic Clinics of North America*, 36(2), 321–336.

Katz, A. E., Gerstman, H. L., Sanderson, R. G., & Buchanan, R. (1982). Stereo earphones and hearing loss. *New England Journal of Medicine*, 307(23), 1460–1461.

Katzenell, U., & Segal, S. (2001). Hyperacusis: Review and clinical guidelines. *Otology & Neurotology*, 22(3), 321–326; discussion 326–327.

Kean, C. (2010, January). MP3 Generation: Noise-induced hearing loss rising among children and adolescents. *ENT Today*. Retrieved from http://www.enttoday.org/details/article/554357/MP3_Generation_Noise-induced_hearing_loss_rising_among_children_and_adolescents.html.

Kim, M. G., Hong, S. M., Shim, H. J., Kim, Y. D., Cha, C. I., & Yeo, S. G. (2009). Hearing threshold of Korean adolescents associated with the use of personal music players. *Yonsei Medical Journal*, 50(6), 771–776.

Kujawa, S. G., & Liberman, M. C. (2006). Acceleration of age-related hearing loss by early noise exposure: Evidence of a misspent youth. *Journal of Neuroscience*, 26(7), 2115–2123.

Kujawa, S. G., & Liberman, M. C. (2009). Adding insult to injury: Cochlear nerve degeneration after "temporary" noise-induced hearing loss. *Journal of Neuroscience*, 29(45), 14077–14085.

Le Prell, C. G., Yamashita, D., Minami, S., Yamasoba, T., & Miller, J. M. (2007). Mechanisms of noise-induced hearing loss indicate multiple methods of prevention. *Hearing Research*, 226, 22–43.

Le Prell, C. G., Hensley, B. N., Campbell, K. C. M., Hall, J. W. I., & Guire, K. (2011). Hearing outcomes in a "normally-hearing" college-student population: Evidence of hearing loss. *International Journal of Audiology*, 50(Supplement 1), S21–31.

Maria, A., Zocoli, F., Morata, T. C., Marques, J. M., & Corteletti, L. J. (2009). Brazilian young adults and noise: Attitudes, habits, and audiological characteristics. *International Journal of Audiology*, 48(10), 692–699.

May, B. J., Little, N., & Saylor, S. (2009). Loudness perception in the domestic cat: Reaction time estimates of equal loudness contours and recruitment effects. *Journal of the Association for Research in Otolaryngology*, 10(2), 295–308.

Meyer-Bisch, C. (1996). Epidemiological evaluation of hearing damage related to strongly amplified music (personal cassette players, discotheques, rock concerts)--high-definition audiometric survey on 1364 subjects. *Audiology*, 35(3), 121–142.

Morata, T. C. (2007). Young people: Their noise and music exposures and the risk of hearing loss. *International Journal of Audiology*, 46(3), 111–112.

Murphy, W., & Tak, S. W. (2009, Posted 11/24/09.). NIOSH science blog: Workplace hearing loss Retrieved April 27, 2010 from http://www.cdc.gov/niosh/blog/nsb112409_hearingloss.html.

National Institute on Deafness and Other Communication Disorders. (2002). Noise-induced hearing loss (NIH Pub. No. 97–4233). Bethesda, MD.

Nelson, J. J., & Chen, K. (2004). The relationship of tinnitus, hyperacusis, and hearing loss. *Ear, Nose, and Throat Journal*, 83(7), 472–476.

Niskar, A. S., Kieszak, S. M., Holmes, A., Esteban, E., Rubin, C., & Brody, D. J. (1998). Prevalence of hearing loss among children 6 to 19 years of age: The Third National Health and Nutrition Examination Survey. *JAMA*, 279(14), 1071–1075.

Niskar, A. S., Kieszak, S. M., Holmes, A. E., Esteban, E., Rubin, C., & Brody, D. J. (2001). Estimated prevalence of noise-induced hearing threshold shifts among children 6 to 19 years of age: The Third National Health and Nutrition Examination Survey, 1988–1994, United States. *Pediatrics*, 108(1), 40–43.

Nondahl, D. M., Shi, X., Cruickshanks, K. J., Dalton, D. S., Tweed, T. S., Wiley, T. L., & Carmichael, L. L. (2009). Notched audiograms and noise exposure history in older adults. *Ear and Hearing*, 30(6), 696–703.

Osei-Lah, V., & Yeoh, L. H. (2010). High frequency audiometric notch: An outpatient clinic survey. *International Journal of Audiology*, 49(2), 95–98.

Schmuziger, N., Patscheke, J., & Probst, R. (2006). Hearing in nonprofessional pop/rock musicians. *Ear and Hearing*, 27(4), 321–330.

Sendowski, I., Braillon-Cros, A., & Delaunay, C. (2004). CAP amplitude after impulse noise exposure in guinea pigs. *European Archives of Oto-Rhino-Laryngology*, 261(2), 77–81.

Shah, S., Gopal, B., Reis, J., & Novak, M. (2009). Hear today, gone tomorrow: An assessment of portable entertainment player use and hearing acuity in a community sample. *Journal of the American Board of Family Medicine*, 22(1), 17–23.

Shargorodsky, J., Curhan, S. G., Curhan, G. C., & Eavey, R. (2010). Change in prevalence of hearing loss in U.S. adolescents. *JAMA*, 304(7), 772–778.

Spankovich, C., Hood, L., Silver, H., Lambert, W., Flood, V., & Mitchell, P. (2011). Associations between diet and both high and low pure tone averages and transient evoked otoacoustic emissions in an older adult population-based study. *Journal of the American Academy of Audiology*, 22, 49–58.

Suter, A. H. (2007). Development of standards and regulations for occupational noise. In M. Crocker (Ed.), *Handbook of noise and vibration control*. Hoboken, NJ: John Wiley & Sons.

Turner, J. G., & Parrish, J. (2008). Gap detection methods for assessing salicylate-induced tinnitus and hyperacusis in rats. *American Journal of Audiology*, 17(2), S185–192.

United States Army Center for Health Promotion and Preventive Medicine (USACHPPM). (2007). U.S. Army Center for Health Promotion and Preventative Medicine. 2006 Veterans Compensation Charts and VA Disability Reports Retrieved October 26, 2007, from http://chppm-www.apgea.army.mil/hcp/comp_reports.aspx.

U.S. Department of Veterans Affairs. (2010). 2009 Annual Benefits Report. Retrieved October 5, 2010 from http://www.vba.va.gov/REPORTS/abr/index.asp.

Vogel, I., Brug, J., Hosli, E. J., van der Ploeg, C. P., & Raat, H. (2008). MP3 players and hearing loss: Adolescents' perceptions of loud music and hearing conservation. *Journal of Pediatrics*, 152(3), 400–404.

Vogel, I., Verschuure, H., van der Ploeg, C. P., Brug, J., & Raat, H. (2009). Adolescents and MP3 players: Too many risks, too few precautions. *Pediatrics*, 123(6), e953–958.

Wang, Y., Hirose, K., & Liberman, M. C. (2002). Dynamics of noise-induced cellular injury and repair in the mouse cochlea. *Journal of the Association for Research in Otolaryngology*, 3(3), 248–268.

Williams, W. (2009). Trends in listening to personal stereos. *International Journal of Audiology*, 48(11), 784–788.

Yoshida, N., Hequembourg, S. J., Atencio, C. A., Rosowski, J. J., & Liberman, M. C. (2000). Acoustic injury in mice: 129/SvEv is exceptionally resistant to noise-induced hearing loss. *Hearing Research*, 141(1–2), 97–106.

Part I
Noise-Induced Hearing Loss: Scope of the Problem

Chapter 2
The Public Health Significance of Noise-Induced Hearing Loss

Peter M. Rabinowitz

1 Introduction

Recognition of hearing loss resulting from noise exposure dates back at least as far as Ramazzini's (1713) classic occupational medicine treatise *De Morbis Artificum* (Diseases of Workers). Ramazzini's vivid discussion of noise-induced hearing loss (NIHL) is notable for its recognition that exposure to both occupational and environmental noise can lead to hearing loss in individuals and entire populations. In describing the coppersmiths of Venice, he compared them to an environmentally exposed population in Egypt:

> ...at Venice, these workers are all congregated in one quarter and are engaged all day in hammering copper to make it ductile so that with it they may manufacture vessels of various kinds. From this quarter there rises such a terrible din that only these workers have shops and homes there; all others flee from that highly disagreeable locality. One may observe these men as they sit on the ground, usually on small mats, bent double while all day long they beat the newly mined copper, first with wooden then with iron hammers till it is as ductile as required. To begin with, the ears are injured by that perpetual din, and in fact the whole head, inevitably, so that workers of this class become hard of hearing and, if they grow old at this work, completely deaf. For that incessant noise beating on the eardrum makes it lose its natural tonus; the air within the ear reverberates against its sides, and this weakens and impairs all the apparatus of hearing. In fact the same thing happens to them as to those who dwell near the Nile in Egypt, for they are all deaf from the excessive uproar of the falling water.

Given the inclusion of NIHL in the first major textbook on occupational diseases, it is surprising that, 300 years later, there is still significant controversy about the true prevalence and public health importance of this condition. This chapter reviews

P.M. Rabinowitz (✉)
Yale Occupational and Environmental Medicine Program, Yale University School of Medicine, 135 College Street, New Haven, CT 06510, USA
e-mail: peter.rabinowitz@yale.edu

C.G. Le Prell et al. (eds.), *Noise-Induced Hearing Loss: Scientific Advances*, Springer Handbook of Auditory Research 40, DOI 10.1007/978-1-4419-9523-0_2,

Table 2.1 Relationship between prevalence, severity, and public health impact of a medical condition

		Severity	
		Low	High
Prevalence	Low	Low public health impact	Potential public health impact
	High	Potential public health impact	High public health impact

the evidence for the public health impact of NIHL and provides a framework for viewing NIHL as a public health issue.

Assessing the public health impact of NIHL involves consideration of both its prevalence in a particular population, as well as the severity of impact of the condition on affected individuals and populations as a whole. As Table 2.1 shows, diseases that are highly prevalent and severe, such as cardiovascular disease and cancer, obviously have a high public health impact, whereas rare or mild diseases do not. Yet even if medical conditions are relatively mild in terms of individual morbidity, they can have a significant public health impact if they are highly prevalent. Therefore, an analysis of the public health importance of NIHL needs to assess carefully the evidence regarding the prevalence and severity of the condition.

2 Estimates of the Public Health Impact of NIHL

Over the years, estimates of the prevalence and severity of NIHL have varied widely. Key reasons behind this variability seem to include the lack of a common case definition for NIHL, the difficulty of distinguishing NIHL from age-related hearing loss (presbycusis), uncertainty about the size of the population that is exposed to harmful levels of noise, and the many ways to assess the impact of the condition on individuals. As this chapter discusses, recent research findings suggest both that older adults are retaining good hearing longer in life (suggesting that previous assumptions about the contribution of aging to adult hearing loss may be flawed) and that NIHL may be increasing as a problem in children and adolescents. Both of these findings, if confirmed, could enhance our appreciation of the relative contribution of NIHL to the overall burden of hearing loss in the population.

2.1 Lack of a Common Case Definition for NIHL

As an example of the diverse ways that NIHL is defined and tracked, current regulatory practice in the United States regarding NIHL employs several different definitions of hearing loss. These definitions include a certain degree of audiometric

Table 2.2 Hearing loss metrics in use in the United States

Hearing loss metric	Criteria
Occupational Safety and Health Administration (OSHA) Standard Threshold Shift (STS)	10-dB change from the baseline audiogram in the average of hearing threshold levels at 2, 3, and 4 kHz, with age correction allowed
OSHA "recordable" hearing loss	10-dB shift from baseline as described above with the average of absolute hearing threshold levels at 2, 3, and 4 kHz greater than or equal to 25 dB HL
American Medical Association (AMA) Hearing Impairment	Hearing threshold average at 0.5 (500 Hz), 1, 2, and 3 kHz greater than 25 dB HL, with 1.5% monaural impairment for each decibel greater than 25 dB

"shift" from a baseline audiogram for a noise-exposed worker being tested in a hearing conservation program [U.S. Occupational Safety and Health Administration (OSHA) STS (OSHA 1983) and OSHA recordable hearing loss], as well as absolute value cutoffs for hearing impairment (AMA hearing impairment: American Medical Association 2008). These definitions are shown in Table 2.2.

Many other governmental definitions of hearing loss are in use in different countries, there are no agreed upon international standards for tracking NIHL, and even across different states in the United States there are varying definitions of compensable hearing loss (Dobie and Megerson 2000). The research literature is similarly diverse, with some studies using governmental definitions to define outcomes and others using hearing threshold levels at single noise-sensitive frequencies, or other combinations of frequencies.

Another method of defining NIHL has been through the use of "notch definitions" determining the presence or absence of a high-frequency "notching" of the audiogram. Such a notch is typically centered around 3,000 or 4,000 Hz with recovery at 8 kHz, (ACOEM Noise and Hearing Conservation Committee 2003), as shown in Fig. 2.1.

The definition of a noise notch provided by Niskar et al. (2001) requires all of the following criteria to be met: (1) thresholds < 15 dB HL at 0.5 and 1.0 kHz; (2) 3, 4, or 6 kHz threshold at least 15 dB worse than thresholds at 0.5 and 1 kHz; and (3) 3, 4, or 6 kHz threshold at least 10 dB worse than 8-kHz threshold. Coles et al. (2000) offered an alternative medicolegal definition of a noise-notch, including a hearing threshold at 3, 4, or 6 kHz that is at least 10 dB greater than at 1 or 2 kHz, and at least 10 dB greater than at 8 kHz. To date, however, there remains no commonly agreed upon definition for an audiometric notch (McBride and Williams 2001), although some published criteria demonstrate good agreement with expert judgment (Rabinowitz et al. 2006a). It is clear that the diagnosis of NIHL, and the estimates regarding the prevalence of NIHL, will vary as a function of the definition selected; thus, comparisons across studies must carefully compare the specific criteria used in each one.

In addition to audiometric definitions, other studies of NIHL prevalence may rely on individual self-report of hearing difficulty in surveys, or use other testing modalities

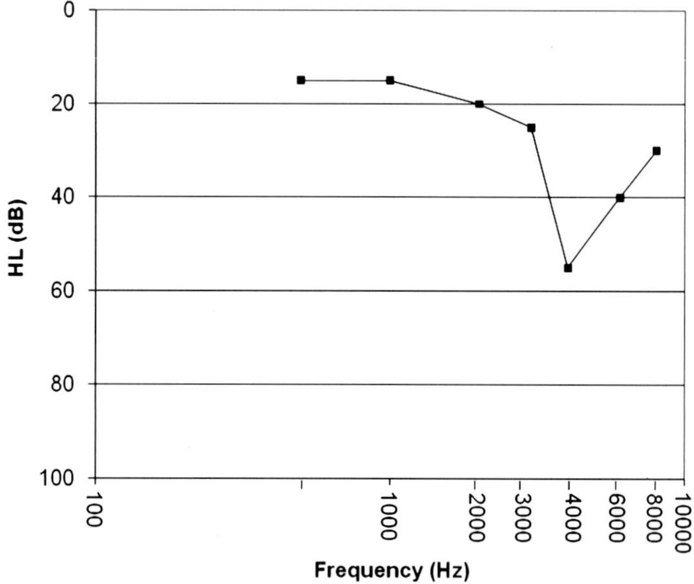

Fig. 2.1 Audiogram showing typical high-frequency "notching" of NIHL

such as otoacoustic emissions that may be sensitive indicators of noise-induced cochlear damage (Korres et al. 2009). As a result of the heterogeneity of these case definitions, comparisons between published studies of NIHL prevalence are often difficult.

2.2 *Differentiation Between NIHL and Presbycusis*

One of the dilemmas in assessing the importance of NIHL as a public health problem is the clinical similarity between presbycusis and NIHL. A central feature of both conditions is sensorineural hearing loss involving predominantly the higher audiometric frequencies. The presence of an audiometric "notch," as described earlier, is thought to be suggestive of noise-induced damage rather than presbycusis (ACOEM Noise and Hearing Conservation Committee 2003). However, as Taylor demonstrated in his studies of noise-exposed weavers (Taylor et al. 1965), over time the effects of noise and aging may superimpose and make a noise notch less evident, thus making it harder to distinguish the relative contributions of noise and aging (for review of age/noise interactions, see Bielefeld, Chap. 10).

Some recommend the use of age-standardized tables of the amount of hearing loss that would be expected in the absence of noise exposure in order to adjust, or "age-correct," for the effects of age and separate out the effects of noise. Examples of such tables include the annexes of the American National Standards Institute (ANSI) standard 3.44 (ANSI 1966). The OSHA noise standard uses similar methods to create an age correction table that can be applied to individual audiograms. Such a process of age standardization can have validity when applied to a population, but will inevitably misclassify some individuals by either under- or overestimating the relative effects of aging and noise for that individual. As a result, some agencies discourage the use of age correction for individual audiograms (NIOSH 1998).

Of course, the use of such aging tables to determine the amount of NIHL occurring in either an individual or a population is based on the dual assumptions that a population can be found without significant noise exposure in order to display the effects of aging alone, and second that the pattern of hearing loss in the reference population can be applied to other populations. In terms of the first issue, although it is possible to find populations who deny working at jobs with significant noise exposure, the ubiquitous environmental noise exposures of daily life as well as non-occupational noise exposures such as power tools and motorized vehicles may be present in the "nonindustrial noise-exposed population" (NINEP) used to develop some of the age tables. In the context of this discussion, it is worth considering the published cross-sectional surveys of the hearing status of populations living in the absence of significant occupational or nonoccupational noise exposures, such as the Mabaan tribe of Sudan; those studies report only minimal changes in hearing with advancing age (Bergman 1966). Therefore, by assuming that a certain proportion of an individual's hearing loss is due to "aging," it is possible that one is underestimating the chronic effects of nonoccupational noise exposure (socioacusis) on his or her hearing status.

In terms of whether the background rate of age-related loss in the population is accurately reflected by standard aging tables such as those in ANSI 3.44, there is evidence suggesting that the general U.S. population may now be exhibiting less "age-related loss" than did previous generations. A recent study explored the effect of age cohort on hearing loss risk among 5,725 adults living in Beaver Dam, Wisconsin, by looking at the rates of hearing impairment in each 5-year birth cohort (Zhan et al. 2010). This study found that after controlling for age, every 5-year increase in birth year lowered the odds of having a hearing impairment by 13% in men and 6% in women. In other words, in the population under study, decreased rates of hearing impairment appear to correlate with more recent dates of birth. Although this could partially reflect differences in noise exposure between birth cohorts, it also suggests that as a result of population-wide changes in medical care, nutrition, and general health, present-day adults may be retaining good hearing longer than previous generations. If this is truly the case, then it would be inappropriate to apply aging tables developed more than 40 years ago to estimates of the current burden of NIHL in the population.

2.3 *Estimates of the Number of Noise-Exposed Persons*

A key method in many estimates of the public health importance of NIHL has been to calculate the approximate size of the population that is exposed to potentially dangerous levels of noise. Usually such estimates have focused on working populations and occupational noise exposure. In 1981, OSHA estimated that 7.9 million manufacturing workers were exposed to noise at daily levels at or above 80 dBA, while the U.S. Environmental Protection Agency (U.S. EPA) in the same year determined that nine million U.S. workers, mostly in manufacturing or utilities industries, were exposed through their occupation to noise levels in excess of 85 dBA (NIOSH 1998). Apparently, there have been no comparable surveys of U.S. workers in recent decades; thus, any current estimates of the burden of NIHL must continue to rely on these probably outdated data. At the same time, some recent studies clearly suggest that a large segment of the U.S. population reports past or present occupational noise exposure. Tak et al. (2009) analyzed 1999–2004 data from the National Health and Nutrition Examination Survey (NHANES) to determine the prevalence of occupational exposure from self-report. They found that 22 million workers (17% of the population-weighted survey) reported exposure to hazardous occupational noise, and among these, 34% reported nonuse of hearing protective devices (HPDs).

Also notable about most estimates of noise-exposed persons used to determine the burden of hearing loss due to noise is that they fail to approximate meaningfully the size of the population exposed to potentially damaging noise outside of work. These noise sources include firearms; power tools; motorcycles, snowmobiles, and other loud vehicles; and amplified music listened to at concerts, nightclubs, or through personal music players. For many of these sources, noise surveys have documented sound levels in a range in which even relatively brief exposures could be damaging over time. Yet rigorous epidemiological studies measuring either the true prevalence of potentially damaging nonoccupational noise exposures or their associations with documented hearing loss remain scarce and sometimes contradictory. For example, the association between reported recreational firearm use and adult hearing loss has been shown in a number of studies (Beckett et al. 2000), but the degree of hearing loss risk from amplified music, although much discussed, especially with reference to adolescent age groups, remains controversial (Zhao et al. 2010). In a study of construction apprentices, Neitzel et al. (2004) found that when compared with the high levels of occupational noise to which they were exposed, nonoccupational noise exposures presented little additional exposure for most workers, although they may contribute significantly to overall exposure in the subset of workers who frequently engaged in noisy activities. Recent studies suggesting increased rates of hearing loss in adolescents have focused greater attention on these issues (Shargorodsky et al. 2010a), and further studies may confirm that the risk of nonoccupational noise exposure is greater than previously believed. In the meantime, any use of occupational noise exposure estimates alone to calculate the burden of NIHL on the U.S. population will inevitably neglect the impact of nonoccupational noise because this impact remains largely unknown.

2.4 Estimates of the Prevalence of Occupational NIHL

In 1972, NIOSH assessed the excess risk of material hearing impairment (defined as binaural average threshold levels in excess of 25 dB HL at 0.5, 1, and 2 kHz) for persons exposed to noise over a 40-year working lifetime as 3% at 80 dBA, 15% at 85 dBA, and 29% at 90 dBA. OSHA used these estimates to set an action level for hearing conservation programs at 85 dBA and a permissible exposure for noise of 90 dBA (OSHA 1983). This type of risk assessment has been part of some estimates of the prevalence of occupational NIHL.

The World Health Organization (WHO) bases many of its public health decisions and policies on studies of the global burden of illness caused by particular diseases or hazards (Terry and Rijt 2010). Applying this approach to occupational NIHL, Nelson et al. (2005) used NIOSH estimates of the prevalence of noise exposure adjusted by data on the distribution of the workforce by occupational category and economic sector, and economic activity rates in each WHO subregion. They defined disabling hearing loss according to WHO criteria as hearing thresholds of greater than 41 dB HL, and extrapolated the risk due to noise from studies of U.S. and British populations. Using these estimates for the worldwide population exposed to noise and the risk of NIHL loss due to such exposure, Nelson et al. (2005) calculated estimates of the attributable fraction (AF) of adult-onset hearing loss resulting from occupational noise exposure. They then applied AFs to WHO estimates of total disability adjusted life years (DALYs) from adult-onset hearing loss to estimate the DALYs due to occupational noise. This modeling exercise found that occupational noise accounts for 16% of the disabling hearing loss in adults (more than four million DALYs), with estimates of disease burden ranging from 7% in developed countries to 21% in underdeveloped and developing WHO subregions. The estimate of the effect of occupational noise on hearing loss burden was greater for males than for females in all subregions. Although this study helped bring NIHL in line with other occupational diseases causing disability worldwide (Driscoll et al. 2005), the analysis failed to consider the impact of lesser degrees of hearing loss, as well as the public health effect of nonoccupational noise.

Dobie (2008), in an alternative analysis, also attempted to estimate the burden of occupational NIHL, and to compare this impact to that of age-related hearing loss. This study considered hearing impairment as a continuous variable, using the AMA hearing impairment criteria of average thresholds at 0.5, 1, 2, and 3 kHz of greater than 25 dB HL with the percentage monaural hearing impairment (MHI) calculated as 1.5% × the pure tone average at 0.5, 1, 2, and 3 kHz – 25 (PTA5123 – 25). Using U.S. Census data, Dobie divided the U.S. population into subgroups based on age, gender, and occupational noise exposure, using NIOSH/OSHA estimates of the size of the noise exposed working population (see earlier). For each subgroup, the burden of hearing loss, in "units of hearing impairment" (UHI), was estimated as the product of MHI and the number of persons in the subgroup. Using these methods to model burden of hearing loss, Dobie (2008) found a result similar to that of the WHO study: that less than 10% of the burden of hearing loss in the United States was due to occupational noise. He concluded that scientific efforts to explore possible preventive treatments for presbycusis such as the role of folate supplementation (see Shargorodsky et al. 2010b) might have a far greater impact on the burden of hearing loss in the United States compared to efforts to reduce noise exposure. Again, as in the WHO analysis, Dobie did not consider either the impact of nonoccupational noise exposures on the population risk of hearing loss or the possible health impact of hearing loss below the level of AMA impairment or occurring at high frequencies (such as 4 kHz) not included in AMA impairment calculations. He also relied, as did the WHO study, on older estimates of both the size of the noise-exposed working population, as well as the expected hearing loss due to aging.

Other smaller studies suggest that the impact of NIHL may be greater than estimated by Nelson et al. (2005) or Dobie (2008). In a Michigan telephone survey study of active surveillance for hearing loss and occupational NIHL using questions added to the U.S. Centers for Disease Control and Prevention (CDC) Behavioral Risk Factor Surveillance System (BRFSS), a high prevalence of self-reported hearing loss was found (44% of respondents ages 75 or older; Stanbury et al. 2008). In that study, approximately 30% of persons whose hearing loss began at age 16 or later associated the hearing loss with occupational noise exposure (Stanbury et al. 2008). A Michigan occupational health surveillance initiative asking audiologists and otolaryngologists to report cases of work-related cases of NIHL identified 1,378 cases between 1992 and 1997, producing evidence that the number of patients with identified occupational NIHL is likely a gross underestimate of the prevalence of the disease (Reilly et al. 1998). Finally, in another study of NHANESIII data, Tak and Calvert (2008) found evidence of hearing difficulty in 11% of individuals ages 18–65, and based on questionnaire responses estimated that 24% of this hearing loss could be attributed to occupational noise, a much higher proportion than found by either the WHO or Dobie analyses. Although this study was limited by its basis in self-reports, the possibility remains that the true burden of illness from occupational noise exposure alone is greater than the modeling studies would suggest.

2.5 *Estimates of the Prevalence of Nonoccupational NIHL*

As mentioned previously, much of the work to date assessing the public health impact of NIHL has focused on the occupational setting. The true extent of noise effects in the general population remains poorly understood. Niskar et al. (2001) studied audiograms of children and adolescents in the NHANESIII survey, and reported that among U.S. children 6–19 years old, 12.5% (~5.2 million) had evidence of audiometric notching suggestive of NIHL. Similarly, a study of audiograms of young adults ages 17–25 entering an industrial workforce found that 16% showed evidence of high-frequency hearing loss at noise-sensitive frequencies, and that even in these young adults, the risk increased with each year of age (Rabinowitz et al. 2006b). At the same time, the rates of high-frequency loss were not increasing over a two-decade period, suggesting that although NIHL may be a widespread problem, it may not be increasing among young people today compared to previous generations, despite increasing use of personal music players and other electronic devices (Rabinowitz et al. 2006b). These studies of young people before entry into the workforce indicate, however, that nonoccupational noise exposure plays a significant role in the overall burden of hearing loss in the population. As further studies are performed, the widespread hearing losses that some have predicted in adolescents due to noise exposure may yet materialize. For example, a recent analysis comparing hearing thresholds of adolescents in the 1988–1994 and 2005–2006 NHANES surveys found that the prevalence of hearing loss increased from 14.9% to 19.5%, often involving the higher (noise-sensitive) frequencies (Shargorodsky et al. 2010a).

2.6 *Estimates of the Severity of NIHL*

As Table 2.1 depicts, the public health impact of a condition involves both the condition's prevalence and the severity. It is clear that, as a medical disorder, hearing loss can affect the quality of life for adults (Dalton et al. 2003). Conventional measures such as the AMA impairment calculations may not capture all of the true morbidity of NIHL. To begin with, NIHL may start to affect overall function at a younger age than age-related loss, and the overall impact of hearing loss in a younger, more active person may be relatively greater than in an older person, such as the impact on learning and communication at work, although this has not been extensively studied. However, one aspect of this premature loss could be the effect of NIHL on risk of accidents in a working age population. Recent studies suggest that NIHL does indeed predispose an individual to the risk of work-related accidents. One such study in British Columbia found that the severity of hearing impairment, calculated as average bilateral hearing threshold levels at 3, 4, and 6 kHz, increases the relative risk of single and multiple work accident events when threshold levels exceed 15 dB HL (Girard et al. 2009). Girard et al. also suggested that loss at audiometric frequencies not included in AMA impairment definitions (such as 4 and 6 kHz) and subtle loss with hearing thresholds less than 25 dB HL can have a significant impact on functioning in working adults, neither of which condition is considered by the Nelson et al. (2005) or Dobie (2008) models of disease burden due to NIHL.

Despite such evidence of the impact of NIHL on younger persons, a review of the published literature reveals that there have been very few studies assessing the true severity and cost of illness of NIHL on individuals. Consequently, current estimates continue to rely on crude measures such as the cost of a worker's compensation claim (Bertsche et al. 2006) or the AMA impairment percentages. Another unknown in the determination of the severity of NIHL as an illness is whether the hearing loss caused by noise damage has any different impact on an individual than the loss due to presbycusis. For example, does a noise-damaged ear process speech or other stimuli differently than an ear affected by presbycusis, with the same audiometric thresholds? (For discussion, see Shrivastav, Chap. 7). The severity of NIHL as a medical condition therefore remains an area for further research and policy discussion, which could add to our understanding of the impact of NIHL on health and function.

One basic dogma that drives current assessment of the impact of NIHL (in addition to the assumption that the effects of noise and aging are additive) is that noise damage stops when noise exposure stops. In other words, according to this dogma, noise damage to the cochlea does not predispose the ear to lose hearing at a faster rate once the person leaves the noisy environment (ACOEM Noise and Hearing Conservation Committee 2003). If, however, noise exposure early in life does change the natural history of the aging ear process and contributes to accelerated loss later, the real impact of NIHL on hearing loss rates would be much greater than currently thought. Several studies in animals suggest this effect of prior noise on subsequent hearing loss could occur in some mammalian species, possibly including humans, which would be a disturbing outcome if confirmed. Specifically, Kujawa and Liberman (2009) reported that a detailed histological examination of the ears of mice exposed to noise levels that caused moderate (~40 dB) temporary hearing loss (which was completely reversible) revealed acute loss of afferent nerve terminals and delayed degeneration of the cochlear nerve. These outcomes provide one potential explanation for the increased age-related changes observed in an earlier study in which mice that were exposed to temporary threshold shift inducing noise were subsequently allowed to age (Kujawa and Liberman 2006). Human epidemiological evidence to support such an acceleration of hearing loss as a result of prior noise exposure remains inconclusive. Gates et al. (2000), in an examination of audiograms of older individuals in the Framingham Study, found that those with evidence of noise notches appeared to have accelerated rates of high-frequency loss over time compared to individuals without such noise notching, but few studies have confirmed this finding. As further research explores the relationship between the size of the temporary threshold shift and the later impact on hearing during aging, as well as the extent to which this translates from rodents to humans, it could radically change our assessment of the long-term impact of noise exposures on the auditory system, and the relative importance of noise and aging in the development of acquired sensorineural hearing loss.

2.7 Future Trends

The lack of certainty in estimates of the current public health impact of NIHL makes it even more difficult to speculate about whether NIHL is increasing or decreasing in importance. However, a few trends are worth noting as areas for future attention. Rapid changes in communication technology are placing new demands on an individual's speech perception abilities, such as hearing a cell phone ring or conducting a conversation in the presence of background noise. The impact of noise-induced cochlear damage on these communication settings may be greater than currently appreciated. In addition, the pace of globalization has increased the amount of communication taking place between persons with multilingual backgrounds, who often need to communicate in a language other than their own first language. There is some evidence that persons communicating in a second language, such as English as a second language for native Spanish speakers, may be more likely to report speech communication difficulties at a given level of hearing loss than native English speakers (Rabinowitz et al. 2005).

3 Summary

Hearing loss is one of the most common chronic conditions in adults (Cruickshanks et al. 1998), and yet it often fails to receive the recognition it deserves, often remaining underdiagnosed and undertreated by healthcare professionals (Bogardus et al. 2003). NIHL, as a subset of hearing loss in general, is also likely to be overlooked. It appears to be a condition that is quite prevalent, but that often exhibits only a mild degree of severity. As such, it may have a significant public health importance but be susceptible to routine underestimation. Attempts to assess the public health impact of NIHL have focused on occupational noise-exposed persons and have used measures of hearing impairment that may not capture the true burden of disease in the general population. Evidence that the U.S. population as a whole may be experiencing less age-related hearing loss than in previous generations suggests that the relative importance of NIHL versus presbycusis may actually be increasing, and that the use of standard tables based on populations norms of 50 years ago to adjust audiograms for the effect of aging may not be appropriate in the future. Provocative new research findings suggest that noise exposure may exert greater long-term damage on the cochlea than previously thought, and that the impact of NIHL on accident risk and other functional abilities in younger adults may be significant. If such findings are confirmed with further study, it will further force a reappraisal of the public health significance of NIHL.

References

ACOEM Noise and Hearing Conservation Committee. (2003). ACOEM evidence-based statement: Noise-induced hearing loss. *Journal of Occupational and Environmental Medicine*, 45(6), 579–581.

American Medical Association (AMA). (2008). *Guides to the evaluation of permanent impairment* (6th ed.). Chicago: American Medical Association.

ANSI. Standard (1966). *Determination of occupational noise exposure and estimation of noise-induced hearing impairment*. Washington, DC: American National Standards Institute.

Beckett, W. S., Chamberlain, D., Hallman, E., May, J., Hwang, S. A., Gomez, M., Eberly, S., Cox, C., & Stark, A. (2000). Hearing conservation for farmers: Source apportionment of occupational and environmental factors contributing to hearing loss. *Journal of Occupational and Environmental Medicine*, 42(8), 806–813.

Bergman, M. (1966). Hearing in the Mabaans. *Archives of Otolaryngology*, 84(4), 411–415.

Bertsche, P. K., Mensah, E., & Stevens, T. (2006). Complying with a corporate global noise health surveillance procedure—Do the benefits outweigh the costs? *AAOHN Journal*, 54(8), 369–378.

Bogardus, S. T., Jr., Yueh, B., & Shekelle, P. G. (2003). Screening and management of adult hearing loss in primary care: Clinical applications. *JAMA*, 289(15), 1986–1990.

Coles, R. R., Lutman, M. E., & Buffin, J. T. (2000). Guidelines on the diagnosis of noise-induced hearing loss for medicolegal purposes. *Clinical Otolaryngology and Allied Sciences*, 25(4), 264–273.

Cruickshanks, K. J., Wiley, T. L., Tweed, T. S., Klein, B. E., Klein, R., Mares-Perlman, J. A., & Nondahl, D. M. (1998) Prevalence of hearing loss in older adults in Beaver Dam, Wisconsin.

The Epidemiology of Hearing Loss Study. *American Journal of Epidemiology*, 148(9), 879–886.

Dalton, D. S., Cruickshanks, K. J., Klein, B. E., Klein, R., Wiley, T. L., & Nondahl, D. M. (2003). The impact of hearing loss on quality of life in older adults. *Gerontologist*, 43(5), 661–668.

Dobie, R. A. (2008). The burdens of age-related and occupational noise-induced hearing loss in the United States. *Ear and Hearing*, 29(4), 565–577.

Dobie, R. A., & Megerson, S. C. (2000). Workers Compensation. In *The noise manual*, (5th ed). Fairfax, VA: American Industrial Hygiene Association.

Driscoll, T., Takala, J., Steenland, K., Corvalan, C., & Fingerhut, M. (2005). Review of estimates of the global burden of injury and illness due to occupational exposures. *American Journal of Industrial Medicine*, 48(6), 491–502.

Gates, G. A., Schmid, P., Kujawa, S.G., Nam, B., & D'Agostino, R. (2000). Longitudinal threshold changes in older men with audiometric notches. *Hearing Research*, 141, 220–228.

Girard, S. A., Picard, M., Davis, A. C., Simard, M., Larocque, R., Leroux, T., & Turcotte, F. (2009). Multiple work-related accidents: Tracing the role of hearing status and noise exposure. *Occupational and Environmental Medicine*, 66(5), 319–324.

Korres, G. S., Balatsouras, D. G., Tzagaroulakis, A., Kandiloros, D., Ferekidou, E., & Korres, S. (2009). Distortion product otoacoustic emissions in an industrial setting. *Noise & Health*, 11(43), 103–110.

Kujawa, S. G., & Liberman, M. C. (2006). Acceleration of age-related hearing loss by early noise exposure: Evidence of a misspent youth. *Journal of Neuroscience*, 26(7), 2115–2123.

Kujawa, S. G., & Liberman, M. C. (2009). Adding insult to injury: Cochlear nerve degeneration after "temporary" noise-induced hearing loss. *Journal of Neuroscience*, 29(45), 14077–14085.

McBride, D. I., & Williams, S. (2001). Audiometric notch as a sign of noise induced hearing loss. *Occupational and Environmental Medicine*, 58(1), 46–51.

Neitzel, R., Seixas, N., Goldman, B., & Daniell, W. (2004). Contributions of non-occupational activities to total noise exposure of construction workers. *Annals of Occupational Hygiene*. 48(5), 463–473.

Nelson, D. I., Nelson, R.Y., Concha-Barrientos, M., & Fingerhut, M. (2005). The global burden of occupational noise-induced hearing loss. *American Journal of Industrial Medicine*, 48(6), 446–458.

National Institute of Occupational Safety and Health (NIOSH). (1998). *Criteria for a recommended standard: Occupational noise exposure revised criteria 1998*. Cincinnati: DHHS. 105.

Niskar, A. S., Kieszak, S. M., Holmes, A. E., Esteban, E., Rubin, C., & Brody, D. J. (2001). Estimated prevalence of noise-induced hearing threshold shifts among children 6 to 19 years of age: The Third National Health and Nutrition Examination Survey, 1988–1994, United States. *Pediatrics*, 108(1), 40–43.

Occupational Safety and Health Administration (OSHA). (1983). 1910.95 CFR *Occupational noise exposure: Hearing Conservation Amendment (Final Rule)*, In: 48 Federal Register, pp. 9738–9785.

Rabinowitz, P. M., Sircar, K. D., Tarabar, S., Galusha, D., & Slade, M. D. (2005). Hearing loss in migrant agricultural workers. *Journal of Agromedicine*, 10(4), 9–17.

Rabinowitz, P. M., Galusha, D., Slade, M. D., Dixon-Ernst, C., Sircar, K. D., & Dobie, R. A. (2006a). Audiogram notches in noise-exposed workers. *Ear and Hearing*, 27(6), 42–50.

Rabinowitz, P. M., Slade, M. D., Galusha, D., Dixon-Ernst, C., & Cullen, M. R. (2006b). Trends in the prevalence of hearing loss among young adults entering an industrial workforce 1985 to 2004. *Ear and Hearing*, 27(4), 369–375.

Ramazzini B. (1713– revised edition 1940). *De morbis artificum* (*Diseases of workers*), the Latin text of 1713, revised by WC Wright. Chicago: University of Chicago Press. Page 437.

Reilly, M. J., Rosenman, K. D., & Kalinowski, D. J. (1998). Occupational noise-induced hearing loss surveillance in Michigan. *Journal of Occupational and Environmental Medicine*, 40(8), 667–674.

Shargorodsky, J., Curhan, S. G., Curhan, G. C., & Eavey, R. (2010a). Change in prevalence of hearing loss in US adolescents. *JAMA*, 304(7), 772–778.

Shargorodsky, J., Curhan, S. G., Eavey, R., & Curhan, G. C. (2010b). A prospective study of vitamin intake and the risk of hearing loss in men. *Otolaryngology Head and Neck Surgery*, 142(2), 231–236.

Stanbury, M., Rafferty, A. P., & Rosenman, K. (2008). Prevalence of hearing loss and work-related noise-induced hearing loss in Michigan. *Journal of Occupational and Environmental Medicine*, 50(1), 72–79.

Tak, S., & Calvert, G. M. (2008). Hearing difficulty attributable to employment by industry and occupation: An analysis of the National Health Interview Survey—United States, 1997 to 2003. *Journal of Occupational and Environmental Medicine*, 50(1), 46–56.

Tak, S., Davis, R. R., & Calvert, G. M. (2009). Exposure to hazardous workplace noise and use of hearing protection devices among US workers—NHANES, 1999–2004. *American Journal of Industrial Medicine*, 52(5), 358–371.

Taylor, W., Pearson, J., Mair, A., & Burns, W. (1965). Study of noise and hearing in jute weaving. *Journal of the Acoustic Society of America*, 38, 113–120.

Terry, R. F., & Rijt, T. V. (2010). Overview of research activities associated with the World Health Organization: Results of a survey covering 2006/07. *Health Research Policy and Systems*, 8(1), 25.

Zhao, F., Manchaiah, V. K., French, D., & Price, S. M. (2010). Music exposure and hearing disorders: An overview. *International Journal of Audiology* 49(1), 54–64.

Chapter 3
Noise-Induced Hearing Loss and Tinnitus: Challenges for the Military

Marjorie A.M. Grantham

1 Introduction

Noise-induced hearing loss (NIHL) and tinnitus present special challenges for the military. Soldiers, sailors, airmen, Marines, and the civilians who serve beside them are exposed to noise levels that are higher than most individuals in industrial operations, putting them at increased risk of hearing loss. Yet these military populations rely on their hearing to a much greater extent than others do. Not only do military personnel, and their civilian counterparts serving in training and combat environments, require hearing for clear communication, but they also need their hearing for optimal survival and lethality. The dangers of miscommunication on the battlefield are clear. For example, imagine hearing "Attack!" instead of "Get back!" in the middle of a firefight. Indeed, the U.S. Marine Corps Center for Lessons Learned concluded that command and control during the battle of Fallujah was significantly degraded when exposure to high-intensity combat operations caused NIHL (Marine Corps Center for Lessons Learned 2005). Although many military troops and government civilians are not directly involved in firing upon and defeating the enemy, many serve in the same noise hazardous environments as these war fighters, both during training and in combat theaters. This chapter focuses on military-specific noise exposure, the effects of NIHL and tinnitus on military operations, hearing conservation programs within the military, and future directions for NIHL and tinnitus research specific to our military and the civilians who support them.

M.A.M. Grantham (✉)
US Army Public Health Command (Provisional),
5158 Blackhawk Road, Aberdeen Proving Ground, MD 21010-5403, USA
e-mail: marjorie.grantham@us.army.mil

C.G. Le Prell et al. (eds.), *Noise-Induced Hearing Loss: Scientific Advances*,
Springer Handbook of Auditory Research 40, DOI 10.1007/978-1-4419-9523-0_3,

2 Military-Specific Noise Exposure

One of the most significant challenges for developing and implementing an effective hearing conservation program in a military environment is that military-related noise levels are uniquely variable. Weapons fire, improvised explosive device (IED) blasts, fixed- and rotary-winged aircraft, armored vehicle, and aircraft carrier noise exposure often exceed the U.S. Occupational Safety and Health Administration (OSHA) daily allowable noise dose in seconds (see Table 3.1 for OSHA noise dose limits and Table 3.2 for examples of sound levels produced during use of military technology). For example, just one round from common small-arms rifles fired regularly by U.S. troops reaches 157 dBP (decibels peak sound pressure level), while another commonly fired weapon exceeds 183 dBP. Loud sounds are not limited to impulsive

Table 3.1 Permissible noise exposure time over work day, with sound level measured in dBA, slow response

Duration	dBA
8 h	85
4 h	88
2 h	91
1 h	94
30 min	97
15 min	100
7.5 min	103
3.75 min	106
1.875 min	109

Army time intensity exchange rates. When the daily noise exposure is composed of two or more periods of noise exposure of different levels, their combined effect is considered, rather than the individual effect of each. If the sum of the following fractions: $C(1)/T(1)+C(2)/T(2)+\ldots+C(n)/T(n)$ exceeds unity, then the mixed exposure is considered to exceed the limit value. $C(n)$ indicates the total time of exposure at a specified noise level, and $T(n)$ indicates the total time of exposure permitted at that level. Exposure to impulse or impact noise should not exceed 140 dB peak sound pressure level. (29 Code of Federal Regulations (CFR) 1910.95, 2009)

Table 3.2 Sound levels measured for different military technologies (dBA)

Technology	Noise type	Decibel level
M2A2 Bradley Tank	Continuous	120 dBA
UH 60 Helicopter	Continuous	110 dBA
155 mm Howitzer	Impulse	181 dBP
9 mm Pistol	Impulse	156 dBP
HMMWV	Continuous	88 dBA
60 mm Mortar	Impulse	180 dBP
MAAWS weapon system	Impulse	184–190 dBP

weapons noise. Naval aircraft carrier flight deck noise levels reach 152 dBA (steady-state, continuous noise, A-weighted decibels). Aircraft carrier flight deck noise levels can exceed the protective capability of even double hearing protection, i.e., both earplugs and cranial noise muffs, sometimes within a single launch for the personnel most exposed to aircraft launch noise (Wilt and Bjorn 2006). Military and civilian personnel operating within these noise levels often must not only communicate using speech, but also rely on their hearing to survive. Their ability to survive may be based on their ability to detect, determine their distance from, and the direction of, enemy versus friendly fire, vehicles, aircraft, and warning signals, as well as the presence and location of possible noncombatant civilians (Ohlin 2005; Wilt and Bjorn 2006).

3 Effects of Noise and Related Hearing Loss on Military Operations

3.1 Impact of Hearing Loss on Troop Readiness

Limited research has shown negative effects of temporary threshold shift (TTS) on the ability to conduct military operations. TTS refers to temporary changes in hearing, which occur due to short-term, intense noise exposure. Temporary changes in hearing are often of lesser concern than lasting, permanent threshold shifts (PTS) given that they are reversible, but even a temporary compromise in hearing status can be devastating for military personnel because these TTS deficits compromise suprathreshold function, such as speech intelligibility (for detailed discussion of suprathreshold function, see Shrivastav, Chap. 7). One study relevant to this issue showed that as speech intelligibility delivered over earphones decreased from 93.5% to 7.1% during tank skills training, the percentage of time only a single round was required to destroy a target decreased from 90% to 62% (Peters and Garinther 1990). This resulted in friendly fire incidents increasing from 0% to 8.1%, and enemy success targeting friendly tanks increased from 7% to 28% (Peters and Garinther 1990). Limited information is available regarding the immediate impact of traumatic brain injury (TBI) and hearing loss on mission success. A recent study of Operation Iraqi Freedom (OIF) and Operation Enduring Freedom (OEF) blast-exposed patients at military medical facilities indicated that more than 50% demonstrated significant hearing loss, while half reported tinnitus (Cave et al. 2007). In addition, blast trauma and TBI accounted for one quarter of all Marine injuries from the start of OIF through 2004 (Fausti et al. 2009).

3.2 Impact of Hearing Loss on Situational Awareness

Both oral communication and normal hearing are assumed to be requirements for maximum survivability and lethality, yet auditory situational awareness is not clearly defined. Currently, there exists no database of sounds critical to troop survivability and lethality available for use in research. Little is known about the importance of stationary and moving sound localization in combat environments (Scharine and Letowski 2005; Grantham et al. 2010). Research examining the contributions of judgments of sound distance and direction is also minimal (Mershon et al. 1981). Clearly, troops need to have normal hearing as defined using the traditional audiogram, i.e., pure tones within the range of speech frequencies at "normal" intensities, in order to communicate on the battlefield. However, they must also be able to hear and distinguish weapons signatures, direction and distance of sniper fire, warning signals, and translation devices. In addition to existing diagnostic audiometric test batteries, new tools are needed to measure baseline suprathreshold troop hearing performance, as well as minimum hearing requirements for successful completion of specific military tasks. Once "normal hearing" for military personnel is defined, the effects of hearing loss, hearing protection devices, communications technology use (e.g., air, bone, tactile), changing auditory environments based on signal-to-noise or direct-to-reflected sound, and time history of noise insult on mission performance must be identified, evaluated, and understood.

3.3 *Impact of Tinnitus on Troop Readiness*

Research directly examining the impact of tinnitus on military operations is not currently available. Tinnitus has, however, increased over the years, as evidenced by its surpassing hearing loss as the number one disability for Gulf War veterans and total fiscal year (FY) 2009 veterans (U.S. Department of Veterans Affairs 2010). Tinnitus, which can interfere with sleep and the ability to focus, may place troops at risk in environments where staying alert equates to staying alive (Tyler 2000). Many questions basic to finding solutions to the problems of NIHL and tinnitus remain. The Institute of Medicine's (IOM) 2005 report on noise and military service recommended that future research investigate "the acoustic parameters associated with," and "the mechanisms, natural history, epidemiology, measurement, and treatment of noise-induced hearing loss and tinnitus" (Institute of Medicine 2005, p. 9). Noise-induced tinnitus is discussed in detail by Kaltenbach, Chap. 8.

3.4 *Financial Impact*

The Department of Veterans Affairs issues an annual benefits report each fiscal year. The most recent data available are for FY 2009 (U.S. Department of Veterans Affairs 2010). This report indicates that tinnitus and hearing loss are the two most prevalent service-connected disabilities for veterans receiving compensation at the end of FY 2009, respectively affecting 639,029 (5.7% of all veterans) and 570,966 (5.1% of all veterans). Indeed, a total of 1,350,484 veterans received compensation for the broad class of "Impairment of Auditory Acuity" during FY 2009; that category includes not only tinnitus and hearing loss, but also other sources of hearing loss such as otitis media (which affected 21,067 veterans). The report further establishes that the number of veterans affected with auditory impairments has been increasing every year, with 88,366 first receiving compensation in 2005, 92,407 first receiving compensation in 2006, 112,421 first receiving compensation in 2007, 118,935 first receiving compensation in 2008, and 135,701 first receiving compensation in 2009. The financial impact of rehabilitation and compensation is significant. Single-year disability costs related to hearing loss exceeded $900 million for 2006, with increasing costs largely related to the noise insults experienced by Soldiers serving in current OEF and OIF military operations (USACHPPM 2007). A more recent report states, "In 2007, [veteran] hearing loss compensation reached over one billion dollars for a predominantly preventable injury" (McIlwain et al. 2008).

Disability compensation is a benefit paid due to injuries or diseases occurring while personnel were serving on active duty or that were made worse by military service. Determinations of eligibility for hearing health benefits through the Department of Veterans Affairs are complex. Veterans' impairment of auditory acuity is calculated separately for each ear, as detailed in 38 CFR 4.85, based on pure tone threshold average and speech discrimination scores, using the Maryland Consonant-Nucleus-Consonant (CNC) test. For example, a pure tone average across 1,000, 2,000, 3,000, and 4,000 Hz of 65 dB HL, with 76% discrimination in the poorer ear, with a PTA of 30 dB HL and 90% discrimination in the better ear, results in 0% disability, but the hearing loss may still be considered service-connected, and the veteran may receive a hearing aid and rehabilitative services, if one of several other criteria are met. Service-connected NIHL has resulted in a significant financial and clinical burden for the VA. In FY 2007, VA audiologists dispensed more than $141 billion in hearing aids and provided related hearing health services costing more than $147 million. The National Center for Rehabilitative Audiology Research has developed a veteran-focused, kiosk-based hearing loss prevention program, to reduce the impact of hearing loss on overall veteran quality of life, including cognitive decline, reduced social interaction, depression, low self-esteem, anxiety, paranoia, frustration, and anger (Saunders and Griest 2009; Folmer et al. 2010).

4 Military Hearing Conservation Programs

4.1 Historical Perspective

From 1862 to 1920, one third of Union Army veterans suffered from hearing loss, most likely from weapons fire noise exposure (Sewell et al. 2004). Hearing loss was considered a disability, and Union Army veterans with unilateral or bilateral hearing loss received pensions ranging from approximately 33% to 66% of the average annual income for workers at that time. Audiology as a military profession did not emerge until later; it found its roots in caring for the hearing-impaired veterans of

World Wars I and II (Bergman 2002; Jerger 2009). At that time, audiology was largely involved in aural rehabilitation, primarily including hearing aid fitting and "lipreading" training at one of only a few Army and Navy installations.

Early research activities were focused on developing better measures of hearing loss and improving hearing aid technology (Bergman 2002). Hearing conservation quickly emerged as a key issue, however, and the U.S. Air Force formalized the first military hearing conservation program in 1948, with Air Force Regulation (AFR) 160-3, "Precautionary Measures Against Noise Hazards" (Nixon 2002). In 1970, OSHA set standards for total allowable workday noise exposure without hearing protection, and in 1978, Department of Defense Instruction (DoDI) 6055.12 required all armed services to meet or exceed the OSHA 1970 standard. In addition, DoDI 6055.12 required adoption of the seven elements of OSHA's industrial-noise exposure-based hearing conservation program: engineering controls, noise monitoring, audiometric testing, hearing conservation training, hearing protection, record-keeping, and program evaluation [29 Code of Federal Regulations (CFR) 1910.95, 2009]. OSHA standards are discussed in detail by Casali, Chap. 12. From 1970 through 1995, the DoD, through the efforts of each branch of the armed services working together, adopted common business practices for completing automated hearing tests and accessing data through common program management capabilities. Program management tools identify troops requiring annual, periodic, and follow-up testing; facilitate evaluation of local program effectiveness; and track both individual and unit deployment hearing readiness. Currently, the Army, Navy, and Air Force use one database management tool, the Defense Occupational and Environmental Health Readiness System – Hearing Conservation (DOEHRS-HC), to conduct and monitor audiometric testing. Audiometric information from this system will soon become readily available to the Department of Veterans Affairs.

Hearing protection and communication systems still have a long way to go to meet the comprehensive needs of military personnel. Military hearing protection must be flexible enough to provide intelligible, clear speech and transmit auditory warning signals, in either a quiet or tactical setting where what is heard by friendly forces does not give their position away to the enemy. On the other hand, military hearing protection must also work in the presence of varying levels of both steady-state and impulse noise, providing necessary attenuation, while not negatively limiting auditory signals contributing to troop situational awareness such as determining the location of the enemy. In addition to these diverse requirements, differences in individual troop hearing acuity and impairment, as well as hearing aid use, must also be considered.

4.2 Modern Approaches

In 2005, a Congress-directed and Department of Veterans Affairs-sponsored study by the Institute of Medicine (IOM), National Academies of Science, evaluated the DoD hearing conservation programs to determine whether current evidence allows the prediction of who will suffer from NIHL or tinnitus, based on time in service or military occupational specialty (Institute of Medicine 2005). The IOM committee's recommendations included the need for both service entry and retirement audiograms for all military personnel. As with all individuals who experience hearing loss, military and civilian DoD personnel with hearing loss face challenges and impacts on their quality of life. Military personnel with either permanent or temporary hearing loss, due to hazardous noise exposure, in the absence of hearing protection use, often do not hear the high-frequency consonants in speech or weapons signatures clearly, or soft sounds such as approaching enemy footsteps, and thus place their lives and those of their comrades at risk.

Military hearing conservation programs continue to change with new research evidence-based processes, technological advances, and metrics for ensuring that hearing loss is not inevitable on retirement from a military career. The current Army Hearing Program developed out of a need to track not only such metrics as significant threshold shifts and overall permanent hearing loss, but to also ensure that commanders know the Army is deploying and mobilizing hearing-ready Soldiers. The Army Hearing Program now consists of (1) hearing readiness, (2) operational hearing, (3) clinical hearing services, and (4) hearing conservation (Department of the Army 2008; McIlwain et al. 2008; Cleveland 2009).

The hearing readiness component "ensures that Soldiers have the required hearing capability to perform their job-specific duties and have the correct personal protective equipment" for each mission, while providing leaders with a hearing classification code designating each Soldier's hearing status, for example, class I indicates that all required audiometric testing is complete and a Soldier demonstrates no more than a mild high-frequency hearing loss, while class III indicates that all required testing is not complete, a significant hearing loss has been identified, or a medical review board determination of ability to remain on active duty status in a current job has not yet been made (ST 4-02.501, Department of the Army 2008). Operational hearing includes the application of engineering and administrative measures to reduce noise exposure and the negative impact of noise on mission success, while ensuring that Soldiers are trained to use current best hearing protection devices and communication systems to maintain their best hearing. Operational hearing services also include deployment of a military audiologist with the deployed troops. Army audiologists have deployed to Iraq since 2004, and the first will soon deploy to Afghanistan. Hearing conservation refers to the traditional, seven-element program based on OSHA (29 CFR 1910.95) and continues to serve as the standard for industrial-type military support operations such as conducted by Department of the Army civilians.

4.3 Challenges for Hearing Conservation Programs

Among the seven steps to conserving hearing, as designated by both OSHA and the Department of Defense (DoD), hearing protection is meant to serve as a *late* step in the prevention of NIHL. Engineering controls, such as the development and implementation of quieter technologies, plus administrative controls, such as requirements for periods of quiet away from aircraft carrier flight deck noise, help to conserve hearing without relying on the proper insertion, fit, and use of hearing protection by a noise-exposed individual. Hearing protection has its limits. As described in the preceding text, aircraft carrier flight deck noise levels can exceed the protective capability of even double hearing protection (Wilt and Bjorn 2006). While some DoD research laboratories are currently researching improved hearing protectors, others are focusing efforts on using technology to reduce the amount of time humans are required to spend time in the high-intensity levels on deck. Currently, little is known about the effects of traumatic noise exposure on the central nervous system (McIlwain et al. 2009). Finally, the military is currently participating in studies measuring the potential for protection against NIHL using novel drug agents such as *N*-acetylcysteine (Kopke et al. 2007) and ebselen (Lynch and Kil 2009), and other candidate therapeutics are emerging (for detailed review and discussion, see Le Prell and Bao, Chap. 13).

However compelling the research findings, equally important is the way in which the evidence and outcomes are communicated to troops. Hearing protection must not only attenuate hazardous noise levels while facilitating mission-critical hearing, but it must also not add significant weight to the troop's load or interfere with movement or other necessary equipment. An often made assumption is that increasing auditory input (i.e., the amount of auditory information presented to the ears and brain) will facilitate mission success. However, what is currently understood about the number of auditory signals humans easily process indicates that we may be limited to clearly tracking four to seven auditory inputs at once, and this number may vary with added visual or other input information (Cowan 1998). Adding visual and tactile information changes our perception of auditory stimuli (Mershon et al. 1980; Crum and Hafter 2001; Koelewijn et al. 2009). In addition to the influence of visual and tactile stimuli, covering the ears with a noise muff type headset or hearing protection device interferes with our ability to receive binaural cues required for localization, but little research indicates how much interference is too much. In light of the need for options that do not cover the outer ears, bone conduction and tactile displays are being studied (Myles and Kalb 2009). Introducing reflective surfaces and unfamiliar, irrelevant sounds, such as those found in urban operations overseas, often interferes with ground troop ability to detect, recognize, identify, and localize mission-relevant sounds (Scharine et al. 2009).

Perceived "overprotection" from hearing protectors is also a challenge faced by the military. Military personnel are more likely to accept attenuation when it is comfortable and not perceived to eliminate what must be heard for mission effectiveness (Ohlin 2010). To find a user-accepted hearing protector that allows military personnel to hear critical sounds at low levels, while providing protection from intense noise, scientists have developed nonlinear hearing protectors and communication systems, such as the Tactical Communication and Protective Systems (TCAPS). "The Army conducts warfare in a specific manner. We shoot, move, and communicate. However, we have placed far too little emphasis on the communication aspect of war fighting. We design radios that will communicate from the other side of the globe and ignore the last few inches it takes to be heard and understood." (LTC Eric Fallon, Army Audiology Consultant, Army Research Laboratory, in Ohlin 2009, p. 23). In addition to providing essential hearing while attenuating hazardous noise, new hearing protectors must be interoperable with all helmets, communication systems, headsets, night vision devices, radios, and other gear worn and used by troops (Powell et al. 2003; Casali et al. 2009) Common sense dictates that military personnel will use, and leaders will require their troops to use, hearing protection and communications technology when presented with solid evidence that the equipment protects them while not interfering with mission-critical hearing and when given a choice of protective devices. Currently, standard evaluation criteria are in development for approving hearing protection devices and tactical communication and protective systems, such as communications headsets, that also provide noise attenuation.

Two goals of every hearing conservation and readiness program should be to reduce hazardous noise exposure and facilitate communication ability in all hazardous noise environments. Currently, hearing protectors do not attenuate noise traveling by bone conduction, for example, directly to the cochlea. Also, hearing protection does not prevent the inhalation or absorption of chemical ototoxins, which also adversely affect hearing. Troops may be exposed to xylene, toluene, organic lead, diesel fuel, kerosene fuel, jet fuel, or organophosphate pesticides (for discussion of chemical interactions with noise insult, see Morata and Johnson, Chap. 11). In addition, convincing military personnel that using hearing protection will not negatively impact their situational awareness poses a real challenge. Resistance to using hearing protection has existed for as long as the military has required its use. Hearing protectors are perceived as uncomfortable or as interfering with mission-critical sound perception (Ohlin 2010). Troops express concern that they cannot hear low-level sounds such as the click of someone moving the lever of a rifle from "safe" to "fire" or the speech commands of a squad leader in the presence of vehicle, aircraft, or weapons fire background noise, all of which impact survivability and lethality. Wearable weapons fire localization systems may use a microphone array to determine the direction and distance of firing and have been developed in response to troop hearing needs in combat, for example, Soldier Wearable Acoustic Targeting System (SWATS).

4.4 Implementing Modern Military Hearing Conservation Programs

Once operationally supportive hearing protection and communication systems are designed, successful use is not automatic. Troops must receive regular, mission-focused training with the systems, in order to trust them in a combat environment. Just as a troop would not be expected to deploy without weapons or protective mask training, so, too, troops must not be expected to deploy without hearing protection and communication systems training. OSHA requires that hearing conservation programs provide annual training, including "the effects of noise on hearing; the purpose, advantages, disadvantages, and attenuation of various types of hearing protectors; instructions on selection, fit, use, and care of hearing protectors; the purpose of audiometric testing; and audiometric test procedures" (CFR 1910.95). DoD hearing programs must add communicating in hazardous noise while using hearing protection or TCAPS, firing a weapon while using hearing protection, and conducting tactical and support operations while using hearing protection to their training programs. A limited number of uniformed audiologists are available to support this hearing protection and communication systems training requirement, as well as to provide comprehensive hearing program support to the total military. To provide effective training, troops must train troops. For example, Soldiers listen to Soldiers, and Marines to Marines. In the past, when uniformed audiologists were not given the opportunity to train troops before deployment, or to train other military medical specialists to train troops during pre-deployment exercises, troops either did not use, or failed to properly use, their hearing protection during deployment (Nemes 2005).

5 Summary

To meet the continuing challenges of a complex, comprehensive hearing program, DoD hearing conservation programs are shifting their focus from solely preventing hearing loss to determining which sounds are critical to troop survivability and lethality; determining the importance of sound localization, direction, and distance to mission success; defining baseline troop hearing performance; and evaluating the effects of hearing loss, hearing protection devices, communications technology, changing auditory environments, and noise exposure on mission performance. Understanding these basic principles will help in the development of the attenuation, intelligibility, and environmental awareness criteria needed to determine whether a hearing protector or tactical communications and protective system is acceptable for use by military personnel. Also challenging is the current separation of advanced weapons technology research and development efforts from medical research related to health and survivability in combat. Shifts in research and program implementation processes place greater emphasis and focus on readiness and conservation for a complete hearing program.

In summary, ongoing and future research efforts will certainly increase our general understanding of the hearing, communication, and auditory situational awareness needs of troops in all environments. Those efforts, combined with the DoD's continued collective efforts to further improve hearing testing and evaluation processes, hearing protection, and training, will improve the services provided to war fighters and assist in reducing noise-induced hearing loss and tinnitus among military and civilian personnel.

Acknowledgments Thank you to Dr. Colleen LePrell and Ms. Leeann Domanico for their careful editing. Thank you to Dr. Kyle Dennis for his expertise regarding Veterans Affairs. Thank you to Dr. Lynne Marshall and CDR Joel Bealer for their expertise regarding Naval hearing conservation efforts. Thank you to the U.S. Army Public Health Command for their continued support.

References

Bergman, M. (2002). On the origins of audiology: American wartime military audiology. *Audiology Today*, Monograph 1, 1–28.

Casali, J. G., Ahroon, W. A., & Lancaster, J. A. (2009). A field investigation of hearing protection and hearing enhancement in one device: For soldiers whose ears and lives depend upon it. *Noise Health*, 11(42), 69–90.

Cave, K. M., Cornish, E. M., & Chandler, D. W. (2007). Blast injury of the ear: Clinical update from the global war on terror. *Military Medicine*, 172(7), 726–730.

Cleveland, L. (2009). Fort Carson: An Army Hearing Program success story. *US Army Medical Department Journal*, 67–75.

Cowan, N. (1998). Visual and auditory working memory capacity. *Trends in Cognitive Sciences*, 2(3), 77–78.

Crum, P. A. C., & Hafter, E. (2001). The residual effects of visual capture on auditory localization. *Proceedings of the Association for Research in Otolaryngology Midwinter Research Meeting*, 24, 259.

Department of the Army. (2008). ST 4–02.501: Army Hearing Program.

Fausti, S. A., Wilmington, D. J., Gallun, F. J., Myers, P. J., & Henry, J. A. (2009). Auditory and vestibular dysfunction associated with blast-related traumatic brain injury. *Journal of Rehabilitation Research and Development*, 46(6), 797–810.

Folmer, R. L., Saunders, G. H., Dann, S. M., Griest, S. E., Leek, M. R., & Fausti, S. A. (2010). Development of a computer-based, multi-media hearing loss prevention education Program for veterans and military personnel. *Perspectives on Audiology*, 6, 6–19.

Grantham, M. A. M., Gaston, J. R., & Letowski, T. R. (2010). Auditory recognition of the direction of walking. Paper presented at the presented at ICSV 17: The 17th International Congress on Sound & Vibration, Cairo, Egypt.

Institute of Medicine. (2005). *Noise and military service: Implications for hearing loss and tinnitus*. Washington, DC: The National Academies Press.

Jerger, J. (2009). *Audiology in the USA*. San Diego: Plural Publishing.

Koelewijn, T., Bronkhorst, A., & Theeuwes, J. (2009). Auditory and visual capture during focused visual attention. *Journal of Experimental Psychology and Human Perception Performance*, 35(5), 1303–1315.

Kopke, R. D., Jackson, R. L., Coleman, J. K. M., Liu, J., Bielefeld, E. C., & Balough, B. J. (2007). NAC for noise: From the bench top to the clinic. *Hearing Research*, 226, 114–125.

Lynch, E. D., & Kil, J. (2009). Development of ebselen, a glutathione peroxidase mimic, for the prevention and treatment of noise-induced hearing loss. *Seminars in Hearing*, 30, 47–55.

Marine Corps Center for Lessons Learned. (2005). Command and control and hearing protection.

McIlwain, D. S., Cave, K., Gates, K., & Ciliax, D. (2008). Evolution of the Army Hearing Program. *US Army Medical Department Journal*, 62–66.

McIlwain, S., Sisk, B., & Hill, M. (2009). Cohort case studies on acoustic trauma in Operation Iraqi Freedom. *US Army Medical Department Journal*, 14–23.

Mershon, D. H., Desaulniers, D. H., Amerson, T. L., Jr., & Kiefer, S. A. (1980). Visual capture in auditory distance perception: Proximity image effect reconsidered. *Journal of Auditory Research*, 20(2), 129–136.

Mershon, D. H., Desaulniers, D. H., Kiefer, S. A., Amerson, T. L., Jr., & Mills, J. T. (1981). Perceived loudness and visually-determined auditory distance. *Perception*, 10(5), 531–543.

Myles, K., & Kalb, J. T. (2009). Vibrotactile sensitivity of the head (Report ARL-TR-4696). Aberdeen Proving Ground, MD: Army Research Laboratory.

Nemes, J. (2005). As their ranks shrink, military audiologists' mission expands. *The Hearing Journal*, 58(9), 19–27.

Nixon, C. W. (2002). A glimpse of history: The origin of hearing conservation was in the military? United States Air Force Research Laboratory, Report Number AFRL-HE-WP-SR-1998-0005. (AFRL-HE-WP-SR-1998–0005). Retrieved from http://handle.dtic.mil/100.2/ADA355531.

Ohlin, D. (2005). Sound identification training: Auditory armament for the battlefield. *CAOHC Update: The Newsletter of the Council for Accreditation in Occupational Hearing Conservation*, 17(2), 1.

Ohlin, D. (2009). Strategic and tactical thinking in the hearing conservation mindset: A military perspective. *Noise Health*, 11(42), 22–25.

Ohlin, D. (2010). Hearing protection: It's not just about noise reduction. *EHS Today: The Magazine for Environment, Health, and Safety Leaders*. Retrieved from http://ehstoday.com/ppe/hearing-protection-not-about-noise-1342/index1.html.

Peters, L. J., & Garinther, G. R. (1990). The effects of speech intelligibility on crew performance in an M1A1 tank simulator (Report A604822). Aberdeen Proving Ground, MD: Human Engineering Laboratory.

Powell, J. A., Kimball, K. A., Mozo, B. T., & Murphy, B. A. (2003). Improved communications and hearing protection in helmet systems: The communications earplug. *Military Medicine*, 168(6), 431–436.

Saunders, G. H., & Griest, S. E. (2009). Hearing loss in veterans and the need for hearing loss prevention programs. *Noise & Health*, 11(42), 14–21.

Scharine, A. A., & Letowski, T. R. (2005). Factors affecting auditory localization and situational awareness in the urban battlefield. (Report A369134). Army Research Lab, Aberdeen Proving Ground, MD: Human Research and Engineering Directorate.

Scharine, A. A., Letowski, T. R., & Sampson, J. B. (2009). Auditory situation awareness in urban operations. *Journal of Military and Strategic Studies*, 1–24.

Sewell, R. K., Song, C., Bauman, N. M., Smith, R. J., & Blanck, P. (2004). Hearing loss in Union Army veterans from 1862 to 1920. *Laryngoscope*, 114(12), 2147–2153.

Tyler, R. S. (2000). *Tinnitus handbook*. San Diego: Thomson Learning.

U.S. Department of Veterans Affairs. (2010). 2009 Annual Benefits Report. Retrieved from http://www.vba.va.gov/REPORTS/abr/index.asp.

USACHPPM. (2007). US Army Center for Health Promotion and Preventative Medicine. 2006 Veterans Compensation Charts and VA Disability Reports Retrieved October 26, 2007 from http://chppm-www.apgea.army.mil/hcp/comp_reports.aspx.

Wilt, J., & Bjorn, V. (2006). Noise and advanced hearing protection. Paper presented at the 45th Navy Occupational Health & Preventive Medicine Conference.

Part II
Relationship Between Noise Exposure and Resulting Anatomical, Physiological, and Perceptual Changes in Hearing

Chapter 4
The Use of Kurtosis Measurement in the Assessment of Potential Noise Trauma

Donald Henderson and Roger P. Hamernik

1 Introduction

The current noise standards for the United States were originally formulated in 1968/1969 as part of the Walsh–Healy Act for Federal Contractors (U.S. Department of Labor 1969). The standards for a permissible noise exposure are based on data from a number of large-scale demographic studies of hearing loss in industrial settings in the United States and Europe (Burns and Robinson 1970; Baughn 1973; Passchier-Vermeer 1974). Table 4.1 illustrates the permissible noise exposures based on the average 8-h measurement of dBA (L_{eq}8h)*. Note three key points: (1) 90 dBA for 8 h or a time-weighted average* of 90 dBA is the permissible noise dose; (2) the *time* and *dBA trading relation** (50% decrease in time for each 5-dB increase in level); and (3) the maximum permissible exposure (115 dBA). The standards were modified in 1999 (MSHA 1999) to create the action level at 85 dBA; however, the permissible noise exposures remain the same, i.e., 90 dBA for 8 h and a 5-dB trading relation.*

Since the original formulation in 1968/1969, a number of laboratory and epidemiological studies have been conducted that have implications on what constitutes a safe exposure (Kryter 1973; Bruel 1980; Davis et al. 2009), but the official noise standards remain essentially the same as they were when published in 1968. In a review of the issues pertaining to noise standards, Eldredge (1976) made the point that for a noise standard to be used it has to be easily understood and easily implemented.

*Note: Words with an asterisk after are defined in the glossary at the end of the text.

D. Henderson (✉)
Department of Communication Disorders and Science, Center for Hearing Research, College of Arts and Sciences, State University of New York at Buffalo, 3435 Main Street, Buffalo, NY 14214, USA
e-mail: donaldhe@buffalo.edu

C.G. Le Prell et al. (eds.), *Noise-Induced Hearing Loss: Scientific Advances*, Springer Handbook of Auditory Research 40, DOI 10.1007/978-1-4419-9523-0_4,

Table 4.1 Permissible noise exposures (Walsh–Healy Act)

Duration per day, hours	Sound level dBA slow response
8	90
6	92
4	95
3	97
2	100
1.5	102
1	105
0.5	110
0.25 or less	115

For the original noise standards, the noise level was the key parameter and was easily measured with available sound level meters. Later, dosimeters were developed.

This chapter reviews the current noise standards and examines the appropriateness of the L_{eq}8h *time-weighted average* (TWA) as an estimate of the traumatic potential of an exposure and introduces the statistical concept of *kurtosis* in the evaluation of non-Gaussian (or complex noise*) exposures as a refinement to our current noise measurements. Animal model data suggest that the *kurtosis metric* (defined in Sect. 5) may serve as a refinement to our current noise measurement practice for the purpose of hearing conservation.

2 Trading Relation: Time and Intensity

The U.S. standards are based on a 5-dB trading relation, whereas European noise exposure standards use a 3-dB trading relation. The European standards are based on the concept of the equal energy hypothesis (EEH). The EEH postulates that the potential noise trauma is directly related to the total A-weighted energy of the exposure (energy = power × time) (Ward et al. 1961; Eldred 1976). Consequently, each doubling of the power of the noise exposure (3 dB) is offset by a 50% reduction in the permissible exposure time. A permissible 8-h exposure is 85 dBA in Europe and 90 dBA in the United States. The difference between *American and European* standards is more dramatic at higher sound levels (i.e., for 100-dBA exposures, the permissible exposure time is 2 h in the United States but only 15 min in Europe). The 5-dB trading relation of the U.S. standard reflects the assumption that *breaks* (lunch, restroom, etc.) provide an opportunity for the ear to begin to recover from the traumatic effects of noise. It is interesting to note that in 1998, the National Institute of Occupational Safety and Health (NIOSH) proposed an amendment to the U.S. noise regulations that included an 85-dBA criterion level for 8 h and a 3-dB trading relation (NIOSH 1998). The NIOSH Amendment was never adopted and U.S. noise standards have remained essentially the same for the last 50 years in spite of many new studies that question conventional theories of the effects of noise on hearing (Bruel 1980; Dunn et al. 1991; Lataye and Campo 1996; Harding and Bohne 2004).

3 Non-Gaussian Noise Exposures

An underlying assumption between both the U.S. and European standards is that hearing loss is related to the total energy of the exposure and the distribution of sound energy over an 8-h period is not a critical variable. There are some interesting observations and experiments to the contrary. For example, a comparison of the hearing loss data of Burns and Robinson (1970) versus that of Passchier-Vermeer (1974) shows that for noise exposures that have the same L_{eq} 8 h, hearing loss at 4 kHz in the Passchier-Vermeer study grows faster and to a greater level. Interestingly, the subjects in the Burns and Robinson study were exposed to a stable, continuous noise (i.e., a Gaussian noise), whereas Passchier-Vermeer's subjects were exposed to a combination of continuous and impact noise (i.e., a non-Gaussian noise or a complex noise).

Experimental studies of complex noise * exposures illustrate the potential hazard of complex noises composed of continuous noise with addition of high-level impacts or noise bursts or impulses. For example, Hamernik and Henderson (1976) reported a substantial synergistic interaction when chinchillas were exposed to a combination of continuous noise (1 h at 95 dBA) and impulses (50 at 158 dBA) compared to either noise dose. The subjects receiving either the impulse or continuous experiment developed only an average 5–10 dB hearing loss from 500 to 8,000 Hz. However, there was substantially more permanent hearing loss [permanent threshold shift (PTS)] and cochlear damage (hair cell loss) in the combination group than the simple addition of the PTS or hair cell loss of both the impulse and continuous exposures. It should be noted that the addition of the two components adds less than 1 dB to the total energy to the combination exposure, but produces 20–30 dB more hearing loss. The acoustic requirements for the interaction effect are interesting, i.e., if the background noise is turned off for 2 s around the impulse, the interaction effect is lost. The addition of the impulse to the background noise adds less than a decibel of total energy of the exposure, conversely turning off the noise for 100 s (2 s × 50 impulses) reduces the total energy by less than a decibel, but the interaction effect of the impulse and continuous noise is 30 dB larger than either component or the sum of the components.

The potential for high-level transients imbedded in Gaussian noise to produce greater hearing loss may be a reflection of two different pathological processes underlying Gaussian and non-Gaussian noise exposures. With exposure to continuous noise, there is an increase in free radical formation in the cochlea (see review, Henderson et al. 2006; Le Prell, Chap. 13). When the accumulation of free radicals exceeds the endogenous antioxidant capacity of the cochlea, the free radicals damage the cochlea, leading to hair cell death. With exposure to impact noise above a *critical level*, the inner ear may be damaged as a result of mechanical failure (Hamernik et al. 1985).

3.1 Critical Level

Critical level refers to the peak level of an exposure at which the cochlea begins to suffer mechanical damage, i.e., the breaking of tight cell junctions and dislocation of tissue (Ward et al. 1961; Henderson and Hamernik 1986). The *critical level* is not a fixed decibel level but rather depends on the type of transient. For short-duration impulses (i.e., gun fire) the *critical level* is higher; for larger duration impacts it is lower.

For industrial types of impact noise the *critical level* was estimated to be between 115 and 123 dB peak sound pressure level (SPL). This range of levels was obtained from a series of equal energy impact noise exposures in which the peak level of the impact was increased in 3-dB steps from 107 to 143 dB and the number of impacts is decreased by 50% for each 3-dB step. Interestingly, for exposures between 107 and 119 dB peak SPL or below the "critical level," the hearing loss and hair cell loss remain relatively constant, independent of the peak level, but as the impact level was increased further, the hearing loss and hair cell damage increased dramatically with each increase in peak level (Henderson and Hamernik 1986).

One possible mechanism to explain the additional hazard associated with non-Gaussian exposures is that the continuous noise renders the cochlea more vulnerable, i.e., generates toxic free radicals and lowers the *critical level*, making the high level transients in the complex noise more traumatic.

4 Alternative Approaches to Evaluating Noise Exposures

Current noise standards evaluate an exposure on the basis of the average noise level over an 8-h work day. Certain occupations, i.e., carpenters, are characterized by above average hearing loss (NIOSH 2007), but their daily exposure is well below the 90 dBA time-weighted average. A key acoustic feature of the exposure is its intermittentency and irregular high noise levels. The average noise levels may be below 90 dBA, but for short periods the noise from hammers and saws can reach the 110–120 dB SPL levels. Thus the hearing loss is not related to average level, but to the intermittent higher levels of the transients. Energy considerations alone ignore the temporal features of a noise exposure.

If the public health goal of having noise standards is to prevent noise-induced hearing loss (NIHL), then it is necessary to develop a noise metric that reflects the additional hazards associated with exposure to noises with widely varying peak levels. There is evidence that complex non-Gaussian noise, i.e., a combination of background noise and impacts or high-level noise bursts, are more dangerous to the ear then continuous Gaussian noise exposures. One approach to estimating the traumatic potential for exposure to non-Gaussian noise is to measure the *kurtosis** of the noise in addition to the L_{eq}.

5 The Use of the Kurtosis* Statistic in Evaluating a Noise

Kurtosis is one of the statistical parameters used to describe a distribution. Wikipedia defines kurtosis as: "In probability theory and statistics, kurtosis (from the Greek word κυρτός, *kyrtos* or *kurtos*, meaning bulging) is a measure of the "peakedness" of the probability distribution of a real-valued random variable. Higher kurtosis means more of the variance is the result of infrequent extreme deviations, as opposed to frequent modestly sized deviations" (http://en.wikipedia.org/wiki/Kurtosis).

or as defined in the Handbook of Engineering Statistics (Pham 2006):

"Kurtosis is a measure of whether the data [i.e. amplitude of noise] are peaked or flat relative to a normal distribution. That is, data sets with high kurtosis tend to have a distinct peak near the mean, decline rather rapidly, and have heavy tails. Data sets with low kurtosis tend to have a flat top near the mean rather than a sharp peak. A uniform distribution would be the extreme case," (see Fig. 4.1).

Defined mathematically, kurtosis is the ratio of the fourth-order central moment to the square of the second order central movement of a sample. It should be noted that Gaussian (G) noise has a $\beta = 3$. The kurtosis measurement can be applied to the history of stocks or hedge funds to illustrate whether its price moves within a relatively narrow range or if there is wide swing in its cost.

5.1 Kurtosis as a Contributing Factor to NIHL

There are a number of examples showing that exposures with the same total A-weighted energy can produce different degrees of hearing loss. Reports from the National Institute for Occupational Safety and Health (2002) indicated that there is a need for a more reliable predictor of NIHL (Dunn et al. 1991; Lei et al. 1994; Lataye and Campo 1996; Hamernik and Qiu 2001; Harding and Bohne 2004; Qiu et al. 2006).

Hamernik and colleagues performed a series of experiments showing that exposures with the same total A-weighted energy and spectrum but different degrees of kurtosis can produce very different amounts of hearing loss and cochlear damage. Lei et al. (1994) exposed chinchillas to either the Gaussian noise (G) or non-Gaussian noise (NG) composed of background noise in combination with either impact noise (up to 126 dB peak SPL) or bursts of noise (106 dB peak SPL) (see Fig. 4.2). All exposures lasted for five consecutive days, and all exposures had the same total energy and the same spectral distribution of energy. The exposures differed only in the value of the kurtosis. Note Fig. 4.2a is an example of broad band noise with kurtosis $\beta(t)=3$, while Figs. 4.2b and c have a continuous noise waveform with either impacts or noise bursts added. In the former case, $\beta(t)=84$ and in the latter $\beta(t)=21$. Figure 4.3 illustrates the distribution of amplitudes of three waveforms similar to those of Fig. 4.2. Note the peakedness of the higher kurtotic noise compared to the Gaussian noise with a β factor$=3$. It is important to note that all three distributions have the same total acoustic energy and spectral distribution.

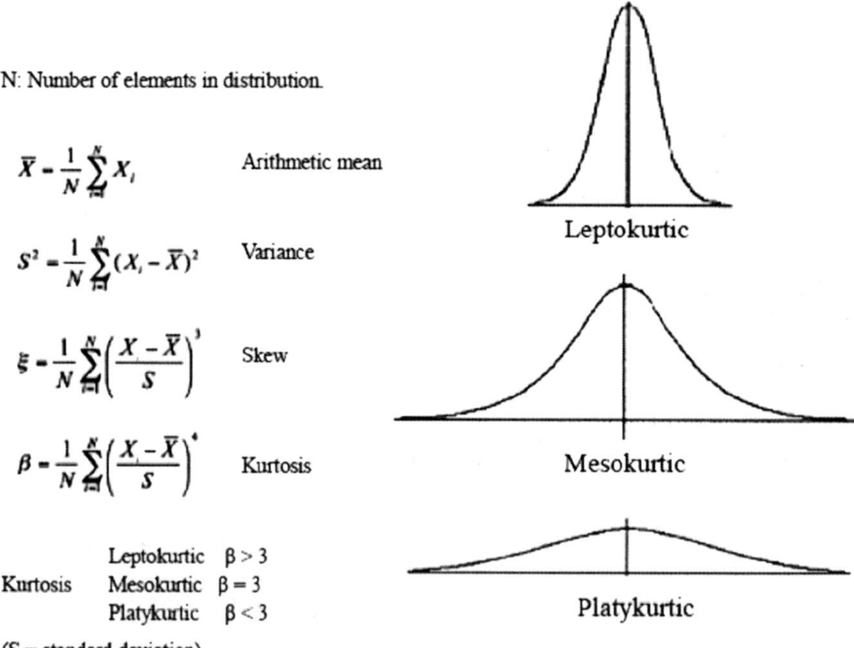

Fig. 4.1 Illustration of the descriptive statistical measures for a distribution of a variable x. For the current paper the x's are the amplitudes of the pressure–time waveform at discrete points over some interval of time. A complex (non-Gaussian) noise exposure consisting of a combination of continuous and impact noise having variable peaks that reach 125 dB SPL as illustrated in Fig. 4.2 would have a value of the kurtosis statistic greater than $\beta = 3$ and an amplitude distribution (i.e., the x's) that would be more peaked than that of a Gaussian distribution. This indicates that the peaks in the waveform are more concentrated around the mean, resulting in "fatter" tails of the distribution

The changes in hearing sensitivity during the 5 days of the exposure and the 3 days of recovery for test frequencies of 0.5, 2, and 8 kHz are shown in Fig. 4.4. Note that during the first 5 days of exposure, the thresholds are at an asymptotic level of approximately 50 dB at 0.5 kHz and 80 dB at 8.0 kHz. The effect of kurtosis does not seem to be a factor during the period of asymptotic threshold shift. However, at 30 days post exposure the permanent threshold shift (PTS) at the 0.5-, 2.0-, and 8.0-kHz test frequencies for the $\beta(t)=3$ exposure caused the least PTS while the $\beta(t)=84$ exposure caused the most. In fact, at 2 kHz, for the same total energy there was approximately 40 dB more PTS for the $\beta(t)=84$ exposure than the $\beta(t)=3$ Gaussian noise exposure.

Figure 4.5 summarizes the PTS and hair cell losses (HCL) for the three groups. The clear implication of the data in this figure is that changing the kurtosis factor $\beta(t)$ from 3 to 84 leads to a regular but dramatic increase in the traumatic potential of an

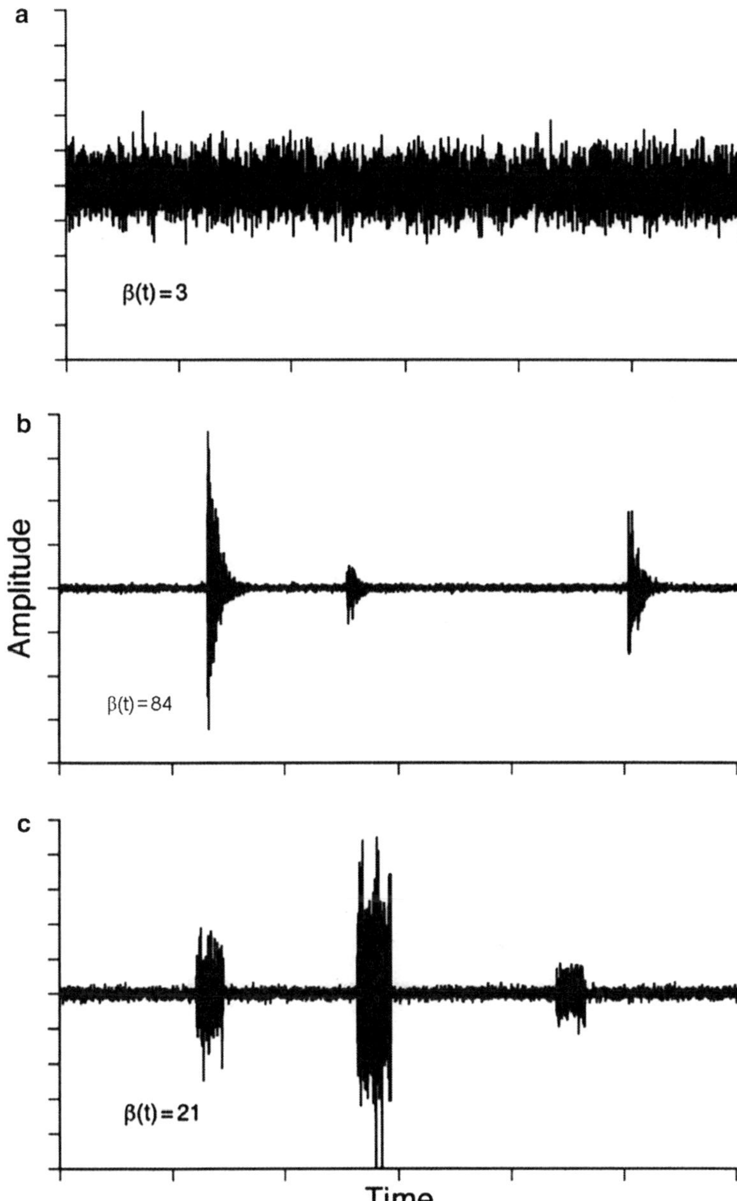

Fig. 4.2 Waveforms of traumatizing noise used in Lei et al. (1994). Note that the three waveforms are from noises that had the same spectrum and total exposure energy but differed in the value of the kurtosis statistic. A Gaussian noise has a kurtosis of $\beta(t) = 3$. (**a**) Gaussian noise. (**b**) Background noise plus impacts. (**c**) Background noise plus noise bursts

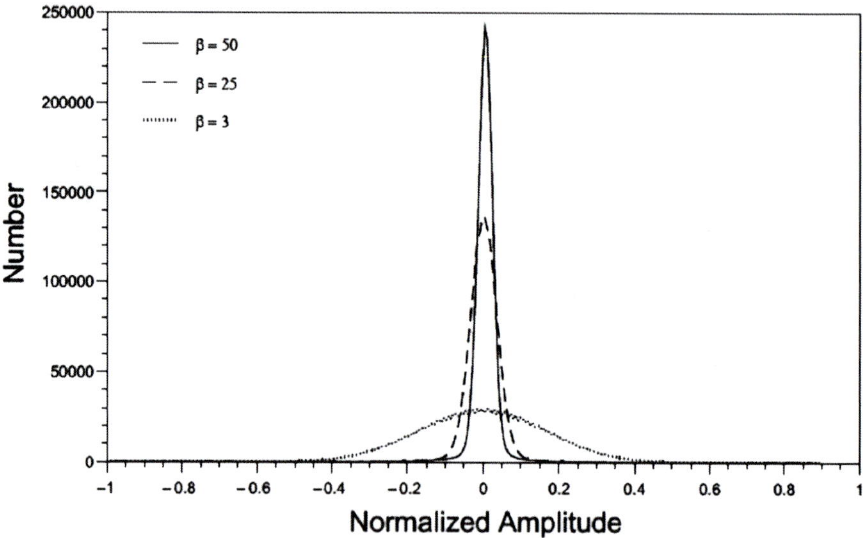

Fig. 4.3 A comparison of three amplitude distributions of actual noises that have the indicated values of the kurtosis $\beta(t)$. Multiple 40-s samples of the noise were used in the calculations. All three of these noises had the same spectrum and energy; they differed only in the value of the kurtosis. The increased "peakedness" of the distributions at $\beta(t)$ increases is clear. The vertical axis refers to the distribution of amplitude measures taken from multiple 40-s samples and the figure represents several million samples

exposure, i.e., PTS difference as great as 40 dB with 70–80% greater HCL. It is important to recognize that these traumatic differences could not be accounted for in the current L_{eq} measurement approach to noise.

The results of Lei et al. (1994) clearly show that the energy distribution over time can be a critical factor in determining the hearing loss from an exposure and raise three questions: (1) Does the kurtosis effect influence all levels of potentially traumatizing sound? (2) Are there classes of noise exposures wherein the equal energy principle operates? (3) Does the kurtosis factor play role in industrial noise environments?

The first two questions are partially answered in a set of experiments by Qiu et al. (2006), Hamernik et al. (2003), and Hamernik et al. (2007). Qiu et al. (2006) exposed chinchillas to either Gaussian noise or non-Gaussian noise with average levels of 90, 100, or 110 dBA. Several clear trends emerge: "At the lowest level [L_{eq} = 90 dB(A)] there were no differences in the trauma produced by G and non-G exposures. However, for L_{eq} > 90 dB(A) nonG exposures produced increased trauma relative to equivalent G exposures" (Qiu et al. 2006).

Hamernik and Qiu (2001) explored the applicability of the "equal energy principle." Sixteen groups of chinchillas (N = 140) were exposed to various equivalent energy noise paradigms at 100–103 dBA. Eleven groups received an interrupted, intermittent, and time-varying (IITV) non-Gaussian exposure quantified by the

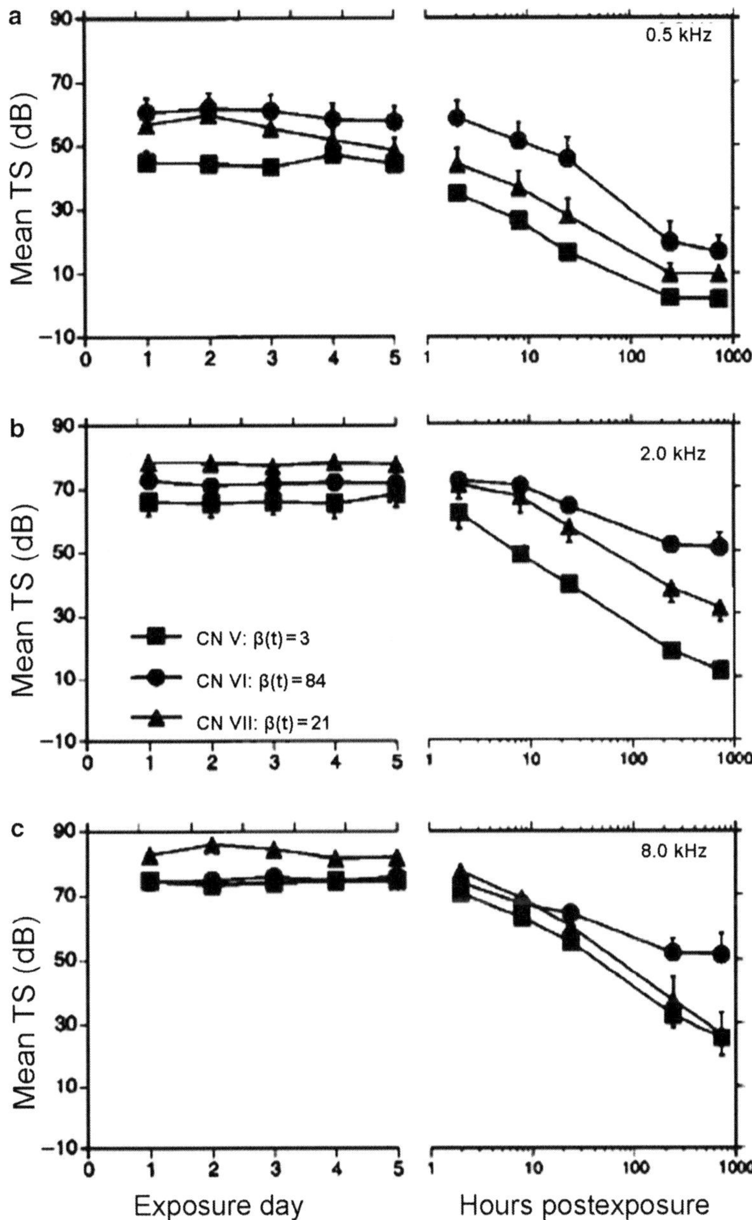

Fig. 4.4 Auditory threshold changes during and after noise exposed. The left side of the figure shows the mean threshold shift (TS) measured daily at the 0.5-, 2.0-, and 8.0-kHz test frequencies during the 5-day Gaussian (β=3) and non-Gaussian (high kurtosis, β=21 and β=84) noise exposures. The right side of the figure shows the corresponding recovery of threshold over a 30-day interval (Lei et al. 1994). Note that by 30 days postexposure the hearing loss is greatest in the higher kurtosis groups despite the same energy and spectrum of the exposure

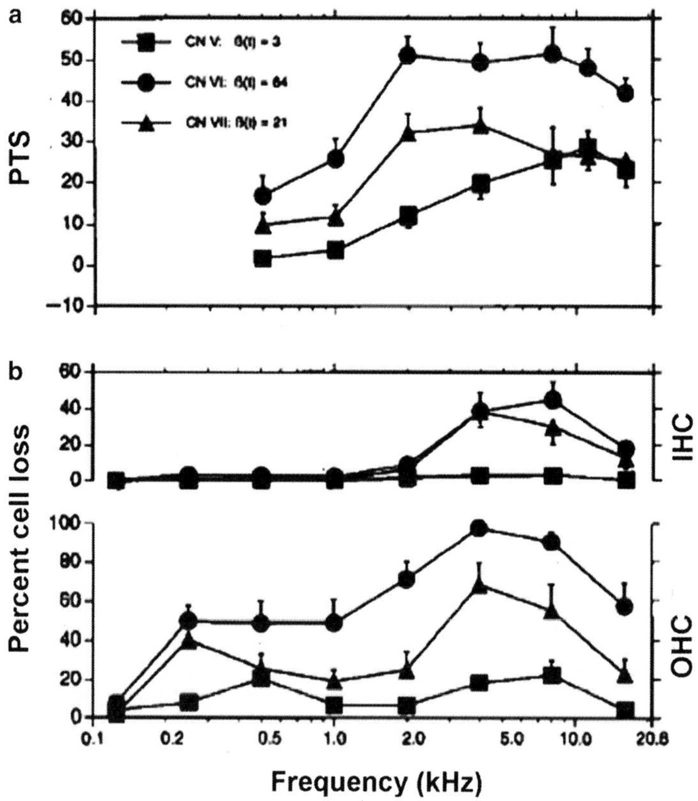

Fig. 4.5 (**a**) The mean permanent threshold shift (PTS) as a function of test frequency and (**b**) the mean percent sensory cell loss as a function of frequency location on the basilar membrane for the three noise-exposed groups shown in Fig. 4.3. Note the increase in PTS and sensory cell loss as the kurtosis of the exposure increasesg (Lei et al. 1994)

kurtosis statistic. The IITV exposures, which lasted for 8 h/day, 5 days/week, for 3 weeks, were designed to model some of the essential features of an industrial work week. Five equivalent energy reference groups were exposed to either a Gaussian or non-Gaussian continuous noise for 5 days, 24 h/day. For IITV exposures at a *fixed equivalent energy* and *fixed kurtosis*, the temporal variations in level did not alter trauma and in some cases the IITV exposures produced results similar to those found for 5-day continuous exposures. However, any increase in kurtosis at a fixed energy was accompanied by an increase in noise-induced trauma. Collectively, these results suggest that the equal energy hypothesis is an acceptable approach to evaluating noise exposures for hearing conservation purposes provided that the kurtosis of the amplitude distribution is taken into consideration.

6 Kurtosis Analysis in an Industrial Setting

Zhao et al. (2010) reported on the prevalence of hearing loss in two industrial plants. Plant A was a textile plant with a Gaussian type noise that had an L_{eq}(A)8 h ranging from 95 dBA to 105 dBA. Plant B was a metal fabricating plant with L_{eq}(A)8 h of 95 dBA with non-Gaussian noise including impacts up the 125 dB peak SPL. Each subject from both plants answered a questionnaire on their otologic and noise history and was examined by a physician to determine if there were any medical or work or recreational activities that would compromise a subject's hearing.

Pure tone, air conduction hearing threshold levels at 0.5, 1, 2, 3, 4, 6, and 8 kHz were measured in each ear. Hearing threshold levels* (HTLs) at each frequency were adjusted for age and gender using the 50th percentile values found in the International Standard Organization (ISO-1999, (1990)) Annex B. The comparison of hearing function between the two plants was based on the incidence of NIHL. The prevalence of an adjusted high-frequency noise-induced hearing loss (AHFNIHL) in workers exposed to non-Gaussian or Gaussian noise was defined as having one or more hearing thresholds, in either ear, at 3, 4, 6 kHz equal to or higher than 30 dB HL.

Unlike laboratory-based experimental conditions, epidemiological studies of NIHL in industry have the additional variable of the different exposure durations across workers. Thus an industrial's exposure was defined as cumulative noise exposure (CNE) (Earshen 1986). The CNE is the product of the average daily noise exposure ($L(A)_{eq,}$8 h) and the log of the years of exposure (T). Specifically,

CNE = $L(A)_{eq,}$8 h + 10 log T (Zhao et al. 2010). Because all subjects in this study never changed their working environment, T was simplified as the number of years each of the 195 subjects worked in the factories.

Table 4.2 shows the L_{eq} levels at each of the two plants. Note that the exposures are 7 dB higher for plant A with Gaussian noise when the CNE is calculated for each plant. Plant B with non-Gaussian noise has an average CNE value of 103.2 ± 4.2 dB, whereas plant A has an average CNE of 110.6 ± 6.0 dB. The prevalence of hearing loss (greater than 30 dB HL at 2, 4, 6, or 8 kHz) is the same for both plants in spite of the fact that plant A has on average a 7-dBA higher noise level.

The kurtosis of the noise in the two plants was calculated. Plant B had a $\beta = 40$; plant A had a $\beta = 3.3$. Figure 4.6 shows the relation between CNE and prevalence of HL (greater than 30 dB HL at 2, 4, 6, and 8 kHz) for the two populations of noise-exposed subjects. Note that the function describing the prevalence of HL from exposure to the non-Gaussian noise of plant A has a steeper slope then the Gaussian noise of plant B. Even though plant B has an average 7 dBA CNE, there is a substantially greater likelihood of developing a HL.

The Zhao et al. (2010) experiment is important for the eventual evolution of noise standards because it shows that the distribution of noise power over time is an important acoustic parameter in determining the noise-induced hearing loss. Data of the type presented by Zhao et al. (2010) are difficult to obtain because of the complexity of doing human research in industry. For example, in most industrial settings the workers are wearing some form of personal protection device (PPD).

Table 4.2 L_{eq}, CNE, and β for the subjects of two factories

Noise type	Total N	Noise source	$N1$	$L_{eq}(A)_{8h}$dB	CNE	$x\beta$
Gaussian	163					30
		Loom ZA 205i	24	98.1 ± 2.1	110 ± 6	
		Loom 1511	75	105.4 ± 2.2		
		Spinner FA507A	23	99.5 ± 2.2		
		Spinner 1301	41	96.1 ± 2.7		
Complex	32					40
		Punch press	17	95.3 ± 2.5	103 ± 4	
		Plate clipper	15	95.2 ± 3.5		

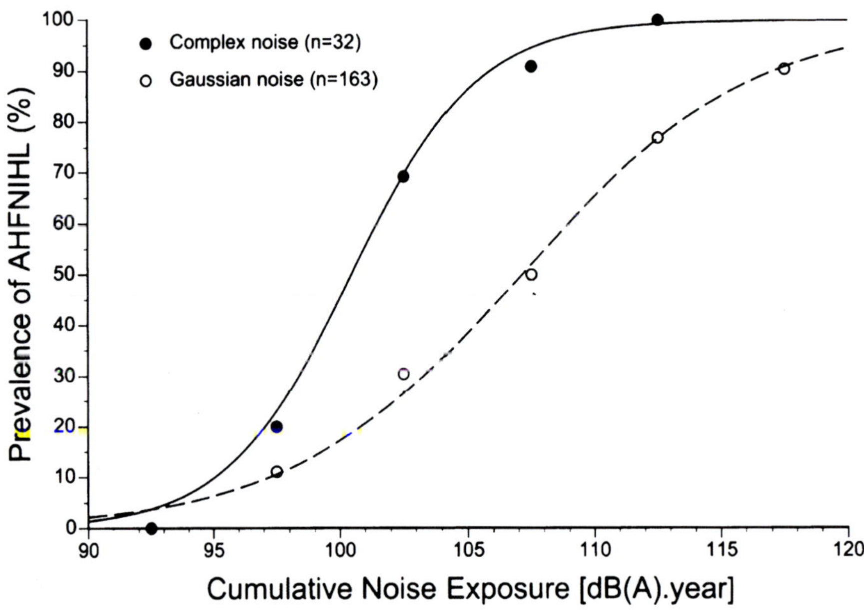

Fig. 4.6 The does–response relation for a population of workers exposed to Gaussian and non-Gaussian noise over a period of many years. The prevalence of age- and gender-adjusted high-frequency noise-induced hearing loss (AHFNIHL) is shown as a function of the cumulative noise exposure (CNE = $L_{Aeq,}$ 8 h +10 log T where T is the exposure time in years). Note that the prevalence of hearing loss is greater for the non-Gaussian exposures and that it starts to occur at a lower CNE than for the Gaussian exposures

The performance characteristics of PPD used in industry leads to a large degree of variance in the relation between noise exposure and hearing loss (Gerges and Casali 2007; Casali, Chap. 12). The Zhao et al. (2010) experiment is rare today because most of the subjects did not wear PPDs. The size of the non-Gaussian subject pool is small, and a larger one could allow comparison between hearing loss from Gaussian and different levels of non-Gaussian noise (correlation analysis). It would also be useful to evaluate kurtosis influence on different levels of L_{eq}. More data are needed with human subjects to be able to determine the range and correction factor associated with degrees of kurtosis.

7 Summary

Our current noise standards require the measurement of the noise level (dBA) and the duration of the exposure. There are dramatic experimental and epidemiological examples of noise exposures with ostensibly the same total energy leading to quite different levels of hearing loss. One factor in these differences is the way the ear responds to a moderate level of noise versus higher level impulses, impacts, and noise blasts. There is convincing evidence that complex noise composed of higher level transient embedded in background noise is particularly dangerous. The point of the chapter is that a noise measurement that includes both time-weighted average or L_{eq}8h and the degree of kurtosis provides a potentially more accurate assessment of the hearing loss expected from the exposure than current energy-based assessments. When the noise standards were first introduced, the addition of a kurtosis evaluation would have been very difficult because of the additional data processing required to compute β as well as dBA or L_{eq}8h. The advancements of noise measurement devices, and the increased computational potential of modern instruments, make measuring both TWA and kurtosis a reasonable possibility.

Glossary

Complex noise Non-Gaussian noise composed of background lower level noise and intermittent higher level impacts or impulses or short bursts of higher level noise.

Decibel (dB) A unit is composed of 10 × sound (x)/reference sound. Note dB is an undefined term unless it has a suffix stating the reference, i.e.,

dB SPL: dB sound pressure level and the reference is 20 μbar pressure
dB SL: dB sensation level and the reference is the listener's threshold for that specific sound.

dBA Sound measurements made with an A-scale weighting on the sound level meter. Low-frequency sounds (less than 500 Hz) are negatively weighted with the A scale because low-frequency sound energy is not as damaging to the ear as sounds above 500 Hz.

Hearing threshold level (HTL) Hearing threshold levels are expressed in decibels relative to appropriate clinical norms.

Kurtosis A statistical parameter used in describing a distribution of a variable. It is this 4th moment of a distribution (see Fig. 4.1) and as kurtosis increases, more of the variance is due to infrequent extreme deviations (i.e., the impact and noise bursts in Fig. 4.2). Kurtosis is used in finance to describe the pattern of variation associated with a given stock or hedge fund.

L_{eq} 8 h Refers to the average sound level over an 8-h period. The L_{eq} 8 h is the basic unit for noise assessment. A 4-h exposure of *x* dBA would be averaged over 8 h to determine the equivalent L_{eq} 8 h.

Trading relation Noise exposures are defined by both the intensity of noise (*I*) and the duration of the exposure (*T*) because hearing loss is related to both *I* and *T*. The trading relation refers to how *I* and *T* are combined. In the current U.S. standards (Table 4.1), for each 5-dB increase in exposure level there is a 50% decrease in time; therefore the trading relation is 5 dB. It should be noted that European noise standards have a 3-dB trading relation.

References

Baughn, W. L. (1973). Relation between daily noise exposure and hearing loss based on evolution of 6,835 industrious noise exposure cases. Report No. AMRL-TR-73.53, Wright-Patterson Air Force Base, OH: Aerospace Mechanical Research Laboratory.

Bruel, P. V. (1980). The influence of high crest factor noise on hearing damage. *Scandinavian Audiological Supplement*, 12, 25–32.

Burns, W., & Robinson, D. W. (1970). An investigation of the effects of occupational noise on hearing. In J. J. Knight (Ed.), *Hearing and noise in industry* (pp. 177–192). London: H.M.S.O.

Davis, R. I., Qiu, W., & Hamernik, R. P. (2009). Role of the kurtosis statistic in evaluating complex noise exposures for the protection of hearing. *Ear & Hearing*, 30(5), 628–634.

Dunn, D. E., Davis, R. R., Merry, C.J., & Franks, J. R. (1991). Hearing loss in the chinchilla from impact and continuous noise exposure. *Journal of the Acoustical Society of America*, 50, 1979–1985.

Earshen, J. J. (1986). Sound measurement: Instrumentation and noise descriptors. In E. H. Berger, W. D. Ward, J. C. Morrill & L. H. Royster (Eds.), *Noise and hearing conservation manual.* Akron, OH: American Industrial Hygiene Association.

Eldred, K. (1976). Demographics of noise pollution with respect to potential hearing loss. In D. Henderson, R. Hamernik, D. Dosanjh, & J. Mills (Eds.), *Effects of noise on hearing.* (pp. 3–20). New York: Raven Press.

Eldredge, D. I. (1976). The problems of criteria for noise exposure. In D. Henderson, R. Hamernik, D. Dosanjh, & J. Mills (Eds.), *Effects of noise on hearing.* (pp. 3–20). New York: Raven Press.

Gerges, S., & Casali, J. G. (2007). Hearing protectors. In M. Crocker (Ed.), *Handbook of noise and vibration control.* Hoboken, NJ: John Wiley & Sons.

Hamernik, R. P., & Henderson, D. (1976). The potentiation of noise by other ototraumatic agents. In D. Henderson, R. P. Hamernik, D. Dosanjh, & J. Mills (Eds.), *Effects of noise on hearing.* (pp. 291–308). New York: Raven Press.

Hamernik, R. P., & Qiu, W. (2001). Energy-independent factors influencing noise-induced hearing loss in the chinchilla model. *Journal of the Acoustical Society of America*, 110, 3163–3168.

Hamernik, R. P., Turrentine, G., & Roberto, M. (1985). Mechanically induced morphological changes in the organ of Corti. In R. J. Salvi, D. Henderson, R. P. Hamernik, & V. Colletti (Eds.), *Basic and applied aspects of noise-induced hearing loss* (pp. 69–84). New York: Plenum Press.

Hamernik, R. P., Qiu, W., & Davis, B. (2003). The effects of the amplitude distribution of equal energy exposures on noise-induced hearing loss: The kurtosis metric. *Journal of the Acoustical Society of America*, 110, 3163–3168.

Hamernik, R. P., Qiu, W., & Davis, B. (2007). Hearing loss from interrupted, intermittent, and time-varying non-Gaussian noise exposures: The applicability of the equal energy hypothesis. *Journal of the Acoustical Society of America*, 122, 2245–2254.

Harding, G. W., & Bohne, B. A. (2004). Noise-induced hair-cell loss and total exposure energy: Analysis of a large data set. *Journal of the Acoustical Society of America*, 115, 2207–2220.

Henderson, D., & Hamernik, R. P. (1986). Impulse noise. Critical review. *Journal of the Acoustical Society of America*, 80, 569–584.

Henderson, D., Bielefeld, E. C., Harris, K. C., & Hu, B. H. (2006). The role of oxidative stress in noise induced hearing loss. *Ear and Hearing*, 27, 1–19.

International Organization for Standardization, ISO 1999, Acoustics: Determination of occupational noise exposure and estimate of noise-induced hearing impairment. Geneva, Switzerland, 1990.

Kryter, K. D. (1973). Impairment to hearing from exposure to noise. *Journal of the Acoustical Society of America*, 53, 1211–1254.

Lataye, R., & Campo, P. (1996). Applicability of the L_{eq} as a damage-risk criterion: An animal experiment. *Journal of the Acoustical Society of America*, 96, 1435–1444.

Lei, S. F., Ahroon, W. A., & Hamernik, R. P. (1994). The application of frequency and time domain kurtosis to the assessment of hazardous noise exposures. *Journal of the Acoustical Society of America*, 99, 1621–1632.

MSHA, Health Standards for Occupational Noise Exposure; Final Rule. U.S. Department of Labor, Mine Safety and Health Administration, 30 CFR Part 62. 64 *Federal Register*, 1999, 49548–49634, 49636–49637.

National Institute of Occupational Safety and Health (NIOSH). (1998). *Criteria for a recommended standard: Occupational noise exposure – revised criteria.* Cincinnati, OH: U.S. Department of Health, Education, and Welfare, National Institute for Occupational Safety and Health, Publication no. 98–126.

National Institute of Occupational Safety and Health (NIOSH). (2002). *Exposure Assessment Methods: Research Needs and Priorities.* Cincinnati, OH: U.S. Department of Health and Human Services, Public Health Service, Centers for Disease Control and Prevention, National Institute for Occupational Safety and Health, DHHS (NIOSH) Publication no. 2002–126.

National Institute of Occupational Safety and Health (NIOSH). (2007). Best practices workshop on impulsive noise and its effect on hearing. National Institute for Occupational Safety and Health. *Journal of Noise Control Engineering*, 53(2)

Passchier-Vermeer, W. (1974). Hearing loss due to continuous exposure to steady-state broad-band noise. *Journal of the Acoustical Society of America*, 56, 1585–1593.

Pham, H. (2006). *Springer handbook of engineering statistics*. New York: Springer-Verlag.

Qiu, W., Hamernik, R. P., & Davis, R. I. (2006). The kurtosis metric as an adjunct to energy in the prediction of trauma from continuous, non-Gaussian noise exposures. *Journal of the Acoustical Society of America*, 120, 3901–3906.

U.S. Department of Labor (1969). Occupational noise exposure. *Federal Register*, 34, 7891–7954.

Ward, W. D., Fleer, R. E., & Glorig, A. (1961). Characteristics of hearing loss produced by gun fire and by steady state noise. *Journal of Auditory Research*, 1, 325–356.

Zhao, Y., Qiu, W., Zeng, L., Chen, S., Cheng, X., Davis, R. I., & Hamernik, R. P. (2010). Application of the kurtosis statistic to the evaluation of the risk of hearing loss in workers exposed to high-level complex noise. *Ear and Hearing*, 31(4), 527–532.

Chapter 5
Noise-Induced Structural Damage to the Cochlea

Bohua Hu

1 Introduction

Noise-induced hearing loss (NIHL) is the most common cause of acquired hearing loss among persons younger than age 40. The major pathological basis of NIHL is mechanical stress to cochlear structures. During noise exposure, mechanical forces drive the basilar membrane to oscillate. Excessive motion of the basilar membrane causes a cascade of structural changes in cochlear sensory cells and their supporting cells, which, in turn, compromise cochlear function. Although acoustic overexposure affects both the peripheral and the central auditory systems, the peripheral organ is the primary target of acoustic exposure.

The generation of noise-induced cochlear damage has been attributed to two basic damaging factors: direct mechanical stress and secondary metabolic disruption. Direct mechanical stress results from the physical forces of acoustic stimuli and occurs during the course of noise exposure. In cochleae subjected to an intense noise exposure, the manifestation of direct mechanical impacts can be detected immediately after the noise exposure. Further metabolic disruption is initiated during the course of noise exposure and continues to develop days or even weeks after the termination of noise exposure. The mechanisms of metabolic disruption have been linked to numerous pathological conditions, including ischemia (Hultcrantz et al. 1979; Axelsson and Dengerink 1987; Nuttall 1999), excitotoxic damage (Puel et al. 1998), metabolic exhaustion (Ishii et al. 1969; Ishida 1978; Omata et al. 1979; Chen et al. 2000), and the intermixing of cochlear fluids (Duvall and Rhodes 1967; Duvall et al. 1969; Bohne and Rabbitt 1983).

Depending on the level of damage, structural changes in the cochlea can be either reversible or irreversible. A mild structural defect results in only a temporary threshold

---

B. Hu (✉)
Department of Communicative Disorders and Sciences, Center for Hearing and Deafness, State University of New York at Buffalo, 3435 Main Street, Buffalo, NY 14214, USA
e-mail: bhu@buffalo.edu

C.G. Le Prell et al. (eds.), *Noise-Induced Hearing Loss: Scientific Advances*, Springer Handbook of Auditory Research 40, DOI 10.1007/978-1-4419-9523-0_5,

shift. Severe structural damage, in contrast, causes permanent defects or even hair cell death, leading to permanent hearing loss.

Investigation into the morphological impacts of acoustic exposure relies on microscopy. In this regard, both light and electron microscopic technologies have been used extensively. Light microscopy allows the examination of a large section of cochlear tissues and is suitable for a quantitative analysis of the overall extent of cochlear damage. By examining the surface preparation of the organ of Corti, the numbers of missing or dying hair cells can be quantified, and the data can be assembled to cochleograms. Recent advances in the technique of confocal microscopy have substantially expanded the capacity and application of light microscopy. Confocal microscopy allows viewing a specific layer of cellular structures in a tissue block without physically sectioning the tissue. This ability enables researchers to examine a series of optical sections and to obtain sequential images along the depth of the specimen. Because each type of cell in the organ of Corti has a unique position in the three-dimensional structure of the organ of Corti, these cells can be individually identified and illustrated. Electron microscopy is ideal for examining fine structures of tissues and cells. Scanning electron microscopy permits the inspection of the surface structures of the cells, which is particularly valuable for the examination of the stereocilia of the hair cell. Transmission electron microscopy is an ideal tool for the examination of fine structures of intracellular organelles, but the specimen preparation is cumbersome and time consuming.

Over the years, our knowledge of the anatomical consequences of acoustic trauma has been steadily improving. A better understanding of structural changes provides new insights into the biological mechanisms of noise-induced cochlear damage. This chapter presents a description of the major structural changes in the organ of Corti resulting from acoustic overstimulation. Because hair cells are particularly susceptible to acoustic trauma, the focus of the chapter is on the structural changes in hair cells. In the past, several excellent reviews on the structural impacts of acoustic trauma have been published (Saunders et al. 1985; Henderson and Hamernik 1986; Slepecky 1986; Axelsson and Dengerink 1987; Borg et al. 1995). Interested readers are referred to these publications for further details.

2 Hair Cell Lesions

2.1 Spontaneous Hair Cell Loss in Normal Cochleae

Before addressing the issue of hair cell damage resulting from acoustic trauma, it is necessary to describe the level of spontaneous hair cell loss in the normal cochlea because preexisting hair cell loss can interfere with the interpretation of the pathological analysis of noise-traumatized cochleae. In cochleae without a history of acoustic insults, loss of hair cells has been observed in various species, even when subjects have normal hearing sensitivity. Erlandsson et al. (1980) quantified the spontaneous hair cell loss in young pigmented guinea pigs and found that the average percentage of spontaneous hair cell loss is 0.58% ± 0.27% (mean ± SD). Bohne et al. (1987) counted the numbers of missing hair cells in normal chinchilla cochleae and found that the average number of missing cells for the outer hair cells (OHCs) is 1.0% ± 0.6% and for the inner hair cells (IHCs) is 0.5% ± 0.2%. In general, spontaneous loss is more frequent in OHCs (Coleman 1976). Although the sites of missing cells are scattered throughout the organ of Corti, there is a tendency for the spontaneous degeneration of hair cells to start from the apical region of the organ of Corti (Coleman 1976; Erlandsson et al. 1980). For the rest of the organ of Corti, spontaneously degenerated hair cells are evenly distributed (Erlandsson et al. 1980).

2.2 *Hair Cell Lesions in Noise-Traumatized Cochleae*

The most pronounced change in cochlear morphology after exposure to intense noise is the degeneration of hair cells in the organ of Corti. Hair cell degeneration commonly occurs in clusters, forming hair cell lesions involving either a few or a large group of cells, depending on the level and the duration of acoustic overexposure. Bohne and Clark (1982) defined the focal lesion as a section of the organ of Corti where hair cell loss involves equal to or greater than 50% of the hair cell population in that section.

Depending on the stage of cochlear pathogenesis, hair cell lesions can be active or inactive. In a lesion for which an active death process is still occurring, the lesion can exhibit several distinct, but overlapping, pathological zones. In the center of the lesion, hair cell degradation has been largely completed, and hair cell loss is evident. The areas adjacent to the center of the lesion are the transition area, where the active disintegration of hair cells is still occurring. Specifically, actin filaments in the cuticular plates of hair cells are degrading. The plasma membrane of the cell is compromised, resulting in an increase in membrane permeability. The formation of large membrane gaps leads to the release of cellular contents to extracellular spaces. The nuclei of the cells are dislocated and become either swollen or condensed. Further away from the center of the lesion, hair cell pathogenesis is at its early stage. In this section, the shape of the cell body becomes irregular and the location of cell nuclei is often elevated. However, the structural integrity of the cuticular plate and the plasma membrane is usually preserved. Further away from the center of the lesion, most hair cells are viable, although a slight disarrangement of the cell alignment is evident. The ultimate fate of these cells varies, as some die and others survive. Figure 5.1 shows an example of a hair cell lesion in a chinchilla cochlea exposed to an intense noise.

As time elapses after noise exposure, cochlear pathogenesis gradually diminishes and hair cell lesions become stable. Inactive lesions are characterized by scar tissue in the regions where dead cells have degraded. The cells that survived can be normal or can display permanent structural defects. The irreversible changes in the cochlea are the pathological basis for permanent hearing loss.

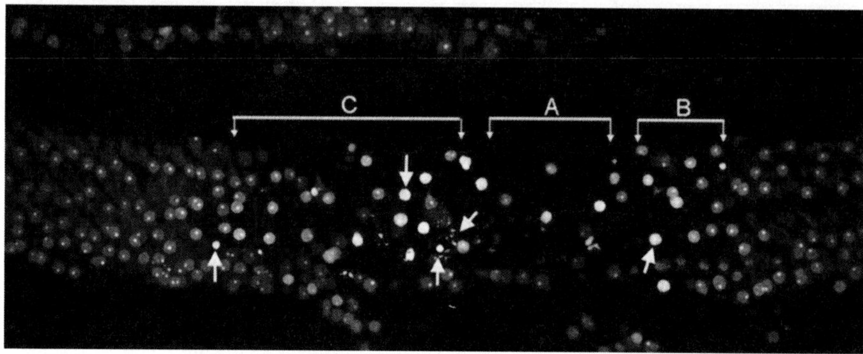

Fig. 5.1 An example of an active hair cell lesion from a chinchilla cochlea exposed to a 4-kHz narrowband noise at 110 dB SPL for 30 min. The cochlea was stained with propidium iodide for the illustration of nuclear morphology. The center of the lesion (**a**) is characterized by loss of hair cell nuclei. Outside the center (**b** and **c**), active cell death is evident. The nuclei of hair cells display an increased propidium iodide fluorescence. Some of the nuclei appear condensed or fragmented (*arrows*). Further away from the center of the lesion, most hair cell nuclei appear normal but are slightly disarranged

2.2.1 Distribution of Hair Cell Lesions

The site of hair cell lesions in the organ of Corti is associated with the frequency component of noise. A high-frequency noise selectively traumatizes the basal end of the organ of Corti and a low-frequency noise preferentially injures the apical portion of the organ of Corti. The distribution pattern of hair cell lesions is also associated with the intensity of noise. For example, Harding and Bohne (2009) reported that when a cochlea was exposed to a high-level of a 4-kHz octave-band noise, focal lesions spread over the entire basal half of the organ of Corti. In contrast, when a cochlea was exposed to the same noise, but at a moderate-level, focal lesions occur in the region of the organ of Corti corresponding to the site activated by the 4-kHz octave-band noise.

Hair cell lesions are either scattered among surviving cells over a large section or are focused in a small section of the organ of Corti (Bohne et al. 1990; Bohne and Harding 2000; Harding and Bohne 2007). Whether lesions are located in a confined area or spread out in a broad region has been found to be associated with the locations of the lesions. When lesions appear in the apical half of the organ of Corti, as a result of exposure to a low-frequency noise, the lesions tend to spread over a broad region. In contrast, when lesions appear in the basal half of the organ of Corti, resulting from exposure to a high-frequency noise, they often appear as a small, concentrated loss of hair cells (Dolan et al. 1975; Bohne and Harding 2000; Harding and Bohne 2009). The broad distribution of hair cell damage in the cochleae subjected to a low-frequency noise is likely to be related to the fact that a low-frequency sound travels further toward the apex of the cochlea than does a high-frequency sound. Consequently, a low-frequency sound induces mechanical stress to a broad region of the organ of Corti.

2.2.2 The Severity of Hair Cell Lesions

The severity of hair cell damage in the cochlea is associated with the nature and conditions of noises, including the intensity, duration, and type of the noises. In regard to intensity, there is a general trend for an increase in the noise level to be associated with an increase in the level of hair cell damage. However, the two events are not parallel and thus the correlation is not direct. Physiological and pathological observations have revealed two phases of lesion potentiation separated by a critical level of noise exposure (Erlandsson et al. 1980; Vertes et al. 1982; Cody and Robertson 1983). Below the critical level, the expansion of hair cell lesions increases slowly. However, as the noise exceeds the critical level, hair cell damage increases substantially. An example of this two-phase increase in hair cell lesions was provided by Erlandsson et al. (1980), who exposed guinea pigs to an intense pure-tone (3.85 kHz), with a level ranging from 102 to 120 dB SPL, for 6 h. As the noise level increased from 102 to 117 dB SPL, the level of hair cell damage remained largely unchanged, with approximately 5–8% cell loss. However, with a further increase in the noise level from 117 to 120 dB SPL, OHC loss suddenly jumped to 37%. This pattern of lesion development suggests that the critical level of the 3.85 kHz pure-tone for guinea pigs is between 117 and 120 dB SPL.

The actual value of the critical level varies with the condition of noise. For an impact noise, the level is 125 dB SPL (Henderson et al. 1991). The critical level is also related to the duration of noise exposure. For a noise with a shorter duration, the critical level is higher than that with a longer duration (Erlandsson et al. 1980). Moreover, the magnitude of the increase in hair cell damage when a noise exceeds its critical level has been found to be related to the frequency of the noise. Erlandsson et al. (1980) reported that the rate of hair cell loss resulting from a high-frequency noise was greater than that caused by a low-frequency noise.

The biological mechanism of the sudden increase in cochlear damage is not clear. Spoendlin (1976) suggested that, when the level of a noise exceeds a critical level, the mechanism of damage shifts from metabolic to mechanical. Several investigations have revealed a shift of cell death modes with an increase in noise levels (Hu et al. 2000; Yang et al. 2004). In a study using a range of noise levels, Hu et al. (2000) reported that, when the noise was at 110 or 115 dB SPL, hair cell damage was mild, and the mode of hair cell death was mainly necrotic. In contrast, when the noise was at 120 dB SPL, hair cell damage increased dramatically, with a shift of the cell death mode from necrosis to apoptosis. This shift in the cell death mode may contribute to the sudden increase in the level of hair cell pathogenesis.

The frequency composition of noise also affects the severity of hair cell damage. High-frequency noise-generated damage is more severe than that resulting from the same noise but with a low-frequency composition. Erlandsson et al. (1980) compared the growth patterns of hair cell lesions caused by exposure to a 1.33-kHz or a 3.85-kHz pure-tone noise in guinea pig cochleae. In the cochleae exposed to the low-frequency tone, the amount of OHC loss increased from 4% to 12% as the intensity of the noise was increased from 102 to 120 dB SPL. In contrast, the OHC loss jumped from 5–8% to 37% when the high-frequency noise was used. The study suggests that a high-frequency noise is more detrimental than a low-frequency noise.

Providing a resting period during exposure to an otherwise continuous noise, as compared with a continuous noise exposure of equal energy, reduces the extent of hair cell damage (Bohne et al. 1985, 1987; Fredelius and Wersall 1992). Bohne et al. (1985, 1987) reported that, when cochleae were exposed to a high-frequency noise (4-kHz octave-band noise), the basal-turn lesions were reduced as compared with the lesions resulting from exposure to the same noise, but uninterrupted. In contrast, when a low-frequency noise (0.5 kHz octave-band noise) was used, the amount of resting-period mediated protection in the basal half of the organ of Corti was not as great as that seen in the apical portion of the organ of Corti. This study suggests that the pattern of lesion reduction is associated with the frequency of noise.

2.2.3 Growth of Hair Cell Lesions

Hair cell damage starts as small lesions in the organ of Corti. Lesions expand during the course of noise exposure and continue to develop after the termination of the exposure. The initial lesion starts in the section of the organ of Corti with the characteristic frequency corresponding to the dominant frequency of the noise. The initial damage can involve one or a few lesions and the lesion expands toward both the apical and the basal ends of the cochlea, leading to formation of a large lesion (Harding and Bohne 2009). Figure 5.2 shows an example of expansion pattern of a hair cell lesion.

The dynamic development of hair cell lesions is difficult to study because a given pathological change can be examined only at one time point. So far, most of the research aimed at elucidating the expansion pattern of hair cell lesions was derived from sequential observations of cochlear pathologies at different time points postexposure. The accurate interpretation of these results is a challenge because there is so much individual variability in cochlear responses to a given exposure, both among animals, between the two cochleae of one animal, and among the sensory cells within a single cochlea.

To investigate the development of hair cell pathogenesis generated during the course of noise exposure, Pye (1981, 1984) exposed guinea pigs to an intense noise (20 kHz) with a duration ranging from 3.25 to 120 min. The exposed cochleae were examined at various times postexposure. For the subjects receiving the exposure for 7.5 min or shorter, the size of the lesions appeared to remain unchanged. The major difference seen with the prolongation of noise duration was the increase in the number of the cochleae that displayed hair cell lesions. A significant increase in the size of hair cell lesions was seen only when the animals were exposed to the noise for at least 60 min. Interestingly, the difference appeared only in the ears examined at 3 weeks postexposure. At 12 weeks postexposure, the difference was no longer significant.

The rate of lesion expansion is related to the phases of cochlear pathogenesis. Early expansion of hair cell lesions is rapid and significant. Thorne and Gavin (1985) exposed guinea pigs to a 5-kHz noise at 125 dB SPL for 30 min. They examined

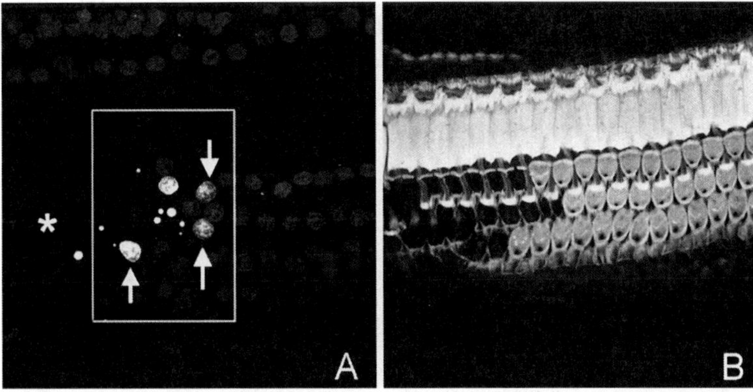

Fig. 5.2 Images showing the expansion pattern of a hair cell lesion in a cochlea exposed to a 4-kHz narrowband noise at 110 dB SPL. The cochlea was doubly stained with propidium iodide for the illustration of nuclei (**a**) and FITC-phalloidin for the illustration of the cuticular plates of hair cells (**b**). (**a**) Nuclear morphology. The asterisk marks the area of the center of the lesion where the outer hair cell nuclei have been degraded. The frame encloses the transition area between the center of the lesion and the relatively normal section of the organ of Corti. Note that cells within the frame (*arrows*) show increased propidium iodide fluorescence. These cells are dying cells. (**b**) The conditions of the cuticular plates of the hair cells illustrated in (**a**). In the area where the hair cells have lost their nuclear staining (see the corresponding nuclear morphology in the area marked by the asterisk in (**a**)), the cuticular plates lack the FITC-phalloidin fluorescence, indicating that these cells have lost their F-actin in the cuticular plates. In contrast, the viable cells with the normal nuclear morphology show strong F-actin staining

the cochlear pathology via scanning electron microscopy at 1 h, or at 1, 7, 14, or 28 days postexposure. They found a significant increase in the number of damaged hair cells within 1 day postexposure. Although the rate of cell death decreases substantially in the later phases of cochlear pathogenesis, chronic cell death has been found several weeks after the termination of noise exposure (Bohne 1976; Hamernik et al. 1984; Yang et al. 2004). The occurrence of acute hair cell death emphasizes the importance of early intervention for patients with acute acoustic trauma. Chronic hair cell damage calls for a prolonged treatment for several weeks, so that the outcome of the treatment can be improved.

Hair cell lesions expand toward both the apical and basal ends of the organ of Corti. However, the basal expansion appears to be more pronounced than does the apical expansion (Thorne et al. 1984; Bohne et al. 1987). By tracking the expansion pattern of active cell death, Hu et al. (2002a) reported that hair cell lesions during the first 2 days after exposure to an intense noise at 110 dB SPL expanded from the junction of first and second cochlear turns to the basal end of the cochlea. The propensity of lesion expansion toward the basal end of the organ of Corti may reflect a structural and metabolic difference between the apical and the basal portions of the organ of Corti. In this regard, the basal end has been shown to have less antioxidant capacity, and, consequently, cells in this region may suffer from greater oxidative stress (Sha et al. 2001).

3 Damage to Hair Cell Structures

3.1 Damage to Stereocilia

Both the IHCs and OHCs have well organized hair bundles on their tops. In mammalian hair cells, these hair bundles are stereocilia, which are organized in W- or V-shaped bundles, consisting of several rows, with their height increasing toward the direction of the basal body. The roots of stereocilia are embedded in the cuticular plate of the cell. The internal skeleton of the stereocilia, consisting of parallel filaments of actin, gives the stereocilia a high stiffness.

The bundles of stereocilia are interconnected by an array of fine fibers, called cross-links. Based on their locations, cross-links can be classified into two groups: the side links and the tip links (Pickles et al. 1984). Side links reside in the lateral side of the stereocilia bundles. Their function has been suggested to be purely physical, linking the stereocilia as a unit, so that they can move together during deflection of the bundle (Pickles et al. 1984). Tip links reside in the top of the stereocilia, running upward to their taller counterpart in the next row (Pickles et al. 1984). The function of tip links has been associated with the sensory transduction function of the hair cell. Stretching of the tip links during acoustic stimulation opens transduction channels, which, in turn, activate the cell. Similar to side links, tip links physically hold the stereocilia together. As a result, the stereocilia bundles can move synchronous in response to acoustic stimuli.

3.1.1 Morphological Changes in Stereocilia of Hair Cells After Acoustic Exposure

Stereocilia are susceptible to acoustic trauma. Defects have been observed in noise-exposed hair cells without showing detectable intracellular abnormalities (Cody and Robertson 1983). After acoustic overexposure, a variety of forms of stereocilium changes has been described. These changes include disarray, separation, falling, fusion, collapsing, shrinking, breaking, detaching, and loss (Robertson et al. 1980; Thorne et al. 1986; Gao et al. 1992). Disarray is an early sign of stereocilium defects (Tsuprun et al. 2003), appearing in different forms: splaying, flopping, and bending. Within stereocilia, the microfilaments lose their normal parallel arrangement and have a wavy appearance (Thorne et al. 1986). Figure 5.3 shows a typical example of stereocilium defects resulting from acoustic trauma.

Fusion is a common abnormality seen in stereocilia after acoustic overstimulation. This pathology is often seen in the late stage of stereocilium pathogenesis (Engström and Borg 1981; Engström 1984). However, there are reports of an early fusion of stereocilia (Engström and Borg 1981; Thorne et al. 1986). Fusion of stereocilia is much more common in IHCs than in OHCs (Mulroy and Curley 1982). Congregation of stereocilia after noise exposure is likely due to the sticky adjoining stereocilium membrane (Cho et al. 1991). The fused stereocilia can develop into a single giant

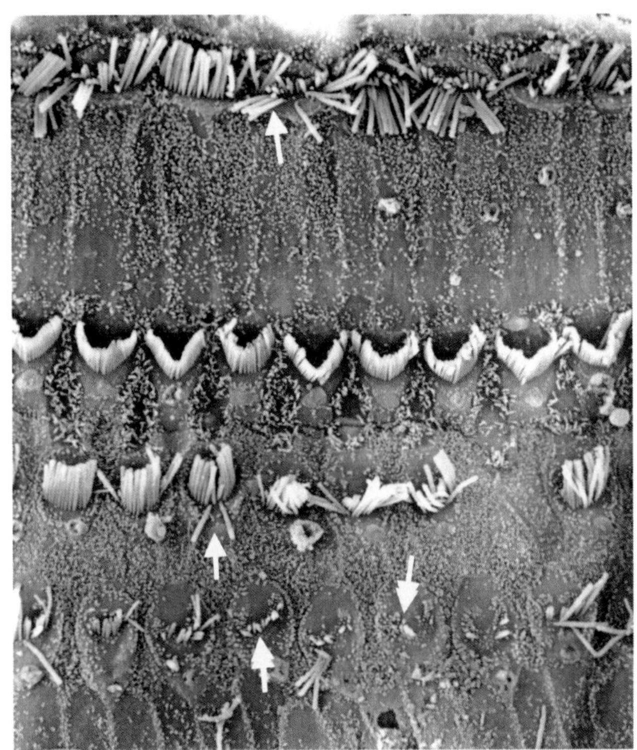

Fig. 5.3 An example of stereocilium damage in a guinea pig cochlea exposed to an impulse noise at 166 dB SPL. Double arrows point to the third row hair cells, where loss or shortening of the stereocilia is visible. Arrows point to the hair cells displaying the splaying of stereocilia (Courtesy of Mr. Jianhe Sun, Institute of Otolaryngology, PLA General Hospital)

hair bundle or irregular masses (Engström and Borg 1981; Slepecky et al. 1981; Engström 1983; Engström et al. 1983). Inside the giant stereocilium, actin cores of two or more stereocilia are packed (Tsuprun et al. 2003). They lose contact with the plasma membrane and are enclosed by a common plasma membrane (Slepecky et al. 1982; Engström et al. 1983). Again, giant stereocilia are more common in IHCs.

In the severely damaged section of the organ of Corti, loss of stereocilia can be seen in both IHCs and OHCs. In this case, the cuticular plates can be completely void of stereocilia, or, in their place, small, rounded protrusions can be found (Thorne et al. 1984). Although hair cells in mammalian cochlea can survive the loss of stereocilia (Jia et al. 2009), these cells contribute to the loss of hearing sensitivity.

Some of the morphological changes observed using scanning electrode microscopy have been attributed to the decrease in the stiffness of the stereocilia due to changes in the structural integrity of actin filaments (Tilney et al. 1982; Canlon 1988). Actin is a dynamic molecule, and its normal structure relies on the balance of polymerization and depolymerization of the molecule. Acoustic trauma causes the depolymerization of actin filaments (Tilney et al. 1982), which alters the core structure of stereocilia. In addition, loss of cross-bridges between adjacent actin filaments in the stereocilium can dramatically affect the rigidity of the stereocilium (Tilney et al. 1982).

The stereociliary rootlets within the cuticular plates are another vulnerable site. A decrease in the length of the supracuticular portion of the stereocilia rootlets has been observed in cochleae with mild damage (Liberman and Dodds 1987). Widespread loss, fracture, and attenuation of the intracuticular portion of the stereocilia rootlets have also been found in cochleae sustained more severe damage (Liberman 1987). These changes in the stereocilium rootlets may affect the orientation of stereocilia.

3.1.2 Differential Vulnerabilities of Stereocilia in Different Hair Cells

OHCs are organized in three rows, and the stereocilia in three rows of OHCs exhibit different vulnerabilities to acoustic trauma. Many studies have shown that the stereocilia in the first row of OHCs often exhibit the most severe damage compared with those in the other two rows of OHCs (Hunter-Duvar 1977; Robertson et al. 1980; Robertson 1981; Thorne et al. 1984).

The stereocilia of OHCs and IHCs also display different vulnerabilities to acoustic trauma. Although OHCs are more likely to die than are IHCs in response to acoustic trauma, their stereocilia are not more vulnerable than those in IHCs. In fact, many studies have shown the opposite pattern, that is, there are more stereocilium defects in IHCs (Engström and Borg 1981; Mulroy and Curley 1982; Robertson 1982; Kaltenbach et al. 1992; Clark and Pickles 1996; Tsuprun et al. 2003). Even without detectable changes in OHC stereocilia in the cochlea, stereocilium defects in IHCs have been observed (Engström 1983). OHCs are responsible for cochlear amplification and contribute to the hearing sensitivity of the cochlea. Therefore, it is expected that OHCs modulate IHC damage in response to acoustic overstimulation. However, Borg and Engström's study showed that damage to the stereocilia of IHCs can still occur even when OHCs are wiped out with a pretreatment of kanamycin (Borg and Engström 1983), suggesting that OHCs are not involved in the damaging process of IHC stereocilia.

The high vulnerability of IHC stereocilia has been attributed to the physical organization of the stereocilia (Clark and Pickles 1996). The stereocilium bundles of OHCs are more tightly packed and are directly coupled with the tectorial membrane. These anatomic properties might render the stereocilia of the OHC more resistant to mechanical stress. One important note here is that, despite the occurrence of greater structural changes in the stereocilia of IHCs, these cells are more likely to survive than are OHCs, suggesting that stereocilium damage is not a fatal event of cell injury and cannot be used as an indicator for the ultimate fate of hair cells (Thorne et al. 1984).

3.1.3 Damage to Cross-Links of the Stereocilia

Like stereocilia, cross-links of stereocilia are susceptible to acoustic trauma. The most common defect associated with cross-links is the breakage of the links resulting from stretching of the links as the stereocilia become disarrayed. Breakage of the cross-links leaves remnant masses on the surface of the stereocilia. In IHCs, it has been found that the tip links running between the tallest stereocilia and the next row of shorter stereocilia are broken, while the tip links running between the other shorter rows of stereocilia remain intact. These findings suggest that tip links between stereocilia are preserved as long as the other links between the stereocilia and the cytoskeleton of the stereocilium remain intact (Pickles et al. 1987).

Although the cross-link is much thinner than is the stereocilium, it appears not to be more vulnerable than the stereocilium itself. Morphological observation has shown that the links remain intact if the bundles of stereocilia show no, or only a very slight degree of, disorganization (Pickles et al. 1987; Clark and Pickles 1996). Even in the hair cells showing stereocilium disarray, tip links can still be preserved in the stereocilia where the cilia remain in a bundle. In terms of the difference between side links and tip links, Clark and Pickles (1996) reported that the side links between the stereocilia often remain intact, even during overstimulation that was able to fracture the stereocilia. This result suggests that the side links are less vulnerable than are the tip links.

3.1.4 Recovery of Stereocilia Damage

The capacity of stereocilium recovery after acoustic trauma is not fully understood. Because temporary hearing loss, which has been linked to stereocilium defects, is recoverable, damage to stereocilia and their cross-links has been thought to be partially reversible. Indeed, there is some morphological evidence showing the recovery of stereocilia after acoustic overstimulation (Tonndorf 1981; Tsuprun et al. 2003). Tonndorf (1981) reported that when animals were sacrificed within minutes after noise exposure, stereocilia of hair cells in the damaged region of the organ of Corti appeared limp, suggesting the loss of the stiffness of the stereocilia. However, 24 h postexposure, the stereocilia returned to their upright position, suggesting the recovery of the lost stiffness. The biological basis of the recovery has been attributed to the recovery of actin cytoskeleton in the stereocilia (Tilney et al. 1982). The restoration of the length of supracuticular rootlets (Liberman and Dodds 1987) may also contribute to the recovery of stiffness. At present, it is not completely clear whether the loss of stereociliary bundles can be regenerated in mammalian hair cells. However, in chick cochlea, there is evidence showing the regeneration of stereociliary bundles of hair cells after acoustic trauma (Cotanche 1987b).

3.2 *Defects in the Reticular Lamina and the Cuticular Plates*

The reticular lamina is the top layer of the organ of Corti, consisting of the apical structures of both hair cells (the cuticular plates) and supporting cells. The reticular lamina separates the endolymph and the cortilymph (Gulley and Reese 1976) and plays an essential role in the maintenance of hair cell homeostasis. The loss of the barrier function of the reticular lamina causes endolymph and cortilymph to mix, creating a cytotoxic environment for hair cells. The generation of large breaks in the reticular lamina also causes the release of cellular debris into the endolymph.

Structural defects in the reticular lamina are commonly observed at two anatomic sites: the cuticular plates of hair cells and the cell–cell junctions between hair cells and supporting cells as well as between supporting cells. The defects in the cuticular plates are often associated with hair cell degeneration, whereas damage to cell–cell junctions is commonly caused by direct mechanical stress due to exposure to a high level of impulse noise or blasts.

Following noise exposure, the cuticular plates may appear to "soften" (Lim and Melnick 1971; Lim 1976) or bulge outward into the subtectorial space. Loss of the structural integrity in the cuticular plates leads to the extrusion of cellular contents into the scala media. Bohne and Rabbitt (1983) reported the formation of a series of holes in the reticular lamina, each with the size and shape that resembles the cuticular plate of the missing OHC, suggesting that these holes are generated due to loss of hair cells. In addition, large holes in the reticular lamina created by the split of the cell–cell junction and sensory cell loss are present (Lim 1976). These structural defects allow communication between the endolymphatic space and the fluid spaces of the organ of Corti before the formation of phalangeal scars by neighboring supporting cells (Bohne and Rabbitt 1983; Ahmad et al. 2003).

A major structural molecule in the cuticular plates is actin. Loss of structural integrity has been linked to the rapid degradation of actin. Hu et al. (2002b) reported that the melting of actin in the cuticular plates of hair cells is an early sign of hair cell death. Loss of the F-actin integrity leads to weakness of the cuticular plate. In addition, the rupture of the cuticular plate can occur in the site of rudimentary kinocilium (Lim and Melnick 1971). Because this anatomic site lacks actin molecules, it is likely to be a weak point for rupture.

Cell–cell junctions are another site of vulnerability. The cell junction in the mammalian organ of Corti consists of tight junctions, gap junctions, adherens junctions, and desmosome (Gulley and Reese 1976; Nadol 1978; Raphael and Altschuler 1991; Kikuchi et al. 2000). Damage to the cell–cell junction is often seen after exposure to a high level of noise. Acoustic overstimulation stretches the structures of the organ of Corti, leading to the detachment of hair cells from their anchorage (Hamernik et al. 1984; Saunders et al. 1985; Henderson and Hamernik 1986). Radial tears in the organ of Corti have been found in animals subjected to impulse-noise exposure (Vertes et al. 1984). Rupture of the organ of Corti can extend across all rows of hair cells and supporting cells, including pillar and Deiters' cells (Thorne et al. 1984).

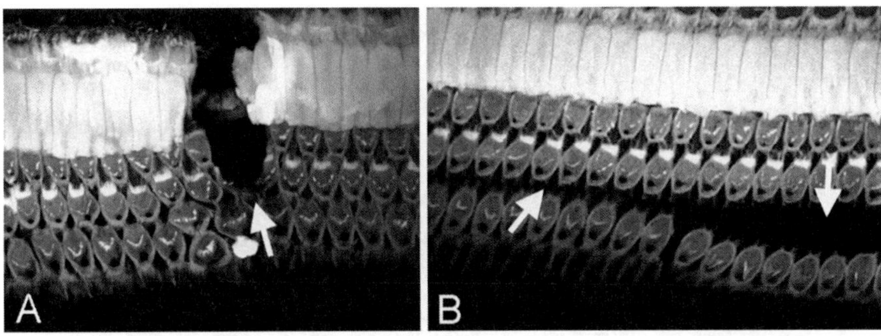

Fig. 5.4 Examples of the separation of cell–cell junctions in the reticular lamina of chinchilla cochleae exposed to an impulse noise at 155 dB pSPL. The cochleae were stained with FITC-phalloidin. (**a**) A radial defect involving both supporting cells and hair cells (*arrow*). (**b**) A split between the second and the third row of outer hair cells (*arrows*)

A large cleft can also occur between Hensen cells and Claudius cells (Spongr et al. 1998) as well as between Hensen–Deiters cell junctions (Lim and Melnick 1971). Figure 5.4 shows examples of the separation of cell–cell junctions in the reticular lamina following exposure to an intense noise.

3.3 Damage to the Plasma Membrane

The plasma membrane is an important cellular structure with multiple functions: enclosing the cell boundary, forming cell–cell adhesions, facilitating extra- and intracellular communication, and maintaining intracellular homeostasis. These functions can be compromised by acoustic overstimulation, primarily through two damaging mechanisms: direct mechanical stress and subsequent metabolic disturbance. Direct mechanical injury is caused by excessive motion of the basilar membrane, which causes a stretching injury to the plasma membrane. Metabolic disturbance induces damage via oxidative stress and energy exhaustion. Lipid peroxidation, a consequence of oxidative stress, causes the malfunction of the plasma membrane, including membrane permeabilization. Energy exhaustion compromises the ATP-dependent enzymes in the plasma membrane.

An increase in the membrane permeability alters hair cells' homeostasis. Malfunction of ion channels leads to an imbalance of intracellular ion concentrations. For example, an aberrant opening of calcium channels causes calcium influx in noise-traumatized hair cells and supporting cells (Fridberger et al. 1998; Lahne and Gale 2008). Blocking such channels has been proven to be protective (Maurer et al. 1999; Shen et al. 2007). Acoustic trauma also results in the formation of membrane pores of different sizes. Mild membrane damage is reversible (Mulroy et al. 1998), whereas severe damage leads to the formation of large membrane

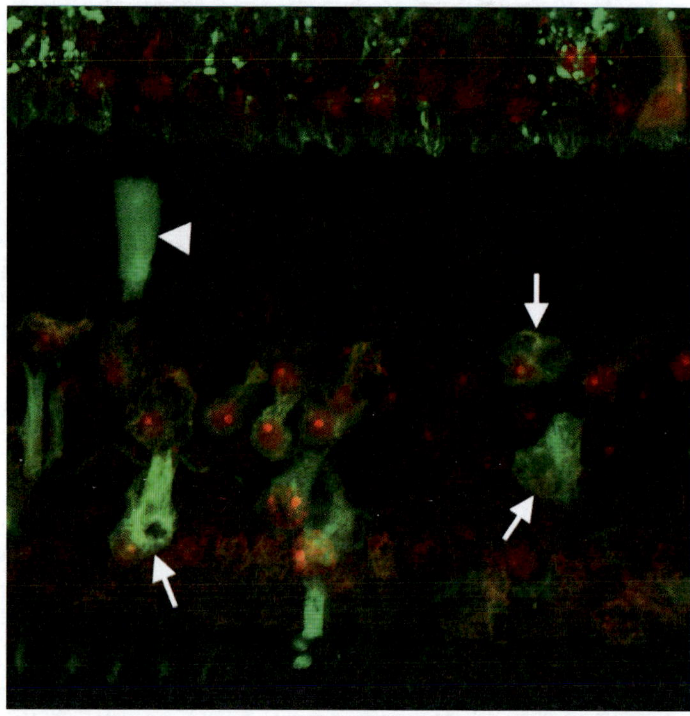

Fig. 5.5 Image showing the accumulation of FITC-dextran fluorescence in hair cells after exposure to an impulse noise at 155 dB pSPL. The cochlea was doubly stained with 40-kDa FITC-dextrans and propidium iodide. Arrows point to the outer hair cells displaying strong FITC-dextran fluorescence. The arrowhead points to an outer pillar cell showing the accumulation of FITC-dextran fluorescence. The intracellular uptake of FITC-dextran molecules indicates an increase in membrane permeability, a sign of the loss of the membrane barrier function

breakers that allows the entry of large carbon spheres into the hair cells (Ahmad et al. 2003). Hu and Zheng (2008) examined the permeability of hair cell membrane with graded sizes of fluorescein isothiocyanate (FITC)-labeled dextrans in chinchilla cochleae exposed to a brief impulse noise at 155 dB peakSPL. Shortly after the noise exposure, there was strong accumulation of the 3-kDa and the 40-kDa FITC-dextran fluorescence in the noise-traumatized hair cells (Fig. 5.5). However, the tissues lacked the 500-kDa FITC-dextran fluorescence. This observation suggests that the size of membrane breaks is somewhere between the 40-kDa and 500-kDa FITC-dextran molecules. In the hair cells at late stages of degeneration, large membrane gaps appear. The occurrence of a large membrane breakdown leads to the release of cellular contents to extracellular spaces, either in the endolymph side or in the cortilymph side of the hair cells. The release of cell debris can cause an inflammatory response, further compromising the cochlear environment for the survival of neighboring viable cells.

3.4 Cytoplasmic Changes

Morphological changes in intracellular organelles have been observed via transmission electron microscopy. Occurrence of vesiculation in the cytoplasm, which apparently involves the endoplasmic reticulum, is evident after acoustic trauma (Lim and Melnick 1971). Increased numbers of Hensen bodies and lysosomes are also evident (Canlon 1988). In the advanced stages of hair cell degeneration, vesiculation is transformed to vacuolization. Vacuolization involves not only the endoplasmic reticulum but also the outer nuclear membrane and the subsynaptic cistern. Cytoplasmic vacuolization causes significant cell-body swelling.

The inner and the OHCs exhibit different levels of degenerative changes. Slepecky et al. (1981, 1982) reported that, in response to an impulse noise exposure, IHCs showed only an increase in lysosomes and multivesicular bodies. Outer hair cells, in comparison, display not only an increase in lysosomes and multivesicular bodies but also the vacuolization of subsurface cisternae and the proliferation of Hensen bodies. This observation is consistent with the notion that OHCs are more susceptible to acoustic trauma than are IHCs. Moreover, Omata et al. (1979) found a decrease in the number of small vesicles, free ribosomes, and coated vesicles in the subnuclear region of hair cells. At present, it is not clear how changes in each of these intracellular organelles affect the molecular signal transduction of the damaging process of hair cells.

4 Variability of Cochlear Damage and Susceptibility to Acoustic Overexposure

The presence of large individual variability is a characteristic of noise-induced cochlear damage (Hunter-Duvar 1977; Lipscomb et al. 1977; Thorne et al. 1984; Saunders et al. 1985). This variability is more prominent between subjects (Dolan et al. 1975) but less obvious between two ears of the same animal (Lipscomb et al. 1977). The level of the variability has been found to be related to noise conditions, including frequency, intensity, and duration. For the factor of noise frequency, Dolan et al. (1975) reported that exposure to a low-frequency sound (125 Hz) generated greater variability than that generated by a high-frequency sound (4 kHz). For the factor of noise intensity, Cody and Robertson (1983) observed differential impacts of three levels of noise exposure (112, 115, and 118 dB SPL) and reported that the variation was smallest for animals exposed at 118 dB SPL and largest at 112 dB SPL, indicating that variability is inversely related to the exposure level. For the factor of noise duration, Thorne et al. (1986) showed that the pattern and extent of damage showed significant variation in animals subjected to short noise exposure but became more consistent as the duration of the exposure lengthened.

The cause of individual variation in the level of cochlear damage is not fully understood but has been linked to many extrinsic and intrinsic factors. The extrinsic factors are related to the variation in the level of noise imposed on the inner ear. This variation can be caused by the unstable performance of noise generation equipment or the difference in outer and middle ear acoustics. To assess the influence of the variation of the noise level in the ear canal, Cody and Robertson (1983) carefully controlled the ear canal sound pressure level at the level of the tympanic membrane. Even with a consistent exposure level, significant variation in the level of inner ear damage was observed, suggesting the presence of intrinsic regulation of the cellular response to acoustic trauma. The possible intrinsic factors include the age, sex, nutritional state, and genetic background of subjects. Future investigations into genetic variations among individuals may shed light on the biological mechanisms responsible for individual variation associated with acoustic trauma.

Different types of cells in the organ of Corti exhibit different levels of susceptibility to acoustic trauma. Hair cells are more vulnerable to acoustic trauma than are supporting cells. Within hair cells, OHCs are more vulnerable than are IHCs (Lim 1976; Lipscomb et al. 1977; Liberman and Kiang 1978; Thorne and Gavin 1985). Evidence for these differences comes from the following three observations. First, pathological analysis of the organ of Corti made immediately or shortly after noise exposure shows that the onset of OHC damage often precedes the onset of IHC damage. Second, the minimal noise level that is capable of inducing OHC death is lower than that needed for the generation of IHC death. Third, physiological assessments reveal that the first 30–50 dB of permanent hearing loss is attributed to loss of OHCs (Stebbins et al. 1979; Hamernik et al. 1989). Detectable IHC loss does not begin to appear until the permanent threshold shift exceeds approximately 30–50 dB. The high vulnerability of OHCs appears to be consistent in both human (Lim 1976) and animal subjects. It is also consistent in cochleae exposed to continuous noise, impulse noise, or even a blast (Yokoi and Yanagita 1984).

Three rows of OHCs often exhibit similar patterns of damage but display different susceptibilities to acoustic trauma. Some studies show the first row of OHCs to be the most vulnerable, as evident by the early onset of damage (Liberman and Kiang 1978; Thorne et al. 1984, 1986; Fredelius et al. 1988), whereas other studies indicate the third row of OHCs to be most vulnerable (Lipscomb et al. 1977; Rydmarker and Nilsson 1987). There is also evidence that the three rows of OHCs are equally susceptible (Poche et al. 1969). It is likely that the difference in these observations is related to the differences in the species, noise conditions, and measures used for the assessment of hair cell damage.

The cause of high susceptibility of OHCs is not clear but has been attributed to differences in their anatomic location (Saunders et al. 1985). OHCs reside in the middle section of the basilar membrane, where sound-induced motion is maximal. In contrast, IHCs reside in the region of the junction between the basilar membrane and the osseous spiral lamina, where the vibration is minimal. Therefore, OHCs sustain greater mechanical stress than do IHCs. Another possible explanation for the difference in the susceptibility is related to the difference in metabolic activity between OHCs and IHCs.

It should be noted that, although OHCs are more likely to die after noise exposure, not all their cellular structures are particularly vulnerable to acoustic overstimulation. Using scanning electron microscopy, researchers have founded that the stereocilia of

IHCs are more vulnerable than those of OHCs (Robertson 1982; Engström 1983; Kaltenbach et al. 1992; Clark and Pickles 1996). However, despite the high vulnerability of stereocilia, IHCs are more likely to survive acoustic trauma, suggesting that stereocilia damage does not necessarily lead to cell death (Thorne et al. 1984).

5 Cell Death Modes and Pathways

There is strong evidence that noise-induced hair cell death is a complex process involving multiple cell death pathways. Based on morphological and biological characteristics, the mode of cell death has been divided into three types: apoptosis, necrosis, and an atypical pathway. Apoptosis is an active cell death pathway, requiring a persistent energy supply during the death process, particularly in the early phase of the degenerative process. Necrosis, in contrast, is a passive cell death pathway that is commonly linked to energy deprivation. In addition to typical apoptotic and necrotic cell death, a third death pathway (an atypical pathway) has been described for the noise-traumatized cochlea (Bohne et al. 2007).

Different cell death modes exhibit different morphologies. Apoptosis is characterized by shrinkage of the cell, whereby the cell body becomes irregular and smaller, and the intracellular organelles become highly packed. This reduction in the cell volume has been attributed to the dehydration of the cell. In addition to a reduction in cell volume, the nuclei of cells shrink and become condensed or fragmented. In fact, the nuclear condensation and fragmentation have been noticed in noise-traumatized hair cells for many years (Thorne et al. 1986). However, only in the past 10 years has this type of morphology been recognized as an apoptotic phenotype. The condensation of nuclear DNA leads to an increase in the density of nuclear staining. During the early phase of nuclear fragmentation, the nuclear membrane is intact and is able to hold the nuclear fragments in place. During the later phase of nuclear fragmentation, rupture of the nuclear membrane leads to spillage of nuclear fragments into the cytosol. Because the nuclear change is an early sign of the cell death and because this morphological change takes place in a step-by-step fashion, it has been used as an important measure to assess apoptotic activity in the cochlea (Hu et al. 2002a; Yang et al. 2004).

In many pathological insults, it has been well established that membrane integrity is preserved at the early stage of apoptosis (Lo et al. 1995; Majno and Joris 1995). The preservation of membrane integrity prevents the early release of cellular contents. At late stages of apoptosis, cells lose their membrane integrity because of a secondary degradation of the membrane. The development of membrane permeability to large molecules has been considered a sign of the shift of cell death from apoptosis to necrosis (Lo et al. 1995; Majno and Joris 1995). Surprisingly, acoustic trauma has been shown to provoke a rapid induction of membrane permeability (Hu and Zheng 2008). At the same time, the acute apoptosis is initiated (Hu et al. 2006). These findings suggest that the increase in membrane permeability to macromolecules is a cofactor for promotion of acute hair cell apoptosis.

Necrotic cell death is characterized by swelling of the cell body, and this enlargement of the cell body has been linked to the vesiculation and vacuolization in the cytoplasm

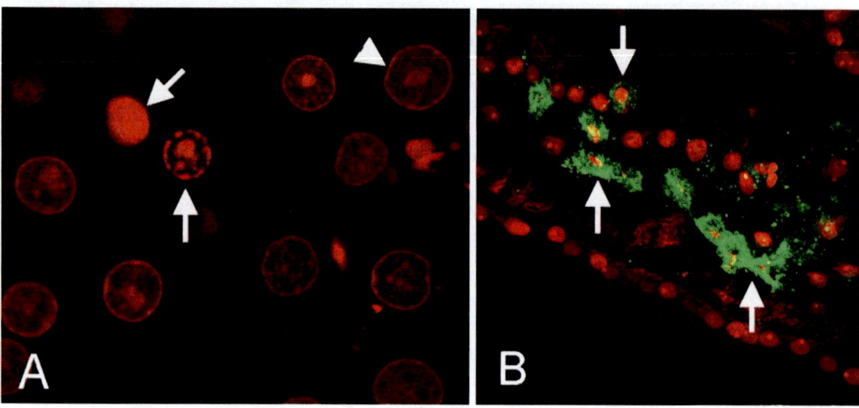

Fig. 5.6 Images showing cell death modes in a chinchilla cochlea exposed to a 4-kHz narrowband noise at 110 dB for 1 h. (**a**) Nuclear morphology illustrated by propidium iodide staining. The arrows point to condensed or fragmented nuclei. The arrowhead points to a swollen nucleus. (**b**) Caspase-9 activity in a noise-traumatized cochlea. Activation of caspase is a signature change of apoptosis. Arrows point to hair cells showing condensed or fragmented nuclei. These cells exhibit strong caspase activity

(Liberman and Dodds 1987). Nuclear swelling is also a phenotype of necrosis. Swelling of nuclei leads to the loss of the normal intranuclear texture. It should be noted that, although necrosis features the swelling of the cell body and the nucleus, the appearance of these changes does not provide a definitive indication of necrosis because swollen cells can still be viable. In this regard, the functional integrity of the plasma membrane has been used as a measure to distinguish necrotic cells from viable cells. Loss of membrane integrity is considered a sign of irreversible changes. Figure 5.6 shows typical phenotypes of apoptotic and necrotic death initiated by acoustic trauma.

Bohne et al. (2007) described a cell death mode with a morphology that is distinct from both apoptosis and necrosis. Cells in this death mode lack all or nearly all of their basolateral plasma membrane. Surprisingly, these cells maintain their normal shape, with cellular debris enclosed within the cell boundary. The nuclei are weakly stained and have chromatin clumped along the nuclear membrane. The authors termed this mode of cell death "the third death pathway." Although the biological basis of this death mode is not clear, its occurrence appears to be related to low or moderate levels of noise exposure. The generation of typical apoptosis, in contrast, is related to high levels of noise exposure.

The propensity of hair cell death toward apoptosis and necrosis appears to be related to the level of noise exposure. A higher level of noise exposure preferentially provokes apoptosis as compared to a lower level noise (Hu et al. 2000; Yang et al. 2004). Hu et al. (2000) compared the patterns of hair cell death in guinea pig cochleae exposed to one of three noise levels (110, 115, or 120 dB SPL). When the cochleae were exposed to the noise at 120 dB SPL, apoptosis dominated the hair cell pathogenesis. In contrast, when the cochleae were exposed to the noise at 110 or 115 dB SPL, the feature change was necrosis.

In noise-damaged cochleae, studies of biological mechanisms underlying apoptosis have emerged. Recent studies have revealed the involvement of multiple apoptotic events, including activation of caspases-3, -8, and -9 (Nicotera et al. 2003), translocation of cytochrome *c* from the mitochondria to the cytosol (Nicotera et al. 2003), and release of EndoG and AIF from the mitochondria to the nuclei (Yamashita et al. 2004; Han et al. 2006). In addition to these apoptotic events, several apoptotic molecules have been identified in noise-traumatized cochlea, including c-Jun-N-terminal kinase (Pirvola et al. 2000), transcriptional factor activator protein-1 (Matsunobu et al. 2004), BCL2-associated agonist of cell death (BAD; Vicente-Torres and Schacht 2006), Bcl-xL and Bak (Yamashita et al. 2008), and tumor necrosis factor-alpha (TNF-α; Fujioka et al. 2006). Using gene-array techniques, several studies have screened the expression of a large number of genes and found differential gene expression in noise-traumatized cochleae (Taggart et al. 2001; Cho et al. 2004; Kirkegaard et al. 2006; Hu et al. 2009; for review see Gong and Lomax, Chap. 9). In a recent study, Hu et al. (2009) screened expression of a large number of apoptosis-related genes in noise-traumatized cochleae and documented a time-dependent expression change. A better understanding of the molecular mechanisms responsible for different cell death pathways will aid our efforts in exploring effective therapeutic strategies for the reduction of noise-induced hearing loss.

6 Damage to the Tectorial Membrane

The tectorial membrane, which consists of a complex matrix of fine fibers, is an acellular structure covering the sulcus spiralis internus and the organ of Corti (Kronester-Frei 1978). Microscopic observation of its upper surface reveals three distinct sections: limbal, middle, and marginal zones. The lower surface (hair-cell side) of the tectorial membrane also has the three zones. This surface features Hensen's stripe, a dark band running parallel to the length of the tectorial membrane, and W-shaped or V-shaped concavities, which are the imprints of the stereocilia. The stereocilia of OHCs are coupled with the tectorial membrane, whereas the stereocilia of IHCs reside beneath the tectorial membrane, without touching the membrane.

Canlon (1987, 1988) observed the morphological changes in the tectorial membrane after acoustic overstimulation and described three major changes. First, the fibers in the middle zone become disordered and occasionally were clumped with an increased waviness. Second, discontinuities and sudden breaks in the Hensen's stripe were visible. Third, the thickness of the tectorial membrane decreased. The tectorial membrane abnormalities were restricted to the regions of the cochlea displaying significant hearing loss (40–50 dB). In another study, deformation of stereocilium imprints was seen in severely damaged cochleae (Morisaki et al. 1991). Inside the imprints, there were remnant hairs from sensory cells, suggesting the detachment of stereocilia from their anchorage.

The regeneration of damaged tectorial membranes in the chick cochlea has been reported by Cotanche (Cotanche 1987a), who found that 1 day after the noise exposure, the supporting cells began to secrete a substance for the regeneration of the tectorial membrane. By 10 days postexposure, a new honeycomb-like matrix had replaced the segment of damaged tectorial membrane. However, the regenerated matrix was not completely normal because it contains only the structure of the lower layer of the tectorial membrane. Currently, it is not clear whether the tectorial membrane of mammalian cochleae can be regenerated following acoustic trauma.

7 Damage to Supporting Cells

Supporting cell damage has not received greater attention, possibly because its damage is not as obvious as that seen in hair cells and because the damage often occurs subsequently to hair cell generation. Although not as susceptible as sensory cells, supporting cells are indeed a target of acoustic overexposure. Using scanning electron microscopy, Thorne et al. (1984) observed the morphological changes in supporting cells after acoustic trauma. Inner sulcus cells close to IHCs appeared ruptured, and cellular debris protruded into the scala media. The Deiters' cells and outer pillar cells often bulged above the reticular lamina. The supporting cells were devoid of, or showed a reduced number of, microvilli.

Pillar cells play an important role in supporting the structure of the organ of Corti. Bohne et al. (1987) described the pattern of pillar cell loss in chinchilla cochleae subjected to a 4-kHz octave-band noise. Pillar cell loss appears mainly in high-frequency lesions. The severity of pillar cell loss appears to be parallel to the level of OHC loss, that is, the greater the outer hair cell loss, the greater the pillar cell loss. Although outer pillar cells and inner pillar cells exhibit a similar pattern of damage, the level of outer pillar cell loss often exceeds that of inner pillar cell loss (Bohne et al. 1987).

At present, the interaction between sensory cell damage and supporting cell damage is poorly understood. Studies of cochlear apoptosis have demonstrated the involvement of extrinsic signal transduction in apoptotic hair cells (Nicotera et al. 2003), suggesting that extracellular signals, possibly from supporting cells, participate in the initiation of hair cell death. Further investigation into the interplay between sensory cell injury and supporting cell injury will provide new insight into the complex process of hair cell damage.

8 Repair Process

8.1 Clearance of Dead Cells

Once hair cells die, clearance of dead cells is an essential step for the restoration of a survival environment for remaining cells. Rapid removal of cell debris also facilitates the reparative processes of the organ of Corti. Based on the morphological analysis of noise-traumatized cochleae, three routes of debris clearance have been proposed.

The first route of clearance is by macrophages or other scavenger cells. Fredelius and Rask-Andersen (1990) analyzed the cochlear pathology at various times (5 min, 4 h, 24 h, 5 days, and 28 days) after acoustic trauma. Phagocytic cells were found at 5 days post-noise exposure. Dendritic macrophages containing engulfed degenerating cells and cell debris were seen in the tunnel of Corti and in the region of the OHCs. The accumulation of leukocytes was also seen in the spiral lamina vessels. Accumulation of these scavenger cells, according to the authors, may contribute to local disposal of degeneration products and thus to the healing of the organ of Corti. In another study, Hirose et al. (2005) found the migration of mononuclear phagocytes to the region of spiral ligament and spiral limbus. These cells may be involved in the recovery process in these regions.

The second route is by protrusions of cellular debris into the endolymph. This change is the consequence of the rupture of the reticular lamina or the formation of cuticular plate holes (Bohne and Rabbitt 1983; Thorne et al. 1984, 1986). The released cellular debris is likely to be absorbed by macrophages (Fredelius and Rask-Andersen 1990; Hirose et al. 2005) or by Reissner's membrane (Hunter-Duvar 1978).

The third route of clearance is via supporting cells. By examining the cochlear distribution of prestin, a motor protein of the OHC, Abrashkin et al. (2006) assessed the fate of OHCs after their breakdown. Under normal conditions, prestin immunoreactivity is localized exclusively in OHCs. In noise-traumatized cochleae, however, prestin immunoreactivity is observed within certain supporting cells, suggesting that these cells start to take up the debris of degenerated OHCs. This observation indicates that supporting cells participate in the disposal of hair cell remains.

8.2 Scar Formation

In the mammalian cochlea, once hair cells die, no new hair cells are regenerated to replace them. Degradation of dead hair cells leads to defects in the reticular lamina, which will be repaired by scar tissue. The pattern of scar formation is dependent on the extent of structural damage. If the sensory loss is sporadic and the surrounding supporting cells are intact, the defect will be sealed by the neighboring supporting cells (Hawkins et al. 1976; Bohne and Rabbitt 1983; Roberto and Zito 1988; Ahmad et al. 2003). Loss of the first row of OHCs can be replaced by outer pillar cells and loss in other rows by the enlarged phalangeal processes of Deiters' cells. Missing IHCs can be replaced by either enlarged inner pillar cells or border cells (Thorne et al. 1984). Severe damage to the organ of Corti results in massive sensory cell loss, which is accompanied by the loss of supporting cells. If an entire section of the organ of Corti collapses, it will be replaced by a flat epithelium. There are transition areas on either side of the flat epithelium where supporting cells, including inner sulcus, Hensen's, and Claudius's cells, are present (Mulroy and Curley 1982; Thorne et al. 1984). As expected, this severe cochlear damage leads to significant permanent hearing loss. Figure 5.7 shows an example of a phalangeal scar and an epithelium scar in noise-traumatized organs of Corti.

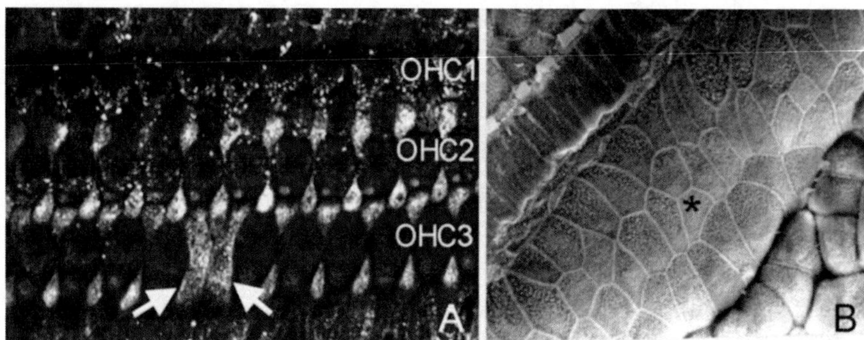

Fig. 5.7 Images showing scar tissue in the organ of Corti. (**a**) A phalangeal scar in a site of a missing outer hair cell (*arrows*) in a chinchilla cochlea exposed to a 4-kHz noise at 110 dB SPL. The enlarged phalangeal processes of two Deiters' cells have sealed the opening left by the missing outer hair cells (OHCs). OHC1, OHC2, and OHC3 represent the first, second, and third row of OHCs. (**b**) The surface view of the reticular lamina using scanning electron microscopy in a guinea pig cochlea exposed to an impulse noise at 166 dB pSPL. The asterisk marks the regions of completely missing OHCs where a flat epithelium is visible (Courtesy of Mr. Jianhe Sun, Institute of Otolaryngology, PLA General Hospital)

9 Alteration of Cochlear Innervation

Hair cells are innervated by the afferent and efferent fibers of the cochlear nerve (Spoendlin 1985). The afferent projection consists of myelinated (about 90–95%) fibers from type I neurons and unmyelinated (about 5–10%) fibers from type II neurons. The fibers from type I neurons innervate IHCs and the fibers from type II neurons are connected to OHCs. Like the afferent projection, the efferent projection also consists of two types of fibers. The fibers originated from the lateral superior olivary nucleus project to the IHC area. They make synaptic contacts with the afferent dendrites associated with the IHCs. The fibers from the medial nucleus of the trapezoid body and periolivary nucleus provide the efferent nerve supply to OHCs, primarily in the contralateral cochlea.

The mechanical stress due to acoustic overstimulation compromises the structural and functional integrity of nerve innervations to hair cells. The resulting degenerative changes are generally considered secondary, occurring after hair cell damage. The degeneration can occur in both the synaptic region and in nerve fibers.

Morphological examination of synapses shows large clear spaces beneath IHCs (Fig. 5.8). These clear spaces are identified as swollen afferent dendrites (Goulios and Robertson 1983; Robertson 1983; Canlon 1988). Swelling of cochlear nerve terminals occurs mainly in the IHC area. Its onset precedes the onset of nerve fiber degeneration, usually within 24–48 h after noise overexposure (Liberman and Mulroy 1982). The synaptic swelling has been linked to the excessive release of glutamate, the principal neurotransmitter of IHCs, during noise exposure (Puel et al. 1998).

The number and the distribution of IHCs showing synaptic swelling are related to noise conditions. In general, the longer the exposure duration and the higher the

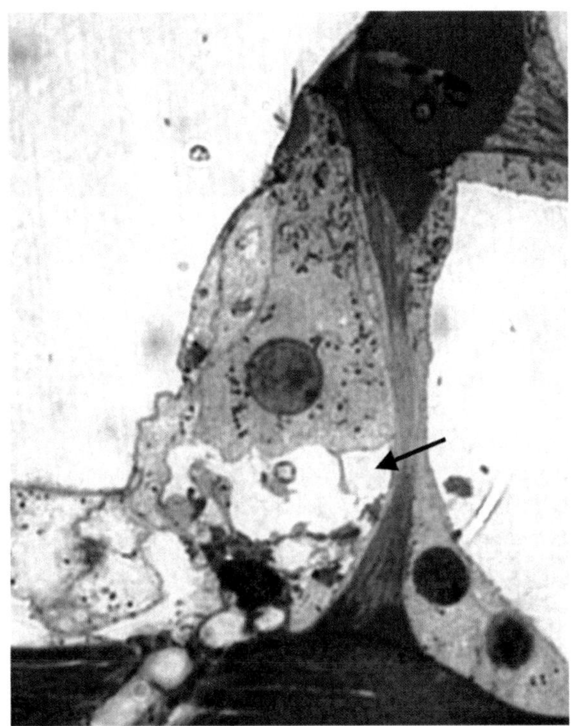

Fig. 5.8 An example of synapse swelling beneath an inner hair cell (IHC) of a chinchilla cochlea exposed to a 4-kHz narrowband noise at 110 dB SPL. The arrow marks the swollen synapse spaces. The IHC membrane at the basal pole is distorted

noise intensity, the greater the number of IHCs that exhibit synaptic swelling (Goulios and Robertson 1983; Puel et al. 1998). In terms of the interaction between the noise duration and the noise intensity, greater swelling appears when cochleae are subjected to a noise with a long duration but a low intensity. In contrast, the swelling is less obvious in cochleae subjected to an exposure of a short duration but a high intensity. With impulse noise, no swelling in the synaptic region is found (Slepecky et al. 1981), suggesting that this synaptic pathology is caused by sustained activation of hair cells. For the distribution pattern of synaptic swelling, a short duration of noise exposure causes the swelling of afferent dendrites in and below the section of the organ of Corti with the characteristic frequency corresponding to the noise frequency. With a longer duration of exposure, swelling expands to the region apical to the exposure frequency. Swelling can occur even when threshold shifts are ultimately reversible (Liberman and Mulroy 1982).

In contrast to the rapid induction of synaptic swelling, the nerve fibers undergo a slow and prolonged degeneration (Liberman and Kiang 1978; Pye 1981, 1984; Prijs et al. 1993; Kujawa and Liberman 2009). The degeneration starts from the distal end of the fibers and expands to their proximal end. The degenerated fibers are commonly seen in the cochlear sites showing hair cell damage. Liberman and Kiang (1978) found that degeneration of afferent nerve fibers follows the pattern of IHC loss. In the region where IHCs remain intact, regardless of OHC conditions, there is no significant nerve fiber degeneration. Prijs et al. (1993) also reported the degeneration of myelinated nerve fibers in the areas where the organ of Corti was destroyed. However, the authors also reported the degeneration of myelinated fibers in some cases without significant hair cell loss. Bohne et al. (1987) found the loss of myelinated nerve fibers in the high-frequency region of the cochlea. Again, the degeneration appears mainly in the region of IHC damage.

The degenerative process of nerve fibers persists for a long period of time postexposure. Pye (1981, 1984) reported that the degeneration occurred only peripherally from the habenula perforate at 3 weeks postexposure. Around 12 weeks postexposure, the degeneration progresses further toward the modiolus. Kujawa and Liberman (2006, 2009) reported that delayed loss of spiral ganglion cells months after noise exposure. This change slowly progresses for years.

There is some evidence of regeneration of nerve fibers into regions where the sensory epithelium was destroyed (Wright 1976; Prijs et al. 1993). Swelling of afferent dendrites has also been found to be reversible (Robertson 1983). However, a recent study suggested that acute noise-induced damage to cochlear nerve terminals is irreversible in adults and that there is minimal nerve regeneration or renewed synaptogenesis after acoustic trauma (Kujawa and Liberman 2009). Further studies on the long-term effects of the degenerative process of hair cell innervation are necessary so that the clinical impacts of nerve degeneration can be better understood.

10 Summary

Noise-induced hearing loss is caused by complex structural changes in the cochlea. The structural damage is initiated during the course of noise exposure and continues to develop after the termination of noise exposure. Over the years, the patterns of morphological changes have been thoroughly characterized. The relationships between noise conditions and morphological changes and between morphological changes and functional impacts have been elucidated. However, the molecular bases and impacts of the structural changes are poorly understood. A better understanding of the interaction between the cellular structural changes and molecular signal transduction will provide significant new insights into the biological mechanism of noise-induced cochlear damage and, more importantly, will aid efforts to explore new therapeutic strategies for preventing noise-induced hearing loss.

References

Abrashkin, K. A., Izumikawa, M., Miyazawa, T., Wang, C. H., Crumling, M. A., Swiderski, D. L., Beyer, L. A., Gong, T. W., & Raphael, Y. (2006). The fate of outer hair cells after acoustic or ototoxic insults. *Hearing Research*, 218(1–2), 20–29.

Ahmad, M., Bohne, B. A., & Harding, G. W. (2003). An in vivo tracer study of noise-induced damage to the reticular lamina. *Hearing Research*, 175(1–2), 82–100.

Axelsson, A., & Dengerink, H. (1987). The effects of noise on histological measures of the cochlear vasculature and red blood cells: A review. *Hearing Research*, 31(2), 183–191.

Bohne, B. A. (1976). Mechanisms of noise damage in the inner ear. In D. Henderson, R. P. Hamernik, D. S. Dosanjh, & J. H. Mills (Eds.), *Effects of noise on hearing* (pp. 41–68). New York: Raven Press.

Bohne, B. A., & Clark, W. W. (1982). Growth of hearing loss and cochlear lesion with increasing duration of noise exposure. In R. Hamernik, D. Henderson & R. Salvi (Eds.), *New perspectives on noise-induced hearing loss* (pp. 283–302). New York: Raven Press.

Bohne, B. A., & Harding, G. W. (2000). Degeneration in the cochlea after noise damage: Primary versus secondary events. *American Journal of Otology*, 21(4), 505–509.

Bohne, B. A., & Rabbitt, K. D. (1983). Holes in the reticular lamina after noise exposure: Implication for continuing damage in the organ of Corti. *Hearing Research*, 11(1), 41–53.

Bohne, B. A., Zahn, S. J., & Bozzay, D. G. (1985). Damage to the cochlea following interrupted exposure to low frequency noise. *Annals of Otology, Rhinology, and Laryngology*, 94(2 Pt 1), 122–128.

Bohne, B. A., Yohman, L., & Gruner, M. M. (1987). Cochlear damage following interrupted exposure to high-frequency noise. *Hearing Research*, 29(2–3), 251–264.

Bohne, B. A., Gruner, M. M., & Harding, G. W. (1990). Morphological correlates of aging in the chinchilla cochlea. *Hearing Research*, 48(1–2), 79–91.

Bohne, B. A., Harding, G. W., & Lee, S. C. (2007). Death pathways in noise-damaged outer hair cells. *Hearing Research*, 223(1–2), 61–70.

Borg, E., & Engström, B. (1983). Damage to sensory hairs of inner hair cells after exposure to noise in rabbits without outer hair cells. *Hearing Research*, 11(1), 1–6.

Borg, E., Canlon, B., & Engström, B. (1995). Noise-induced hearing loss. Literature review and experiments in rabbits. Morphological and electrophysiological features, exposure parameters and temporal factors, variability and interactions. *Scandanavian Audiology Supplementum*, 40, 1–147.

Canlon, B. (1987). Acoustic overstimulation alters the morphology of the tectorial membrane. *Hearing Research*, 30(2–3), 127–134.

Canlon, B. (1988). The effect of acoustic trauma on the tectorial membrane, stereocilia, and hearing sensitivity: Possible mechanisms underlying damage, recovery, and protection. *Scandanavian Audiology Supplementum*, 27, 1–45.

Chen, G. D., McWilliams, M. L., & Fechter, L. D. (2000). Succinate dehydrogenase (SDH) activity in hair cells: A correlate for permanent threshold elevations. *Hearing Resesarch*, 145(1–2), 91–100.

Cho, H., Sakamoto, H., Hachikawa, K., & Nakai, Y. (1991). Electron microscopic observation of communication between inner ear stereocilia under normal and noise stimulated conditions. *Acta Oto-Laryngologica*, 486(Supplementum), 13–18.

Cho, Y., Gong, T. W., Kanicki, A., Altschuler, R. A., & Lomax, M. I. (2004). Noise overstimulation induces immediate early genes in the rat cochlea. *Molecular Brain Research*, 130(1–2), 134–148.

Clark, J. A., & Pickles, J. O. (1996). The effects of moderate and low levels of acoustic overstimulation on stereocilia and their tip links in the guinea pig. *Hearing Research*, 99(1–2), 119–128.

Cody, A. R., & Robertson, D. (1983). Variability of noise-induced damage in the guinea pig cochlea: Electrophysiological and morphological correlates after strictly controlled exposures. *Hearing Research*, 9(1), 55–70.

Coleman, J. W. (1976). Hair cell loss as a function of age in the normal cochlea of the guinea pig. *Acta Oto-Laryngologica*, 82(1–2), 33–40.

Cotanche, D. A. (1987a). Regeneration of the tectorial membrane in the chick cochlea following severe acoustic trauma. *Hearing Research*, 30(2–3), 197–206.

Cotanche, D. A. (1987b). Regeneration of hair cell stereociliary bundles in the chick cochlea following severe acoustic trauma. *Hearing Research*, 30(2–3), 181–195.

Dolan, T. R., Ades, H. W., Bredberg, G., & Neff, W. D. (1975). Inner ear damage and hearing loss after exposure to tones of high intensity. *Acta Oto-Laryngologica*, 80(5–6), 343–352.

Duvall, A. J., 3 rd, & Rhodes, V. T. (1967). Ultrastructure of the organ of Corti following intermixing of cochlear fluids. *Annals of Otology, Rhinology, and Laryngology*, 76(3), 688–708.

Duvall, A. J., 3 rd, Sutherland, C. R., & Rhodes, V. T. (1969). Ultrastructural changes in the cochlear duct following mechanical disruption of the organ of Corti., *Annals of Otology, Rhinology, and Laryngology* 78(2), 342–357.

Engström, B. (1983). Stereocilia of sensory cells in normal and hearing impaired ears. A morphological, physiological and behavioural study. *Scandanavian Audiology Supplementum*, 19, 1–34.

Engström, B. (1984). Fusion of stereocilia on inner hair cells in man and in the rabbit, rat and guinea pig. *Scandanavian Audiology*, 13(2), 87–92.

Engström, B., & Borg, E. (1981). Lesions to cochlear inner hair cells induced by noise. *Archives of Oto-Rhino-Laryngology*, 230(3), 279–284.

Engström, B., Flock, A., & Borg, E. (1983). Ultrastructural studies of stereocilia in noise-exposed rabbits. *Hearing Research*, 12(2), 251–264.

Erlandsson, B., Hakanson, H., Ivarsson, A., Nilsson, P., & Wersall, J. (1980). Hair cell damage in the guinea pig due to different kinds of noise. *Acta Oto-Laryngologica*, 357 (Supplementum), 1–43.

Fredelius, L., & Rask-Andersen, H. (1990). The role of macrophages in the disposal of degeneration products within the organ of corti after acoustic overstimulation. *Acta Oto-Laryngologica*, 109(1–2), 76–82.

Fredelius, L., & Wersall, J. (1992). Hair cell damage after continuous and interrupted pure tone overstimulation: A scanning electron microscopic study in the guinea pig. *Hearing Research*, 62(2), 194–198.

Fredelius, L., Rask-Andersen, H., Johansson, B., Urquiza, R., Bagger-Sjoback, D., & Wersall, J. (1988). Time sequence of degeneration pattern of the organ of Corti after acoustic overstimulation. A light microscopical and electrophysiological investigation in the guinea pig. *Acta Oto-Laryngologica*, 106(1–2), 81–93.

Fridberger, A., Flock, A., Ulfendahl, M., & Flock, B. (1998). Acoustic overstimulation increases outer hair cell Ca^{2+} concentrations and causes dynamic contractions of the hearing organ. *Proceedings of the National Academy of Sciences of the USA*, 95(12), 7127–7132.

Fujioka, M., Kanzaki, S., Okano, H. J., Masuda, M., Ogawa, K., & Okano, H. (2006). Proinflammatory cytokines expression in noise-induced damaged cochlea. *Journal of Neuroscience Research*, 83(4), 575–583.

Gao, W. Y., Ding, D. L., Zheng, X. Y., Ruan, F. M., & Liu, Y. J. (1992). A comparison of changes in the stereocilia between temporary and permanent hearing losses in acoustic trauma. *Hearing Research*, 62(1), 27–41.

Goulios, H., & Robertson, D. (1983). Noise-induced cochlear damage assessed using electrophysiological and morphological criteria: an examination of the equal energy principle. *Hearing Research*, 11(3), 327–341.

Gulley, R. L., & Reese, T. S. (1976). Intercellular junctions in the reticular lamina of the organ of Corti. *Journal of Neurocytology*, 5(4), 479–507.

Hamernik, R. P., Turrentine, G., Roberto, M., Salvi, R., & Henderson, D. (1984). Anatomical correlates of impulse noise-induced mechanical damage in the cochlea. *Hearing Research*, 13(3), 229–247.

Hamernik, R. P., Patterson, J. H., Turrentine, G. A., & Ahroon, W. A. (1989). The quantitative relation between sensory cell loss and hearing thresholds. *Hearing Research*, 38(3), 199–211.

Han, W., Shi, X., & Nuttall, A. L. (2006). AIF and endoG translocation in noise exposure induced hair cell death. *Hearing Research*, 211(1–2), 85–95.

Harding, G. W., & Bohne, B. A. (2007). Distribution of focal lesions in the chinchilla organ of Corti following exposure to a 4–kHz or a 0.5–kHz octave band of noise. *Hearing Research*, 225(1–2), 50–59.

Harding, G. W., & Bohne, B. A. (2009). Relation of focal hair-cell lesions to noise-exposure parameters from a 4- or a 0.5-kHz octave band of noise. *Hearing Research*, 254(1–2), 54–63.

Hawkins, J. E., Jr., Johnsson, L. G., Stebbins, W. C., Moody, D. B., & Coombs, S. L. (1976). Hearing loss and cochlear pathology in monkeys after noise exposure. *Acta Oto-Laryngologica*, 81(3–4), 337–343.

Henderson, D., & Hamernik, R. P. (1986). Impulse noise: critical review. *Journal of the Acoustical Society of America*, 80(2), 569–584.

Henderson, D., Subramaniam, M., Gratton, M. A., & Saunders, S. S. (1991). Impact noise: the importance of level, duration, and repetition rate. *Journal of the Acoustical Society of America*, 89(3), 1350–1357.

Hirose, K., Discolo, C. M., Keasler, J. R., & Ransohoff, R. (2005). Mononuclear phagocytes migrate into the murine cochlea after acoustic trauma. *Journal of Comparative Neurology*, 489(2), 180–194.

Hu, B. H., & Zheng, G. L. (2008). Membrane disruption: An early event of hair cell apoptosis induced by exposure to intense noise. *Brain Research*, 1239, 107–118.

Hu, B. H., Guo, W., Wang, P. Y., Henderson, D., & Jiang, S. C. (2000). Intense noise-induced apoptosis in hair cells of guinea pig cochleae. *Acta Oto-Laryngologica*, 120(1), 19–24.

Hu, B. H., Henderson, D., & Nicotera, T. M. (2002a). Involvement of apoptosis in progression of cochlear lesion following exposure to intense noise. *Hearing Research*, 166(1–2), 62–71.

Hu, B. H., Henderson, D., & Nicotera, T. M. (2002b). F-actin cleavage in apoptotic outer hair cells in chinchilla cochleas exposed to intense noise. *Hearing Research*, 172(1–2), 1–9.

Hu, B. H., Henderson, D., & Nicotera, T. M. (2006). Extremely rapid induction of outer hair cell apoptosis in the chinchilla cochlea following exposure to impulse noise. *Hearing Research*, 211(1–2), 16–25.

Hu, B. H., Cai, Q., Manohar, S., Jiang, H., Ding, D., Coling, D. E., Zheng, G., & Salvi, R. (2009). Differential expression of apoptosis-related genes in the cochlea of noise-exposed rats. *Neuroscience*, 161(3), 915–925.

Hultcrantz, E., Angelborg, C., & Beausang-Linder, M. (1979). Noise and cochlear blood flow. *Archives of Oto-Rhino-Laryngology*, 224(1–2), 103–106.

Hunter-Duvar, I. (1977). Inner ear correlates in acoustic trauma. *Transactions. Section on Otolaryngology. American Academy of Ophthalmology and Otolaryngology*, 84(2), 422.

Hunter-Duvar, M. (1978). Reissner's membrane and endocytosis of cell debris. *Acta Oto-Laryngologica*, 351(Supplementum), 24–32.

Ishida, M. (1978). [Lactate dehydrogenase (LDH) in the inner ear during acoustic stimulation]. *Acta Otolaryngologica*, 85(1–2), 17–23.

Ishii, D., Takahashi, T., & Balogh, K. (1969). Glycogen in the inner ear after acoustic stimulation. A light and electron microscopic study. *Acta Oto-Laryngologica*, 67(6), 573–582.

Jia, S., Yang, S., Guo, W., & He, D. Z. (2009). Fate of mammalian cochlear hair cells and stereocilia after loss of the stereocilia. *Journal of Neuroscience*, 29(48), 15277–15285.

Kaltenbach, J. A., Schmidt, R. N., & Kaplan, C. R. (1992). Tone-induced stereocilia lesions as a function of exposure level and duration in the hamster cochlea. *Hearing Research*, 60(2), 205–215.

Kikuchi, T., Kimura, R. S., Paul, D. L., Takasaka, T., & Adams, J. C. (2000). Gap junction systems in the mammalian cochlea. *Brain Research Reviews*, 32(1), 163–166.

Kirkegaard, M., Murai, N., Risling, M., Suneson, A., Jarlebark, L., & Ulfendahl, M. (2006). Differential gene expression in the rat cochlea after exposure to impulse noise. *Neuroscience*, 142(2), 425–435.

Kronester-Frei, A. (1978). Ultrastructure of the different zones of the tectorial membrane. *Cell Tissue Research*, 193(1), 11–23.

Kujawa, S. G., & Liberman, M. C. (2006). Acceleration of age-related hearing loss by early noise exposure: Evidence of a misspent youth. *Journal of Neuroscience*, 26(7), 2115–2123.

Kujawa, S. G., & Liberman, M. C. (2009). Adding insult to injury: Cochlear nerve degeneration after "temporary" noise-induced hearing loss. *Journal of Neuroscience*, 29(45), 14077–14085.

Lahne, M., & Gale, J. E. (2008). Damage-induced activation of ERK1/2 in cochlear supporting cells is a hair cell death-promoting signal that depends on extracellular ATP and calcium. *Journal of Neuroscience*, 28(19), 4918–4928.

Liberman, M. C. (1987). Chronic ultrastructural changes in acoustic trauma: Serial-section reconstruction of stereocilia and cuticular plates. *Hearing Research*, 26(1), 65–88.

Liberman, M. C., & Dodds, L. W. (1987). Acute ultrastructural changes in acoustic trauma: Serial-section reconstruction of stereocilia and cuticular plates. *Hearing Research*, 26(1), 45–64.

Liberman, M. C., & Kiang, N. Y. (1978). Acoustic trauma in cats. Cochlear pathology and auditory-nerve activity. *Acta Oto-Laryngologica*, 358(Supplementum), 1–63.

Liberman, M. C., & Mulroy, M. J. (1982). Acute and chronic effects of acoustic trauma: Cochlear pathology and auditory-nerve pathophysiology. In R. Hamernik, D. Henderson & R. Salvi (Eds.), *New perspectives on noise-induced hearing loss* (pp. 105–135). New York: Raven Press.

Lim, D. J. (1976). Ultrastructural cochlear changes following acoustic hyperstimulation and ototoxicity. *Annals of Otology, Rhinology and Laryngology*, 85(6 Pt. 1), 740–751.

Lim, D. J., & Melnick, W. (1971). Acoustic damage of the cochlea. A scanning and transmission electron microscopic observation. *Archive of Otolaryngology*, 94(4), 294–305.

Lipscomb, D. M., Axelsson, A., Vertes, D., Roettger, R., & Carrol, J. (1977). The effect of high level sound on hearing sensitivity, cochlear sensorineuroepithelium and vasculature of the chinchilla. *Acta Oto-Laryngologica*, 84(1–2), 44–56.

Lo, A. C., Houenou, L. J., & Oppenheim, R. W. (1995). Apoptosis in the nervous system: Morphological features, methods, pathology, and prevention. *Archives of Histology and Cytology*, 58(2), 139–149.

Majno, G., & Joris, I. (1995). Apoptosis, oncosis, and necrosis. An overview of cell death. *The American Journal of Pathology*, 146(1), 3–15.

Matsunobu, T., Ogita, K., & Schacht, J. (2004). Modulation of activator protein 1/DNA binding activity by acoustic overstimulation in the guinea-pig cochlea. *Neuroscience*, 123(4), 1037–1043.

Maurer, J., Heinrich, U. R., Hinni, M., & Mann, W. (1999). Alteration of the calcium content in inner hair cells of the cochlea of the guinea pig after acute noise trauma with and without application of the organic calcium channel blocker diltiazem. *Journal of Oto-Rhino-Laryngology and Its Related Specialties*, 61(6), 328–333.

Morisaki, N., Nakai, Y., Cho, H., & Shibata, S. (1991). Imprints of the tectorial membrane following acoustic overstimulation and kanamycin treatment. *Acta Oto-Laryngologica*, 486 (Supplementum), 486, 19–31.

Mulroy, M. J., & Curley, F. J. (1982). Stereociliary pathology and noise-induced threshold shift: A scanning electron microscopic study. *Scanning Electron Microscopy* (Pt 4), 1753–1762.

Mulroy, M. J., Henry, W. R., & McNeil, P. L. (1998). Noise-induced transient microlesions in the cell membranes of auditory hair cells. *Hearing Research*, 115(1–2), 93–100.

Nadol, J. B., Jr. (1978). Intercellular junctions in the organ of Corti. *Annals of Otology, Rhinology and Laryngology*, 87(1 Pt. 1), 70–80.

Nicotera, T. M., Hu, B. H., & Henderson, D. (2003). The caspase pathway in noise-induced apoptosis of the chinchilla cochlea. *Journal of the Association for Research in Otolaryngology*, 4(4), 466–477.

Nuttall, A. L. (1999). Sound-induced cochlear ischemia/hypoxia as a mechanism of hearing loss. *Noise and Health*, 2(5), 17–32.

Omata, T., Ohtani, I., Ohtsuki, K., Ogawa, Y., & Ouchi, J. (1979). Electron microscopical and histochemical studies of outer hair cells in acoustically exposed rabbits. *Archives of Oto-Rhino-Laryngology*, 222(2), 127–132.

Pickles, J. O., Comis, S. D., & Osborne, M. P. (1984). Cross-links between stereocilia in the guinea pig organ of Corti, and their possible relation to sensory transduction. *Hearing Research*, 15(2), 103–112.

Pickles, J. O., Osborne, M. P., & Comis, S. D. (1987). Vulnerability of tip links between stereocilia to acoustic trauma in the guinea pig. *Hearing Research*, 25(2–3), 173–183.

Pirvola, U., Xing-Qun, L., Virkkala, J., Saarma, M., Murakata, C., Camoratto, A. M., Walton, K. M., & Yikoski, J. (2000). Rescue of hearing, auditory hair cells, and neurons by CEP-1347/KT7515, an inhibitor of c-Jun N-terminal kinase activation. *Journal of Neuroscience*, 20(1), 43–50.

Poche, L. B., Jr., Stockwell, C. W., & Ades, H. W. (1969). Cochlear hair-cell damage in guinea pigs after exposure to impulse noise. *Journal of the Acoustical Society of America*, 46(4), 947–951.

Prijs, V. F., Keijzer, J., Versnel, H., & Schoonhoven, R. (1993). Recovery characteristics of auditory nerve fibres in the normal and noise-damaged guinea pig cochlea. *Hearing Research*, 71(1–2), 190–201.

Puel, J. L., Ruel, J., Gervais d'Aldin, C., & Pujol, R. (1998). Excitotoxicity and repair of cochlear synapses after noise-trauma induced hearing loss. *NeuroReport*, 9(9), 2109–2114.

Pye, A. (1981). Acoustic trauma effects with varying exposure times. *Archives of Oto-Rhino-Laryngology*, 230(3), 265–271.

Pye, A. (1984). The effects of short noise exposures in the guinea pig. *Archives of Oto-Rhino-Laryngology*, 240(2), 107–114.

Raphael, Y., & Altschuler, R. A. (1991). Reorganization of cytoskeletal and junctional proteins during cochlear hair cell degeneration. *Cell Motility and the Cytoskeleton*, 18(3), 215–227.

Roberto, M., & Zito, F. (1988). Scar formation following impulse noise-induced mechanical damage to the organ of Corti. *The Journal of Laryngology and Otology*, 102(1), 2–9.

Robertson, D. (1981). Combined electrophysiology and ultrastructure of acoustic trauma in the guinea pig cochlea. *Archives of Oto-Rhino-Laryngology*, 230(3), 257–263.

Robertson, D. (1982). Effects of acoustic trauma on stereocilia structure and spiral ganglion cell tuning properties in the guinea pig cochlea. *Hearing Research*, 7(1), 55–74.

Robertson, D. (1983). Functional significance of dendritic swelling after loud sounds in the guinea pig cochlea. *Hearing Research*, 9(3), 263–278.

Robertson, D., Johnstone, B. M., & McGill, T. J. (1980). Effects of loud tones on the inner ear: a combined electrophysiological and ultrastructural study. *Hearing Research*, 2(1), 39–43.

Rydmarker, S., & Nilsson, P. (1987). Effects on the inner and outer hair cells. *Acta Oto-Laryngologica*, 441(Supplementum), 25–43.

Saunders, J. C., Dear, S. P., & Schneider, M. E. (1985). The anatomical consequences of acoustic injury: A review and tutorial. *Journal of the Acoustical Society of America*, 78(3), 833–860.

Sha, S. H., Taylor, R., Forge, A., & Schacht, J. (2001). Differential vulnerability of basal and apical hair cells is based on intrinsic susceptibility to free radicals. *Hearing Research*, 155(1–2), 1–8.

Shen, H., Zhang, B., Shin, J. H., Lei, D., Du, Y., Gao, X., Wang, Q., Ohlemiller, K. K., Piccirillo, J., & Bao, J. (2007). Prophylactic and therapeutic functions of T-type calcium blockers against noise-induced hearing loss. *Hearing Research*, 226(1–2), 52–60.

Slepecky, N. (1986). Overview of mechanical damage to the inner ear: Noise as a tool to probe cochlear function. *Hearing Research*, 22, 307–321.

Slepecky, N., Hamernik, R., Henderson, D., & Coling, D. (1981). Ultrastructural changes to the cochlea resulting from impulse noise. *Archives of Oto-Rhino-Laryngology*, 230(3), 273–278.

Slepecky, N., Hamernik, R., Henderson, D., & Coling, D. (1982). Correlation of audiometric data with changes in cochlear hair cell stereocilia resulting from impulse noise trauma. *Acta Oto-Laryngologica*, 93(5–6), 329–340.

Spoendlin, H. (1976). Anatomical changes following various noise exposures. In D. Henderson, R. P. Hamernik, D. S. Dosanjh, & J. H. Mills (Eds.), *Effects of Noise on Hearing* (pp. 69–89). New York: Raven Press.

Spoendlin, H. (1985). Anatomy of cochlear innervation. *American Journal of Otolaryngology*, 6(6), 453–467.

Spongr, V. P., Henderson, D., & McFadden, S. L. (1998). Confocal microscopic analysis of the chinchilla organ of Corti following exposure to high-level impact noise. *Scandanavian Audiology Supplementum*, 48, 15–25.

Stebbins, W. C., Hawkins, J. E., Jr., Johnson, L. G., & Moody, D. B. (1979). Hearing thresholds with outer and inner hair cell loss. *American Journal of Otolaryngology*, 1(1), 15–27.

Taggart, R. T., McFadden, S. L., Ding, D. L., Henderson, D., Jin, X., Sun, W., & Salvi, R. (2001). Gene expression changes in chinchilla cochlea from noise-induced temporary threshold shift. *Noise and Health*, 3(11), 1–18.

Thorne, P. R., & Gavin, J. B. (1985). Changing relationships between structure and function in the cochlea during recovery from intense sound exposure. *Annals of Otology, Rhinology and Laryngology*, 94(1 Pt. 1), 81–86.

Thorne, P. R., Gavin, J. B., & Herdson, P. B. (1984). A quantitative study of the sequence of topographical changes in the organ of Corti following acoustic trauma. *Acta Oto-Laryngologica*, 97(1–2), 69–81.

Thorne, P. R., Duncan, C. E., & Gavin, J. B. (1986). The pathogenesis of stereocilia abnormalities in acoustic trauma. *Hearing Research*, 21(1), 41–49.

Tilney, L. G., Saunders, J. C., Egelman, E., & DeRosier, D. J. (1982). Changes in the organization of actin filaments in the stereocilia of noise-damaged lizard cochleae. *Hearing Research*, 7(2), 181–197.

Tonndorf, J. (1981). Stereociliary dysfunction, a case of sensory hearing loss, recruitment, poor speech discrimination and tinnitus. *Acta Oto-Laryngologica*, 91(5–6), 469–479.

Tsuprun, V., Schachern, P. A., Cureoglu, S., & Paparella, M. (2003). Structure of the stereocilia side links and morphology of auditory hair bundle in relation to noise exposure in the chinchilla. *Journal of Neurocytology*, 32(9), 1117–1128.

Vertes, D., Nilsson, P., Wersall, J., Axelsson, A., & Bjorkroth, B. (1982). Cochlear hair cell and vascular changes in the guinea pig following high level pure-tone exposures. *Acta Oto-Laryngologica*, 94(5–6), 403–411.

Vertes, D., Axelsson, A., Hornstrand, C., & Nilsson, P. (1984). The effect of impulse noise on cochlear vessels. *Archives of Otolaryngology*, 110(2), 111–115.

Vicente-Torres, M. A., & Schacht, J. (2006). A BAD link to mitochondrial cell death in the cochlea of mice with noise-induced hearing loss. *Journal of Neuroscience Research*, 83(8), 1564–1572.

Wright, C. G. (1976). Neural damage in the guinea pig cochlea after noise exposure. A light microscopic study. *Acta Oto-Laryngologica*, 82(1–2), 82–94.

Yamashita, D., Miller, J. M., Jiang, H. Y., Minami, S. B., & Schacht, J. (2004). AIF and EndoG in noise-induced hearing loss. *NeuroReport*, 15(18), 2719–2722.

Yamashita, D., Minami, S. B., Kanzaki, S., Ogawa, K., & Miller, J. M. (2008). Bcl-2 genes regulate noise-induced hearing loss. *Journal of Neuroscience Research*, 86(4), 920–928.

Yang, W. P., Henderson, D., Hu, B. H., & Nicotera, T. M. (2004). Quantitative analysis of apoptotic and necrotic outer hair cells after exposure to different levels of continuous noise. *Hearing Research*, 196(1–2), 69–76.

Yokoi, H., & Yanagita, N. (1984). Blast injury to sensory hairs: A study in the guinea pig using scanning electron microscopy. *Archives of Oto-Rhino-Laryngology*, 240(3), 263–270.

Chapter 6
Neural Coding of Sound with Cochlear Damage

Eric D. Young

1 Introduction

The cochlea is a delicate and complex structure designed to transduce sound into the electrical activity of neurons. Damage to any of the components of the cochlea can result in hearing impairment. The development, structure, and vulnerability of the cochlea are the subject of the other chapters in this volume. Here, the focus is on the consequences of cochlear malfunction for the representation of sound in the brain. To make the task feasible in the face of the many known physiological causes of hearing impairment, this chapter discusses the well-studied effects of damage to inner (IHCs) or outer hair cells (OHCs) and to spiral ganglion neurons (SGNs). For most cochlear malfunctions, the hair cells and the SGNs are the final common path for the effects of the damage, so that the implications of a particular type of damage can usually be understood in terms of its effects on these cells and the consequent changes in the brain. Most of the chapter focuses on the effects of acoustic trauma (i.e., exposure to a sustained loud sound), with some discussion of the effects of ototoxic substances, because these have most often been used in experimental studies to produce controlled cochlear lesions.

1.1 Perceptual Effects of Cochlear Damage

Wherever possible, the perceptual aspects of hearing impairment (Moore 2007) are discussed in relation to their physiological correlates. Some important perceptual

E.D. Young (✉)
Department of Biomedical Engineering, Johns Hopkins School of Medicine,
720 Rutland Ave, Baltimore, MD 21205, USA
e-mail: eyoung@jhu.edu

C.G. Le Prell et al. (eds.), *Noise-Induced Hearing Loss: Scientific Advances*,
Springer Handbook of Auditory Research 40, DOI 10.1007/978-1-4419-9523-0_6,

effects of cochlear damage are listed below as an introduction to the subject matter of this chapter.

- The most obvious effect of hearing impairment is a *loss of audibility*, wherein soft sounds cannot be heard because the auditory threshold is elevated. The elevated perceptual threshold, of course, corresponds to the elevated thresholds of neurons in the auditory nerve (AN), usually because of hair cell damage.
- Less obvious but no less challenging is a *loss of frequency selectivity*, so that a person's ability to separate and analyze sounds of different frequencies is impaired (Tyler et al. 1984; Glasberg and Moore 1986; Dubno and Dirks 1989). Thus, even when sounds are amplified to make them audible, lack of frequency resolving power may interfere with perception of complex stimuli like speech or music. The loss of perceptual resolving ability seems to be directly related to a similar loss of frequency tuning in neurons of the auditory system (Liberman and Dodds 1984b).
- Persons with a hearing loss often show *loudness recruitment*, meaning an unnaturally rapid growth of the perceptual loudness of a stimulus as its physical intensity is increased (Moore 2007). Two possible neural sources of recruitment are discussed: mechanical changes in basilar membrane movements (Moore and Glasberg 1997; Ruggero et al. 1997; Schlauch et al. 1998); and changes in synaptic strength of neurons in the brain (Salvi et al. 1990; Syka et al. 1994; Szczepaniak and Moller 1996; Cai et al. 2009).

The aforementioned perceptual deficits are usually attributed to hair-cell damage. However, it is increasingly clear that damage to SGNs or reorganization of neural circuits in the brain can play a part in hearing impairment. Three aspects of neural damage are discussed in this context.

- In some cases, the summated electrical potentials produced by the neural responses of the AN or the brain are missing or abnormal, but the signals associated with the hair cells seem to be normal (Starr et al. 1996; Rance 2005; Zeng et al. 2005). The associated perceptual deficits are often deficits in timing auditory stimuli (e.g., poor gap detection or modulation sensitivity) as opposed to sound-level-related deficits like detection thresholds, tuning, intensity perception, and so forth. Such deficits are classed as *auditory neuropathy*, because the deficit is thought to be in the SGN or the hair-cell/SGN synapse rather than in hair cell transduction.
- The ability to use information encoded in the high-frequency temporal fluctuations in sounds, so-called *temporal fine structure*, is apparently compromised by cochlear hearing loss (Moore 2008). This deficit has been shown to affect pitch perception, binaural processing, masking, and speech perception. Little is known of the underlying physiological correlates of this phenomenon.
- Damage to the cochlea produces secondary changes in the organization of the central nervous system, including degeneration of synapses and neurons, changes in the strength or pattern of connections among neurons, changes in the excitability of neurons, and reorganization of sound representations (Syka 2002).

These central changes may underlie phenomena such as *tinnitus*, a phantom percept of sound (see Kaltenbach, Chap. 8), but they may also contribute to a variety of other perceptual deficits.

Deficits in *temporal processing*, either in the ability to analyze a rapid sequence of sounds or the ability to integrate sound energy across time (Moore 2007; Reed et al. 2009) are not discussed. The effects of hearing loss on temporal processing are complex and variable; their physiological correlates have not been studied in detail except for hearing loss accompanying aging (Walton 2010). Temporal deficits seem to be produced by disorders in central processing, including the deficits produced by developmental damage to hearing (e.g., Sanes and Constantine-Paton 1985; Kilgard and Merzenich 1998), and deserve a more lengthy treatment than is possible here.

Hearing loss that results from conductive problems in the middle or outer ear is also not discussed. Conductive problems prevent sound from being efficiently coupled to the cochlea and do not necessarily involve damage to the neural elements of the cochlea, the focus of this chapter.

1.2 The Components of Auditory Transduction

The schematic diagram in Fig. 6.1 shows the major functional components of cochlear transduction. This figure provides an overview of the physiological processes most often implicated in hearing impairment; these are the processes discussed in this chapter. A good overall introduction to cochlear physiology is provided by Pickles (2008).

Sound is coupled into the cochlea by the outer and middle ears, not shown in Fig. 6.1 (Rosowski 1994). The sound pressure fluctuations produce motion of the basilar membrane (BM), in the direction of the dashed blue arrow (number 2; Robles and Ruggero 2001). The BM motion is coupled to the cilia of the IHCs and OHCs through a shearing motion (solid blue arrow, number 3) of the tectorial membrane (TM; Richardson et al. 2008) with respect to the epithelium holding the hair cells, the organ of Corti. The resulting deflection of the stereocilia of the hair cells (number 3) leads to cochlear transduction. Transduction actually occurs at the tips of the stereocilia, where transduction channels are opened and closed by the relative mechanical motion of the cilia as they are displaced (Gillespie and Müller 2009). This allows a transducer current (red lines), carried mainly by potassium ions, to flow into the hair cells, depolarizing them. The energy for hair-cell transduction is provided by the stria vascularis (SV, number 1), which transports potassium into the scala media (SM, shown yellow in Fig. 6.1), a process that also produces a positive potential as high as 100 mV, the endolymphatic potential (EP), in the SM (Zdebik et al. 2009). Hair cell depolarization by the transduction current drives the synapse from the IHCs to SGNs (number 4), initiating action potentials in AN fibers (Fuchs et al. 2003). The AN carries information about sound from the cochlea to the auditory circuits in the brain (number 5).

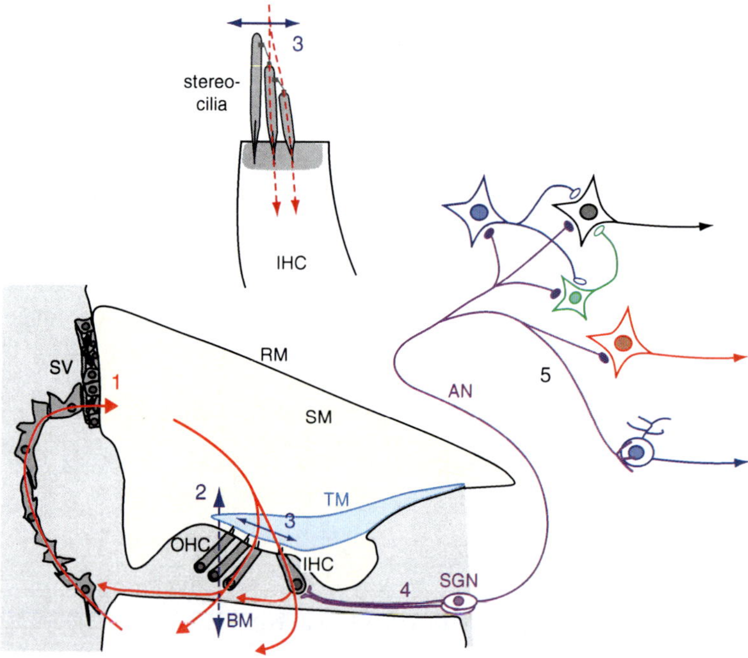

Fig. 6.1 A schematic of the major components of cochlear transduction shown in a cross-sectional view. The yellow region is the scala media (SM), which contains a high-potassium solution, endolymph, and a significant positive potential with respect to the other cochlear fluid spaces, the EP. The other spaces of the cochlea, scala vestibuli and scala tympani, are vertically above and below the SM in this drawing, but are not shown. Reissner's membrane (RM) and the BM separate the SM from the other two spaces, which contain perilymph with the usual composition of extracellular fluid. The red arrows show the potassium cycling in the cochlea (1); potassium is actively transported through the stria vascularis (SV) into the SM to produce the endolymph and the EP. The transduction current in IHC and OHC is carried by potassium, which is recycled to the SV via specialized transport systems formed by cells in the lateral wall of the cochlea (at left in the figure) and by other circuits in the supporting cells around the hair cells. Sound drives BM motion (2), resulting in shear displacement of the stereocilia of hair cells (3). The cilia of an IHC are shown in more detail in the inset. The OHCs amplify BM motion and the IHCs activate SGNs (4) through a chemical synapse. AN fibers innervate neurons in the cochlear nucleus in the brain (5)

2 The Endolymph and EP

Because potassium is the predominant ion both inside the hair cells and in the SM, the transducer current in hair cells (red lines in Fig. 6.1) is carried by potassium. Potassium is probably also the charge carrier in other hair cell systems, including the vestibular system (Wangemann 1995) and nonmammalian hearing or vestibular organs, because these structures also have a high-potassium endolymphatic spaces like the SM (e.g., chick, Runhaar et al. 1991). A potassium transducer current confers the advantage that the hair cells are freed from the metabolic load of transporting potassium to maintain their intracellular environment.

Figure 6.1 shows that the potassium currents associated with transduction are part of a loop of potassium transport (red arrows) in which potassium is secreted into the SM by the SV, flows through the hair cells as transducer current, and is then recycled to the SV by a transport system in the lateral wall of the cochlea (Konishi et al. 1978; Zidanic and Brownell 1990).

2.1 *Role of the EP in Cochlear Transduction*

A large EP (80–100 mV) is a unique feature of the mammalian cochlea. In fact, the driving force moving potassium current through the hair-cell transduction channels is provided primarily by the EP, since there is little concentration gradient for potassium between the cytoplasm of hair cells and the endolymph. The EP is important to normal cochlear function in vivo. If the EP is reduced by systemic administration of the drug furosemide, which blocks one of the active transport mechanisms that produce the EP, the spontaneous activity of AN fibers is reduced proportionally to the EP reduction (by ~0.02 log units of rate/mV) and the thresholds of the fibers for responses to sound increase (by ~0.9 dB/mV; Sewell 1984c). Furosemide also changes the rate-level functions and tuning curves of AN fibers in a way qualitatively similar to the effects of acoustic trauma (Sewell 1984a, b). These effects are consistent with a decrease in transduction current in the hair cells when the EP is reduced, thereby attenuating the effects of the OHCs on BM motion (discussed in Sect. 3) and the transducer action of the IHCs in activating the SGNs.

2.2 *Aging and the Degeneration of the Endolymph System*

Degeneration of the endolymphatic system is thought to play a role in presbycusis, the hearing loss observed in aging (Schuknecht and Gacek 1993). In humans and experimental animals, degeneration of the lateral wall of the SM is observed with aging of normal animals and animals with mutations that accelerate age-related hearing loss (Schulte and Schmiedt 1992; Hequembourg and Liberman 2001; Wu and Marcus 2003). The degeneration consists of loss of cellular elements, especially the fibrocytes which are important in potassium recycling and a decreased staining for enzymes such as Na^{+},K^{+}-ATPase that play a role in endolymph secretion. In the most studied animal model, the gerbil, the result of the degeneration is a decrease in the EP and eventually a decrease in the potassium concentration in the SM (Schmiedt 1996). Interestingly, the potassium concentration often stays near normal values after the EP has begun to decrease. In the absence of hair-cell or SGN lesions, the decrease in EP in aged gerbils is sufficient to account for the auditory threshold shifts produced by aging. This was demonstrated in two ways. First, gerbils were raised to old age in a quiet environment; such animals have little or no degeneration of hair cells or SGNs but have hearing loss. Second, young animals were treated with furosemide to reduce the EP to values that matched the EP of aged animals (Schmiedt et al. 2002; Lang et al. 2010); those young animals showed the same threshold shifts as quiet-reared aged animals.

Although EP degeneration seems to be involved in presbycusis, it is certainly not the only problem faced by the aging cochlea. The pattern of hearing loss in aging animals varies by species, often determined by the presence of genes predisposing to age-related hearing loss (reviewed in Bielefeld et al. 2008). In addition, a number of changes occur in the representation of sound in the brain that may be as important as peripheral degeneration to hearing impairment in the elderly (Walton 2010).

EP degeneration does not seem to play a role in hearing loss due to acoustic trauma. Acoustic trauma in young animals can produce substantial threshold shifts without a change in the EP (Hirose and Liberman 2003). The EP is reduced only for severe exposures that produce permanent hearing loss and substantial damage to the hair cells. Even in those cases, the EP usually recovers to normal values in a few days. This occurs despite substantial degeneration of the fibrocytes in the lateral wall of the SM. Thus the endolymphatic system seems to have significant reserve capacity that reduces its vulnerability to acoustic trauma.

2.3 *Generation of the Endolymph and Relevant Deafness Genes*

The endolymph is generated in the SV by a two-stage active transport process (Zdebik et al. 2009, which should be consulted for details). Anatomically, the SV consists of two layers of epithelial cells separated by an extracellular space, called the intrastrial space. One cell layer (the marginal cells) lies between the intrastrial space and the SM; the second (the intermediate and basal cells) lies between the intrastrial space and the connective tissue beneath the SV. The intermediate and basal cells are coupled together via gap junctions to form a continuous intracellular compartment that is also connected to the fibrocytes in the tissue of the lateral wall (Kikuchi et al. 2000). These are shown schematically in Fig. 6.1 by the gray arc of cells in the lateral wall tissue. In the first stage of transport, potassium is transported from the extracellular space into the intracellular compartment of the fibrocytes, intermediate and basal cells. The energy is provided by ATP via Na^+,K^+-ATPase and the sodium-potassium-chloride cotransporter (Spicer and Schulte 1991; Crouch et al. 1997; Nin et al. 2008). This system extracts potassium from the perilymph and the extracellular space, thus picking up potassium as it flows from the hair cells. The potassium flows to the intermediate and basal cells in the SV through the gap junctions that couple all these cells together. The concentrated potassium in this intracellular space diffuses through the apical membranes of the intermediate cells into the intrastrial space. The cells bounding the intrastrial compartment are linked by tight junctions, so that the intrastrial compartment is isolated from other fluid compartments for both diffusion of ions and electrical current. The apical membranes of intermediate cells are permeable only to potassium, so there is a potassium diffusion potential across this membrane, giving the intrastrial fluid a positive potential of several tens of millivolts (Salt et al. 1987; Takeuchi et al. 2000; Nin et al. 2008; Quraishi and Raphael 2008). The second transport stage is in the marginal cells

A similar two-membrane transport occurs here: active transport into the marginal cells from the intrastrial space followed by diffusion of potassium into the SM. The result is a high concentration of potassium in the SM and the positive EP, which is the sum of the potassium diffusion potentials across the apical membranes of the intermediate and marginal cells.

The model described in the preceding text for SV function is directly supported by direct physiological studies of ion concentrations and membrane potentials (Wangemann and Schacht 1996). The model also depends on evidence from pharmacological analysis, immunocytochemical localization of molecules (Spicer and Schulte 1991; Kikuchi et al. 2000), transport of dye among cells (Jagger and Forge 2006), and mutations that affect hearing (Mistrik and Ashmore 2009; Zdebik et al. 2009).

Mutations of several ion transport proteins expressed in the SV and the lateral wall of the cochlea affect endolymph function, resulting either in reduced potassium concentration or reduced EP or both. For example, the potassium channels in the apical membranes of the transporting cells are KCNJ10 in the intermediate cells and KCNQ1/KCNE1 in the marginal cells. Mutations of either should block the potassium transport function of the SV and reduce the EP; in fact, such mutations cause deafness in humans and loss of cochlear function in experimental knockout mice (Vetter et al. 1996; Splawski et al. 1997; Marcus et al. 2002; Scholl et al. 2009). Another group of mutations that has been instructive are those in connexin genes (Angeli et al. 2000; Cohen-Salmon et al. 2002), which code for the molecules that form the gap junctions between cells. In the cochlea, connexins form the continuous intracellular compartment through fibrocytes and intermediate/basal cells by which potassium flows to the SV. Mutations in connexin genes cause deafness and cochlear degeneration. Several other gene defects that affect hearing through the endolymph system are known (Zdebik et al. 2009; Gong and Lomax, Chap. 9).

3 Tuning and Frequency Resolution

The first step in cochlear transduction of sound is displacement of the BM (number 2 in Fig. 6.1; Robles and Ruggero 2001). The motion is frequency specific so that each location on the BM responds maximally to sound of a particular frequency, called the best frequency (BF), with smaller responses to sound at nearby frequencies. This frequency tuning is critical to normal hearing because it is the basis for the analysis of complex auditory stimuli (Kim et al. 1980; Moore 2004b). Frequency tuning and cochlear sensitivity turn out to be different aspects of the same mechanism, called the *cochlear amplifier*, which increases the sensitivity of cochlear transduction and sharpens the frequency tuning of BM responses. The idea of the cochlear amplifier has driven much of the work on the cochlea for the last 50 years (reviewed in Ashmore 1987). Several of the causes of hearing impairment, including acoustic trauma, ototoxic antibiotics, loss of the EP due to aging, and genetic defects in the cochlea have their effects primarily by interfering with the cochlear amplifier. This section discusses the evidence for the cochlear amplifier and its characteristics.

Note that the term *characteristic frequency* is often used along with or instead of BF. The two terms sometimes refer to slightly different aspects of tuning. Because this usage is not standard, BF is used here to mean the frequency at which responses are strongest at low sound levels (near threshold) and characteristic frequency will not be used.

3.1 Basilar Membrane Tuning

Schematics of typical frequency responses of two parts of the basilar membrane are shown in Fig. 6.2a, b. These plots show the gain of the BM response plotted versus stimulus frequency. Gain is defined as BM velocity divided by the sound pressure at the eardum. The behavior of the high-frequency part of the BM is well established (Fig. 6.2b; Cooper and Rhode 1997; Ruggero et al. 1997). The responses are tuned, in that maximum BM motion or gain is observed at the BF of the location (the vertical dashed line). However, the tuning varies with sound level. At low sound levels (20 dB in this example) the gain is large and the tuning is sharp, in that the BM velocity decreases rapidly as the stimulus frequency moves away from BF. At higher sound levels (e.g., 100 dB), the tuning is broader, the gain is smaller, and the frequency of maximum response moves to lower frequencies. The level-dependent amplification is confined to frequencies near BF. At low frequencies (below 0.7 in Fig. 6.2b), the gain is not level dependent and the BM filtering is linear. The level-dependent amplification is an expression of the function of the cochlear amplifier and represents an amplifier in the strict sense of the word, in that the power in the motion of the basilar membrane is larger than the power in the acoustic stimulus over the range of levels where level-dependent gain is expressed (Diependaal et al. 1987; Dallos et al. 2006; Ashmore 2008).

It is technically more difficult to make BM velocity measurements at the low-frequency end of the cochlea, so there is more uncertainty about those responses (Robles and Ruggero 2001). Figure 6.2a shows the behavior of one set of measurements (Cooper and Rhode 1997) that are similar to the high-frequency data in some ways. A similar level-dependent behavior of the gain is observed, although the sharpness of tuning does not vary as much and the gain decrease with level is smaller. Support for the behavior in Fig. 6.2a is provided by indirect psychophysical experiments in which masking (Plack et al. 2008) or measurements of otoacoustic emissions (OAE, a sound measured in the ear canal that is produced by OHCs; Gorga et al. 2008) are used to estimate BM response characteristics, giving a result consistent with Fig. 6.2a.

The level-dependent gain and tuning shown in Fig. 6.2a, b are seen only in the best physiological preparations. Indeed, acoustic trauma or other damage to the cochlea eliminates the level-dependent gain, giving a BM response like the dashed line in Fig. 6.2b (labeled *D*; Nuttall and Dolan 1996; Ruggero et al. 1996); this response is also observed after the death of the animal. There is no change in gain with sound level and the tuning is as broad as at the highest sound levels. There is

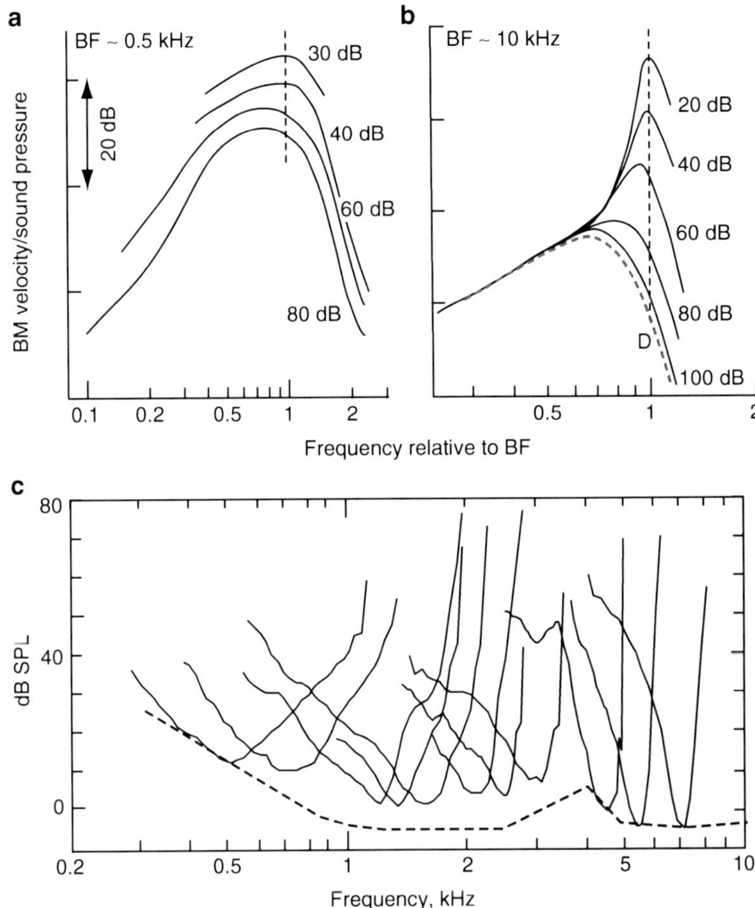

Fig. 6.2 Normal tuning of the BM and AN fibers. (**a, b**) Gain functions for the BM at sites with low (**a**) and high (**b**) BFs. The plots show response *gain* of a point on the basilar membrane, the measured velocity of BM displacement divided by sound pressure at the eardrum. The abscissa is stimulus frequency, plotted normalized by the BF. The gains were determined at several sound pressure levels, indicated by the labels on the curves (in dB SPL). The plot marked *D* in (**b**) shows the gain after death or complete loss of the OHCs. The plots are schematics drawn after data from the apex of the guinea pig BM (**a**, Cooper and Rhode 1995) and from data typical of the base in guinea pigs, chinchillas, and gerbils (**b**, Cooper and Rhode 1997; Ruggero et al. 1997; Overstreet et al. 2002). (**c**) Tuning curves from 11 AN fibers from one normal cat. The curves show the sound pressure necessary to produce a 20 spike/s increase in response (actually 1 spike in 50 ms) at various frequencies. The line at bottom is the lowest threshold observed across a number of animals (Redrawn from Miller et al. 1997 with permission)

also no evidence of a level-dependent cochlear amplifier, in that the gain of the response does not vary with stimulus level (i.e., the dashed curve *D* applies at all sound levels). The dashed curve in Fig. 6.2b apparently reflects the response of a fully passive BM, that is, one in which the cochlear amplifier does not function.

3.2 Roles of the IHCs and OHCs in Cochlear Transduction

It is clear from several decades of research that the IHCs and OHCs serve different roles in cochlear transduction. The OHCs have been shown to be the source of the level-dependent gain, the cochlear amplifier, and thus participate in the generation of the motion of the BM. The IHCs, by contrast, are transducers which sense BM motion and activate SGNs to convey information about sound to the brain. This separation of function means that cochlear lesions that damage OHCs have different characteristics from those that damage IHCs.

The innervation of the IHCs and OHCs provides one source of evidence for their different roles. The IHCs are innervated by the majority (~95%) type of SGNs (so-called type 1 neurons), which have myelinated axons (Spoendlin 1971; Kiang et al. 1982) and send their axons to innervate the principal cells of the cochlear nucleus (Rouiller et al. 1986). The OHCs are innervated by a different group of SGNs, type 2, which have unmyelinated axons (Spoendlin 1971; Kiang et al. 1982) and project to granule-cell, not principal cell, areas of the cochlear nucleus (Brown and Ledwith 1990). The type 1 SGNs respond strongly to sound, consistent with their role as the main afferent auditory pathway. The type 2 SGNs receive functional synaptic inputs from OHCs at low rates (Weisz et al. 2009), but have not been shown to respond to sound in vivo (e.g., Robertson 1984). The principal innervation of OHCs seems to be the efferent olivocochlear bundle, consisting of axons of neurons in the brain stem that project to the cochlea; the medial olivocochlear bundle innervates the OHCs and has a generally inhibitory effect on afferent activity in the type 1 SGNs (reviewed by Warr 1992; Guinan 1996). Because there is no direct connection between medial olivocochlear efferents and IHCs or SGNs, the effect of the efferents must be to reduce the BM motion, and therefore the input to the IHCs, via the efferent connection to OHCs (Brown et al. 1983).

A variety of evidence suggests that the OHCs form the cochlear amplifier. First, these cells are motile, meaning that they contract when their membrane is depolarized (Brownell et al. 1985; Ashmore 2008), so the OHCs can provide a source of energy for amplification. Moreover, their contraction is rapid enough to follow high-frequency auditory stimuli. Both characteristics are needed (Ashmore 1987). Second, furosemide causes a reduction or loss of BM level-dependent responses, producing the same effect as damage to the cochlea, that is, making the BM gain like the dashed line in Fig. 6.2b (Ruggero and Rich 1991). The effect is reversible and has been interpreted as due to a reduction of OHC responses by the reduction of the EP caused by the drug. Third, the level-dependent gain of the BM is reduced or eliminated by stimulation of the medial olivocochlear efferents (Murugasu and Russell 1996), which affect the cochlea only through their innervation of OHCs, as discussed above. Fourth, ototoxic antibiotics often produce a hair-cell lesion in which regions of the cochlea have no OHCs, but apparently intact IHCs. AN fibers innervating these IHCs show substantial elevations of threshold and broadening of tuning, consistent with a loss of level-dependent tuning (Evans and Harrison 1975; Dallos and Harris 1978; Liberman and Dodds 1984b). Finally, chimeric mice with varying levels of expression of the molecule *prestin* were analyzed for the sensitivity and tuning sharpness of AN responses (Cheatham et al. 2009). Prestin is the molecule that produces somatic motility in OHCs (Zheng et al. 2000; Liberman et al. 2002) and seems to be a necessary component of the cochlear amplifier. In these chimeric mice, the degree of threshold sensitivity and sharpness of tuning was linearly related to the expression of prestin in the OHCs.

3.3 Tuning of AN Fibers in Normal Cochleas

Tuning curves of normal fibers in the cat AN are shown in Fig. 6.2c. These curves plot threshold versus stimulus frequency for tones, where threshold is the sound level necessary to produce a criterion rate increase. Similar tuning curves are obtained in other species (e.g., guinea pig: Evans 1972; mouse: Taberner and Liberman 2005; gerbil: Ohlemiller and Echteler 1990; chinchilla: Temchin et al. 2008). They show a V-shape with the lowest threshold at the BF, which corresponds to the BF of the point on the BM innervated by the fiber (Liberman 1982). There is a change in shape of the curves across the BF axis, meaning that the curves are more symmetric and broader (on a log frequency scale) at low BFs. High BF tuning curves (BF>3 kHz) have a sharply tuned and low-threshold "tip" centered on BF and a higher threshold "tail" at frequencies below BF.

AN tuning curves should be related to BM tuning curves. The BM and neural data shown in Fig. 6.2 are not directly comparable because different measures are plotted on the ordinates of the figures. However, BM tuning curves can be derived from data like those in Fig. 6.2a, b. In such comparisons, the tip characteristics compare well with AN responses (reviewed by Temchin et al. 2008), although the thresholds in the tails are often systematically different. In another approach, neural tuning can be measured in a way directly comparable to the basilar membrane data using the method of Wiener kernels (Recio-Spinoso et al. 2005). This method yields neural gain functions with properties essentially the same as the BM gain; in particular, the properties of low and high BF tuning are similar and the level-dependent gain is similar.

3.4 Tuning After Acoustic Trauma

Exposure to intense noise or other sound produces a range of lesions in the cochlea, depending on the severity of the exposure. The effects of such an acoustic trauma are usually defined in terms of the threshold shift immediately after the exposure (temporary threshold shift, TTS) and the steady-state threshold shift after a period of time (permanent threshold shift, PTS). Typically TTS declines over a period of a days or weeks to PTS. Significant effects on hearing can occur for exposures that produce little or no PTS (e.g., Noreña and Eggermont 2005). However, most of the examples given here show the effects of moderate PTS (~40–60 dB). This degree of hearing loss is usually sufficient to motivate the use of a hearing aid, but it is not so severe that it is difficult to analyze the response properties of AN fibers.

The tuning curves of AN fibers from a cat exposed to a narrow band of noise centered on 2 kHz are shown in Fig. 6.3a. These are typical of tuning curves seen in the AN following a variety of insults, including ototoxic exposure, acoustic trauma, furosemide poisoning, or anoxia (Evans 1975; Evans and Harrison 1975; Kiang et al. 1976; Dallos and Harris 1978; Robertson 1982; Liberman and Dodds 1984b; Sewell 1984a). For acoustic trauma, the degree of threshold shift and the BF region affected are variable from exposure to exposure, but generally the effects are largest at or slightly above the frequency of the exposure (shown by the black bar in Fig. 6.3a). Comparing these tuning curves with Fig. 6.2c, several changes can be noted. First, the thresholds are elevated over a wide BF range centered on the exposure frequency. Second, the shapes of the tuning curves are different; the tip portion is smaller or missing and the tuning curves appear to be composed mainly of a broad low-frequency tail. The sharp high-frequency edge of the tuning curve remains, but for more severe damage, the slope of this part of the tuning curve also decreases (Liberman 1984). From the BM data, one expects that the BF of the fiber should decrease after acoustic trauma. Although this decrease cannot be seen from data like Fig. 6.3a, it has been shown to occur by filling AN fibers with a dye and reconstructing them to determine the point of innervation of the BM; the fibers' expected BFs in a normal animal can then be determined from the known BM frequency map (Liberman 1982). In this case, neurons with threshold shifts larger than approximately 40–50 dB showed lower BFs than expected from the cochlear map (Liberman 1984), suggesting a downward shift in BF with trauma.

The changes in tuning have also been studied in animals using systemic injection of furosemide to produce a temporary decrease in hair cell function. Tuning curves before (*n*) and during (*f*) furosemide effects are shown in Fig. 6.3b; these are schematics showing typical results obtained by Sewell (1984a). Tuning curves are shown for a low BF neuron (left) and a high BF neuron (right). The high BF neuron shows effects similar to the tuning curves in Fig. 6.3a, a threshold shift and a loss of the tip of the tuning curve; there is also a decrease in the BF during the furosemide. The low BF neuron shows a shift in threshold, but the tuning curve does not resolve into a tip and tail portion. In this case, the BF shifts toward higher frequencies during the furosemide. These differences between high and low BF fibers resemble the differences in BM tuning (Fig. 6.2) and suggest that there are differences in BM physiology between the apex and base of the cochlea (Robles and Ruggero 2001). The dividing line between low BF and high BF behavior varies with species, but is approximately 1–3 kHz (Shera et al. 2010).

3.5 *IHCs Versus OHCs Again*

The effects of lesions that damage IHCs versus OHCs are systematically different and are important for understanding the effects of acoustic trauma. The critical experimental step in characterizing these differences was the analysis of the state of

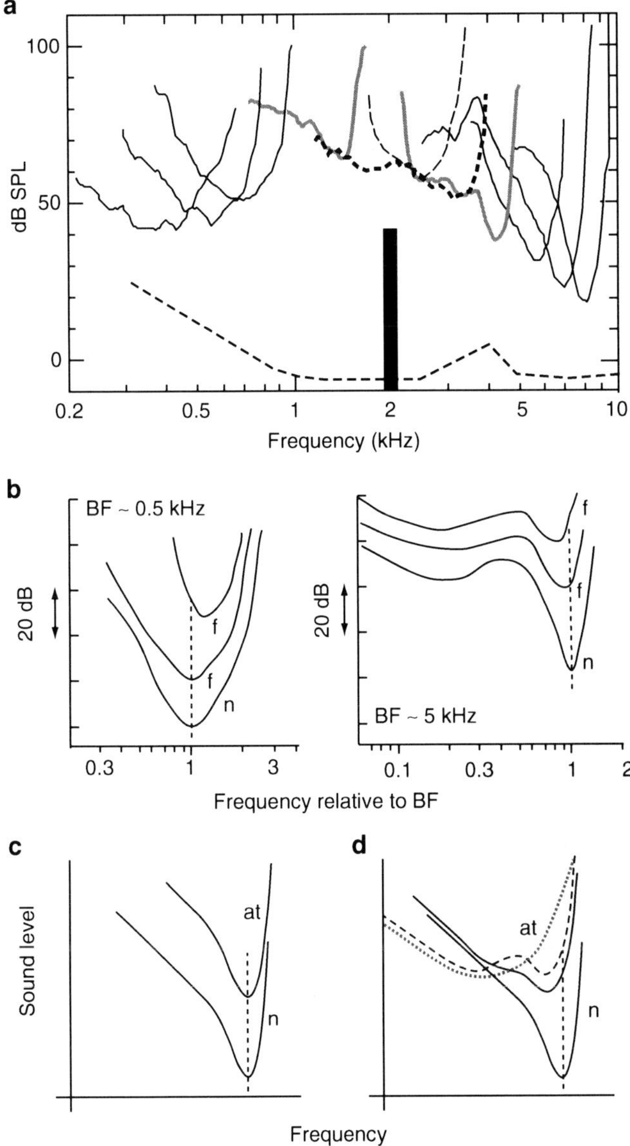

Fig. 6.3 Tuning of AN fibers after damage to hair cells. (**a**) Threshold tuning curves for 10 AN fibers from a cat with PTS from an exposure to a 2 kHz band of noise (bandwidth 50 Hz) at 110 dB SPL for 2 h. The *black bar* shows the frequency range of the noise. Four of the tuning curves with large threshold shifts are plotted with different line styles to distinguish them. The *dashed line* at bottom shows the lowest thresholds of fibers across a number of normal cats. (Reprinted from Miller et al. 1997 with permission.) (**b**) Schematics showing typical tuning curves before (*n*) and during (*f*) the effects of systemic furosemide. Shown are examples for a low BF (*left*) and a high BF (*right*) fiber. The two furosemide tuning curves in each case are from different injections, and the effects were reversible. The *vertical dashed lines* show BF in the normal tuning curve. (Redrawn after Sewell 1984a with permission.) (**c, d**) Schematics summarizing the effects on tuning curves of damage to IHC stereocilia (**c**) and to OHCs (**d**), either a loss of OHCs or damage to OHC stereocilia. Normal tuning curves (*n*) and tuning curves following acoustical trauma (*at*) are shown. In the *at* cases, the lesion was a PTS. The three *at* tuning curves in D represent varying degrees of OHC lesion, see text (Based on Fig. 14 of Liberman and Dodds 1984b)

the stereocilia on hair cells (Liberman and Beil 1979; Liberman and Dodds 1984b). Hair cells can survive acoustic trauma with their stereociliary bundles damaged. Because transduction depends on the stereociliary bundles, such hair cells do not function normally, which can lead to functional impairment in regions of the cochlea with most or all of the hair cells still present. Thus, simply documenting the presence or absence of hair cells, as is often done, is not an adequate anatomical characterization of the lesion.

Damage to IHC stereocilia with normal OHCs leads to a threshold elevation that is roughly constant across frequency, so that the tuning-curve shape does not change (Fig. 6.3c). This conclusion from the analysis of stereocilia is consistent with the effects of carboplatin (Wang et al. 1997), an ototoxic cancer drug. In chinchillas, it can produce a preparation with damage to IHCs and SGNs but not to OHCs (Trautwein et al. 1996); in such cases, the remaining active AN fibers are sharply tuned but often have elevated thresholds.

Damage to OHC stereocilia or loss of OHCs (in the absence of IHC damage) leads to both a threshold shift and a broadening of tuning curves (Fig. 6.3d). The shape of the resulting tuning curve depends on the degree of OHC damage. Moderate damage to OHC stereocilia usually results in a reduced tip (Fig. 6.3d, solid *at* curve), with less change in the tail region. More severe damage of OHC cilia or loss of the OHCs can cause hypersensitivity of the tail (Fig. 6.3d, dashed *at* curve), and a wipeout of OHCs gives a bowl-shaped tuning curve with no sign of a tip (Fig. 6.3d, dotted *at* curve).

Generally in acoustic trauma both IHCs and OHCs are damaged (Liberman and Dodds 1984b; but see Hamernik et al. 1984 for examples of the variability of the effects of acoustic trauma). Thus, the tuning curves show a sum of the effects of IHC and OHC damage. Such tuning curves show a threshold shift at all frequencies with a loss of tuning sharpness and a reduced tip of the tuning curve (Liberman and Dodds 1984b; Miller et al. 1997); there is usually little sign of tail hypersensitivity because the threshold shift due to IHC damage moves the tail threshold upward. Notice that this behavior is consistent with the lack of tail hypersensitivity with furosemide poisoning (Fig. 6.3b), where the responses of both IHCs and OHCs are affected by the reduction in the EP.

Tail hypersensitivity seems to be related to the so-called component 2 (C2) response of AN fibers (Liberman and Kiang 1984). C2 is a poorly tuned response mode of fibers that appears only at high sound levels. It is not subject to acoustic trauma and thus has the characteristics needed to explain bowl shaped tuning curves with tail hypersensitivity. C2 is discussed in more detail in Sect. 6.2.

The behavior of threshold shift and tuning width of AN fibers following acoustic trauma is shown in Fig. 6.4. Schematic tuning curves before and after trauma are shown in Fig. 6.4a, where the 10 dB bandwidth and the standard Q_{10} measure of tuning are defined. The thresholds at BF of a population of AN fibers following acoustic trauma are shown in Fig. 6.4b. The threshold shift is the distance from the best threshold line (dashed) to the fiber thresholds (Heinz and Young 2004). The widths of tuning (Q_{10}) of the same population of fibers are shown in Fig. 6.4d. The range of Q_{10}s in normal animals is shown by the two solid lines. Following acoustic trauma, the Q_{10}s are below the range of normal (meaning broader tuning than normal) over an approximately

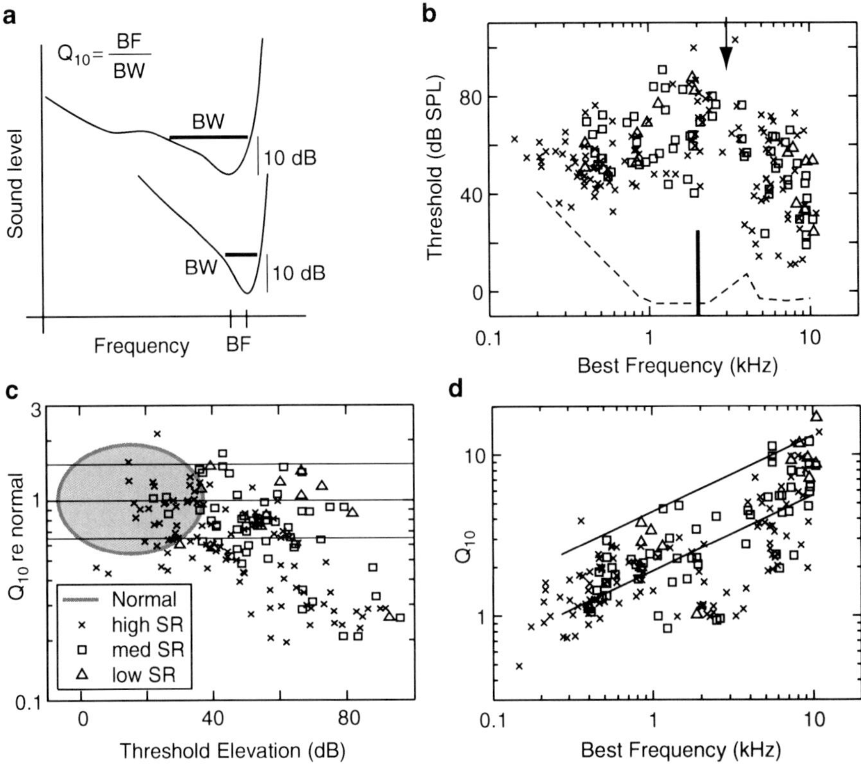

Fig. 6.4 Tuning data from AN fibers in cats with PTS from acoustic trauma. (**a**) Schematic tuning curves showing the bandwidth (*BW*) 10 dB above threshold and the formula for Q_{10}; note that Q_{10} gets smaller as the *BW* increases. (**b**) Thresholds at BF of tuning curves for AN fibers from the exposed cats. The *vertical gray bar* shows the exposure stimulus (a 50 Hz wide band of noise centered at 2 kHz at 103 or 108 dB SPL for 4 h). The *dashed line* shows the lowest thresholds in a group of normal cats. The arrow at top points to a BF where few fibers were encountered, explained in the text. (**c**) Q_{10} values normalized by the mean Q_{10} at the same BF in normal animals (ordinate) plotted against threshold shift (abscissa), meaning threshold relative to the *dashed line* in (**b**). The *gray oval* shows the location of data from normal animals. The *horizontal lines* show the mean Q_{10} and 95% confidence limits for the normal data. The points are calculated from the data in (**b**) and (**d**). (**d**) Tuning bandwidth (Q_{10}) plotted versus BF for the same data as in (**b**). The lines show the range of data in normal animals. In all plots, the symbol style identifies the spontaneous rate (SR) category of the fiber, identified in the inset of (**c**) (Reprinted from Heinz and Young 2004 with permission)

2-octave range centered on the exposure frequency. Generally the tuning is broader in cases with larger threshold shifts (Fig. 6.4c). However, the neurons with elevated thresholds (>40 dB) in Fig. 6.4c fall into two groups: those with near normal Q_{10}s (normalized Q_{10}s near 1) and those with subnormal Q_{10}s. The former group are mostly neurons with little or no spontaneous activity (open symbols). An interpretation of these data is that the neurons with near-normal Q_{10}s innervate IHCs with enough damage to have elevated thresholds and no spontaneous activity (Liberman and

Dodds 1984a; Wang et al. 1997), which lie in a region with functioning OHCs to produce sharp tuning. By contrast, the neurons with subnormal Q_{10}s innervate regions with damaged OHCs and either normal or damaged IHCs.

The spontaneous rates (SR) of AN fibers vary from near 0 to greater than 100/s. SR correlates with the size and location of the synapse the fiber makes on its IHC (Liberman 1980) and with several suprathreshold properties, discussed in Sects. 4, 5 and 6. Fibers are classified as low ($SR < 1$), medium ($1 \leq SR < 18$), and high ($SR \geq 18$). For the present discussion, it is most important that SRs decrease with IHC damage, presumably because of the decrease in transducer current in damaged IHCs (Liberman and Dodds 1984a).

The understanding of the role of IHCs and OHCs summarized in the preceding text has been formalized in computational models that reproduce the response properties of AN fibers with varying degrees of accuracy (e.g., Deng and Geisler 1987; Zhang et al. 2001; Sumner et al. 2003; reviewed by Lopez-Poveda 2005; Heinz 2010). One model that specifically incorporates IHC and OHC components (Bruce et al. 2003; Zilany and Bruce 2006) can be applied to hearing impairment by independently "lesioning" the IHC and OHC components of the model consistent with the summary of data in Fig. 6.3b, d. The OHCs are lesioned to match the broadening of tuning curves and the IHCs are lesioned to match the thresholds, taking into account the threshold shift from the OHC damage. This model does well at predicting the phenomena shown in this and subsequent sections (Zilany and Bruce 2007).

3.6 Perceptual Frequency Resolution

The human auditory system often operates as if the first processing stage is a parallel bank of filters (called "auditory filters") with gain functions like the tuning curves shown in Fig. 6.2c (reviewed in Moore 2004b); such filters also show a level-dependent gain similar to the BM (Glasberg and Moore 2000). The listener has available only the signals at the outputs of the filters, so the listener's ability to separate or analyze sounds by frequency is limited by the bandwidths of the filters. The signals at the output of the filters form an *excitation pattern* (Moore and Glasberg 1983), which is a smoothed or smeared version of the stimulus frequency spectrum. Models of this excitation pattern have been used to explore several perceptual phenomena, like the loudness of sounds (Moore and Glasberg 1997) and the representation of spectral shape (e.g., Leek and Summers 1996).

In hearing impairment, the bandwidths of the auditory filters become larger, as shown for AN fibers in Fig. 6.4d (reviewed in Moore 2007) and the level-dependence is reduced. The filter broadening is larger as the threshold shift (and presumably the degree of damage to hair cells) increases, similar to the AN data in Fig. 6.4c. Consistent with wider auditory filters, hearing-impaired listeners show a reduced ability to analyze sounds for their frequency content (Summers and Leek 1994) and an increased susceptibility to interfering sounds in situations like speech perception (ter Keurs et al. 1992; Turner 2006).

One approach to analyzing speech perception in normal and impaired ears is the articulation index (e.g., Ching et al. 1998). This model attempts to predict speech perception performance on the basis of the audibility of sounds. Because of elevated thresholds in a hearing-impaired listener, information about some frequency regions is not available, thus reducing the ability of the auditory system to identify and discriminate sounds with energy at those frequencies. Although this model often works well for mild hearing impairment, it generally overestimates performance with moderate to severe hearing loss (e.g. Pavlovic 1984). Presumably, the prediction error occurs because the articulation index model does not take into account the effects of broadened auditory filters, as well as damage to other components of the auditory system such as the SGNs and neural circuits in the brain, discussed in Sects. 5 and 7.

4 Responses to Speech-like Stimuli

The changes in the neural representation of stimuli like speech when the cochlea is damaged are largely accounted for by the direct effects of the damage, threshold shift, and degraded tuning. However, there are also important nonlinear effects from the interaction of the multiple frequency components of the stimulus. This section discusses the neural representation of speech in normal and damaged ears.

4.1 Analysis of AN Responses with Phase-Locking: Synchrony Capture

Hearing impairment reduces the quality of the neural representation of auditory stimuli, studied most carefully for speech stimuli following acoustic trauma (Palmer and Moorjani 1993; Miller et al. 1997). The analysis of neural responses to a speech stimulus presents difficulties because the stimulus contains a number of different frequency components and the neuron may respond to one or more of them (reviewed by Young 2008). An informative approach to this problem has been to analyze the *phase-locking* of the neuron to the stimulus waveform; this method allows the responses to be teased apart into responses to individual stimulus components. Figure 6.5a shows the time waveform of a vowel from a natural speech stimulus. This waveform contains frequencies across a wide range, shown in the magnitude of the Fourier transform of the signal (Fig. 6.5c). There are prominent peaks of energy in the signal called *formants*; the first three (lowest-frequency) formants are identified in Fig. 6.5c as F_1, F_2, and F_3 (Stevens 1998). The formants correspond to the resonant frequencies of the vocal tract and are the dominant characteristic of speech sounds, for both perception and production (Fant 1970). Figure 6.5b shows the instantaneous firing rate of an AN fiber in response to the waveform in Fig. 6.5a. The firing rate fluctuates in a fixed relationship to the waveform of the stimulus. In this case, the fiber's BF is

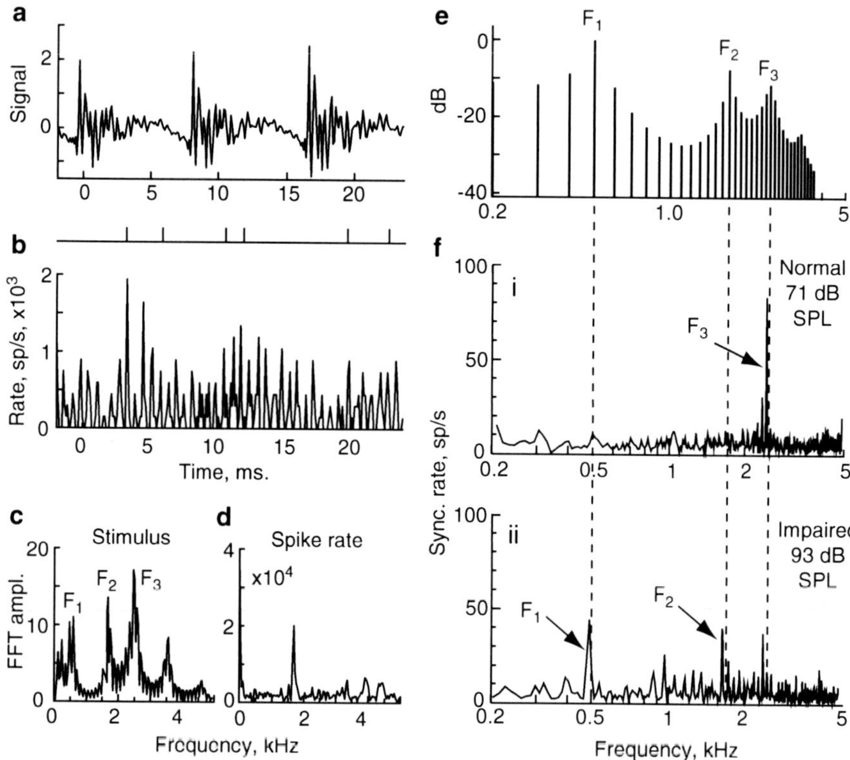

Fig. 6.5 Examples of the responses of AN fibers to vowels. (**a**) Short portion of the waveform of the first vowel in the word *basketball*. (**b**) A phase-locked response of an AN fiber to the vowel in A. The plot shows the instantaneous discharge rate, or the instantaneous probability of response, of the fiber, on the same time axis as the vowel. The inset between (**a**) and (**b**) shows a typical spike train of a fiber, where the vertical ticks are the times of occurrence of the spikes. The rate plot is computed from approximately 50 repetitions of the stimulus, that is, from approximately 50 such spike trains, by binning the spikes into 0.2 ms bins to form the instantaneous rate histogram. (Phase-locking is discussed again in Fig. 6.7.) (**c**) Magnitude of the Fourier transform of the stimulus in (**a**). The peaks occur at the formants of the vowel, the first three of which are marked F_1, F_2, and F_3. (**d**) Fourier transform of the rate plot from (**b**). Note that the fiber responds only to the F_2 component of the vowel, which is the largest energy peak in the vowel near the BF of the fiber (1.77 kHz). (**a–d** reprinted from Young 2008 with permission.) (**e**) Spectrum of a different vowel, /eh/, responses to which are shown in (**f**) and in Fig. 6.6. The vowel is synthesized and periodic, so it has energy at the harmonics of 100 Hz, the *vertical lines* shown in this figure. (**f**) Magnitude of the Fourier transform of responses of two neurons to the/eh/, a normal fiber (i) and a fiber from an animal with an approximately 55 dB threshold shift due to acoustic trauma (ii). Both fibers have BFs near the F_3 frequency (2.5 kHz). The *dashed lines* show the formant frequencies (**e**, **f** reprinted from Miller et al. 1997 with permission)

1.77 kHz and its temporal response pattern follows the F_2 component of the stimulus, which is approximately 1.7 kHz. This "locked" response is evident in Fig. 6.5d, which shows the magnitude of the Fourier transform of the response rate in Fig. 6.5b. There is a peak of response at the F_2 frequency and little response at other frequencies.

The behavior shown in Fig. 6.5d is typical of responses to speech sounds in normal AN fibers (Young and Sachs 1979; Miller and Sachs 1983; Sinex and Geisler 1983; Delgutte and Kiang 1984; Palmer et al. 1986) in that the fibers generally respond exclusively or nearly so to the formant frequency nearest the fiber's BF, referred to as *synchrony capture*. Another example is shown in Fig. 6.5f-i, which shows the magnitude of the Fourier transform of a normal-hearing AN fiber's responses to the vowel-like stimulus whose frequency content is shown in Fig. 6.5e. Again, the response is primarily to one formant, in this case to F_3, which is near the fiber's BF (2.5 kHz).

Synchrony capture depends on the tuning of the fiber, which selects the stimulus components that are near the fiber's BF. However, it also depends on cochlear *suppression*, an important nonlinear property by which energy at one frequency suppresses responses to other frequencies, both in neural responses (Abbas and Sachs 1976; Delgutte 1990; Cai and Geisler 1996b) and BM responses (Ruggero et al. 1992; Cooper 1996). One evidence for a role of suppression in synchrony capture is that the neurons whose responses are shown in Fig. 6.5d, f-i would respond strongly to a tone at the frequency of F_1 presented by itself at the same sound level as the F_1 component of the vowel (Wong et al. 1998). Apparently the response to the F_1 component of the vowel is suppressed by other components of the vowel, mainly the components at the other formants that drive the neuron more strongly (Sachs and Young 1979; Le Prell et al. 1996). This suppression is the essence of synchrony capture. Suppression has been demonstrated convincingly in the simpler situation of only two tones, wherein the suppression of one tone by the other is easy to see (Kim et al. 1980). Another kind of evidence for a role of suppression in synchrony capture is that a linear model of AN fibers that does not include suppression does not display synchrony capture (Sinex and Geisler 1984), whereas a nonlinear model with suppression does (Deng and Geisler 1987).

4.2 *The Population Representation*

The representation of speech is usually considered to be a collective one, in which the vowel spectrum is encoded by the distribution of activity across the population of AN fibers. In such a *tonotopic representation* each frequency component of the sound is represented by the activity of fibers with BFs near the frequency of the component. In normal animals, the population of AN fibers provides a good tonotopic representation of the vowel, in that the responses to the formants occur in separate BF groups of fibers. Figure 6.6a shows the distribution of responses phase locked to the first three formants of the /eh/ of Fig. 6.5e in a population of AN fibers in normal cats. The three plots show responses to F_1, F_2, and F_3 from top to bottom. The response to each formant is centered near its tonotopic place in the population, i.e., where the formant frequency equals the BF of the fibers. At the F_1 and F_2 places in the population (Fig. 6.6a-i, a-ii), synchrony capture is observed, in that the fraction of the response devoted to the formant is near 1. Responses to F_3 are not as strong as those to the other formants (Fig. 6.6a-iii), but are still the largest response at the F_3

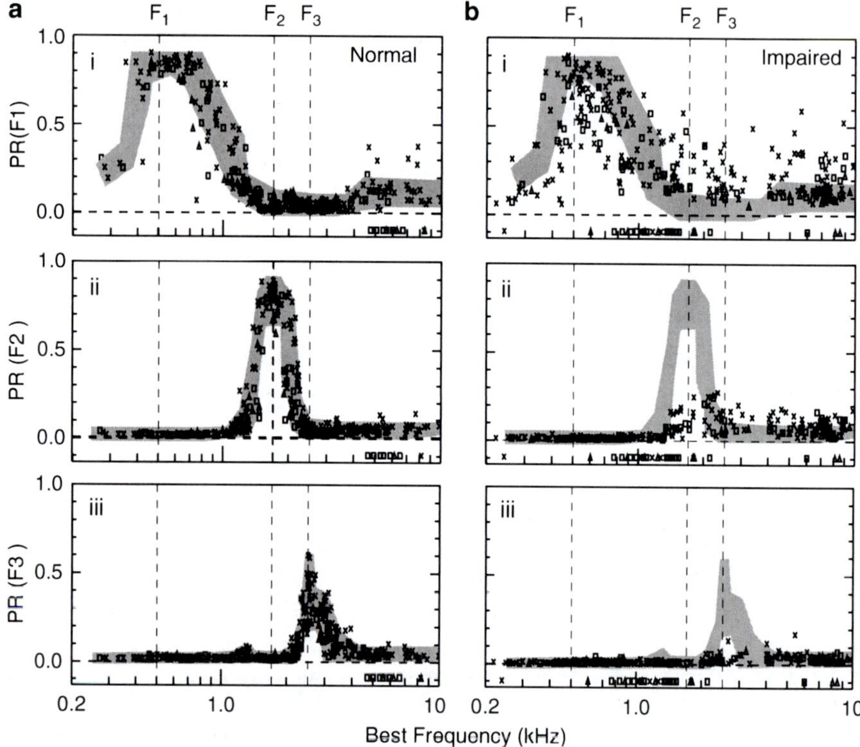

Fig. 6.6 Responses to a vowel of a population of fibers in normal cats (**a**) and in cats exposed to acoustic trauma (**b**). The formant frequencies of the vowel (/eh/as in met, Fig. 6. 5e) are indicated by the *vertical dashed lines*. Each plot shows the fraction of the phase-locked response power of the fibers devoted to one formant of the vowel, that is, the *power ratio PR* for F_1 (i), F_2 (ii), and F_3 (iii). Each data point is the response of one AN fiber and the gray areas show the loci of the data in normal animals. The symbols identify the SRs of the fibers (triangle, low; box medium; and X high). The symbols below the abscissa are fibers for which the stimulus was below threshold. In the normal animals, the stimulus level was 69 dB SPL, in the exposed animals it was 92 dB SPL (Reprinted from Miller et al. 1997 with permission)

place (Fig. 6.5f-i). Responses to other frequency components of the vowels are not shown, but they are small (Schilling et al. 1998), presumably suppressed by the responses to the formants.

4.3 Effects of Acoustic Trauma

After acoustic trauma, synchrony capture is weakened or disappears and the phase locking becomes more broadband. The AN fiber in Fig. 6.5f-ii is typical of fibers in impaired ears in that responses to many frequency components of the stimulus are present. Responses to both F_1 and F_2 are evident in this fiber, which should respond mainly to F_3, as well as responses at nonformant frequencies. A similar effect is seen for fibers with BFs near the F_2 frequency, which respond to F_1. The population representation shown in Fig. 6.6b was constructed for fibers from a group of animals with thresholds and tuning similar to that shown in Fig. 6.4. In particular, there was a threshold shift of approximately 55 dB in the BF range of F_2 and F_3 with broadened tuning. Responses in this population (Fig. 6.6b) showed a loss of the tonotopic order characteristic of normal animals. Responses to F_2 and F_3 were dramatically reduced, except that small responses to F_2 occurred at BFs at and above the F_2 frequency (Fig. 6b-ii). Responses to F_1 spread to higher BFs (Fig. 6b-i); note particularly that there were significant responses to F_1 at the BFs corresponding to F_2 and F_3, where the F_1 response was suppressed to near zero in normal animals (compare to Fig. 6a-i).

The changes in the representation after acoustic trauma are consistent with the changes in tuning curve shape. In particular, the loss of tuning-curve tips in the fibers tuned near F_2 (or F_3) gives them a "low-pass" shape (Fig. 6.3a), which reduces the ability of those fibers to discriminate the formants (Miller et al. 1999b). This change in the tuning curves is consistent with the increase in the responses to F_1 among fibers with higher BFs and also with the increase in the diversity of phase-locking to components of the vowel, including nonformant components (Fig. 6.5f-ii; Schilling et al. 1998).

There is probably also an effect of reduced suppression in the responses after acoustic trauma. So-called two-tone suppression, wherein the discharge rate in response to an exciter tone is reduced by the addition of a suppressor tone (Abbas and Sachs 1976; Delgutte 1990; Cai and Geisler 1996a), is weaker after acoustic trauma, although it is not abolished (Schmiedt and Zwislocki 1980; Salvi et al. 1983; Miller et al. 1997). Reduced suppression most likely contributes to the lack of synchrony capture after acoustic trauma.

The fibers that show small responses to the formants, those with BFs near F_2 and F_3 in Fig. 6.6b-ii, b-iii, appear to be phase locking weakly, suggesting that a loss of the ability to phase lock might be the cause of the poor representation after acoustic trauma. In fact, these fibers are phase locked simultaneously to many components of the stimulus (Fig. 6.5f-ii and Schilling et al. 1998), giving a significant overall phase locking. It seems unlikely that a disorder of phase locking per se is the cause of the poor representation at the higher frequencies. This point is discussed in more detail in Sect. 5.3.

4.4 *Representation of F_2 and F_3 and the Effects of High-Pass Amplification*

The responses in Fig. 6.6b come from AN fibers in cochleas with a high-frequency hearing loss, meaning that the thresholds, and presumably the hair cells, are close to normal at BFs below 0.8–1 kHz, but elevated at higher BFs (as in Fig. 6.4b). The threshold shifts are largest at BFs near the frequencies of F_2 and F_3. Usually a person with such a hearing loss is given a hearing aid with larger gain at higher frequencies, to overcome the loss of audibility at high frequencies and to attempt to smooth out the variations in loudness between low and high frequencies (Moore 2007).

Amplifying the vowel stimulus in a way that is typical of such a hearing aid (without the amplitude compression circuits that are needed in a real aid) improves the population representation shown in Fig. 6.6b by overcoming the spread of F_1 to higher BFs and increasing the responses to F_2 and F_3 in neurons with appropriate BFs (Schilling et al. 1998; Miller et al. 1999a). However, it does not change the broadband phase locking, nor is synchrony capture restored. The responses to F_2 at the F_2 place are partially restored, although the spread of F_2 to higher BFs shown in Fig. 6.6b-ii is not changed (except by getting stronger).

Comparison of Fig. 6.6b and a suggests that the information about F_2 and F_3 present in the normal ear is reduced by acoustic trauma. That is, in the traumatized ear there is no place in the cochlea where F_2 or F_3 dominates the responses of the neurons; thus there are no neurons that provide information to the brain specifically about F_2 or F_3. In fact, the neurons at the F_2 and F_3 places respond as strongly or more strongly to F_1, so their responses are misleading. This impression was quantified by obtaining responses to a series of /eh/ vowels that were identical except for the frequency of F_2 (over the range 1.4–2 kHz). In normal animals, there was a strong rate representation of the F_2 frequency (Miller et al. 1999b) that was sufficient to support behavioral performance in F_2 frequency discrimination (May et al. 1996). After acoustic trauma, the representation of F_2 was lost, as predicted from Fig. 6.6b-ii, in that there was no change in the response rate of neurons in any part of the cochlea to signal the change in F_2 (Miller et al. 1999b). When the high frequencies were amplified as above, the rate changes induced by formant changes were restored, for fibers with BFs near F_2, and discriminability returned to near-normal values (Miller et al. 1999a). However the pattern of response was still abnormal in that responses to F_2 spread widely to frequencies above the F_2 place.

5 Damage to SGNs as a Primary Lesion in Cochlear Damage

The discussion above is based on damage to hair cells as the primary pathophysiological mechanism in acoustic trauma. However, degeneration of SGNs is often observed in cases of presbycusis, even in the absence of loss of hair cells or damage to the stria vascularis (Schuknecht and Gacek 1993; note that stereocilia were not analyzed here). In acoustic trauma, hair cells are presumably damaged by the direct mechanical effects of intense sound on the transduction apparatus of the cells, and also by oxidative stress (Le Prell and Bao, Chap. 13). SGNs are most likely damaged through excitotoxicity, as a consequence of excessive glutamate release from the hair cells during the strong stimulation (Zheng et al. 1997; Ruel et al. 2007). After trauma, swelling of dendritic terminals of SGNs occurs over a period of hours (Liberman and Mulroy 1982; Robertson 1983). The swelling resolves over a few days and may leave intact SGNs and synapses. However, recent experiments, in which counts were made of hair cells, SGNs, and fibrocytes, show that degeneration of SGNs may occur after an acute trauma, over a long time period (~2 years in the mice used in these experiments), even in cases where the hair cells are still present and functional (Kujawa and Liberman 2006).

A similar effect was observed when the synapses between IHCs and SGNs were evaluated (Kujawa and Liberman 2009). An acoustic trauma that led to only a TTS was used. After resolution of the TTS, no degeneration of hair cells was seen and electrical potentials associated with the OHCs were normal [distortion product otoacoustic emissions (DPOAE); for an introduction to otoacoustic emissions and auditory brain stem responses, see Kraus and McGee 1992]. However, significant permanent loss of SGNs and ribbon synapses between IHCs and SGNs occurred, the latter within a week, the former progressively over years (over most of the lifetimes of the mice used in these experiments). Moreover, the *amplitude* of the auditory brain stem response (ABR, the averaged evoked activity of neurons in the AN and the auditory brain stem) was reduced, even though its *threshold* was not elevated. The interpretation of the latter result is that the threshold of the ABR depends primarily on the hair cells, as long as there are some intact SGNs remaining, whereas its amplitude depends primarily on the number of functional SGNs (Kujawa and Liberman 2009).

It is clear from the preceding discussion that loss of SGNs can be a primary lesion after acoustic trauma; thus hearing impairment, especially hearing impairment at suprathreshold levels (Shrivastav, Chap. 7), may often be a result of SGN degeneration rather than or in addition to hair-cell damage.

5.1 Dead Regions

In population studies of AN fibers following acoustic trauma, there is often a gap in the fiber population at a BF approximately a half octave above the exposure frequency (the arrow at about 3 kHz in Fig. 6.4b for example). Given that efforts are made to obtain a complete sample of BFs in these experiments, the gap seems to reflect a loss of AN fibers with BFs over a narrow region where the effect of the exposure stimulus was strongest. For additional examples of BF gaps, see Fig. 6.2a in Miller et al. (1999b) and Fig. 6.4a in Miller et al. (1997).

In psychoacoustic studies of hearing impaired subjects, it is possible to show the existence of frequency regions without functioning inputs to the auditory system (Moore 2004a). These are called "dead regions" and are presumed to correspond to a BF region of missing IHCs or SGNs. Such regions can be demonstrated in up to 50–60% of individuals with moderate to severe hearing loss (Vinay and Moore 2007). The perceptual effect of a dead region is simplest to understand for a region that covers a large BF range extending upward from an edge frequency. In this case, there are no responses to tones at the frequencies within the dead region, so that amplifying the frequencies in the dead region does not lead to an improvement in speech perception (Baer et al. 2002). Another kind of problem occurs for a circumscribed dead region. For example, suppose the BF region from 1 to 2 kHz were a dead region in Fig. 6.3a. Stimulus energy within that region would produce responses in higher BF neurons (e.g., the 3.2-kHz neuron whose tuning curve is plotted with the heavy dashed line), leading to an upward shift of the apparent frequency of the

1–2 kHz components of the stimulus. At best, the shift would lead to confusion about these components; the shift may also explain why sounds with frequency components in such a dead region often sound distorted and unnatural.

5.2 *Neuropathy*

Loss of SGNs does not have to be punctate and discrete as with the neural and perceptual examples discussed in the previous subsection. Aging is often accompanied by a more or less uniform partial loss of SGNs that is progressive (Otte et al. 1978). In the data of Otte and colleagues, there was a progressive loss of approximately 2,500 SGNs per decade from age 10 to age 90. Because of the subtlety of the lesion, the perceptual deficits caused by such a diffuse loss of SGNs might be difficult to measure until the loss becomes large. In particular, it should not lead to an elevation of the threshold for hearing, so common diagnostic hearing evaluations are unlikely to detect the lesion. However, a diffuse lesion could lead to decreased auditory performance by two mechanisms. First, the decrease in the number of active fibers could lead to a decrease in the amount of information passed to the brain about sound. There seem to be more fibers in the AN than are needed for simple discriminations (Siebert 1968; Colburn et al. 2003), so a diffuse loss of SGNs will first be noted in difficult listening situations, like speech in the presence of reverberation or substantial interfering sounds. Second, degeneration of SGNs leads to secondary changes in the cochlear nucleus and other parts of the central auditory system (Syka 2002). These changes can produce problems in auditory processing, discussed in Sect. 7. Both mechanisms are possible explanations for the fact that older listeners typically have more problems in speech perception than younger listeners with similar hearing losses (threshold shifts).

In the introduction to this chapter, auditory neuropathy was briefly discussed. This term refers to a hearing impairment in which hair cell function remains, but neural activity in the brain is abnormal or missing (Starr et al. 1996; Rance 2005; Zeng et al. 2005). In the clearest case, DPOAEs are normal, suggesting intact OHC function, but ABRs are missing or abnormal, suggesting abnormalities in neural responses. Such a condition might be caused by a pathology anywhere in the chain of synaptic transmission from IHCs to the auditory cortex. So far, mutations associated with auditory neuropathy provide the best information. Such mutations are in genes associated with hair cell synapses or the neurons of the auditory pathway (Khimich et al. 2005; Delmaghani et al. 2006; Santarelli et al. 2009; Schoen et al. 2010). However, the physiology of these mutations have not been analyzed sufficiently to show how the defects might produce the symptoms of neuropathy. For example, it has been reported that carboplatin, an ototoxic cancer drug that can produce a specific lesion of IHCs and SGNs, does not produce an adequate animal model of neuropathy (El-Badry and McFadden 2009), implying that the defect in neuropathy is more specific than just damage to IHCs and SGNs.

It is widely thought that neuropathy involves a defect in synaptic function that interferes with the transmission of precisely timed auditory events, thus explaining the loss of potentials that depend on synchronous activation of neurons, like the ABR (Rance 2005). Evidence for a synaptic timing defect has been obtained in one case, the so-called bassoon mutant, which involves a protein important in forming the synaptic ribbon in IHC/SGN synapses (Khimich et al. 2005). Bassoon mutant mice show the general characteristics of neuropathy; in particular, they show degraded precision in their responses to the onsets of sounds, although their ability to phase lock in the steady state of the stimulus is not affected (Buran et al. 2010).

5.3 *Fine Structure*

Acoustic stimuli can be decomposed into an *envelope* and *fine structure* (Smith et al. 2002), illustrated in Fig. 6.7a–d. The waveform of a speech sound (Fig. 6.7a) is filtered by the BM (Fig. 6.7b) into different frequency channels (e.g., fibers with different tuning curves as in Fig. 6.2c). The signal in one frequency channel is shown in Fig. 6.7c (a simulation using a filter of the approximate form of a BM filter in the human ear at 1.2 kHz). The BM signal, shown by the solid line, oscillates at a frequency near 1.2 kHz; in addition, its amplitude fluctuates as shown by the heavy gray line marked *envelope*. The envelope fluctuation represents the instantaneous energy in the original speech sound at the frequencies that pass through this BM filter. The oscillations of the waveform are called the *fine structure*.

Human listeners are very good at using envelope information, especially for speech in quiet; if the fine structure in speech is eliminated (by replacing it with random noise with the same envelope as the speech) then the speech is still intelligible with envelopes in as few as 4–8 frequency bands (Shannon et al. 1995). For such no-fine-structure stimuli, the envelope of BM signals such as that in Fig. 6.7c would be similar to the envelope in the full stimulus but the fine structure would be random and unrelated to the information in the stimulus.

There is evidence that fine structure is used by human listeners in a number of perceptual judgments (Moore 2008), including pitch perception, binaural processing, masking by fluctuating background sounds, and speech perception. A well-known case is interaural-time-difference sensitivity, in which delays in the fine structure from one ear to the other are used to estimate the interaural time delay of the stimulus and thus the position of the sound in space (Wightman and Kistler 1992). An intriguing recent finding is that hearing-impaired listeners have deficits in the use of fine-structure information, for example, in interaural time difference computation (Lacher-Fougère and Demany 2005) or in speech perception (Lorenzi et al. 2006; Hopkins et al. 2008). This is a surprising finding that may be pointing to an aspect of the physiology of hearing impairment that has so far been overlooked.

Both the envelope and the fine structure of sounds are represented by AN fiber spike trains. For illustration, Fig. 6.7d shows simulated spike trains in six fibers that innervate the BM filter of Fig. 6.7c. The discharge rates of the fibers rise and fall with the envelope of the signal, shown by the higher density of spikes where the envelope is large. Note the two 15-ms time periods shown by heavy black bars at the

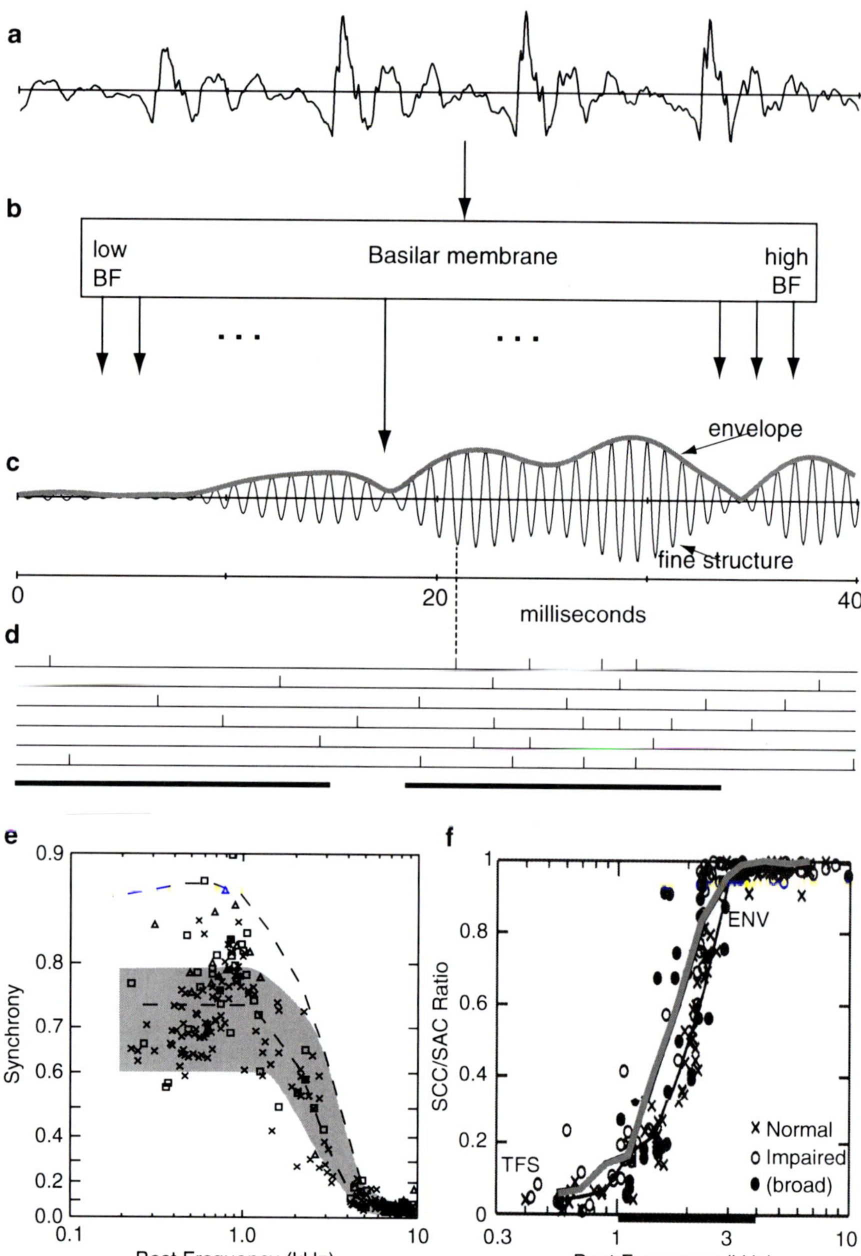

Fig. 6.7 The decomposition of speech into an envelope and fine structure. (**a**) A short segment of speech at the ear. (**b**) BM decomposition by frequency into multiple signals at different BFs. (**c**) The BM signal, the input to IHCs, at a BF of 1.2 kHz. The actual waveform (*solid line*) oscillates at a frequency near the BF (the *fine structure*) and fluctuates in amplitude (the *envelope*, heavy *gray line*). (**d**) Spike trains from six model AN fibers drawn to show the features that represent the fine structure (phase locking to the stimulus waveform) and the envelope (the discharge rate follows the envelope). The *vertical dashed line* shows how one example spike aligns with (is phase locked to) the stimulus waveform. (**e**) Strength of phase locking to BF tones 50 dB above threshold for a population of AN fibers, a subset of the population from cats after acoustic trauma in Fig. 6. 6b. Excluded fibers are ones in which 50 dB above threshold was above 115 dB SPL. The *shaded region* and *lines*

bottom. Only 6 spikes occur during the first bar where the envelope is small, but 19 spikes occur during the second bar where the envelope is large. The average discharge rates of the fibers thus represent the overall energy passing through the BM filter innervated by these fibers. When the envelope amplitude (measured by the discharge rate over a time window with a duration of tens or hundreds of milliseconds) is plotted versus BF, the result is a *rate code*, a neural representation of the spectrum of the stimulus (Sachs and Young 1979). The rate code has a low temporal precision, because it needs to encode only the frequencies of the envelope of the stimulus (below 50 Hz for speech, Drullman et al. 1994). Such a code exists at all levels of the central auditory system.

The fine structure of the stimulus is represented by phase-locking of AN fibers to the stimulus waveform (as in Fig. 6.5). The vertical dashed line in Fig. 6.7d shows how one spike lines up with a negative peak of the BM waveform; all the other spikes in Fig. 6.7d line up at the same stimulus phase, which is phase locking. Phase-locked discharges provide information about speech in AN fibers (for BFs below a few kiloHertz, above which fibers are not phase-locked; Young and Sachs 1979; Delgutte and Kiang 1984; Palmer et al. 1986) and in some neurons in cochlear nucleus (Blackburn and Sachs 1990). However, neurons elsewhere in the auditory system do not phase lock at frequencies above 100 Hz or so. Thus, if phase-locking to fine structure is to be used by the brain, it must be recoded into a non-phase-locked form at a peripheral level. Intriguing models have been proposed for the recoding that depend on the relative phase of the neural response in successive frequency bands (Shamma 1985; Deng and Geisler 1987; Carney et al. 2002), but direct evidence for these coding mechanisms is lacking.

The reasons that hearing-impaired listeners have a deficit in fine-structure perception are not clear. Possibilities include loss of the neural representation of fine structure itself (i.e., loss of phase-locking), changes in its organization (e.g., changes in the relative phase response along the BM; Heinz et al. 2010), or changes in the central nervous system that compromise the ability to extract the information (Moore 2008).

Existing data rule out the first of the three explanations in the previous paragraph, in that phase-locking is as strong after acoustic trauma as before. Figure 6.7e shows the strength of phase-locking to a BF tone in a subset of the fibers for which data were shown in Fig. 6.6 (Miller et al. 1997). The ordinate is the synchrony of spiking to the tone, which varies from 0 for random spike timing to 1 for perfect phase locking with spikes lined up at the same phase of every cycle of the stimulus. Synchrony was determined from responses to BF tones 50 dB above threshold. Points show synchrony values for individual neurons after acoustic trauma, plotted against BF. The shaded region shows data from normal animals using the same

Fig. 6 7 (continued) show synchrony in normal animals; see text. The symbols show the neurons' SRs, as in Figs. 4 and 6. The ordinate scale is chosen to make the variance of the synchrony measure constant along the ordinate (Johnson 1980). (Reprinted from Miller et al. 1997 with permission.) (**f**) A plot of the ratio of the responses to envelope and fine structure versus BF. Data from normal chinchillas (Xs, the thin solid line shows average values) and animals with acoustic trauma (the *circles*; *filled circles* show fibers with significantly broadened tuning; the wide *gray line* shows average values) (Redrawn from Kale and Heinz 2010 with permission)

analysis method and the dashed lines show the range of phase locking in normal cats computed in a way that gives the maximum possible phase locking (Johnson 1980). These data show that fibers are capable of phase locking over a range comparable to normal after acoustic trauma. Similar results have been obtained elsewhere (Harrison and Evans 1979; Kale and Heinz 2010), including studies at high sound levels, at and above 100 dB SPL (Liberman and Kiang 1984; Wong et al. 1998).

A measure of the effects of acoustic trauma on fine-structure in complex stimuli is shown in Fig. 6.7f (Kale and Heinz 2010). The stimuli in this case were sinusoidally amplitude-modulated BF tones, but a similar result was obtained for single-formant vowels. The ordinate shows a measure of the ratio of envelope to fine structure responses (Louage et al. 2004; Heinz and Swaminathan 2009). Values of 0 mean responses are dominated by fine structure (marked *TFS*) and values of 1 mean domination by the envelope (marked *ENV*). As expected, the responses are dominated by fine structure at low BFs where phase locking is strong and transition to domination by envelope over the BF range where phase locking disappears (1–4 kHz). The data for ears with acoustic trauma (circles and the gray line) show slightly stronger relative responses to the envelope than do the normal ears (Xs and the black line), for BF regions with substantial threshold shift (the bar on the abscissa). A separate measure of fine structure responses shows that the difference is not caused by weaker fine structure responses in impaired ears, but rather by slightly larger envelope responses.

Thus, hearing-impaired listeners' difficulty with fine-structure information is not caused by a lack of responses to fine-structure in individual neurons. However, there is evidence that acoustic trauma produces a change in the distribution of the phase of fine-structure responses across neurons with different BFs (Heinz et al. 2010). Such a difference could be responsible for poor fine-structure coding by interfering with the (unknown) central mechanisms for decoding fine structure.

6 Compression, Rate Responses, and Loudness Recruitment

Another important property of level-dependent amplification in BM responses is compression of the dynamic range of the stimulus. Figure 6.8a shows the BM velocity as a function of sound level for a tone at BF (*~BF*) and a tone well below BF (*<BF*); these data are typical of measurements of BM motion (especially in the high BF part of the cochlea; e.g., Ruggero et al. 1997). As shown in Fig. 6.2b, the response at BF is nonlinear, in that the gain of the response changes with sound level. Here, it is convenient to assume that the BM velocity is proportional to P^m, where P is sound pressure in Pascals and m is the slope of the velocity–pressure relationship in the log–log plot of Fig. 6.8a [i.e., $\log(velocity) = (\text{const}) + m \log(P)$]. For the velocity at BF, the response is linear at low sound levels (<30 dB SPL), meaning that $m = 1$. For reference, a dashed line with slope 1 is shown at right (marked *linear*); the BF velocity plot is parallel to this line at low sound levels. At higher sound levels, the slope of the BF velocity line is smaller, around 0.2. The low slope means that the velocity of the BM increases by less than 20 dB between sound levels of 30 and

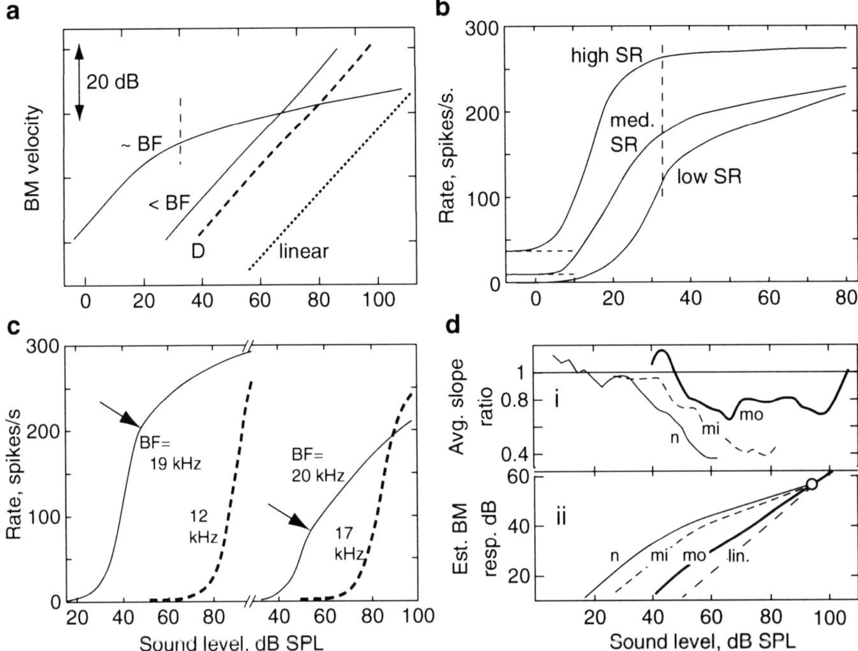

Fig. 6.8 Growth with sound level of BM and AN fiber responses to tones. (**a**) BM velocity versus sound level showing compressive growth for tones near BF (*~BF*) and linear growth for tones at frequencies below BF (*<BF*). The line marked *linear* shows a slope of 1, which means velocity is proportional to sound pressure in this log–log plot. *D* is the postmortem velocity function. (Redrawn after data of Ruggero et al. 1997.) (**b**) Sketches of typical plots of discharge rate versus sound level for cat AN fibers, showing different dynamic-range behavior for fibers of three SR groups (Sachs and Abbas 1974). (**c**) Rate versus sound level for two guinea pig AN fibers comparing rate functions to BF tones (*solid*) and tones well below BF (*dashed*). The *arrows* show where compression is assumed to set in. (Redrawn from Yates et al. 1990 with permission.) (**d-i**) Estimates of the compression ratio from rate functions in normal (*n*) ears and ears with mild (*mi*) and moderate (*mo*) threshold shift following acoustic trauma. (**d-ii**) Estimates of the BM velocity function computed by integrating the compression slopes in the top plot. The responses were assumed equal at 95 dB SPL (*unfilled circle*). The dashed line (*lin*) shows a linear velocity function for comparison (Redrawn from Heinz et al. 2005 with permission)

100 dB SPL. This is a substantial *compression* of the BM response, which is important in dealing with the wide dynamic range of hearing (Oxenham and Bacon 2003). Whereas the sound level in the environment can vary over 80–100 dB, AN fibers have much narrower dynamic ranges, only 20–30 dB (Sachs and Abbas 1974; Palmer and Evans 1982); BM compression helps to match the fibers' dynamic ranges to those of natural sounds.

For tones sufficiently below BF, the response is linear, the curve marked "*<BF*" in Fig. 6.8a. The linear BM input/output curve in Fig. 6.8a corresponds to the fact that the gain of the BM velocity as plotted in Fig. 6.2a does not change with sound level for relative frequencies (frequency/BF) below 0.7. The BM response is also

linear if the cochlea is damaged by acoustic trauma (TTS Ruggero et al. 1996), furosemide (Ruggero and Rich 1991), or the death of the animal. The line marked *D* in Fig. 6.8a shows the postmortem response to tones at BF.

6.1 Compression in AN Responses

The properties of the BM velocity response in Fig. 6.8a can be seen in the responses of AN fibers. Fig. 6.8b shows schematic rate functions for AN fibers (e.g., Sachs and Abbas 1974; Winter et al. 1990; Yates et al. 1990); the rate of discharge of action potentials is plotted as a function of the sound level of an approximately 200 ms tone burst at BF. As sound level and BM velocity increase, the fibers increase their discharge rates. Three classes of fibers have been defined in cats (Liberman 1978), and the results are similar in other animals. The fibers with the lowest thresholds usually have substantial spontaneous activity, meaning ongoing spiking in the absence of a stimulus (40 spikes/s in this case for the *high SR* example). Rate functions of these fibers have a steep dynamic portion (between 0 and 30 dB SPL here) with an approximately constant *saturation* rate at higher sound levels. High SR fibers respond to sound levels low enough to be on the linear portion of the BM input/output function. Thus their rate functions represent the input/output relationship of a neuron in the absence of compression.

Medium- and low-SR fibers have higher thresholds and wider dynamic ranges on the sound level axis. Often there is a break (i.e., a sharp decrease) in the slope of these rate functions at about the sound level (the vertical dashed lines in Fig. 6.8a, b) where the BM slope should change from linear to compressive (Sachs and Abbas 1976; Yates et al. 1990). Thus the medium- and low-SR fibers sample both the linear (at low sound levels) and the compressive (at higher sound levels) portion of the BM response. To the extent that the compressive range is included, their discharge rates change over a wider range of sound levels.

The difference between the compressive BM response at frequencies near BF and the linear response at frequencies below BF can also be seen in AN rate functions. Fig. 6.8c shows rate functions at BF and below BF for two AN fibers (Yates et al. 1990). The arrows point to the change in slope of the BF rate functions (solid lines) that occurs at the transition from linear to compressive BM velocity growth. The dashed lines show rate functions at frequencies below BF that are responses to linear growth of BM velocity. The ratio of the slope of the compressive part of the rate-function (sound levels above the arrows) to the slope of the below-BF rate functions provides a measure of the compression factor (the exponent m in the discussion of Fig. 6.8a) of the BM. The actual calculation is more general than that described in the preceding text and allows slope estimation as a function of the sound level of the tone at BF (Yates et al. 1990), which is how the slope ratio is plotted in Fig. 6.8d-i (Heinz et al. 2005). These data are population averages of fibers in normal animals (*n*) and animals with mild (*mi*) and moderate (*mo*) acoustic trauma. For normal animals (*n*) the slopes begin near 1 at low sound levels, where the BM response is linear at BF, and decrease as compression sets in at higher levels. At and above

60 dB, the slope cannot be reliably estimated for normal animals, so one can say only that the compression slope is 0.4 or less at higher levels. By integrating the slope estimate (with extrapolation of the slope to sound levels higher than 60 dB), the BM response function can be estimated, as shown in Fig. 6.8d-ii. The result is generally similar to directly measured BM velocity responses like that shown in Fig. 6.8a.

After acoustic trauma, the same analysis leads to smaller estimates of compression, that is, slope ratios closer to 1. Estimates were made for mild (*mi*, threshold shifts of 25–30 dB at the exposure frequency) and moderate (*mo*, 45–50 dB shifts) degrees of PTS (Fig. 6.8d-i; Heinz et al. 2005). In the moderate group, the compression ratio estimate is about 0.8 at high sound levels. Correspondingly, the estimated BM velocity response (Fig. 6.8d-ii) is closer to linear in the hearing impaired fibers.

The effects of compression can also be estimated in psychophysical experiments (Oxenham and Plack 1997). A two-tone masking experiment is done that measures the intensity of a low-frequency masker required to just mask a probe tone. It is assumed that the listener uses fibers with BFs at the probe frequency to detect the probe. The neural response to the low-frequency masker should grow linearly with sound pressure while the response to the probe should be compressed. Thus, the masker sound level at masked threshold should grow more slowly than the level of the probe. The ratio of the two growth rates should be the compression ratio. In that experiment, the inferred compression is roughly the same as that measured with direct observation of the BM or AN rate functions.

6.2 AN Responses After Acoustic Trauma, Component 2

The changes in AN rate responses after damage to the cochlea depend on the nature of the hair cell lesion. A pure OHC lesion should modify AN responses through its effects on BM motion. The postmortem curve in Fig. 6.8a shows the expected effect of an OHC lesion, a loss of cochlear amplification reducing the BM velocity (except at high sound levels) and eliminating compression. If the IHCs are intact, then all the AN responses should have shapes similar to the high-SR response in Fig. 6.8b, because the input/output function of the BM is now linear at all sound levels; of course, the thresholds should be elevated because of the loss of cochlear amplification. Nonsaturating rate functions (as for the low- and medium-SR responses) should not occur because there is no longer any compression and the slopes of AN rate functions should become on average steeper. These expectations are largely met for AN fibers in ears where a mainly OHC lesion was produced by an ototoxic antibiotic (Evans 1975; Schmiedt et al. 1980; Harrison 1981).

In the case of damage to IHC stereocilia (with intact OHCs), the primary effects are caused by weaker synaptic activation of SGNs because of reduced transducer currents in the damaged IHCs. The thresholds of AN fibers should be elevated and the maximum discharge rates reduced. Both effects should decrease the slopes of rate functions: the threshold elevation moves the rate functions into the compression range of BM motion (which is still present because the OHCs are assumed normal), decreasing the slope. Thus, the effects of pure OHC and pure IHC lesions should be opposite.

After acoustic trauma, both IHC and OHC damage occur and additional effects due to C2 responses become important (Liberman and Dodds 1984b; Heinz and Young 2004). Figure 6.9a shows schematic rate functions that are typical of a normal high-SR fiber (*n*) and a fiber following acoustic trauma (*at*). This figure also shows some properties of the component 2 (C2) responses at high sound levels (Gifford and Guinnan 1983; Liberman and Kiang 1984; Sewell 1984b). All of the discussion up to this point has referred to the component 1 (C1) responses, at lower sound levels. At high levels, there is a sudden change in behavior of AN fibers; the discharge rate may drop over a range of a few decibels and then recover to a different rate, as shown in Fig. 6.9a. In all cases, there is a 180° change in the phase of response to the stimulus waveform, that is, an inversion of the phase-locked response, that occurs over the same few decibels as the rate dip. The rate change and phase inversion are the *C1–C2 transition*. C1 and C2 are thought to represent two different modes of stimulation of the IHCs. Liberman and Kiang (1984) argued that C1 represents stimulation of the IHCs through their tallest stereocilia; the stimulus for C1 is the BM motion which is affected by BM tuning, the mechanical effects of OHCs, efferent effects through the olivocochlear bundle, and acoustic trauma. C2 is a different stimulus mode, perhaps through the shorter stereocilia, which is much less sensitive to sound, poorly tuned or not tuned at all, and not affected by OHC status. For reasons that are not understood, these two stimuli are in phase opposition. C2 is weaker than C1 at low sound levels, but rises rapidly in amplitude and overwhelms C1 when the transition occurs. The interaction of the two components when their amplitudes are approximately equal explains the rate dip and the phase inversion at the transition point. Recent models of cochlear processing incorporate two parallel paths, one for C1 and the second for C2 (e.g., Sumner et al. 2003; Zilany and Bruce 2006).

Acoustic trauma affects C1 but not C2. Similar effects on C1 are produced by furosemide (Sewell 1984b). Because acoustic trauma produces a mixed IHC/OHC lesion, a variety of rate functions is observed after trauma, including ones expected from the arguments above about pure IHC or OHC lesions and mixtures (Heinz and Young 2004). The slope of the main rising phase of the rate function, the portion of the function just above threshold, is a convenient measure of these functions. The slopes are shown in Fig. 6.9a by the heavy gray lines. The distribution of these slopes is shown in Fig. 6.9b for normal animals and animals with two degrees of threshold shift after acoustic trauma (the same groups of AN fibers as are shown in Fig. 6.8d). Notice that the distributions spread broadly over a decade range (Salvi et al. 1983). The slopes are slightly but significantly smaller in the traumatized populations, suggesting that the IHC effects were slightly larger in these data.

When C1 is attenuated, as in Fig. 6.9a, the rate function at high sound levels can be dominated by the large rate increase at the C1/C2 transition. For normal ears, this rate change is masked by the response to C1, which produces a higher rate than C2 at low sound levels and a comparable rate at the high sound levels. However, the C2 rate increase may be important in impaired ears, especially with hearing aids. Indeed, hearing-impaired listeners are more sensitive to temporal modulation of sounds (Fullgrabe et al. 2003), perhaps because of the increased slope of the rate function near the C2 transition. In the AN, the rate changes associated with the C2 transition

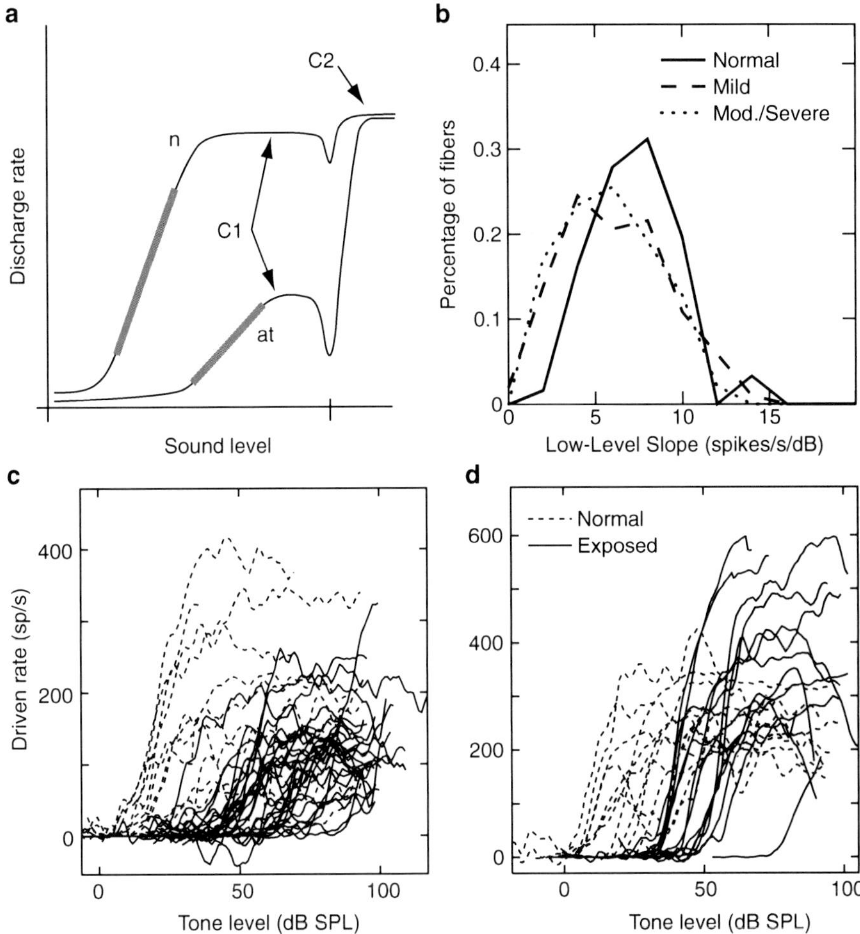

Fig. 6.9 Effects of acoustic trauma on rate functions. (**a**) Schematic rate-versus level functions for BF tones from a fiber in a normal animal (*n*) and following acoustic trauma (*at*) or furosemide. C1 and C2 are labeled by the arrows; the transition from C1 to C2 is shown by the dip in the rate functions, which would occur at 80–100 dB SPL. The rates for C2 are shown to be the same for the two fibers, but this is not necessarily so and a range of C2 rates is observed. However, C2 rates do not change with furosemide poisoning whereas C1 rates do (Sewell 1984b). The heavy shaded lines show the slopes plotted in (**b**). (Based on rate functions in Liberman and Kiang 1984; Sewell 1984b, and Heinz and Young 2004.) (**b**) Distributions of slopes of BF-tone rate functions computed from the low-level rapidly rising portion of the function, the gray bars in (**a**). Note that the steep slopes at the C1/C2 transition are not included. (Reprinted from Heinz and Young 2004 with permission.) (**c**) Rate-versus-level functions for BF tones from primarylike neurons in the cochlear nucleus (see text) of normal (*dashed lines*) cats and cats exposed to acoustic trauma (*solid lines*). (**d**) Same data for chopper neurons in the cochlear nucleus. Note that rates are decreased following trauma in primarylike neurons, similar to (**a**), but increased in choppers (**c**, **d** reproduced from Cai et al. 2009, with permission)

have the effect of increasing the responses of fibers to the envelope of the stimulus, consistent with this idea (Kale and Heinz 2010). Recently, however, it has been found that temporal modulation of AN responses such as onset and offset responses are faster after acoustic trauma, which may also contribute (Scheidt et al. 2010).

In summary, the slopes of the C1 part of rate functions on average decrease somewhat in cases of mixed IHC and OHC lesions typical of acoustic trauma. However, the C2 response at high sound levels produce a steep slope that may be important for loud sounds.

6.3 Rate Functions and Loudness Recruitment

The perceptual equivalent of the data in Figs. 6.8 and 6.9 is loudness, the sense of the volume of a sound. Loudness increases with the physical stimulus intensity as do most peripheral auditory responses, such as the velocity of the BM or the discharge rates of AN fibers. Usually it is assumed that perceptual loudness is proportional to some measure of the total neural activity in the AN, say the summed discharge rate across the nerve or equivalently the excitation pattern in loudness models (Moore and Glasberg 1997; Moore and Glasberg 2004). This assumption has been tested by estimating the total activity in the AN (e.g., Pickles 1983; Relkin and Doucet 1997), but a critical test is difficult because both loudness and neural activity increase together with stimulus level. There are quantitative differences between the two measures, but these can be explained by assuming that there is some compensating nonlinearity in the central auditory system.

In hearing impaired ears, loudness shows *recruitment*, meaning that loudness increases faster with sound level in the impaired ear than in the normal ear (Moore 2007). As a result, sounds of 80–100 dB may be equally loud in a normal ear and an impaired ear, even though the zero of loudness is at approximately 0 dB in the normal ear and at a significantly elevated value (say 50 dB) in the impaired ear. Recruitment can be demonstrated most clearly in a person with one normal ear and one impaired ear. The subject is asked to adjust the intensity of the sound in the normal ear to equate its loudness to that of a series of sounds of different intensities in the impaired ear, or vice versa (Miskolczy-Fodor 1960; Moore et al. 1985; Zeng and Turner 1991; Stillman et al. 1993). The resulting matches are roughly linearly related (on decibel scales) with a slope that tells how fast the loudness increases with level in the impaired ear relative to the normal ear. Typically these slopes are between one and three for ears with approximately 50 dB threshold shifts.

The loudness balance functions have been used to estimate the input/output function of the BM in a way similar to that used with neural data in Fig. 6.8d (Schlauch et al. 1998). In this case, it is assumed that loudness balance occurs when the BM velocity at the frequency of the tone is equated between the normal and impaired ears and further that the BM response is linear in the impaired ear, as in the *D* data in Fig. 6.8a. With these assumptions, the BM velocity in the normal ear can be estimated from the loudness balances (consult the original paper for details of the method). The result is a compressive function very similar to measured BM velocity input/output functions at BF (Fig. 6.8a). Thus, the assumption that loudness is related to BM response is supported; this result was used to argue that loudness is proportional to BM velocity at BFs near the tone frequency, which implies that recruitment is a reflection of the increased slope of the BM input/output function after acoustic trauma, and therefore is a reflection of the loss of compression (Moore 2007).

Of course, the BM velocity must be conveyed to the brain by the AN, and Fig. 6.9a, b show that the C1 rate responses of AN fibers do not increase their slopes after acoustic trauma. To further analyze recruitment in neural data, the perceptual loudness balance experiments were simulated with neural data by matching the discharge rates of neurons in impaired and normal ears (Heinz et al. 2005; Cai et al. 2009). The assumption, again, is that loudness is a function of the total activity in some group of neurons (all of the AN fibers or AN fibers with BFs within some bandwidth of the stimulus tone). With this analysis, the properties of any nonlinearity in the mapping from AN responses to loudness do not matter, as long as it is the same for normal and impaired ears. Simulated loudness balance experiments based on AN fiber rates did not give the same results as experiments in human listeners, in that the slopes of loudness balance functions were usually less than or equal to 1 (instead of greater than 1) in the C1 range of sound levels, mainly because of the smaller discharge rates in animals after acoustic trauma (Fig. 6.9b). This result argues against the interpretation of recruitment as reflecting a change in the input/output slope of the BM. Of course, the slopes increase at high sound levels where C2 responses appear, but the overall shape of the rate-balance curves is very different from loudness-balance curves.

Other explanations for recruitment after acoustic trauma have been suggested, including abnormally fast spread of activity in the AN due to broadened tuning (Kiang et al. 1970; Evans 1975) and a compressed distribution of AN fiber thresholds because the threshold shift reduces the overall dynamic range of the cochlea (Moore et al. 1985; Zeng and Turner 1991). These hypotheses also were not supported by the response properties of AN fibers in cats exposed to acoustic trauma (Heinz et al. 2005; Cai et al. 2009). Yet, identical animals with the same degree of acoustic trauma showed recruitment similar to that seen in human observers (May et al. 2009). It seems likely that recruitment actually results from changes in the central nervous system, discussed in the next section.

6.4 Recruitment in Neurons in the Central Auditory System

Although hypotheses based on the rate of growth of AN responses with sound level do not seem to be an explanation for recruitment, there is substantial evidence of steeper neural response growth after acoustic trauma in neurons of the central auditory system. Studies based on evoked potentials or neural population responses show recruitment-like hyperexcitability after acoustic trauma for neurons in the cochlear nucleus (Saunders et al. 1972; Lonsbury-Martin and Martin 1981), inferior colliculus (Salvi et al. 1990; Szczepaniak and Moller 1996; Wang et al. 2002), and cortex (Popelar et al. 1987; Syka et al. 1994).

Examples of changes in response after acoustic trauma are shown for neurons in the cochlear nucleus in Fig. 6.9c, d (Cai et al. 2009). The cochlear nucleus is an assembly of several different neuron types (reviewed in Young and Oertel 2004); data from two of these are shown here. Figure 6.9c shows rate versus sound level functions for neurons in normal animals (dashed lines) and animals with an approximately 50 dB threshold shifts (solid lines) for one kind of neuron, so-called *primarylike*. These neurons receive large synaptic terminals from AN fibers and generally have response properties similar to those of AN fibers. The effects of acoustic trauma are similar in primarylike neurons and AN fibers: the thresholds are shifted and the discharge rates are decreased. For a second neuron type, so-called *choppers* (Fig. 6.9d), the result is the opposite. Thresholds are increased after trauma, but the discharge rates increase significantly. The causes of the different behavior of primarylike and chopper neurons are not known at present; however, it seems likely that the increase in response in choppers results from adjustments of the strength of the synapses from AN fibers onto the cells. Synaptic strength is often modulated to regulate the overall average level of activity of neurons, a process called homeostatic plasticity (Turrigiano 2008). In this case, plasticity could operate in chopper neurons to overcome the decreased spontaneous and sound-driven activity in AN fibers after trauma. A similar adjustment does not occur in primarylike neurons, perhaps because the large synaptic terminals made by AN fibers on primarylike neurons normally operate significantly above threshold and so are not subject to homeostatic adjustment. Plasticity in the central auditory system is discussed by Kaltenbach (Chap. 8).

When loudness balance experiments are simulated with cochlear nucleus data, primarylike neurons, like AN fibers, do not produce results consistent with recruitment, whereas choppers do (Cai et al. 2009). These data suggest that loudness recruitment may be caused by changes in the strength of synapses in the central nervous system rather than (directly) by changes in the properties of the cochlea.

7 Damage to the Central Nervous System

The changes in response rate shown in Fig. 6.9d are examples of the kinds of changes that occur in the central auditory system after acoustic trauma or other damage to the cochlea. Similar changes in the excitatory synaptic strength of auditory (Bledsoe et al. 2009) and nonauditory (Shore et al. 2008) inputs to the cochlear nucleus have been reported after TTS or PTS. A full discussion of this subject is beyond the scope of this chapter, but it is worthwhile to mention a few well studied examples of central reorganization after cochlear damage. Much of the work on the effects of acoustic trauma in the brain is designed to study the mechanisms of tinnitus. That work is the subject of another chapter in this volume (Kaltenbach, Chap. 7).

Immediately after damage to the cochlea by acoustic trauma, a widespread degeneration of fibers and both excitatory and inhibitory terminals is seen in the cochlear nucleus (Kim et al. 2004). Over a period of up to 32 weeks, the degeneration disappears and is replaced by formation of new terminals, with a tendency toward more excitatory than inhibitory terminals. Analysis of synaptic transmission in animals with cochlear damage also suggests a strengthening of excitatory (Oleskovich and Walmsley 2002; Vale and Sanes 2002; Muly et al. 2004) and a weakening of inhibitory (Suneja et al. 1998; Vale et al. 2004; Takesian et al. 2009) connections, along with changes in ion channel density that increase the excitability of the cells (Francis and Manis 2000; Leao et al. 2004). The increased strength of excitatory relative to inhibitory synaptic responses is consistent with an increased expression of genes associated with excitatory synapses relative to inhibitory synapses (Asako et al. 2005; Cui et al. 2007; Dong et al. 2010). Finally, neurons in brainstem auditory structures show fewer or weaker inhibitory responses following cochlear damage (Salvi et al. 1990; Ma and Young 2006).

A well studied effect of cochlear damage on the auditory parts of the brain is the reorganization of the BF map in the auditory cortex following mechanical damage to the cochlea or acoustic trauma (Rajan et al. 1993; Noreña and Eggermont 2005). The cortical map reorganizes in the region of damage so that the map for the range of BFs where there is cochlear damage is replaced by a region responsive to the low-frequency edge of the lesion. Another dramatic map reorganization phenomenon occurs in neonatally deafened animals in which electrical stimulation of the cochlea is done as a simulation of a cochlear implant (Snyder et al. 1990; Leake et al. 2000). The stimulation induces an apparent change in neural connections from the cochlea to the inferior colliculus such that the point-to-point connectivity that underlies the normal IC BF map is substantially broadened. This would have the effect of decreasing the precision of the representation of the cochlear output in the IC.

The anatomical and biochemical changes that occur in the brain following damage to the cochlea are significant. Like the situation illustrated in Fig. 6.9, it is likely that many perceptual effects of acoustic trauma will ultimately be traced to the indirect effects of reorganization of the central auditory system.

8 Summary

The damage to hearing from noise-induced hearing loss or acoustic trauma can be analyzed in terms of the effects on cochlear hair cells and spiral ganglion neurons. Although the stria vascularis can be damaged by acoustic trauma, that effect does not seem to compromise cochlear sensory function unless the damage is severe. Hair cell damage can be subacute, mainly damage to the stereocilia, which decreases the sensitivity of transduction but does not destroy the cells and sometimes is repaired. More severe exposures can destroy the hair cells, damage that cannot be repaired because hair cells do not regenerate in mammals. OHC damage has its effects by decreasing the sensitivity and sharpness of tuning of the BM response to sound. Subacute IHC damage produces a decrease in the sensitivity of transduction of BM motion and thus a decrease in the sensitivity of activation of the SGNs. Destruction of IHCs, of course, leads to a loss of representation of the frequencies that map to the damaged region of the cochlea, producing a so-called dead zone. Damage to hair cells can be diagnosed clinically by shifts in the threshold of audibility, widening of auditory filters, and loudness recruitment.

An important recent result is that hair cells can often survive an acoustic trauma in a functional state, while the SGNs connected to those hair cells degenerate, presumably from excitotoxic damage. This result has substantial implications for diagnosis of the nature of the lesion in the cochlea. Diagnostic measures like the amplitude of otoacoustic emissions and the threshold of averaged neural responses (e.g., the ABR) are primarily sensitive to hair cell function and may not be sensitive to the degree of loss of SGNs. Information about SGNs can be gained from the suprathreshold amplitude of the averaged neural responses, which is not routinely done presently, and perhaps also by suprathreshold perceptual measures.

An important set of outstanding questions relates to the status of central auditory processing following acoustic trauma. The degeneration of SGNs discussed in the preceding text should have substantial effects by producing degeneration and reorganization of connections in the cochlear nucleus and more central auditory nuclei including the cortex. The analysis of recruitment discussed in connection with Fig. 6.9 shows that central effects can be large and can change the interpretation of long-accepted models of auditory processing. The importance of these effects in hearing impairment have been studied only in connection with a few conditions like auditory neuropathy or aging where they produce effects that cannot be explained by hair cell damage. At present there is little in the way of direct data on the neural representation of sound to test the theories of neuropathy or aging developed by studying hearing impaired listeners. This is an important area for further study.

Acknowledgments Comments from Amanda Lauer, Sean Slee, Brad May, Michael Heinz, and the editors of the SHAR series and of this volume improved the presentation of this chapter. Preparation of this chapter was supported by NIH grants DC00109 and DC0100594.

References

Abbas, P. J., & Sachs, M. B. (1976). Two-tone suppression in auditory-nerve fibers: Extension of a stimulus-response relationship. *Journal of the Acoustical Society of America*, 59(1), 112–122.

Angeli, S., Ultrera, R., Dib, S., Chiossone, E., Naranjo, C., Henriquez, O., & Porta, M. (2000). GJB2 gene mutations in childhood deafness. *Acta Oto-Laryngologica*, 120, 133–136.

Asako, M., Holt, A. G., Griffith, R. D., Buras, E. D., & Altschuler, R. A. (2005). Deafness-related decreases in glycine-immunoreactive labeling in the rat cochlear nucleus. *Journal of Neuroscience Research*, 81, 102–109.

Ashmore, J. F. (1987). A fast motile response in guinea-pig outer hair cells: The cellular basis of the cochlear amplifier. *Journal of Physiology-London*, 388, 323–347.

Ashmore, J. F. (2008). Cochlear outer hair cell motility. *Physiological Reviews*, 88, 173–210.

Baer, T., Moore, B. C. J., & Kluk, K. (2002). Effects of low pass filtering on the intelligibility of speech in noise for people with and without dead regions at high frequencies. *Journal of the Acoustical Society of America*, 112, 1133–1144.

Bielefeld, E. C., Coling, D., Chen, G. D., Li, M., Tanaka, C., Hu, B. H., & Henderson, D. (2008). Age-related hearing loss in the Fischer 344/NHsd rat substrain. *Hearing Research*, 241, 26–33.

Blackburn, C. C., & Sachs, M. B. (1990). The representations of the steady-state vowel sound/ε/in the discharge patterns of cat anteroventral cochlear nucleus neurons. *Journal of Neurophysiology*, 63, 1191–1212.

Bledsoe, S. C., Jr., Koehler, S., Tucci, D. L., Zhou, J., Le Prell, C., & Shore, S. E. (2009). Ventral cochlear nucleus responses to contralateral sound are mediated by commissural and olivocochlear pathways. *Journal of Neurophysiology*, 102, 886–900.

Brown, M. C., & Ledwith, J. V. (1990). Projections of thin (type-II) and thick (type-I) auditory-nerve fibers into the cochlear nucleus of the mouse. *Hearing Research*, 49, 105–118.

Brown, M. C., Nuttall, A. L., & Masta, R. I. (1983). Intracellular recordings from cochlear inner hair cells: Effects of stimulation of the crossed olivocochlear efferents. *Science*, 222, 69–72.

Brownell, W. E., Bader, C. R., Bertrand, D., & de Ribaupierre, Y. (1985). Evoked mechanical responses of isolated cochlear outer hair cells. *Science*, 227, 194–196.

Bruce, I. C., Sachs, M. B., & Young, E. D. (2003). An auditory-periphery model of the effects of acoustic trauma on auditory nerve responses. *Journal of the Acoustical Society of America*, 113, 369–388.

Buran, B. N., Strenzke, N., Neef, A., Gundelfinger, E. D., Moser, T., & Liberman, M. C. (2010). Onset coding is degraded in auditory nerve fibers from mutant mice lacking synaptic ribbons. *Journal of Neuroscience*, 30, 7587–7597.

Cai, Y., & Geisler, C. D. (1996a). Suppression in auditory-nerve fibers of cats using low-side suppressors. I. Temporal aspects. *Hearing Research*, 96, 94–112.

Cai, Y., & Geisler, C. D. (1996b). Suppression in auditory-nerve fibers of cats using low-side suppressors. III. Model results. *Hearing Research*, 96, 126–140.

Cai, S., Ma, W.-L. D., & Young, E. D. (2009). Encoding intensity in ventral cochlear nucleus following acoustic trauma: Implications for loudness recruitment. *Journal of the Association for Research in Otolaryngology*, 10, 5–22.

Carney, L. H., Heinz, M. G., Evilsizer, M. E., Gilkey, R. H., & Colburn, H. S. (2002). Auditory phase opponency: A temporal model for masked detection at low frequencies. *Acustica-Acta Acustica*, 88, 334–347.

Cheatham, M. A., Low-Zeddies, S., Naik, K., Edge, R., Zheng, J., Anderson, C. T., & Dallos, P. (2009). A chimera analysis of *Prestin* knock-out mice. *Journal of Neuroscience*, 29, 12000–12008.

Ching, T. Y. C., Dillon, H., & Byrne, D. (1998). Speech recognition of hearing-impaired listeners: Predictions from audibility and the limited role of high-frequency amplifcation. *Journal of the Acoustical Society of America*, 103, 1128–1140.

Cohen-Salmon, M., Ott, T., Michel, V., Hardelin, J. P.,Perfettini, I., Eybalin, M., Wu, T., Marcus, D.C., Wangemann, P., Willecke, K., & Petit, C. (2002). Targeted ablation of connexin26 in the inner ear epithelial gap junction network causes hearing impairment and cell death. *Current Biology*, 12, 1106–1111.

Colburn, H. S., Carney, L. H., & Heinz, M. G. (2003). Quantifying the information in auditory-nerve responses for level discrimination. *Journal of the Association for Research in Otolaryngology*, 4, 294–311.

Cooper, N. P. (1996). Two-tone suppression in cochlear mechanics. *Journal of the Acoustical Society of America*, 99, 3087–3098.

Cooper, N. P., & Rhode, W. S. (1995). Nonlinear mechanics at the apex of the guinea-pig cochlea. *Hearing Research*, 82(2), 225–243.

Cooper, N. P., & Rhode, W. S. (1997). Mechanical responses to two-tone distortion products in the apical and basal turns of the mammalian cochlea. *Journal of Neurophysiology*, 78, 261–270.

Crouch, J. J., Sakaguchi, N., Lytle, C., & Schulte, B. A. (1997). Immunohistochemical localization of the Na-K-Cl co-transporter (NKCC1) in the gerbil inner ear. *Journal of Histochemistry and Cytochemistry*, 45, 773–778.

Cui, Y. L., Holt, A. G., Lomax, C. A., & Altschuler, R. A. (2007). Deafness associated changes in two-pore domain potassium channels in rat inferior colliculus. *Neuroscience*, *149*, 421–433.

Dallos, P., & Harris, D. (1978). Properties of auditory nerve responses in absence of outer hair cells. *Journal of Neurophysiology*, 41, 365–383.

Dallos, P., Zheng, J., & Cheatham, M. A. (2006). Prestin and the cochlear amplifier. *Journal of Physiology-London*, 576, 37–42.

Delgutte, B. (1990). Two-tone rate suppression in auditory-nerve fibers: Dependence on suppressor frequency and level. *Hearing Research*, 49, 225–246.

Delgutte, B., & Kiang, N. Y. S. (1984). Speech coding in the auditory nerve: I. Vowel-like sounds. *Journal of the Acoustical Society of America*, 75, 866–878.

Delmaghani, S., del Castillo, F. J., Michel, V., Leibovici, M., Aghaie, A., Ron, U., Van Laer, L., Ben-Tai, N., Van Camp, G., Weil, D., Langa, F., Lathrop, M., Avan, P., & Petit, C. (2006). Mutations in the gene encoding pejvakin, a newly identified protein of the afferent auditory pathway, cause DFNB59 auditory neuropathy. *Nature Genetics*, 38, 770–778.

Deng, L., & Geisler, C. D. (1987). A composite auditory model for processing speech sounds. *Journal of the Acoustical Society of America*, 82, 2001–2012.

Diependaal, R. J., de Boer, E., & Viergever, M. A. (1987). Cochlear power flux as an indicator of mechanical activity. *Journal of the Acoustical Society of America*, 82, 917–926.

Dong, S., Mulders, W. H., Rodger, J., Woo, S., & Robertson, D. (2010). Acoustic trauma evokes hyperactivity and changes in gene expression in guinea-pig auditory brainstem. *European Journal of Neuroscience*, 31, 1616–1628.

Drullman, R., Festen, J. M., & Plomp, R. (1994). Effect of temporal envelope smearing on speech reception. *Journal of the Acoustical Society of America*, 95, 1053–1064.

Dubno, J. R., & Dirks, D. D. (1989). Auditory filter characteristics and consonant recognition for hearing-impaired listeners. *Journal of the Acoustical Society of America*, 85, 1666–1675.

El-Badry, M. M., & McFadden, S. L. (2009). Evaluation of inner hair cell and nerve fiber loss as sufficient pathologies underlying auditory neuropathy. *Hearing Research*, 255, 84–90.

Evans, E. F. (1972). The frequency response and other properties of single fibres in the guinea-pig cochlear nerve. *Journal of Physiology-London*, 226, 263–287.

Evans, E. F. (1975). The sharpening of cochlear frequency selectivity in the normal and abnormal cochlea. *Audiology*, 14, 419–442.

Evans, E. F., & Harrison, R. V. (1975). Correlation between cochlear outer hair cell damage and derioration of cochlear nerve tuning properties in the guinea-pig. *Journal of Physiology-London*, 256, 43P–44P.

Fant, G. (1970). *Acoustic theory of speech production*. The Hague: Mouton.

Francis, H. W., & Manis, P. B. (2000). Effects of deafferentation on the electrophysiology of ventral cochlear nucleus neurons. *Hearing Research*, 149, 91–105.

Fuchs, P. A., Glowatzki, E., & Moser, T. (2003). The afferent synapse of cochlear hair cells. *Current Opinion in Neurobiology*, 13, 452–458.

Fullgrabe, C., Meyer, B., & Lorenzi, C. (2003). Effect of cochlar damage on the detction of complex temporal envelopes. *Hearing Research*, 178, 35–43.

Gifford, M. L., & Guinnan, J. J. (1983). Effects of crossed olivocochlear-bundle stimulation on cat auditory nerve fiber responses to tones. *Journal of the Acoustical Society of America*, 74, 115–123.

Gillespie, P. G., & Müller, U. (2009). Mechanotransduction by hair cells: Models, molecules, and mechanisms. *Cell*, 139, 33–44.

Glasberg, B. R., & Moore, B. C. J. (1986). Auditory filter shapes in subjects with unilateral and bilateral cochlear impairments. *Journal of the Acoustical Society of America*, 79, 1020–1033.

Glasberg, B. R., & Moore, B. C. J. (2000). Frequency selectivity as a function of level and frequency measured with uniformly exciting notched noise. *Journal of the Acoustical Society of America*, 108, 2318–2328.

Gorga, M. P., Neely, S. T., Dierking, D. M., Kopun, J., Jolkowski, K., Groenenboom, K., Tan, H., & Stiegemann, B. (2008). Low-frequency and high-frequency distortion product otoacoustic emission suppression in humans. *Journal of the Acoustical Society of America*, 123, 2172–2190.

Guinan, J. J. (1996). Physiology of olivocochlear efferents. In P. Dallos, A. N. Popper, & R. R. Fay (Eds.), *The cochlea* (pp. 435–502). New York: Springer.

Hamernik, R. P., Turrentine, G., Roberto, M., Salvi, R., & Henderson, D. (1984). Anatomical correlates of impulse noise-induced mechanical damage in the cochlea. *Hearing Research*, 13, 229–247.

Harrison, R. V. (1981). Rate-versus-intensity functions and related AP responses in normal and pathological guinea pig and human cochleas. *Journal of the Acoustical Society of America*, 70, 1036–1044.

Harrison, R. V., & Evans, E. F. (1979). Some aspects of temporal coding by single cochlear fribres from regions of cochlear hair cell degeneration in the guinea pig. *Archives of Oto-Rhino-Laryngology*, 224, 71–78.

Heinz, M. G. (2010). Computational modeling of sensorineural hearing loss. In R. Meddis, E. A. Lopez-Poveda, R. R. Fay, & A. N. Popper (Eds.), *Computational models of the auditory system* (Vol. 35, pp. 177–202). New York: Springer.

Heinz, M. G., & Swaminathan, J. (2009). Quantifying envelope and fine-structure coding in auditory nerve responses to chimaeric speech. *Journal of the Association for Research in Otolaryngology*, 10, 407–423.

Heinz, M. G., & Young, E. D. (2004). Response growth with sound level in auditory-nerve fibers after noise-induced hearing loss. *Journal of Neurophysiology*, 91, 784–795.

Heinz, M. G., Issa, J. B., & Young, E. D. (2005). Auditory-nerve rate responses are inconsistent with common hypotheses for the neural correlates of loudness recruitment. *Journal of the Association for Research in Otolaryngology*, 6, 91–105.

Heinz, M. G., Swaminathan, J., Boley, J. D., & Kale, S. (2010). Across-fiber coding of temporal fine-structure: Effects of noise-induced hearing loss on auditory-nerve responses. In E. A. Lopez-Poveda, A. R. Palmer & R. Meddis (Eds.), *The neurophysiological bases of auditory perception* (pp. 621–630). New York: Springer.

Hequembourg, S., & Liberman, M. C. (2001). Spiral ligament pathology: A major aspect of age-related cochlear degeneration in C57BL/6 mice. *Journal of the Association for Research in Otolaryngology*, 2, 118–129.

Hirose, K., & Liberman, M. C. (2003). Lateral wall histopathology and endocochlear potential in the noise-damaged mouse cochlea. *Journal of the Association for Research in Otolaryngology*, 4, 339–352.

Hopkins, K., Moore, B. C. J., & Stone, M. A. (2008). Effects of moderate cochlear hearing loss on the ability to benefit from temporal fine structure information in speech. *Journal of the Acoustical Society of America*, 123, 1140–1153.

Jagger, D. J., & Forge, A. (2006). Compartmentalized and signal-selective gap junctional coupling in the hearing cochlea. *Journal of Neuroscience*, 26, 1260–1268.

Johnson, D. H. (1980). The relationship between spike rate and synchrony in responses of auditory-nerve fibers to single tones. *Journal of the Acoustical Society of America*, 68, 1115–1122.

Kale, S., & Heinz, M. G. (2010). Envelope coding in auditory nerve fibers following noise-induced hearing loss. *Journal of the Association for Research in Otolaryngology, (epub ahead of print)*. doi: 10.1007/s10162-010-0223-6

Khimich, D., Nouvian, R., Pujol, R., tom Dieck, S., Egner, A., Gundelfinger, E. D., & Moser, T. (2005). Hair cell synaptic ribbons are essential for synchronous auditory signalling. *Nature*, 434, 889–894.

Kiang, N. Y., Moxon, E. C., & Levine, R. A. (1970). Auditory-nerve activity in cats with normal and abnormal cochleas. In G. E. W. Wolstenholme & T. Knight (Eds.), *Sensorineural hearing loss* (pp. 241–273). London: Churchill.

Kiang, N. Y. S., Liberman, M. C., & Levine, R. A. (1976). Auditory-nerve activity in cats exposed to ototoxic drugs and high-intensity sounds. *Annals of Otology, Rhinology, and Laryngolgy*, 85, 752–768.

Kiang, N. Y., Rho, J. M., Northrop, C. C., Liberman, M. C., & Ryugo, D. K. (1982). Hair-cell innervation by spiral ganglion cells in adult cats. *Science*, 217, 175–177.

Kikuchi, T., Adams, J. C., Miyabe, Y., So, E., & Kobayashi, T. (2000). Potassium ion recycling pathway via gap junction systems in the mammalian cochlea an its interruption in hereditary nonsyndromic deafness. *Medical Electron Microscopy*, 33, 51–56.

Kilgard, M. P., & Merzenich, M. M. (1998). Plasticity of temporal information processing in the primary auditory cortex. *Nature Neuroscience*, 1, 727–731.

Kim, D. O., Molnar, C. E., & Matthews, J. W. (1980). Cochlear mechanics: Nonlinear behavior in two-tone responses as reflected in cochlear-nerve-fiber responses and in ear-canal sound pressure. *Journal of the Acoustical Society of America*, 67, 1704–1721.

Kim, J. J., Gross, J. S., Morest, D. K., & Potashner, S. J. (2004). Quantitative study of degeneration and new growth of axons an dsynaptic endings in the chinchilla cochlear nucleus after acoustic overstimulation. *Journal of Neuroscience Research*, 77, 829–842.

Konishi, T., Hamrick, P. E., & Walsh, P. J. (1978). Ion transport in guinea pig cochlea: I. Potassium and sodium transport. *Acta Oto-Laryngologica*, 86, 22–34.

Kraus, N., & McGee, T. (1992). Electrophysiology of the human auditory system. In A. N. Popper & R. R. Fay (Eds.), *The mammalian auditory pathway: Neurophysiology* (pp. 335–403). Berlin: Springer-Verlag.

Kujawa, S. G., & Liberman, M. C. (2006). Acceleration of age-related hearing loss by early noise exposure: Evidence of a misspent youth. *Journal of Neuroscience*, 26, 2115–2123.

Kujawa, S. G., & Liberman, A. M. (2009). Adding insult to injury: Cochlear nerve degeneration after "temporary" noise-induced hearing loss. *Journal of Neuroscience*, 29, 14077–14085.

Lacher-Fougère, S., & Demany, L. (2005). Consequences of cochlear damage for the detection of interaural phase differences. *Journal of the Acoustical Society of America*, 118, 19–26.

Lang, H., Jyothi, V., Smythe, N. M., Dubno, J. R., Schulte, B. A., & Schmiedt, R. A. (2010). Chronic reduction of endocochlear potential reduces auditory nerve activity: Further confirmation of an animal model of metabolic presbyacusis. *Journal of the Association for Research in Otolaryngology*, 11, 419–434.

Le Prell, G., Sachs, M. B., & May, B. J. (1996). Representation of vowel-like spectra by discharge rate responses of individual auditory-nerve fibers. *Auditory Neuroscience*, 2, 275–288.

Leake, P. A., Snyder, R. L., Rebscher, S. J., Moore, C. M., & Vollmer, M. (2000). Plasticity in central representations in the inferior colliculus induced by chronic single- vs. two-channel electrical stimulation by a cochlear implant after neonatal deafness. *Hearing Research*, 147, 221–241.

Leao, R. N., Berntson, A., I.D., F., & Walmsley, B. (2004). Reduced low-voltage activated K^+ conductances and enhanced central excitability in a congenitally deaf (dn/dn) mouse. *Journal of Physiology-London*, 559, 25–33.

Leek, M. R., & Summers, V. (1996). Reduced frequency selectivity and the preservation of spectral contrast in noise. *Journal of the Acoustical Society of America*, 100, 1796–1806.

Liberman, M. C. (1978). Auditory-nerve response from cats raised in a low-noise chamber. *Journal of the Acoustical Society of America*, 63, 442–455.

Liberman, M. C. (1980). Morphological differences among radial afferent fibers in the cat cochlea. An electron-microscopic study of serial sections. *Hearing Research*, 3(1), 45–63.

Liberman, M. C. (1982). The cochlear frequency map for the cat: Labeling auditory-nerve fibers of known characteristic frequency. *Journal of the Acoustical Society of America*, 72, 1441–1449.

Liberman, M. C. (1984). Single-neuron labeling and chronic cochlear pathology. I. Threshold shift and characteristic-frequency shift. *Hearing Research*, 16, 33–41.

Liberman, M. C., & Beil, D. G. (1979). Hair cell condition and auditory nerve response in normal and noise-damaged cochleas. *Acta Oto-Laryngologica*, 88, 161–176.

Liberman, M. C., & Dodds, L. W. (1984a). Single-neuron labeling and chronic cochlear pathology. II. Stereocilia damage and alterations of spontaneous discharge rates. *Hearing Research*, 16, 43–53.

Liberman, M. C., & Dodds, L. W. (1984b). Single-neuron labeling and chronic cochlear pathology. III. Stereocilia damage and alterations of threshold tuning curves. *Hearing Research*, 16, 55–74.

Liberman, M. C., & Kiang, N. Y. (1984). Single-neuron labeling and chronic cochlear pathology. IV. Stereocilia damage and alterations in rate- and phase-level functions. *Hearing Research*, 16, 75–90.

Liberman, M. C., & Mulroy, M. J. (1982). Acute and chronic effects of acoustic trauma: Cochlear pathology and auditory nerve pathophysiology. In R. P. Hamernik, D. Henderson, & R. Salvi (Eds.), *New perspectives on noise-induced hearing loss* (pp. 105–135). New York: Raven Press.

Liberman, M. C., Gao, J., He, D. Z. Z., Wu, X., Jia, S., & Zuo, J. (2002). Prestin is required for electromotility of the outer hair cell and for the cochlear amplifier. *Nature*, 419, 300–304.

Lonsbury-Martin, B. L., & Martin, G. K. (1981). Effects of moderately intense sound on auditory sensitivity in rhesus monkeys: Behavioral and neural observations. *Journal of Neurophysiology*, 46, 563–586.

Lopez-Poveda, E. A. (2005). Spectral processing by the peripheral auditory system: Facts and models. *International Review of Neurobiology*, 70, 7–48.

Lorenzi, C., Gilbert, G., Carn, H., Garnier, S., & Moore, B. C. J. (2006). Speech perception problems of the hearing impaired reflect inability to use temporal fine structure. *Proceedings of the National Academy of Sciences of the USA*, 103, 18866–18869.

Louage, D. H., van der Heijden, M., & Joris, P. X. (2004). Temporal properties of responses to broadband noise in the auditory nerve. *Journal of Neurophysiology*, 91, 2051–2065.

Ma, W.-L., & Young, E. D. (2006). Dorsal cochlear nucleus response properties following acoustic trauma: Response maps and spontaneous activity. *Hearing Research*, 216–217, 176–188.

Marcus, D. C., Wu, T., Wangemann, P., & Kofuji, P. (2002). KCNJ10 (Kir4.1) potassium channel knockout abolishes endocochlear potential. American Journal of Physiology-Cell Physiology, 282, C403–407.

May B.J., Huang A., Le Prell G., Hienz R.D. (1996) Vowel formant frequency discrimination in cats: Comparison of auditory nerve representations and psychophysical thresholds. *Auditory Neuroscience*, 3, 135–162.

May, B. J., Little, N., & Saylor, S. (2009). Loudness perception in the domestic cat: Reaction time estimates of equal-loudness contours and recruitment effects. *Journal of the Association for Research in Otolaryngology*, 10, 295–308.

Miller, M. I., & Sachs, M. B. (1983). Representation of stop consonants in the discharge patterns of auditory-nerve fibers. *Journal of the Acoustical Society of America*, 74, 502–517.

Miller, R. L., Schilling, J. R., Franck, K. R., & Young, E. D. (1997). Effects of acoustic trauma on the representation of the vowel/ε/in cat auditory nerve fibers. *Journal of the Acoustical Society of America*, 101, 3602–3616.

Miller, R. L., Calhoun, B. M., & Young, E. D. (1999a). Contrast enhancement improves the representation of/e/-like vowels in the hearing-impaired auditory nerve. *Journal of the Acoustical Society of America*, 106, 2693–2708.

Miller, R. L., Calhoun, B. M., & Young, E. D. (1999b). Discriminability of vowel representations in cat auditory-nerve fibers after acoustic trauma. *Journal of the Acoustical Society of America*, 105, 311–325.

Miskolczy-Fodor, F. (1960). Relation between loudness and diration of tonal pulses. III. Response in cases of abnormal loudness function. *Journal of the Acoustical Society of America*, 32, 486–492.

Mistrik, P., & Ashmore, J. F. (2009). The role of potassium recirculation in cochlear amplification. *Current Opinion in Otolaryngology*, 17, 394–399.

Moore, B. C. J. (2004a). Dead regions of the cochlea: Conceptual foundations, diagnosis and clinical applications. *Ear and Hearing*, 25, 98–116.

Moore, B. C. J. (2004b). *An introduction to the psychology of hearing*. Amsterdam: Elsevier.

Moore, B. C. J. (2007). *Cochlear hearing loss: Physiological, psychological, and technical issues*. Chichester: John Wiley & Sons.

Moore, B. C. J. (2008). The role of temporal fine structure processing in pitch perception, masking, and speech perception for normal-hearing and hearing-impaired people. *Journal of the Association for Research in Otolaryngology*, 9, 399–406.

Moore, B. C. J., & Glasberg, B. R. (1983). Formulae describing frequency selectivity as a function of frequency and level and their use in calculating excitation patterns. *Journal of the Acoustical Society of America*, 74, 750–753.

Moore, B. C. J., & Glasberg, B. R. (1997). A model of loudness perception applied to cochlear hearing loss. *Auditory Neuroscience*, 3, 289–311.

Moore, B. C. J., & Glasberg, B. R. (2004). A revised model of loudness perception applied to cochlear hearing loss. *Hearing Research*, 188, 70–88.

Moore, B. C. J., Glasberg, B. R., Hess, R. F., & Birchall, J. P. (1985). Effects of flanking noise bands on the rate of growth of loudness of tones in normal and recruiting ears. *Journal of the Acoustical Society of America*, 77, 1505–1513.

Muly, S. M., Gross, J. S., & Potashner, S. J. (2004). Noise trauma alters D-[^{3}H] aspartate release and AMPA binding in chinchilla cochlear nucleus. *Journal of Neuroscience Research*, 75, 585–596.

Murugasu, E., & Russell, I. J. (1996). The effect of efferent stimulation on basilar membrane displacement in the basal turn of the guinea pig cochlea. *Journal of Neuroscience*, 16, 325–332.
Nin, F., Hibino, H., Doi, K., Suzuki, T., Hisa, Y., & Kurachi, Y. (2008). The endocochlear potential depends on two K^+ diffusion potentials and an electrical barrier in the stria vascularis of the inner ear. *Proceedings of the National Academy of Sciences USA*, 105, 1751–1756.
Noreña, A. J., & Eggermont, J. J. (2005). Enriched acoustic environment after noise trauma reduces hearing loss and prevents cortical map reorganization. *Journal of Neuroscience*, 25, 699–705.
Nuttall, A. L., & Dolan, D. F. (1996). Steady-state sinusoidal velocity responses of the basilar membrane in guinea pig. *Journal of the Acoustical Society of America*, 99, 1556–1565.
Ohlemiller, K. K., & Echteler, S. M. (1990). Functional correlates of characteristic frequency in single cochlear nerve fibers of the Mongolian gerbil. *Journal of Comparative Physiology A*, 167, 329–338.
Oleskovich, S., & Walmsley, B. (2002). Synaptic transmission in the auditory brainstem of normal and congenitally deaf mice. *Journal of Physiology-London*, 540, 447–455.
Otte, J., Schuknecht, H. F., & Kerr, A. G. (1978). Ganglion cell populations in normal and pathological human cochleae. Implications for cochlear implantation. *Laryngoscope*, 88, 1231–1246.
Overstreet, E. H., Temchin, A. N., & Ruggero, M. A. (2002). Basilar membrane vibrations near the round window of the gerbil cochlea. *Journal of the Association for Research in Otolaryngology*, 3, 351–361.
Oxenham, A. J., & Bacon, S. P. (2003). Cochlear compression: Perceptual measures and implications for normal and impaired hearing. *Ear and Hearing*, 24, 352–366.
Oxenham, A. J., & Plack, C. J. (1997). A behavioral measure of basilar-membrane nonlinearity in listeners with normal and impaired hearing. *Journal of the Acoustical Society of America*, 101, 3666–3675.
Palmer, A. R., & Evans, E. F. (1982). Intensity coding in the auditory periphery of the cat: Responses of cochlear nerve and cochlear nucleus neurons to signals in the presence of bandstop masking noise. *Hearing Research*, 7(3), 305–323.
Palmer, A. R., & Moorjani, P. A. (1993). Responses to speech signals in the normal and pathological peripheral auditory system. *Progress in Brain Research*, 97, 107–115.
Palmer, A. R., Winter, I. M., & Darwin, C. J. (1986). The representation of steady-state vowel sounds in the temporal discharge patterns of the guinea pig cochlear nerve and primary-like cochlear nucleus neurons. *Journal of the Acoustical Society of America*, 79, 100–113.
Pavlovic, C. V. (1984). Use of the articulation index for assessing residual auditory function in listeners with sensorineural hearing impairment. *Journal of the Acoustical Society of America*, 75, 1253–1258.
Pickles, J. O. (1983). Auditory-nerve correlates of loudness summation with stimulus bandwidth, in normal and pathological cochleae. *Hearing Research*, 12, 239–250.
Pickles, J. O. (2008). *An introduction to the physiology of hearing* (3 rd ed.). London: Academic Press.
Plack, C. J., Oxenham, A. J., Simonson, A. M., O'Hanlon, C. G., Drga, V., & Arfifianto, D. (2008). Estimates of compression at low and high frequencies using masking additivity in normal and impaired ears. *Journal of the Acoustical Society of America*, 123, 4321–4330.
Popelar, J., Syka, J., & Berndt, H. (1987). Effect of noise on auditory evoked responses in awake guinea pigs. *Hearing Research*, 26, 239–247.
Quraishi, I. H., & Raphael, R. M. (2008). Generation of the endocochlear potential: A biophysical model. *Biophysical Journal*, 94, L64–66.
Rajan, R., Irvine, D. R., Wise, L. Z., & Heil, P. (1993). The effect of unilateral partial cochlear lesions in adult cats on the representation of lesioned and unlesioned cochleas in primary auditory cortex. *Journal of Comparative Neurology*, 338, 17–49.
Rance, G. (2005). Auditory neuropathy/dys-synchrony and its perceptual consequences. *Trends in Amplification*, 9, 1–43.
Recio-Spinoso, A., Temchin, A. N., van Dijk, P., Fan, Y. H., & Ruggero, M. A. (2005). Wiener-kernel analysis of responses to noise of chinchilla auditory-nerve fibers. *Journal of Neurophysiology*, 93, 3615–3634.

Reed, C. M., Braida, L. D., & Zurek, P. M. (2009). Review of the literature on temporal resolution in listeners with cochlear hearing impairment: A critical assessment of the role of suprathreshold deficits. *Trends in Amplification*, 13, 4–43.

Relkin, E. M., & Doucet, J. R. (1997). Is loudness simply proportional to the auditory nerve spike count? *Journal of the Acoustical Society of America*, 101, 2735–2740.

Richardson, G. P., Lukashkin, A. N., & Russell, I. J. (2008). The tectorial membrane: One slice of a complex cochlear sandwich. *Current Opinion in Otlaryngology*, 16, 458–464.

Robertson, D. (1982). Effects of acoustic trauma on stereocilia structure and spiral ganglion cell tuning properties in the guinea pig cochlea. *Hearing Research*, 7, 55–74.

Robertson, D. (1983). Functional significance of dendritic swelling after loud sounds in the guinea pig cochlea. *Hearing Research*, 9, 263–278.

Robertson, D. (1984). Horseradish peroxidase injection of physiologically characterized afferent and efferent neurons in the guinea pig spiral ganglion. *Hearing Research*, 15, 113–121.

Robles, L., & Ruggero, M. A. (2001). Mechanics of the mammalian cochlea. *Physiological Reviews*, 81, 1305–1352.

Rosowski, J. J. (1994). Outer and middle ears. In R. R. Fay & A. N. Popper (Eds.), *Comparative hearing: Mammals* (pp. 172–247). New York: Springer-Verlag.

Rouiller, E. M., Cronin-Schreiber, R., Fekete, D. M., & Ryugo, D. K. (1986). The central projections of intracellularly labeled auditory nerve fibers in cats: An analysis of terminal morphology. *Journal of Comparative Neurology*, 249, 261–278.

Ruel, J., Wang, J., Rebillard, G., Eybalin, M., Lloyd, R., Pujol, R., & Puel, J. L. (2007). Physiology, pharmacology and plasticity at the inner hair cell synaptic complex. *Hearing Research*, 227, 19–27.

Ruggero, M. A., & Rich, N. C. (1991). Furosemide alters organ of Corti mechanics: Evidence for feedback of outer hair cells upon the basilar membrane. *Journal of Neuroscience*, 11, 1057 1067.

Ruggero, M. A., Robles, L., & Rich, N. C. (1992). Two-tone suppression in the basilar membrane of the cochlea: Mechanical basis of auditory-nerve rate suppression. *Journal of Neurophysiology*, 68, 1087–1099.

Ruggero, M. A., Rich, N. C., & Recio, A. (1996). The effect of intense acoustic stimulation on basilar-membrane vibrations. *Auditory Neuroscience*, 2, 329–345.

Ruggero, M. A., Rich, N. C., Recio, A., & Narayan, S. S. (1997). Basilar-membrane responses to tones at the base of the chinchilla cochlea. *Journal of the Acoustical Society of America*, 101, 2151–2163.

Runhaar, G., Schedler, J., & Manley, G. A. (1991). The potassium concentration in the cochlear fluids of thc embryonic and post-hatching chick. *Hearing Research*, 56, 227–238.

Sachs, M. B., & Abbas, P. J. (1974). Rate versus level functions for auditory-nerve fibers in cats: Tone-burst stimuli. *Journal of the Acoustical Society of America*, 56(6), 1835–1847.

Sachs, M. B., & Abbas, P. J. (1976). Phenomenological model for two-tone suppression. *Journal of the Acoustical Society of America*, 60, 1157–1163.

Sachs, M. B., & Young, E. D. (1979). Encoding of steady-state vowels in the auditory nerve: Representation in terms of discharge rate. *Journal of the Acoustical Society of America*, 66, 470–479.

Salt, A. N., Melichar, I., & Thalmann, R. (1987). Mechanisms of endocochlear potential generation by stria vascularis. *Laryngoscope*, 97, 984–991.

Salvi, R. J., Hamernik, R. P., & Henderson, D. (1983). Response patterns of auditory nerve fibers during temporary threshold shift. *Hearing Research*, 10, 37–67.

Salvi, R. J., Saunders, S. S., Gratton, M. A., Arehole, S., & Powers, N. (1990). Enhanced evoked response amplitudes in the inferior colliculus of the chinchilla following acoustic trauma. *Hearing Research*, 50, 245–257.

Sanes, D. H., & Constantine-Paton, M. (1985). The development of stimulus following in the cochlear nerve and inferior colliculus of the mouse. *Brain Research*, 354, 255–267.

Santarelli, R., Del Castillo, I., Rodriguez-Ballesteros, M., Scimemi, P., Cama, E., Arslan, E., & Starr, A. (2009). Abnormal cochlear potentials from deaf patients with mutations in the otoferlin gene. *Journal of the Association for Research in Otolaryngology*, 10, 545–556.

Saunders, J. C., Bock, G. R., James, R., & Chen, C. S. (1972). Effects of priming for audiogenic seizure on auditory evoked responses in the cochlear nucleus and inferior colliculus of BALB-c mice. *Experimental Neurology*, 37, 388–394.

Scheidt, R. E., Kale, S., & Heinz, M. G. (2010). Noise-induced hearing loss alters the temporal dynamics of auditory-nerve responses. *Hearing Research, (epub ahead of print).*

Schilling, J. R., Miller, R. L., Sachs, M. B., & Young, E. D. (1998). Frequency shaped amplification changes the neural representation of speech with noise-induced hearing loss. *Hearing Research*, 117, 57–70.

Schlauch, R. S., DiGiovanni, J. J., & Ries, D. T. (1998). Basilar membrane nonlinearity and loudness. *Journal of the Acoustical Society of America*, 103, 2010–2020.

Schmiedt, R. A. (1996). Effects of aging on potassium homeostasis and the endocochlear potential in the gerbil cochlea. *Hearing Research*, 102, 125–132.

Schmiedt, R. A., & Zwislocki, J. J. (1980). Effects of hair cell lesions on responses of cochlear nerve fibers. II. Single- and two-tone intensity functions in relation to tuning curves. *Journal of Neurophysiology*, 43, 1390–1405.

Schmiedt, R. A., Zwislocki, J. J., & Hamernik, R. P. (1980). Effects of hair cell lesions on responses of cochlear nerve fibers. I. Lesions, tuning curves, two-tone inhibition, and responses to trapezoidal-wave patterns. *Journal of Neurophysiology*, 43, 1367–1389.

Schmiedt, R. A., Lang, H., Okamura, H. O., & Schulte, B. A. (2002). Effects of furosemide applied chronically to the round window: A model of metabolic presbyacusis. *Journal of Neuroscience*, 22, 9643–9650.

Schoen, C. J., Emery, S. B., Thorne, M. C., Ammana, H. R., Sliwerska, E., Arnett, J., Hortsch, M., Hannan, F., Burmeister, M., & Lesperance, M. M. (2010). Increased activity of diaphanous homolog 3 (DIAPH3)/diaphanous causes hearing defects in humans with auditory neuropathy and in *Drosophila. Proceedings of the National Academy of Sciences of the USA*,, 107, 13396–13401.

Scholl, U. I., Choi, M., Liu, T., Ramackers, V. T., Hausler, M. G., Grimmer, J., Tobe, S. W., Farhi, A., Nelson-Williams, C., & Lifton, R. P. (2009). Seizures, sensorineural deafness, ataxia, mental retardation, and electrolyte imbalance (SeSAME syndrome) caused by mutations in KCNJ10. *Proceedings of the National Academy of Sciences of the USA*, 106, 5842–5847.

Schuknecht, H. F., & Gacek, M. R. (1993). Cochlear pathology in presbyacusis. *Annals of Otology Rhinology and Laryngology*, 102, 1–16.

Schulte, B. A., & Schmiedt, R. A. (1992). Lateral wall Na,K-ATPase and endocochlear potentials decline with age in quiet-reared gerbils. *Hearing Research*, 61, 35–46.

Sewell, W. F. (1984a). The effects of furosemide on the endocochlear potential and auditory-nerve fiber tuning curves in cats. *Hearing Research*, 14, 305–314.

Sewell, W. F. (1984b). Furosemide selectively reduces one component in rate-level functions from auditory-nerve fibers. *Hearing Research*, 15, 69–72.

Sewell, W. F. (1984c). The relation between the endocochlear potential and spontaneous activity in auditory nerve fibers of the cat. *Journal of Physiology-London*, 347, 685–696.

Shamma, S. A. (1985). Speech processing in the auditory system. I: The representation of speech sounds in the responses of the auditory nerve. *Journal of the Acoustical Society of America*, 78, 1612–1621.

Shannon, R. V., Zeng, F. G., Kamath, V., Wygonski, J., & Ekelid, M. (1995). Speech recognition with primarily temporal cues. *Science*, 270, 303–304.

Shera, C. A., Guinan, J. J., & Oxenham, A. J. (2010). Otoacoustic estimation of cochlear tuning: Validation in the chinchilla. *Journal of the Association for Research in Otolaryngology*, 11:343–65.

Shore, S. E., Koehler, S., Oldakowski, M., Hughes, L. F., & Syed, S. (2008). Dorsal cochlear nucleus responses to somatosensory stimulation are enhanced after noise-induced hearing loss. *European Journal of Neuroscience*, 27, 155–168.

Siebert, W. M. (1968). Stimulus transformations in the peripheral auditory system. In A. Kollers & M. Eden (Eds.), *Recognizing patterns* (pp. 104–133). Cambridge, MA: MIT Press.

Sinex, D. G., & Geisler, C. D. (1983). Responses of auditory-nerve fibers to consonant-vowel syllables. *Journal of the Acoustical Society of America*, 73, 602–615.

Sinex, D. G., & Geisler, C. D. (1984). Comparison of the responses of auditory nerve fibers to consonant-vowel syllables with predictions from linear models. *Journal of the Acoustical Society of America*, 76, 116–121.

Smith, Z. M., Delgutte, B., & Oxenham, A. J. (2002). Chimaeric sounds reveal dichotomies in auditory perception. *Nature*, 416.

Snyder, R. L., Rebscher, S. J., Cao, K. L., Leake, P. A., & Kelly, K. (1990). Chronic intracochlear electrical stimulation in the neonatally deafened cat. I: Expansion of central representation. *Hearing Research*, 50, 7–33.

Spicer, S. S., & Schulte, B. A. (1991). Differentiation of inner ear fibrocytes according to their ion transport related activity. *Hearing Research*, 56, 53–64.

Splawski, I., Timothy, K. W., Vincent, G. M., & Atkinson, D. L. (1997). Molecular basis of the long-QT syndrome associated with deafness. *New England Journal of Medicine*, 336, 1562–1567.

Spoendlin, H. (1971). Primary structural changes in the organ of Corti after acoustic overstimulation. *Acta Oto-Laryngologica*, 71(2), 166–176.

Starr, A., Picton, T. W., Sininger, Y., Hood, L. J., & Berlin, C. I. (1996). Auditory neuropathy. *Brain*, 119, 741–753.

Stevens, K. N. (1998). *Acoustic phonetics*. Cambridge, MA: MIT Press.

Stillman, J. A., Zwislocki, J. J., Zhang, M., & Cefaratti, L. K. (1993). Intensity just-noticeable differences at equal-loudness levels in normal and pathological ears. *Journal of the Acoustical Society of America*, 93, 425–434.

Summers, V., & Leek, M. R. (1994). The internal representation of spectral contrast in hearing-impaired listeners. *Journal of the Acoustical Society of America*, 95, 3518–3528.

Sumner, C. J., O'Mard, L. P., Lopez-Poveda, E. A., & Meddis, R. (2003). A nonlinear filter-bank model of the guinea-pig cochlear nerve: Rate response. *Journal of the Acoustical Society of America*, 113, 3264–3274.

Suneja, S. K., Potashner, S. J., & Benson, C. G. (1998). Plastic changes in glycine and GABA release and uptake in adult brain stem auditory nuclei after unilateral middle ear ossicle removal and cochlear ablation. *Experimental Neurology*, 151, 273–288.

Syka, J. (2002). Plastic changes in the central auditory system after hearing loss, restoration of function, and during learning. *Physiological Reviews*, 82, 601–636.

Syka, J., Rybalko, N., & Popelar, J. (1994). Enhancement of the auditory cortex evoked responses in awake guinea pigs after noise exposure. *Hearing Research*, 78, 158–168.

Szczepaniak, W. S., & Moller, A. R. (1996). Evidence of neuronal plasticity within the inferior colliculus after noise exposure: A study of evoked potentials in the rat. *Electroencephalography and Clinical Neurophysiology*, 100, 158–164.

Taberner, A. M., & Liberman, A. M. (2005). Response properties of single auditory nerve fibers in the mouse. *Journal of Neurophysiology*, 93, 557–569.

Takesian, A. E., Kotak, V. C., & Sanes, D. H. (2009). Developmental hearing loss disrupts synaptic inhibition: Implications for auditory processing. *Future Neurology*, 4, 331–349.

Takeuchi, S., Ando, M., & Kakigi, A. (2000). Mechanisms generating endocochlear potential: Role played by intermediate cells in stria vascularis. *Biophysical Journal*, 79, 2572–2582.

Temchin, A. N., Rich, N. C., & Ruggero, M. A. (2008). Threshold tuning curves of chinchilla auditory-nerve fibers. I. Dependence on characteristic frequency and relation to the magnitudes of cochlear vibration. *Journal of Neurophysiology*, 100, 2889–2898.

ter Keurs, M., Festen, J. M., & Plomp, R. (1992). Effect of spectral envelope smearing on speech reception. I. *Journal of the Acoustical Society of America*, 91, 2872–2880.

Trautwein, P., Hofstetter, P., Wang, J., Salvi, R., & Nostrant, A. (1996). Selective inner hair cell loss does not alter distortion product otoacoustic emissions. *Hearing Research*, *96*, 71–82.

Turner, C. W. (2006). Hearing loss and the limits of amplification. *Audiol and Neuro-Otology*, 11(Suppl. 1), 2–5.

Turrigiano, G. G. (2008). The self-tuning neuron: Synaptic scaling of excitatory synapses. *Cell*, 135, 422–435.

Tyler, R. S., Hall, J. W., Glasberg, B. R., Moore, B. C., & Patterson, R. D. (1984). Auditory filter asymmetry in the hearing impaired. *Journal of the Acoustical Society of America*, 76, 1363–1368.

Vale, C., & Sanes, D. H. (2002). The effect of bilateral deafness on excitatory and inhibitory synaptic strength in the inferior colliculus. *European Journal of Neuroscience*, 16, 2394–2404.

Vale, C., Juiz, J. M., Moore, D. R., & Sanes, D. H. (2004). Unilateral cochlear ablation produces greater loss of inhibition in the contralateral inferior colliculus. *European Journal of Neuroscience*, 20, 2133–2140.

Vetter, D. E., Mann, J. R., Wangemann, P., Liu, J., McLaughlin, K. J., Lesage, F., Marcus, D. C., Lazdunski, M., Heinemann, S. F., & Barhanin, J. (1996). Inner ear defects induced by null mutation of the isk gene. *Neuron*, 17, 1251–1264.

Vinay, & Moore, B. C. J. (2007). Prevalence of dead regions in subjects with sensorineural hearing loss. *Ear and Hearing*, 28, 231–241.

Walton, J. P. (2010). Timing is everything: Temporal processing deficits in the aged auditory brainstem. *Hearing Research*, 264, 63–69.

Wang, J., Powers, N. L., Hofstetter, P., Trautwein, P., Ding, D., & Salvi, R. (1997). Effects of selective inner hair cell loss on auditory nerve fiber threshold, tuning and spontaneous and driven discharge rate. *Hearing Research*, 107, 67–82.

Wang, J., Ding, D., & Salvi, R. J. (2002). Functional reorganization in chinchilla inferior colliculus associated with chronic and acute cochlear damage. *Hearing Research*, *168*, 238–249.

Wangemann, P. (1995). Comparison of ion transport mechanisms between vestibular dark cells and strial marginal cells. *Hearing Research*, 90, 149–157.

Wangemann, P., & Schacht, J. (1996). Homeostatic mechanisms in the cochlea. In P. Dallos, A. N. Popper, & R. R. Fay (Eds.), *The cochlea* (pp. 130–185). New York: Springer.

Warr, W. B. (1992). Organization of olivocochlear efferent systems in mammals. In D. B. Webster, A. N. Popper, & R. R. Fay (Eds.), *The mammalian auditory pathway: Neuroanatomy* (pp. 410–448). New York: Springer-Verlag.

Weisz, C., Glowatzki, E., & Fuchs, P. (2009). The postsynaptic function of type II cochlear afferents. *Nature*, 461, 1126–1129.

Wightman, F. L., & Kistler, D. J. (1992). The dominant role of low-frequency interaural time differences in sound localization. *Journal of the Acoustical Society of America*, 91, 1648–1661.

Winter, I. M., Robertson, D., & Yates, G. K. (1990). Diversity of characteristic frequency rate-intensity functions in guinea pig auditory nerve fibres. *Hearing Research*, 45, 191–202.

Wong, J. C., Miller, R. L., Calhoun, B. M., Sachs, M. B., & Young, E. D. (1998). Effects of high sound levels on responses to the vowel/ε/in cat auditory nerve. *Hearing Research*, 123, 61–77.

Wu, T., & Marcus, D. C. (2003). Age-related changes in cochlear endolymphatic potassium and potential in CD-1 and CBA/CaJ mice. *Journal of the Association for Research in Otolaryngology*, 4, 353–362.

Yates, G. K., Winter, I. M., & Robertson, D. (1990). Basilar membrane nonlinearity determines auditory nerve rate-intensity functions and cochlear dynamic range. *Hearing Research*, 45, 203–220.

Young, E. D. (2008). Neural representation of spectral and temporal information in speech. *Philosophical Transactions of the Royal Society B*, 363, 923–945. doi: doi:10.1098/rstb.2007.2151

Young, E. D., & Oertel, D. (2004). The cochlear nucleus. In G. M. Shepherd (Ed.), *The synaptic organization of the brain* (5th Ed., pp. 125–163). New York: Oxford University Press.

Young, E. D., & Sachs, M. B. (1979). Representation of steady-state vowels in the temporal aspects of the discharge patterns of populations of auditory-nerve fibers. *Journal of the Acoustical Society of America*, 66, 1381–1403.

Zdebik, A. A., Wangemann, P., & Jentsch, T. J. (2009). Potassium ion movement in the inner ear: Insights from genetic disease and mouse models. *Physiology*, 24, 307–316.

Zeng, F. G., & Turner, C. W. (1991). Binaural loudness matches in unilaterally impaired listeners. *Quarterly Journal of Experimental Psychology*, 43A, 565–583.

Zeng, F. G., Kong, Y. Y., Michalewski, H. J., & Starr, A. (2005). Perceptual consequences of disrupted auditory nerve activity. *Journal of Neurophysiology*, 93, 3050–3063.

Zhang, X., Heinz, M. G., Bruce, I. C., & Carney, L. H. (2001). A phenomenological model for the responses of auditory-nerve fibers: I. Nonlinear tuning with compression and suppression. *Journal of the Acoustical Society of America*, 109, 648–670.

Zheng, X. Y., Henderson, D., Hu, B. H., & McFadden, S. L. (1997). Recovery of structure and function of inner ear afferent synapses following kainic acid excitotoxicity. *Hearing Research*, 105, 65–76.

Zheng, J., Shen, W., He, D. Z. Z., Long, K. B., Madison, L. D., & Dallos, P. (2000). Prestin is the motor protein of cochlear outer hair cells. *Nature*, 405, 149–155.

Zidanic, M., & Brownell, W. E. (1990). Fine structure of the intracochlear potential field. I. The silent current. *Biophysical Journal*, 57, 1253–1268.

Zilany, M. S. A., & Bruce, I. C. (2006). Modeling auditory-nerve responses for high sound pressure levels in the normal and impaired auditory periphery. *Journal of the Acoustical Society of America*, 120, 1446–1466.

Zilany, M. S. A., & Bruce, I. C. (2007). Representation of the vowel/eh/in normal and impaired auditory nerve fibers: Model predictions of responses in cats. *Journal of the Acoustical Society of America*, 122, 402–417.

Chapter 7
Suprathreshold Auditory Processing in Noise-Induced Hearing Loss

Mini N. Shrivastav

1 Introduction

This chapter reviews auditory processing changes that occur in individuals with sensorineural hearing loss due to noise exposure. The typical patient with noise-induced hearing loss (NIHL) initially presents with a high-frequency sensorineural hearing loss that spreads to other frequencies as the noise exposure continues. Studies on various animal models have shown that the configuration of hearing loss and subsequent recovery depend on the intensity, frequency, and duration of exposure (see Clark 1991 and Salvi and Boettcher 2008, for review). Briefly stated, the sensorineural hearing loss resulting from noise exposure seems to be restricted mostly to the frequency of exposure for low-intensity [83 dB sound pressure level (SPL), Salvi et al. 1978] stimulation, while high-intensity (95 dB SPL) stimulation results in hearing loss at frequencies one half to an octave above the exposure frequency. The half-octave shift has been well described in the seminal work of Davis et al. (1950). For long-term exposures, the amount of temporary threshold shift (TTS) typically increases linearly over the first 24 h of exposure and then plateaus to a level referred to as asymptotic threshold shift (ATS). Recovery after noise exposure is also dependent on the intensity and duration of exposure. Recovery tends to be slowest for audiometric frequencies around 4,000 Hz, regardless of what the frequency of exposure is (Davis et al. 1950). In general, small amounts of ATS due to low-intensity exposures are associated with faster recovery, whereas large amounts of ATS (greater than around 55 dB) due to high-intensity exposures are associated with slower recovery and in some cases, permanent threshold shifts (PTS; Clark 1991). As the auditory system sustains ATS for weeks and months, the amount of

M.N. Shrivastav (✉)
Department of Speech, Language, and Hearing Sciences,
University of Florida, Gainesville, FL 32611, USA
e-mail: mnshriv@ufl.edu

C.G. Le Prell et al. (eds.), *Noise-Induced Hearing Loss: Scientific Advances*,
Springer Handbook of Auditory Research 40, DOI 10.1007/978-1-4419-9523-0_7,

recovery at the eventual cessation of the exposure decreases. At the extreme, ATS is a good predictor of PTS for prolonged exposures.

Although the audiometric configuration in patients with NIHL is well documented, there is less information on the nature of suprathreshold auditory processing deficits in these individuals. A large body of literature on the nature of suprathreshold auditory processing deficits in sensorineural hearing loss, specifically of cochlear origin, does exist. However, many of these studies either fail to mention the specific etiology of the hearing loss, or combine listeners with NIHL along with hearing-impaired individuals with other etiologies into one listener group. Nevertheless, the body of literature on sensorineural hearing loss does provide a model for NIHL. Collectively, these studies indicate that the listening difficulties faced by an individual with high-frequency sensorineural hearing loss are much more than just the loss of audibility resulting from their higher hearing thresholds. Rather, in many cases, it is a pervasive and life-changing difficulty in hearing everyday sounds, the most important of which is speech. These difficulties are exacerbated in less-than-ideal listening conditions such as the presence of background noise or reverberation.

A classic example of a hearing disorder that results in significant listening difficulties is that of age-related hearing loss (ARHL). Some may consider ARHL a vascular, metabolic, and/or neural disorder rather than a true sensorineural hearing disorder of cochlear origin (Mills et al. 1998) due to the questionable involvement of the sensory cells in the cochlea. Nevertheless, for several reasons, ARHL may serve as a good model for describing suprathreshold auditory deficits in cochlear hearing losses such as NIHL. First, although the initial stages of NIHL are characterized by a notch around 4,000 Hz, a patient with a long history of noise exposure can present with a high-frequency sensorineural hearing loss, not unlike a patient with ARHL. Second, and more importantly, age-related changes in suprathreshold auditory processing have been studied extensively, especially when it comes to psychoacoustic abilities such as frequency and temporal resolution. The ARHL literature is not conclusive for some aspects of auditory processing, particularly those that involve complex processing abilities such as across-channel comparisons. In these cases, the few studies that have included hearing-impaired listeners (e.g., Grose and Hall 1996) have tended to use single small groups of listeners that span a wide age range and in which the etiology of hearing loss is often not identified.

The study of changes in suprathreshold auditory processing abilities in NIHL is important for multiple reasons. First, as has been well documented in the ARHL literature, a typical listener with high-frequency sensorineural hearing loss experiences speech understanding difficulties that are often disproportionate to their hearing loss. These patients often experience poor audibility of high-frequency signals such as the consonants in speech. However, compensating for the reduced audibility with appropriate amplification very often does not result in improved speech understanding, especially in noisy conditions. A common complaint of these patients is that "I can hear it, but I can't understand it." In other words, the processing difficulties of these individuals cannot be explained by a simple decrease in audibility. In an attempt to understand the underlying factors related to this phenomenon, a large body of research on suprathreshold changes in hearing in these patients has been compiled over several decades (see Fitzgibbons and Gordon-Salant 2010 for a review). This research has ranged from studies dealing with simple frequency and temporal resolution to those on complex processing abilities involving multiple frequency channels. Findings include broadened auditory filters and hence poor frequency resolution (e.g., Patterson et al. 1982; Peters and Moore 1992) and poor temporal resolution as measured by increased gap detection thresholds (Schneider et al. 1994), and poor duration discrimination (Abel et al. 1990; Fitzgibbons and Gordon-Salant 1994). However, there is evidence that some of these changes may be related to reduced audibility rather than advancing age (e.g., Sommers and Humes 1993). Also, age-related deficits seem to be more evident in temporal processing tasks involving complex rather than simple stimuli (Humes and Christopherson 1991; Fitzgibbons and Gordon-Salant 1995).

It has been the goal of several studies to determine if the speech understanding deficits found in sensorineural hearing loss are related to these auditory processing abilities and other factors such as cognitive abilities (see CHABA 1988 and Humes and Dubno 2010 for reviews). There is evidence that age-related factors such as poor temporal processing are associated with poorer speech recognition in speech temporally altered by compression or reverberation. Some of these factors also come into play when speech perception becomes complicated due to multiple talkers, low contextual cues, and greater memory load (Gordon-Salant and Fitzgibbons 1995a, b, 1997). However, in general, cognitive abilities and most simple psychoacoustic tasks that involve the detection of a subtle change in spectral or temporal characteristics of non-speech stimuli fail to show any association with speech recognition, especially when background noise is present. This is not surprising because the detection and discrimination of simple features in isolation is not very similar to speech recognition in complex listening situations. Hence, more recently, the focus has shifted to more global and complex processing abilities such as spectrotemporal processing involving simultaneous processing across multiple frequency channels. There is evidence that deficits in some of these complex processing tasks may be related to the speech perception problems encountered in sensorineural hearing loss. For example, at least one study has found a significant association between the ability to detect changes in the spectral shape of broadband noise stimuli (an ability that presumably involves across-channel integration of intensity) and nonsense syllable identification among older hearing-impaired listeners for whom audibility was ensured (Shrivastav et al. 2006). Similar associations have been observed by Litvak et al. (2007) for spectral modulation detection thresholds and vowel- and consonant-identification in cochlear implantees.

This chapter focuses on the various suprathreshold changes that occur in cases of NIHL. As far as possible, this chapter attempts to focus on studies in which the etiology of hearing loss of the subjects has been clearly established as NIHL. However, in some cases, owing to the paucity of studies in which the specific etiology of cochlear hearing loss is identified, the approach of this chapter is to use ARHL or general sensorineural hearing loss as a model for NIHL. Changes in basic aspects of auditory processing, such as frequency resolution, temporal resolution, and loudness perception, are reviewed first. This is followed by a discussion of changes in more complex auditory processing abilities such as across-channel spectrotemporal processing.

2 Frequency Resolution

Frequency resolution or selectivity is a measure of the ear's ability to extract the signal of interest from the surrounding background noise. Frequency selectivity is often measured by estimating the characteristics of the internal auditory filter for a given signal frequency using psychophysical tuning curves (see Moore 2003, for an extensive review). A broadened auditory filter would make the signal harder to detect in the presence of more background noise. Several studies have indicated that listeners with NIHL have greater susceptibility to masking, a finding that is related to reduced frequency resolution in these individuals. Zwicker and Schorn (1978) measured psychoacoustical tuning curves in a group of normal ears and in several pathological ears, including those with NIHL. Tuning curves were generated by determining thresholds in quiet for pure tone signals and then determining the level of a pure-tone masker that was just enough to mask the test tone (for maskers of different frequencies). The signal frequency was either 500 or 4,000 Hz, and the masker frequencies encompassed a range surrounding the signal frequency. For the 500-Hz signal, ears with normal hearing and ears with NIHL showed similar tuning curves. However, for the 4,000-Hz signal, tuning curves for the noise-damaged ears at first appeared to be flattened (less sharp) when compared to those with normal hearing. The flattening, however, disappeared upon taking into two factors, especially for hearing losses greater than 55 dB. First, when the tuning curves for the NIHL group were compared to those for normal-hearing listeners at *higher* intensity levels, the two sets of curves looked similar. This indicated that the initially observed flattening probably reflected the effects of recruitment. Second, when listeners were trained to ignore difference tones between the signal and the masker and to focus only on the test tones, the slope of the high-frequency tail of the tuning curve increased markedly. For these two reasons, Zwicker and Schorn (1978) suggested that tuning curve data for ears with higher levels of NIHL should be interpreted with caution. In later work, Schorn and Zwicker (1990) generated psychoacoustical tuning curves and estimates of frequency resolution for a group of normal ears and ears with NIHL in quiet and in background noise, for 500-Hz and 4,000-Hz signal frequencies. For the normal ears, the frequency resolution factor (FRF; see Schorn and Zwicker 1990, for details of measurement) in quiet and noisy conditions were 1 and 0.79, respectively, showing that frequency resolution is slightly impaired in conditions of background noise. For the group with NIHL, frequency resolution for the 500-Hz signal was only slightly impaired in quiet (FRF = 0.75), while that for the 4,000-Hz signal was more severely reduced (FRF = 0.54). In background noise, subjects with NIHL showed more impairment than the normal hearing controls (FRF of 0.54 and 0.28, respectively, for the 500-Hz and 4,000-Hz signals). Similar curves measured in chinchillas for a 2,000-Hz signal (Salvi et al. 1983) showed that normal ears were characterized by psychophysical tuning curves with a low-threshold, narrowly tuned V-shaped tip centered at the signal frequency and a high-threshold, broadly tuned tail region. Ears with approximately 30 dB of NIHL showed tuning curves with high-threshold broadly tuned tips connected to high-threshold tail regions.

Taken together, it seems that even a moderate amount of NIHL, especially at high frequencies, results in significant reductions in frequency resolution in ears with NIHL. It is important to keep in mind that the measurement of frequency resolution using masking is limited in higher degrees of NIHL due to the necessity for higher masker levels and the probability of detecting difference tones, the latter being true for simultaneous masking paradigms. The reader is referred to Tyler (1986) for an excellent review of various earlier and alternate methods to estimate frequency resolution in a variety of hearing impairments including NIHL.

3 Temporal Resolution

Temporal resolution can be measured in a variety of ways including gap detection, difference limens for signal duration and gap duration, temporal modulation transfer functions, temporal order judgment, temporal integration, voice onset times (VOT) for consonants, and masking period patterns. Animal experiments have shown that gap detection thresholds are affected by noise exposure. For example, Giraudi et al. (1980, 1982) showed that chinchillas that had significant noise-induced ATS had higher gap detection thresholds than their normal hearing counterparts, especially for low-level background noises. Tyler et al. (1982) measured temporal integration, gap detection thresholds, duration difference limens, and gap difference limens for a group of normal-hearing listeners and those with cochlear hearing loss of various etiologies including NIHL. In general, the hearing-impaired listeners showed poorer temporal resolution, regardless of whether they and the normal-hearing listeners were compared at equal sensation levels or equal sound pressure levels. Zwicker and Schorn (1982) used masking period patterns to measure the temporal resolution factor (TRF; see Zwicker and Schorn 1982 for details of measurement) of normal ears and those with various pathologies. Briefly stated, the TRF involves the measurement of threshold for a steady-state pure tone in quiet and in modulated and unmodulated noise. The signal threshold is plotted as a function of the temporal position of the tone within the period of the masker. The TRF is then computed as the ratio of the thresholds in the three different conditions and represents a measure of temporal resolution. Zwicker and Schorn (1982) found that a group of listeners with NIHL had a markedly low TRF (0.2) for a 4,000-Hz signal when compared to normal hearing listeners (TRF = 1). Listeners with NIHL showed nearly normal TRFs for 500-Hz and 1,500-Hz signals.

Schorn and Zwicker (1990) measured temporal resolution using masking period patterns for a group of normal hearing listeners and for those with NIHL with and without the presence of background noise for 500-Hz and 4,000-Hz signals. The normal-hearing listeners' TRF increased from 1.0 in quiet to 3.0 in background noise for both signal frequencies, indicating better temporal resolution abilities in the presence of background noise than in quiet. For the listeners with NIHL, TRFs were similar to those for normal-hearing listeners for the 500-Hz signal frequency. However, for the 4,000-Hz signal frequency, listeners with NIHL showed poorer temporal resolution than their normal-hearing counterparts (TRF = 0.5 in quiet and

1.25 in background noise). These results indicate that for higher frequencies at least, listeners with NIHL do not show the improvement in temporal resolution in background noise that normal-hearing listeners exhibit.

There is evidence that noise exposure can affect temporal processing at cortical levels. For example, Tomita et al. (2004) showed that acute acoustic trauma resulting from a 1 h exposure to a high frequency (5 or 6 kHz) tone presented at 120 dB SPL resulted in poor representation of VOT and gap-duration in the primary auditory cortex in cats. Animal studies have also suggested that noise exposure at early ages may adversely affect temporal processing in adulthood. Aizawa and Eggermont (2006) exposed 6-week-old kittens to a 120 dB SPL one-third octave band noise centered at 5,000 Hz for 2 h. When the animals were at least 22 weeks old, cortical neural responses, including responses to VOT and gap duration, were measured. Results indicated that both the minimum detectable VOT and gap duration were increased in the adult cats that had been exposed to noise in early life.

While comparing the temporal resolution abilities in those with cochlear hearing loss and normal hearing, it is important to consider factors such as the presentation level, the audible bandwidth available to the listener, and the effect of widened auditory filters. For example, those with cochlear hearing loss in general may perform worse than their normal hearing counterparts at listening tasks involving low presentation levels, but equally well or better than normal-hearing listeners at higher presentation levels or when the audible bandwidth is controlled for. More recently, the focus has shifted to more global and complex processing abilities such as spectrotemporal processing involving simultaneous processing across multiple frequency channels. Readers are referred to Moore (1996) and Rawool (2006) for reviews of temporal resolution in cochlear hearing loss.

4 Loudness Perception

It has long been recognized that individuals with cochlear hearing loss have abnormal loudness perception (Fowler 1936; Moore 2004). These individuals often experience loudness recruitment, an effect that has traditionally been described as an abnormally rapid growth of loudness at suprathreshold levels (Moore 1998), or more recently, as an "abnormally large loudness at an elevated threshold" (Buus and Florentine 2002). In humans, most studies that have described recruitment have included heterogeneous listener groups with varying etiologies of cochlear hearing loss. Burke and Creston (1966) demonstrated recruitment using Bekesy audiometry in a group of 22 human listeners with noise exposure as the cause of hearing loss. Several animal experiments have reported various behavioral and physiological measures reflecting recruitment after TTS or PTS was induced in the subjects during the experimental protocol. For example, Pugh et al. (1979) induced NIHL in macaque monkeys by exposing them to an hour-long 108 dB SPL octave band at noise centered at 8,000 Hz. They measured both behavioral [using reaction time, based on the work of Moody (1970), as an indirect measure of perceived loudness] and electrical (using whole nerve action potentials) metrics of recruitment and found good agreement between these measures. Recruitment was demonstrated in the monkeys immediately after the noise exposure and during the subsequent recovery period. Monkeys subjected to a PTS resulting from exposure to 160 h of 8,000 Hz octave band noise at 118 dB SPL also demonstrated recruitment using both metrics. Similarly, May et al. (2009) used reaction time measurements to demonstrate behavioral loudness recruitment in domestic cats that were exposed to a 109 dB SPL 50-Hz bandwidth centered at 2,000 Hz for 4 h. They generated equal latency contours that were similar to human equal loudness contours (Fletcher and Munson 1933), especially when differences in absolute sensitivity and audible frequency range between humans and cats were considered.

The physiological correlate of loudness recruitment would be an abnormal growth of response of neurons with increasing sound level (Joris 2009). Until recently, there was a lack of agreement about the lowest level in the auditory pathway at which this pattern is observed. For example, at the level of the auditory nerve, contradictory to what one would expect to find, noise exposure and subsequent acoustic trauma is associated with reduced auditory nerve responses and *slower* growth of discharge rate with increasing level (Heinz et al. 2005). Recently, Cai et al. (2008) measured the response growth of ventral cochlear nucleus (VCN) neurons in cats exposed to a 111–112 dB SPL 50-Hz bandwidth noise centered at 2,000 Hz for 4 h. They found that neurons with chopper type of responses showed abnormally rapid growth with increasing sound levels. This suggests that the VCN may be the lowest level at which the physiological correlate of recruitment may be observed.

5 Spectrotemporal Processing

According to the traditional power spectrum model of masking (Fletcher 1940), signal detection in the presence of a noise masker is based on the output of a single auditory filter. In other words, masker components whose frequencies fall beyond a certain bandwidth, or critical band, do not influence the detectability of the signal. However, the results of many subsequent experiments involving stimuli spanning large bandwidths argue that simultaneous comparisons of distant auditory filters can actually enhance the detectability of signals. The ability to carry out such across-channel comparisons has been tested using a variety of paradigms including profile analysis (Green 1988), comodulation masking release (e.g., Hall et al. 1984), and modulation detection interference (MDI, Yost and Sheft 1989).

Profile analysis involves the measurement of intensity discrimination thresholds for a signal added to one component of a multitonal complex, with the components surrounding the signal component acting as maskers. Green et al. (1983) studied the effect of component spacing on profile analysis. They found that thresholds decreased with increasing component spacing, and beyond a certain frequency separation, for increasing number of components. These results suggested that profile analysis, rather than being a local phenomenon, is based on simultaneous comparisons across a wide range spanning many critical bands.

Comodulation masking release (CMR) experiments demonstrate that listeners' detection of a signal actually *improves* in the presence of a distant comodulated masker compared to conditions when the masker is present but unmodulated. In these experiments, listeners are asked to detect a signal such as a pure tone in the presence of a masker, typically a narrowband noise centered in frequency around the signal. The signal threshold is measured as a function of the masker bandwidth. The noise masker can either be random, with no coherence in the peaks and dips across the bandwidth of the noise, or "comodulated," such that there is a degree of coherence in the peaks and dips of the noise. When the masker is random noise, the results are similar to those of Fletcher (1940), with the masked signal threshold initially increasing with masker bandwidth until the critical bandwidth is reached and then remaining constant. However, when the masker is comodulated, the signal threshold increases initially, but *decreases* as the masker bandwidth exceeds the critical bandwidth. Increases in the bandwidth of this comodulated masker actually result in further improvements in detection, suggesting that comparing across multiple frequency channels is advantageous to the listener. The decrease in the signal threshold is a measure of the comodulation masking release, and suggests that listeners are able to make use of the synchronous fluctuations of distant frequencies in the noise to improve the detection of the signal. However, other experimental paradigms such as MDI (Yost and Sheft 1989) seem to suggest that across-channel comparisons may not always be advantageous to the listener. In these experiments, listeners are asked to detect amplitude modulation (AM) of a signal in the presence of a masker distant in frequency from the signal. The masker can either be modulated or unmodulated. When the masker is unmodulated, there is no effect on the AM detection threshold of the signal. However, when the masker is modulated, the AM detection threshold of the signal *increases*, suggesting that masker modulation interferes with the detection of signal modulation, particularly when the masker and the signal are modulated at the same rate. Introduction of asynchrony in the modulation rates of the signal and interfering components markedly reduces the amount of interference in the AM detection of the signal.

The ability to perform these simultaneous across-channel comparisons may be particularly relevant for speech understanding, which often involves spectral and temporal changes that span wide frequency ranges that are well beyond a single frequency channel. Hence, it is likely that speech perception changes found in cochlear hearing loss are related to changes in the ability to perform these across-channel simultaneous comparisons, especially when listening to speech in fluctuating background noise.

There have been relatively few studies on various aspects of spectrotemporal processing in hearing-impaired listeners, particularly those focusing on ARHL or NIHL. Most studies that have included listeners with hearing loss have tended to not specify the etiology of hearing loss. In paradigms using profile analysis, it is not clear that listeners with cochlear hearing loss perform any worse than normal-hearing individuals (Lentz and Leek 2002, 2003). However, these studies indicate that hearing-impaired listeners may be using different strategies, such as placing more weight on frequencies close to their hearing loss, than normal-hearing listeners do when trying to detect the signal. The presence of large individual differences in both normal-hearing and hearing-impaired listeners also makes it hard to draw any definite conclusions from these studies. Data from experiments involving amplitude modulation detection are also similarly inconclusive. There is evidence that hearing-impaired listeners with sloping hearing loss have greater MDI when compared to normal-hearing listeners (Opie and Bacon 1993), particularly when the signal frequency is in the region of the hearing loss. However, listeners with flat hearing loss configurations (Grose and Hall 1994) and those tested using envelope comparison interference (Grose and Hall 1996) did not seem to perform differently than normal-hearing listeners. Similarly, Hall et al. (2008) reported no differences between younger normal-hearing listeners and hearing-impaired adults with mild-to-moderate sensorineural hearing loss for spectral integration of speech bands.

In contrast with the data just reviewed, experiments using CMR have repeatedly shown that hearing-impaired listeners tend to show less release from masking, or reduced magnitude of CMR, than their normal-hearing counterparts (Hall et al. 1988; Hall and Grose 1989). The results of several experiments suggest that the reduced CMR in cochlear hearing-impairment is related to reduced frequency selectivity in these individuals (Hall et al. 1988; Hall and Grose 1989; Moore et al. 1993). Recent evidence also suggests that reduced CMR in cochlear hearing loss may be related to more central processes such as a lack of two-tone suppression (Ernst et al. 2010).

There are very little published data on the nature of across-channel processing abilities in individuals whose hearing loss has been specifically attributed to noise exposure. Hall et al. (1996) reviewed data on CMR and monaural envelope correlation perception on a group of three normal hearing listeners and four listeners with NIHL. For the CMR experiment, they included conditions in which the signal was centered at regions with relatively normal (1,000 Hz) and poor hearing (2,000 Hz). In the case of the former, they used either a 20-Hz-wide noise centered at 1,000 Hz [on-signal band (OSB)], or the OSB plus two 20-Hz-wide comodulated flanking bands centered at 800 and 1,200 Hz. The CMR would be the difference in detection thresholds between the OSB only and the OSB plus comodulation bands conditions. Although the listeners with NIHL showed higher thresholds in both conditions, both groups of listeners showed substantial amounts of CMR (13 dB and 9–11 dB for the normal-hearing and NIHL listeners, respectively). For the region with poor hearing sensitivity (2,000 Hz), Hall et al. (1996) used three different test conditions (OSB only, OSB plus flanking bands that were spaced closer in frequency, and OSB plus flanking bands that were spaced widely apart). The latter two conditions were included to determine the role of frequency selectivity in the performance of the listeners with NIHL. Listeners with normal-hearing showed 9 and 13 dB masking releases for the wide and narrow frequency separation conditions, respectively. Large individual differences were apparent within the listeners with NIHL. While one of the listeners had a CMR of only 2 dB, the remaining three showed CMR in the range of 8 dB for the condition with wide frequency separation. For the narrow frequency separation condition, one of the listeners had a CMR of 9 dB, while the remaining three showed poor release from masking with CMR in the range of 3–7 dB. Hall et al. (1996) concluded that the poorer performance of the listeners with NIHL for the narrow frequency separation condition was probably related to the poor frequency selectivity in these listeners.

Hall et al. (1996) also reported data on monaural envelope correlation perception from the same four listeners with NIHL. They measured the ability of the listeners to detect correlations between the envelopes of noise stimuli centered at either 1,000 or 2,000 Hz. For the 2,000-Hz condition, they also investigated the effect of the frequency spacing of the noise bands. They found that regardless of what the center frequency was, the listeners with NIHL performed much worse (in some cases at chance level) than their normal-hearing controls. These results suggest that the ability to detect temporal modulation across distant frequency channels may be negatively affected by the presence of NIHL.

6 Summary

It is apparent that hearing problems encountered by individuals with NIHL are not limited to their reduced audibility at selected frequency regions. It is accompanied by deficits in both simple and complex suprathreshold processing, many of which are not attenuated even when factors such as reduced sensation level are accounted for. Owing to the paucity of studies specifically studying individuals with NIHL, there is little information on how these deficits are influenced by factors such as the duration of exposure and amount of TTS/PTS. Further, recent evidence suggests that monitoring simple detection thresholds as a metric of recovery from TTS does not capture the underlying physiological effects of noise exposure. Specifically, Kujawa and Liberman (2009) have shown that TTS in the mouse resulted in acute and progressive physiological changes such as loss of afferent nerve terminals and cochlear nerve degeneration that continued even after detection thresholds have reverted back to normal. This finding highlights the importance of investigating how simple and complex suprathreshold processing changes in NIHL.

Few studies have systematically investigated the association of suprathreshold processing abilities to the speech understanding deficits encountered in NIHL. However, the literature on ARHL suggests that a strong correlation between speech understanding deficits and deficits in basic auditory processing abilities such as frequency and temporal resolution is unlikely, especially if the audibility of the entire speech spectrum is ensured either by selecting subjects whose thresholds fall within predetermined levels or by shaping the stimuli using appropriate amplification. This is especially true because most of the suprathreshold processing deficits in individuals with NIHL seem to be restricted to a narrow frequency range around the region of hearing loss. Recent research on individuals with ARHL has suggested that some of these speech understanding deficits may be linked to underlying problems with complex suprathreshold processing abilities such as spectral modulation detection. These promising findings suggest that similar explanations for the speech understanding deficits in NIHL may be linked to complex suprathreshold abilities such as across-channel processing.

Acknowledgments The author thanks Caleb Williams and Ethan Levien for invaluable help with the preparation of this manuscript.

References

Abel, S. M., Krever, E. M., & Alberti, P. W. (1990). Auditory detection, discrimination and speech processing in ageing, noise-sensitive and hearing-impaired listeners. *Scandinavian Audiology*, 19(1), 43–54.

Aizawa, N., & Eggermont, J. J. (2006). Effects of noise-induced hearing loss at young age on voice onset time and gap-in-noise representations in adult cat primary auditory cortex. *Journal of the Association for Research in Otolaryngology*, 7(1), 71–81.

Burke, K. S., & Creston, J. E. (1966). Recruitment in noise-induced hearing loss. *Acta Oto-Laryngologica*, 62(1–6), 351–361.

Buus, S., & Florentine, M. (2002). Growth of loudness in listeners with cochlear hearing losses: Recruitment reconsidered. *Journal of the Association for Research in Otolaryngology*, 3(2), 120–139.

Cai, S., Ma, W. D., & Young, E. D. (2008). Encoding intensity in ventral cochlear nucleus following acoustic trauma: Implications for loudness recruitment. *Journal of the Association for Research in Otolaryngology*, 10, 5–22.

Committee on Hearing, Bioacoustics, and Biomechanics (CHABA). (1988). Speech understanding and aging. *Journal of the Acoustical Society of America*, 83, 859–895.

Clark, W. W. (1991). Recent studies of temporary threshold shift (TTS) and permanent threshold shift (PTS) in animals. *Journal of the Acoustical Society of America*, 90(1), 155–163.

Davis, H., Morgan, C. T., Hawkins, J. E., Galambos, R. & Smith, F. W. (1950). Final report on temporary deafness following exposure to loud tones and noise. *Acta Oto-Laryngologica*, 88(Supplement), 1–57.

Ernst, S. M., Rennies, J., Kollmeier, B., & Verhey, J. L. (2010). Suppression and comodulation masking release in normal-hearing and hearing-impaired listeners. *Journal of the Acoustical Society of America*, 128(1), 300–309.

Fitzgibbons, P. J., & Gordon-Salant, S. (1994). Age effects on measures of auditory duration discrimination. *Journal of Speech and Hearing Research*, 37(3), 662–670.

Fitzgibbons, P. J., & Gordon-Salant, S. (1995). Age effects on duration discrimination with simple and complex stimuli. *Journal of the Acoustical Society of America*, 98(6), 3140–3145.

Fitzgibbons, P. J., & Gordon-Salant, S. (2010). Behavioral studies with aging humans: Hearing sensitivity and psychoacoustics. In S. Gordon-Salant, R. D. Frisina, A. N. Popper, & R. R. Fay (Eds.), *The aging auditory system* (pp. 111–134). New York: Springer.

Fletcher, H. (1940). Auditory patterns. *Reviews of Modern Physics*, 12, 47–65.

Fletcher, H., & Munson, W. A. (1933). Loudness, its definition, measurement and calculation. *Journal of the Acoustical Society of America*, 5(2), 82–108.

Fowler, E. P. (1936). A method for the early detection of otosclerosis. *Archives of Otolaryngology*, 24, 731–741.

Giraudi, D. M., Salvi, R. J., Henderson, D., & Hamernik, R. P. (1980). Gap detection by the chinchilla. *Journal of the Acoustical Society of America*, 68(3), 802–806.

Giraudi, D. M., Salvi, R. J., Henderson, D., & Hamernik, R. P. (1982). Gap detection in hearing-impaired chinchillas. *Journal of the Acoustical Society of America*, 72(5), 1387–1393.

Gordon-Salant, S., & Fitzgibbons, P. J. (1995a). Comparing recognition of distorted speech using an equivalent signal-to-noise ratio index. *Journal of Speech and Hearing Research*, 38(3), 706–713.

Gordon-Salant, S., & Fitzgibbons, P. J. (1995b). Recognition of multiply degraded speech by young and elderly listeners. *Journal of Speech and Hearing Research*, 38(5), 1150–1156.

Gordon-Salant, S., & Fitzgibbons, P. J. (1997). Selected cognitive factors and speech recognition performance among young and elderly listeners. *Journal of Speech, Language, and Hearing Research*, 40(2), 423–431.

Green, D. M. (1988). *Profile analysis: Auditory intensity discrimination*. London: Oxford University Press.

Green, D. M., Kidd, G., Jr., & Picardi, M. C. (1983). Successive versus simultaneous comparison in auditory intensity discrimination. *Journal of the Acoustical Society of America*, 73(2), 639–643.

Grose, J. H., & Hall, J. W., III. (1994). Modulation detection interference (MDI) in listeners with cochlear hearing loss. *Journal of Speech and Hearing Research*, 37(3), 680–686.

Grose, J. H., & Hall, J. W., III (1996). Cochlear hearing loss and the processing of modulation: Effects of temporal asynchrony. *Journal of the Acoustical Society of America*, 100(1), 519–527.

Hall, J. W., & Grose, J. H. (1989). Spectro-temporal analysis and cochlear hearing impairment: Effects of frequency selectivity, temporal resolution, signal frequency, and rate of modulation. *Journal of the Acoustical Society of America*, 85(6), 2550–2562.

Hall, J. W., III, Haggard, M. P., & Fernandes, M. A. (1984). Detection in noise by spectro-temporal analysis. *Journal of the Acoustical Society of America*, 76(1), 50–56.

Hall, J. W., III, Davis, A. C., Haggard, M. P., & Pillsbury, H. C. (1988). Spectro-temporal analysis in normal-hearing and cochlear-impaired listeners. *The Journal of the Acoustical Society of America*, 84(4), 1325–1331.

Hall, J. W., III, Grose, J. H., & Mendoza, L. (1996). Spectro-temporal processing in cochlear hearing-impaired listeners. In A. Axelsson, H. M. Borchgrevink, R. P. Hamernik, P. A. Hellstrom, D. Henderson, & R. J. Salvi (Eds.), *Scientific basis of noise-induced hearing loss* (pp. 243–251). New York: Thieme.

Hall, J. W., III, Buss, E., & Grose, J. H. (2008). Spectral integration of speech bands in normal-hearing and hearing-impaired listeners. *Journal of the Acoustical Society of America*, 124(2), 1105–1115.

Heinz, M. G., Issa, J. B., & Young, E. D. (2005). Auditory-nerve rate responses are inconsistent with common hypotheses for the neural correlates of loudness recruitment. *Journal of the Association for Research in Otolaryngology*, 6, 91–105.

Humes, L. E., & Christopherson, L. (1991). Speech identification difficulties of hearing-impaired elderly persons: The contributions of auditory processing deficits. *Journal of Speech and Hearing Research*, 34(3), 686–693.

Humes, L. E., & Dubno, J. R. (2010). Factors affecting speech understanding in older adults. In S. Gordon-Salant, R. D. Frisina, A. N. Popper, & R. R. Fay (Eds.), *The aging auditory system* (pp. 211–257). New York: Springer.

Joris, P. X. (2009). Recruitment of Neurons and Loudness. Commentary on "Encoding intensity in ventral cochlear nucleus following acoustic trauma: Implications for loudness recruitment" by Cai et al. *Journal of the Association for Research in Otolaryngology*. doi: 10.1007/s10162-008-0142-y. *Journal of the Association for Research in Otolaryngology*, 10(1), 1–4. doi: 10.1007/s10162-009-0156-0.

Kujawa, S. G., & Liberman, M. C. (2009). Adding insult to injury: Cochlear nerve degeneration after "temporary" noise-induced hearing loss. *Journal of Neuroscience*, 29(45), 14077–14085.

Lentz, J. J., & Leek, M. R. (2002). Decision strategies of hearing-impaired listeners in spectral shape discrimination. *Journal of the Acoustical Society of America*, 111(3), 1389–1398.

Lentz, J. J., & Leek, M. R. (2003). Spectral shape discrimination by hearing-impaired and normal-hearing listeners. *Journal of the Acoustical Society of America*, 113(3), 1604–1616.

Litvak, L. M., Spahr, T., Saoji, A. A., & Fridman, G. Y. (2007). Relationship between perception of spectral ripple and speech recognition in cochlear implant and vocoder listeners. *Journal of the Acoustical Society of America*, 122(2), 982–991.

May, B. J., Little, N., & Saylow, S. (2009). Loudness perception in the domestic cat: Reaction time estimates of equal loudness contours and recruitment effects. *Journal of the Association for Research in Otolaryngology*, 10(2), 295–308.

Mills, J. H., Schulte B. A., Boettcher, F. A., & Dubno, J. R. (1998). A comparison of age-related hearing loss and noise-induced hearing loss. In D. Henderson, D. Prasher, R. Kopke, R. Salvi, &

R. Hamernik (Eds.), *Noise-induced hearing loss: Basic mechanisms, prevention and control* (pp. 497–511). London: Noise Research Network Publications.

Moody, D. B. (1970). Reaction time as an index of sensory function. In W. C. Stebbins (Ed.), *Animal psychophysics: The design and conduct of sensory experiments* (pp. 227–302). New York: Appleton.

Moore, B. C. J. (1996). Effects of noise-induced hearing loss on temporal resolution. In A. Axelsson, H. M. Borchgrevink, R. P. Hamernik, P. A. Hellstrom, D. Henderson, & R. J. Salvi (Eds.), *Scientific basis of noise-induced hearing loss* (pp. 252–263). New York: Thieme.

Moore, B. C. J. (1998). *Cochlear hearing loss*. London: Whurr.

Moore, B. C. J. (2003). Frequency selectivity, masking, and the critical band. In *An introduction to the psychology of hearing* (5th ed., pp. 65–125). San Diego: Academic Press.

Moore, B. C. J. (2004). Testing the concept of softness imperceptions: Loudness near threshold for hearing-impaired ears. *Journal of the Acoustical Society of America*, 115(6), 3103–3111.

Moore, B. C. J., Shailer, M. J., Hall, J. W., & Schooneveldt, G. P. (1993). Comodulation masking release in subjects with unilateral and bilateral hearing impairment. *Journal of the Acoustical Society of America*, 93(1), 435–451.

Opie, J. M., & Bacon, S. P. (1993). Modulation detection interference in subjects with a mild high-frequency hearing loss. *Journal of the Acoustical Society of America*, 94(3), 1812.

Patterson, R. D., Nimmo-Smith, I., Weber, D. L., & Milroy, R. (1982). The deterioration of hearing with age: Frequency selectivity, the critical ratio, the audiogram, and speech threshold. *Journal of the Acoustical Society of America*, 72(6), 1788–1803.

Peters, R. W., & Moore, B. C. (1992). Auditory filter shapes at low center frequencies in young and elderly hearing-impaired subjects. *Journal of the Acoustical Society of America*, 91(1), 256–266.

Pugh, J. E., Jr., Moody, D. B., & Anderson, D. J. (1979). Electrocochleography and experimentally induced loudness recruitment. *Archives of Oto-Rhino-Laryngology*, 224(3–4), 241–255.

Rawool, V. (2006). The effects of hearing loss on temporal processing. *Hearing Review*, 13(6), 42–46.

Salvi, R., & Boettcher, F. A. (2008). Animal models of noise-induced hearing loss. In P. M. Conn (Ed.), *Sourcebook of models for biomedical research* (pp. 289–301). Totowa, NJ: Humana Press.

Salvi, R. J., Hamernik, R. P., & Henderson, D. (1978). Discharge patterns in the cochlear nucleus of the chinchilla following noise-induced asymptotic threshold shift. *Experimental Brain Research*, 32(3), 301–320.

Salvi, R. J., Hamernik, R. P., & Henderson, D. (1983). Response patterns of auditory nerve fibers during temporary threshold shift. *Hearing Research*, 10(1), 37–67.

Schneider, B. A., Pichora-Fuller, M. K., Kowalchuk, D., & Lamb, M. (1994). Gap detection and the precedence effect in young and old adults. *Journal of the Acoustical Society of America*, 95(2), 980–991.

Schorn, K., & Zwicker, E. (1990). Frequency selectivity and temporal resolution in patients with various inner ear disorders. *Audiology*, 29(1), 8–20.

Shrivastav, M. N., Humes, L. E., & Kewley-Port, D. (2006). Individual differences in auditory discrimination of spectral shape and speech-identification performance among elderly listeners. *Journal of the Acoustical Society of America*, 119(2), 1131–1142.

Sommers, M. S., & Humes, L. E. (1993). Auditory filter shapes in normal-hearing, noise-masked normal, and elderly listeners. *Journal of the Acoustical Society of America*, 93(5), 2903–2914.

Tomita, M., Norena, A. J., & Eggermont, J. J. (2004). Effects of an acute acoustic trauma on the representation of a voice onset time continuum in cat primary auditory cortex. *Hearing Research*, 193(1–2), 39–50.

Tyler, R. S. (1986). Frequency resolution in hearing-impaired listeners. In B. C. J. Moore (Ed.), *Frequency selectivity in hearing* (pp. 309–363). London: Academic Press.

Tyler, R. S., Summerfield, Q., Wood, E. J., & Fernandes, M. A. (1982). Psychoacoustic and phonetic temporal processing in normal and hearing-impaired listeners. *Journal of the Acoustical Society of America*, 72(3), 740–752.

Yost, W. A., & Sheft, S. (1989). Across-critical-band processing of amplitude-modulated tones. *Journal of the Acoustical Society of America*, 85(2), 848–857.

Zwicker, E., & Schorn, K. (1978). Psychoacoustical tuning curves in audiology. *Audiology*, 17(2), 120–140.

Zwicker, E., & Schorn, K. (1982). Temporal resolution in hard-of-hearing patients. *Audiology*, 21(6), 474–492.

Chapter 8
The Neurobiology of Noise-Induced Tinnitus

James A. Kaltenbach and Ryan Manz

1 Introduction

Tinnitus is the perception of an ongoing sound that has no external physical source. It is estimated that approximately 50 million people in the United States experience tinnitus (Tyler 2008). The percentage of patients with noise-induced hearing loss who also have tinnitus has been estimated at 35% (Axelsson and Sandh 1985). The incidence of noise-induced tinnitus (NIT) among people with all forms of tinnitus ranges from 20% to 42% (Axelsson and Barrenas 1992; Kowalska and Sulkowski 2001). The Tinnitus Data Archive of the Oregon Health and Sciences University shows that excessive exposure to noise accounted for 45% of cases in which the cause of tinnitus was known or suspected (http://www.tinnitusarchive.org; see also Meikle and Greist 1991). Like other forms of tinnitus, NIT can be either acute or chronic. The acute form lasts for relatively short periods, usually seconds or days, although in some cases can persist for several weeks. Chronic tinnitus represents the more clinically significant form, lasting from months to years.

1.1 Acute NIT

The temporary form of tinnitus usually has its onset immediately following noise exposure (Loeb and Smith 1967; Atherley et al. 1968). Acute tinnitus may be perceived on the same side as the exposed ear or on the opposite side, and in the

J.A. Kaltenbach (✉)
Department of Neurosciences, NE-63, The Cleveland Clinic,
9500 Euclid Avenue, Cleveland, OH 44195, USA
e-mail: kaltenj@ccf.org

C.G. Le Prell et al. (eds.), *Noise-Induced Hearing Loss: Scientific Advances*,
Springer Handbook of Auditory Research 40, DOI 10.1007/978-1-4419-9523-0_8,

vast majority of subjects tested with loud sound exposure, the induced tinnitus was most commonly tonal, with "ringing" as the most common descriptor (Chermak and Dengerink 1987). The pitch of acute NIT is usually matched to high frequencies and is shifted toward frequencies above that of the exposure tone (Loeb and Smith 1967).

1.2 Chronic NIT

Chronic NIT is most commonly associated with hearing loss (Kowalska and Sulkowski 2001), and the severity of tinnitus typically increases with the severity of the hearing loss (Axelsson and Barrenas 1992), although it should be underscored that the term "severity" has multiple meanings in the literature, and can refer to loudness or to the degree of suffering caused by the tinnitus, which may not correspond to loudness. Chronic NIT varies considerably in its psychophysical attributes, but most commonly is described as an ongoing whistling or ringing sound (Nicolas-Puel et al. 2006). The pitch or pitch spectrum of NIT is most often matched to frequencies of 2 kHz or greater and corresponds to the frequency range in which associated hearing losses are maximal (Cahani et al. 1983; Axelsson and Sandh 1985; Noreña et al. 2002). Less frequently, NIT is matched to frequencies below 2 kHz and sometimes as low as a few hundred Hertz. Contrary to common assumptions, the pitch of NIT does not correspond to the edge frequency of the hearing loss (Axelsson and Sandh 1985; Noreña et al. 2002). Konig et al. (2006) found that the pitch of NIT averaged 1.5 octaves above the edge frequency. NIT has been found to be matched to sounds in the range of 10–45 dB sound pressure levels (SPL); however, because hearing thresholds are usually elevated in subjects with NIT, the sound level to which NIT is matched is usually within 10–15 dB of these elevated hearing thresholds (Axelsson and Sandh 1985).

1.3 The Three Faces of Tinnitus

Like all sound percepts, tinnitus has three components: auditory, attentional, and emotional (Jastreboff and Hazell 2004). Tinnitus begins with the auditory sensation, the actual sound percept. By itself, the tinnitus sound would seem to be without particular significance because its level is usually so low (<15 dB SL). Indeed, many, if not most, people with tinnitus are not bothered by it. However, in the more problematic state, the tinnitus sound can capture a person's awareness, leading to the attentional component, the extent to which the subject actually listens to the tinnitus sound(s). To be a clinical problem, tinnitus sounds must attract a great deal of attention, so that the person focuses excessively on the presence of tinnitus. This hyperattentiveness then leads to undesirable emotional reactions that are the clinical hallmarks of tinnitus: annoyance, frustration, depression, anxiety, and, in extreme cases, thoughts of suicide.

Until recently, the clinical approach to tinnitus has been to treat its emotional component. A number of drug therapies are available, and some of these have been reported to have significant benefits for tinnitus patients, greatly reducing the depression or anxiety caused by tinnitus sounds (Robinson 2007). Some people derive benefits from psychological counseling or conditioning designed to help them neutralize the emotional impact of the unwanted noise. Although some patients can, in fact, learn to tolerate the sounds of tinnitus, they nonetheless usually continue to perceive the sounds negatively, wishing that some day the tinnitus sounds will end. Unfortunately, NIT is likely to be a significant clinical problem for a long time to come owing to the prevalence of loud noise in industrialized societies.

1.4 The Importance of Understanding Underlying Mechanisms

The ability to treat the auditory component of tinnitus has been hampered by the lack of understanding of its underlying biological mechanisms. Without this understanding, it is difficult to develop effective therapies, as such therapies ultimately depend on knowing what structures to target for treatment and what functional defect needs to be corrected within those structures. Fortunately, much progress has been made in recent years toward an understanding of tinnitus-producing mechanisms, particularly those underlying NIT. This understanding has come from a combination of anatomical, physiological, pharmacological, and molecular studies in animals and from human studies using brain imaging techniques. In addition, a host of new behavioral techniques have been developed for testing animals for tinnitus following intense noise exposure (Brozoski et al. 2002; Heffner and Harrington 2002; Turner et al. 2006). A major theme that has emerged from these studies is that NIT is a problem induced by damage or overstimulation of the ear, which triggers plasticity of neurons in the central auditory system (Fig. 8.1). This plasticity renders neurons more excitable, and consequently more likely to behave as though they are responding to sound, even in quiet environments. This chapter summarizes anatomical, physiological, pharmacological, imaging, and behavioral data on NIT from studies in animals and humans, with emphasis on the neuroplastic changes caused by noise exposure. Where applicable, a brief overview of other (non-noise) models (cochlear ablation, salicylate) is provided because, though the focus of this chapter is NIT, it is possible that data from other models can have relevance to NIT.

2 Insights into Mechanisms of NIT Gained from Clinical Studies

Tinnitus displays several features that reveal its underlying mechanisms. Most of these features, described in the following section, point to plasticity as an important process in the induction and maintenance of tinnitus.

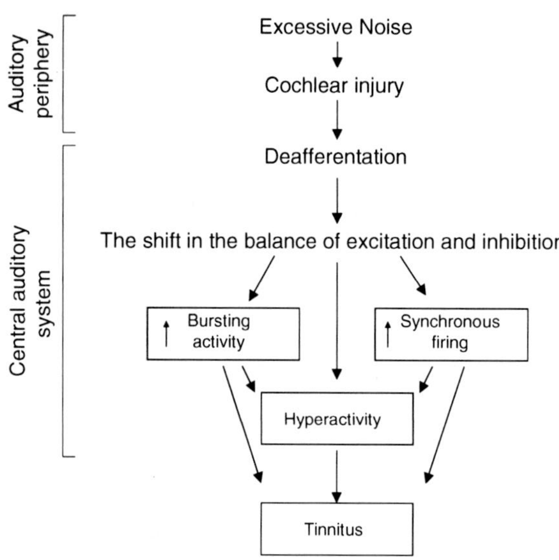

Fig. 8.1 Contemporary view summarizing the major chain of events between exposure to intense sound and the emergence of Tinnitus

2.1 *The Role of the Auditory Periphery in NIT*

NIT shares with many other forms of tinnitus the fact that it begins with cochlear injury. Noise-induced injury to the inner ear usually involves more damage to outer hair cells (OHCs) than to inner hair cells (IHCs). In fact, tinnitus is often associated with alterations in distortion product otoacoustic emissions (DPOAEs), which provide a measure of OHC function (Kowalska and Sulkowski 2001; Bartnik et al. 2009). Shiomi et al. (1997) found that a vast majority of patients with tinnitus displayed alterations of DPOAEs even when there was no accompanying hearing loss.

It does not necessarily follow, however, that the tinnitus-producing signals in the auditory system originate peripherally. The literature suggests that the generators of tinnitus often lie beyond the auditory periphery. This view is based in part on neurophysiological data (see Sect. 3), but there is also a clinical reason: It is commonly reported that patients who have undergone section of either the cochlear portion of the eighth nerve or the entire eighth nerve continue to experience tinnitus postsurgically. This persistence has been observed when the eighth nerve sections are either unilateral or bilateral. Jackson (1985) found no improvement of tinnitus after eighth nerve section in a patient with NIT. The most intensive clinical investigation based on hundreds of surgeries in human subjects with tinnitus of various etiologies was published in 1981 by House and Brackman. They found that 62% of patients undergoing surgical section of the eighth nerve continued to experience their tinnitus postsurgically. In a substantial percentage of these patients, the tinnitus was worse after surgery than before. Similar results have been reported by other surgeons (Silverstein 1976; Gardner 1984). In addition, Berliner et al. (1992) found that 42% of patients with no prior tinnitus who underwent sectioning of the eighth cranial nerve to remove an acoustic neuroma developed tinnitus postoperatively. These findings have led to the commonly held view that various forms of tinnitus, including NIT, are triggered by cochlear hair cell injury but that the actual generators lie centrally.

2.2 The Importance of Plasticity in NIT

2.2.1 Temporal Plasticity

Temporal changes in the pitch of tinnitus are sometimes experienced by persons with NIT. In a study of three subjects with NIT, Penner (1983) determined the pitch of tinnitus by identifying the frequency of a tone that most closely matched the tinnitus percept. The test subjects perceived pitch matches over a wide range of frequencies, suggesting that the tinnitus percept was more like a narrow band of noise than a pure tone. However, in each case, one frequency was identified that more closely matched the tinnitus pitch than the other frequencies. Subsequent retesting of these subjects demonstrated clear variations in the pitch match within the same test sessions, but even larger variations were apparent when retesting was conducted on different days. In a later study, Burns (1984) examined pitch changes in patients with tinnitus of various etiologies, but two of the patients studied reported that their tinnitus was caused by noise exposure. Pitch matches for the tinnitus percept were compared with pitch matches for externally presented tones. Tinnitus pitch matches were found to fluctuate considerably across sessions, with variability in pitch of tinnitus far exceeding variations in perceived pitch of externally presented tones.

Some authors have argued that changes in tinnitus perception are more complicated than can be explained in terms of simple changes in loudness and pitch. Patients with NIT can experience their tinnitus without necessarily experiencing a distinct pitch. Further complicating the picture, tinnitus can have multiple components, each with a different pitch and spread over a broad frequency range (Penner 1983; Noreña et al. 2002). Changes in the loudness of one component without a simultaneous change in the loudness of other components could lead to changes in perceived pitch, even though the corresponding spectral (i.e., frequency) content of the tinnitus may not have changed. Unfortunately, changes such as these have seldom been examined in studies of patients specifically suffering from NIT. For these reasons, the incidence of pitch changes in the general NIT tinnitus population is unknown, but the available data suggest that shifts in pitch are common in the tinnitus population as a whole. The changes indicate that there is probably a high degree of plasticity in the magnitude and distribution of the tinnitus-generating signal within the tonotopic framework of the auditory system.

2.2.2 Stimulus-Dependent Plasticity

Another intriguing feature of tinnitus that represents a distinct form of plasticity is residual inhibition (RI), the transient period during which tinnitus is momentarily suppressed following cessation of a previous stimulus. RI-inducing stimuli include low-level sound, such as nontraumatic masking noise or a tone, or electrical current delivered to various areas in and around the ear or to the cochlea or brain (Feldmann 1971; Dobie et al. 1986; Shi et al. 2009). The incidence of inducible RI in patients with NIT is probably high: Roberts et al. 2006) found that approximately 75% of tinnitus subjects tested experienced some measurable degree of RI. The duration of RI is usually short, on the order of tens of seconds (Terry et al. 1983; Roberts 2007), although RI durations of days to weeks have occasionally been reported (Hazell and Wood 1981; Goldstein et al. 2001). Terry et al. (1983) found that subjects with NIT showed strong evidence of RI lasting more than 10 s; this contrasts with the results from patients with Ménière's disease, who showed little or no evidence of RI. Residual inhibition lasting more than 10 s cannot be related to the phenomenon of forward masking, a poststimulus suppression of auditory sensitivity that has a peripheral origin and generally does not exceed durations of more than a few hundred milliseconds (Wilson and Carhart 1971; Smith 1977). RI implies the existence in the auditory system of a form of inhibition that is activity dependent, which is discussed in Sect. 3.

3 Insights into Mechanisms of NIT from Animal Studies

3.1 The Search for Neural Correlates of NIT in the Auditory Periphery

The term tinnitus is often used interchangeably with "ringing of the ears," suggesting a peripheral origin. However, as discussed in Sect. 2, the clinical literature often reports that tinnitus is not abolished by sectioning the auditory nerve, which points to a central origin of tinnitus in many cases. Animal studies have provided further support for this centralist view. Liberman and Kiang (1978) compared recordings of spontaneous activity of single auditory nerve fibers of noise-exposed and normal hearing cats. Auditory nerve fibers sampled from cats that had been exposed to noise several weeks earlier displayed an upward shift in their neural response thresholds. However, spontaneous rates decreased among neurons with shifted response thresholds. Liberman and Dodds (1984) later showed that the decreased activity occurred among fibers originating from cochlear regions with damaged inner hair cell stereocilia. Fibers originating from neighboring regions with intact hair cells and their stereocilia displayed normal levels of activity. There was no increase in mean spontaneous firing rates. These results would seem to cast doubt on the notion that the generators of tinnitus are located in the auditory periphery.

On the other hand, noise exposure can induce *transient* increases in spontaneous activity of a subset of auditory nerve fibers for periods of 3–20 min (Lonsbury-Martin and Martin 1981). Such increases could represent a neural correlate of the acute form of tinnitus, such as that which follows exposures to intense sound (90–110 dB SPL) for short periods (5–20 min) (Loeb and Smith 1967; Atherley et al. 1968). Moreover, a peripheral origin of *chronic* NIT cannot be ruled out entirely. Liberman and Kiang (1978) observed long-term increases in bursting activity of auditory nerve fibers, a property that was later suggested as a possible basis of tinnitus (Chen and Jastreboff 1995). Although increases in bursting have not yet been shown to be a property displayed by auditory nerve fibers in response to acoustic stimulation, bursting activity could still be an underlying contributor to NIT, even if it does not strongly resemble stimulus-driven activity. The emergence of bursts of spikes could be interpreted by the brain as a high-pitch sound (i.e., tinnitus), as both bursts of spikes and high-frequency sound-evoked activity are characterized by a high incidence of short interspike intervals. These considerations suggest that the acute form of NIT may have a peripheral correlate and that even the chronic form could stem from increases in bursting activity originating at the peripheral level, although, to our knowledge, the latter hypothesis has not been formally addressed.

3.2 Neurophysiological Correlates of NIT at the Central Level

3.2.1 Hyperactivity

One of the most widely reported noise-induced changes to be discovered in the central auditory system with relevance to tinnitus is an increase in the levels of spontaneous activity, or what is usually referred to as hyperactivity. Chronic, noise-induced hyperactivity has now been well documented for several auditory centers, including the dorsal cochlear nucleus (DCN) (Kaltenbach et al. 2000; Brozoski et al. 2002; Shore et al. 2008), the inferior colliculus (IC) (Mulders and Robertson 2009; Dong et al. 2009, 2010a), and the auditory cortex (AC) (Komiya and Eggermont 2000; Seki and Eggermont 2003; Noreña and Eggermont 2006). The most compelling evidence that chronic noise-induced hyperactivity in the central auditory system may result in the perception of tinnitus is that some of the same regions of the auditory system that have been shown to be hyperactive in animals following noise exposure have also been found to be hyperactive in imaging studies of human subjects with tinnitus (for review see Lanting et al. 2009).

In addition, NIT has been demonstrated in animals exposed to similar noise conditions that induce hyperactivity in central auditory centers (Heffner and Harrington 2002; Brozoski et al. 2002; Kaltenbach et al. 2004). When recordings of activity and behavioral tests for tinnitus were conducted in the same animals, the behavioral evidence for tinnitus was found to be either associated with or moderately correlated with the level of hyperactivity in the DCN (Brozoski et al. 2002; Kaltenbach et al. 2004). A more recent study reported a correlation between NIT demonstrated in animals tested behaviorally and the presence of a triad of spontaneous activity changes in the IC that included low interspike interval variation, increased bursting spontaneous activity, and a tendency for burst discharge rates to correspond with the tinnitus pitch (Bauer et al. 2008). Moreover, the tonotopic distribution of hyperactivity following noise exposure has been shown to resemble the tonotopic pattern of increased activity resulting from tonal stimulation, both showing peaks in the high-frequency region of the DCN (Kaltenbach and Afman 2000). This is consistent with the finding that NIT typically has a high pitch (Axelsson and Sandh 1985). Thus, mechanisms of induction of hyperactivity and related changes would seem to be highly relevant to an understanding of the processes underlying tinnitus.

Tinnitus-related hyperactivity emerges by a process involving neural plasticity. For example, the presence of hyperactivity in central auditory centers that lack a counterpart in the auditory nerve implies central adjustments that cannot be explained by a simple relay of information from the periphery. For very intense sound exposures, hyperactivity in the DCN and IC develops slowly over several days. Neural sensitivity to sound during this same period is reduced, suggesting that the increased activity may be a homeostatic compensatory response to lost input (Kaltenbach et al. 1998; Mulders and Robertson 2009). There are changes over time in the tonotopic profile of the induced hyperactivity (Kaltenbach et al. 2000; Mulders and Robertson 2009). For example, in its earliest phase of emergence, the hyperactivity was found to be broadly distributed across most of the range of frequencies composing the tonotopic map of the DCN. Over the next 2 weeks, the hyperactivity became increasingly attenuated in the low- and high-frequency regions. By 1 month after exposure, the initially highest activity remained only in the mid-frequency region near the locus that represents the frequency of the exposure tone. The peak of the hyperactivity profile shifted toward high frequencies between 1 and 6 months postexposure. These changes suggest some flux in the balance of excitation and inhibition in different tonotopic regions over time. Lastly, hyperactivity in the DCN continued to be present even after the cochlea had been ablated (Zacharek et al. 2002; see, however, Mulders and Robertson 2009). The findings are consistent with those of Liberman and Dodds (1984), which did not show any hyperactivity in the auditory nerve following noise exposure. Together these results suggest that, although the induction of hyperactivity may be triggered by cochlear injury, it emerges and continues to be maintained centrally once it is established, and its magnitude and tonotopic distribution can change over time.

It should be noted that increases in spontaneous activity include increases in both bursting and nonbursting discharges (Chang et al. 2002; Finlayson and Kaltenbach 2009). Bursting activity is of special interest in the context of both tinnitus and plasticity. As mentioned earlier, a study by Bauer et al. (2008) suggests a relationship between the incidence of bursting and behavioral evidence for tinnitus in rats. Moreover, bursting discharges have been implicated in the induction of long-term potentiation (LTP) (Thomas et al. 1998; Watanabe et al. 2002). Thus increased bursting could generate plastic changes in other central auditory centers receiving inputs from bursting neurons (Harnett et al. 2009).

3.2.2 Plastic Reorganization

Plastic reorganization of tonotopic maps has for many years been regarded as evidence of changes in underlying circuitry. This type of plasticity has been identified in the auditory cortex (Robertson and Irvine 1989; Rajan et al. 1993; Eggermont 2006) and to a lesser extent in the IC (Irvine et al. 2003; Izquierdo et al. 2008) but not in the DCN (Meleca et al. 1997; Rajan and Irvine 1998) after cochlear injury. When a restricted region of the cochlea is damaged by noise exposure, the corresponding tonotopic map region in the auditory cortex does not lose its sensitivity to sound, but instead, becomes more sensitive to frequencies near the edge of the hearing loss (Eggermont and Komiya 2000; Komiya and Eggermont 2000). As a result, a narrow range of frequencies at the edge of the hearing loss becomes overrepresented. The expanded map region includes the location where neurons become hyperactive (Komiya and Eggermont 2000). The significance of these results is that if neurons representing a single frequency are induced into a state of hyperactivity, they might be interpreted by the brain as signaling the presence of a tone whose pitch is determined by the frequency that falls in the expanded map area. Accordingly, tinnitus having a more or less distinct pitch might be a product of this increase in the number of hyperactive neurons that now encode the same frequency. Despite the appeal of this model, a role of tonotopic map reorganization in the auditory cortex by itself or any other auditory center as a source of tinnitus sensations has recently been called into question (Roberts et al. 2006) on the grounds that the pitch or pitch spectrum of tinnitus does not generally correspond to frequencies bordering on the edge of a hearing loss (i.e., the frequency region that becomes over-represented after reorganization).

3.2.3 Increased Neural Synchrony

Another important change that has been proposed as a mechanism of tinnitus is increased neural synchrony (Eggermont 1995; Roberts 2011). This concept derives from the hypothesis that neurons that are synchronized are likely to have a strong impact in driving postsynaptic targets and recruiting cortical and downstream neurons into a tinnitus percept (Roberts et al. 2010). Indeed, animals exposed to noise develop an increase in neural synchrony at the cortical level (Seki and Eggermont 2003; Noreña and Eggermont 2006). The increased synchrony is found in primary auditory cortex in the portion of the tonotopic map that shows edge frequency expansion (see Sect. 3.2.2). Interestingly, increases in neural synchrony are observed in the secondary auditory cortex (areas AII) but not in primary auditory cortex in animals treated with another inducer of tinnitus, quinine (Ochi and Eggermont 1997). It is possible that noise-induced increases in synchrony in the cortex originate subcortically, as increases in synchrony have also recently been found in the central nucleus of the IC after noise exposure (Bauer et al. 2008).

4 Mechanisms of Induction of NIT

If tinnitus results from one or more of the changes in neural properties described in Sect. 3, then the natural question to follow is: What are the processes or mechanisms by which such changes are induced? This section reviews the anatomical and molecular changes that seem most likely to reflect the mechanisms underlying the tinnitus-related changes described in Sect. 3.

4.1 Evidence for Synaptic Plasticity Causing Shifts in the Balance of Excitation and Inhibition

4.1.1 Anatomical Findings

Several researchers (Bilak et al. 1997; Muly et al. 2002; Kim et al. 2004a,b,c) have examined short and long-term effects of exposure to octave band or white noise on the innervation of the anteroventral (AVCN) or posteroventral (PVCN) cochlear nucleus of adult chinchillas. They have shown that noise-induced injury to the cochlea is followed by degeneration of axonal inputs to corresponding regions of the PVCN and AVCN that can persist for at least 8 months. During this period of degeneration, there is also regrowth of axons that replace the lost inputs. Electron microscopic evaluation of the endings in the PVCN reveal substantial losses of both excitatory and inhibitory synapses 1 week after exposure; these losses become progressively greater between 16 and 24 weeks, but at 24–32 weeks, there was nearly complete recovery of excitatory endings (Kim et al. 2004a). In contrast, the number of inhibitory endings had only partially recovered. Thus, the net long-term effect of noise exposure was a gain in excitation in the PVCN. Some of the increase in excitatory synapses may involve sprouting of collaterals from preexisting inputs to the CN. Staining of GAP-43, a marker for synaptic remodeling, has also been found to increase in the VCN after noise exposure (Illing et al. 2005; Meidinger et al. 2006).

4.1.2 Decreased Inhibitory Neurotransmission (Glycine and GABA)

Wang et al. (2009) reported alterations in subunit composition of glycine receptors in the DCN of rats with behavioral evidence of NIT. Another study (Dong et al. 2010a) found evidence for down-regulation of inhibitory transmitter receptors in the central auditory system of noise-exposed guinea pigs. In the CN ipsilateral to the exposure, expression of the γ-aminobutyric acid-A (GABA-A) receptor subunit α1 (GABRA1) and glycine receptor subunit α1 (GLRA1) were decreased at 2 weeks postexposure. Decreases in these same receptor subunits as well as in the expression of glutamate decarboxylase (GAD), an enzyme required for the synthesis of GABA, were all decreased in the contralateral IC at 2 and 4 weeks after unilateral noise exposure. These decreases in GAD were associated with hyperactivity in the IC in the same animals, providing a link between loss of inhibition and increased spontaneous activity. Further supporting this link, noise-induced decreases in GABRA1 were found to be restricted to the tonotopic region where noise-induced hyperactivity had previously been shown to develop (Dong et al. 2010b).

4.1.3 Increased Excitatory Transmission (Glutamate and Acetylcholine)

There appears to be much plasticity in glutamatergic transmission in central auditory nuclei after noise exposure. In general, changes depend on the postexposure time, the type of manipulation, and the nucleus studied. In the ventral part of the DCN, glutamatergic transmission was increased at some postexposure times but was decreased at other times (Muly et al. 2004). In contrast, noise exposure caused a decrease in glutamatergic transmission in the dorsal DCN, suggesting that the increases are confined to the low-frequency half of the DCN. Both the AVCN and PVCN showed significant changes in glutamatergic transmission after exposure, with increases occurring during the first week after exposure and decreases or return to control levels occurring at later postexposure times. Dong et al. (2010a) found that noise exposure induced an increase in the expression of the *N*-methyl-D-aspartate (NMDA) receptor subunit 1 in the ipsilateral CN over the 4-week period of observation. This is potentially significant because NMDA receptors are potent mediators of plastic changes, such as long-term potentiation (LTP), which is generally viewed as a facilitator of neural activity. Mechanisms for LTP have been identified in the DCN (Fujino & Oertel 2003), and it has been hypothesized that LTP might be a mechanism by which hyperactivity is induced in the DCN by noise exposure (Tzounopoulos 2008). Taken together, the dynamic changes in glutamatergic neurotransmission observed over many weeks are consistent with large changes in the level of excitation in the CN.

Zeng et al. (2009) recently presented evidence indicating that loss of primary afferent input to the CN due to cochlear trauma leads to an up-regulation in the expression of VGlut2 in the granule cell domain of the CN. VGlut2 is a vesicular glutamate transporter that is associated with nonauditory inputs to the CN, such as that from somatosensory nuclei. This finding suggests an increase in descending modulatory influence on CN cells from the somatosensory system. Consistent with this hypothesis, Shore et al. (2008) found that the bimodally sensitive cells in the DCN (likely fusiform cells), which respond to sound as well as stimulation of somatosensory nuclei, develop an increased sensitivity to trigeminal ganglion stimulation after noise exposure. The increased sensitivity was apparent as a decrease in mean response threshold.

Noise exposure leads to long-term changes in cholinergic transmission in the CN. Acetylcholine is released in the CN by collateral branches of the olivocochlear bundle, which originate in the periolivary nuclei of the superior olivary complex (SOC) and terminate in the granule cell domain (Godfrey et al. 1990, 1997).

The effect of acetylcholine on granule cells is probably excitatory (Chen et al. 1998, 1999). Chang et al. (2002) found that bursting neurons in the superficial DCN became more sensitive to the cholinergic agonist carbachol between 7 and 39 days after noise exposure. Increased sensitivity to cholinergic input has also been observed in the fusiform cell layer of the DCN, in vivo after noise exposure (Kaltenbach and Zhang 2007). Further, noise exposure causes an up-regulation of choline acetyltransferase (ChAT), the enzyme of acetylcholine synthesis found in cholinergic synapses (Jin et al. 2006). The largest increases were found in the granule cell domain, although increases in the fusiform cell layer were also observed. Because ChAT is synthesized by presynaptic terminals, the increased expression of this enzyme was interpreted as possibly reflecting an increase in the number of cholinergic synapses. If this interpretation is correct, tinnitus induction may involve a compensatory increase in the input from the olivocochlear bundle. However, the role of the olivocochlear system is likely to be complex, as some cholinergic inputs are received by projection neurons of the CN, whereas others are received by granule cells, which exert an excitatory effect on inhibitory interneurons. Further work is needed to clarify the net effect of the altered cholinergic input to the CN on neural activity.

4.1.4 Other Relevant Changes in Neurotransmission: Norepinephrine and Serotonin

Noise exposure can induce up-regulation of serotonin and noradrenergic activity in the auditory brain stem. These transmitters are released by nonauditory inputs coming mainly from the dorsal raphe nucleus and locus coeruleus, respectively (Klepper and Herbert 1991; Thompson et al. 1995; Thompson and Thompson 2001). Their role in sensory processing is generally believed to be modulatory, meaning that they likely affect global activity, both spontaneous and stimulus driven. After moderate to intense noise exposure (70, 90, and 110 dB SPL), serotonergic activity was increased in the DCN and PVCN in an intensity-dependent manner (Cransac et al. 1998). Noradrenergic activity was increased at an exposure level of 70 dB SPL but not at higher levels (Cransac et al. 1998). Finally, there were no changes in either of these transmitters in the VCN, IC, or auditory cortex (Cransac et al. 1998). Serotonin has been reported to have either excitatory or inhibitory effects on CN neurons (Ebert and Ostwald 1992). The effects on DCN neurons, however, have not been systematically studied. If the effect of serotonin on DCN is found to be excitatory, it could also contribute to the emergence of tinnitus-related activity in this nucleus.

Collectively, the results of these studies indicate that noise exposure has strong effects on neurotransmission, although the magnitude and direction of the effects clearly depend on the recovery time when the measurements are performed. This suggests that the balance of excitation and inhibition in the central auditory system is in a state of flux after noise exposure. In this regard, the molecular data are consistent with the electrophysiological data, which show temporal variations in the degree and tonotopic distribution of hyperactivity and other tinnitus-related changes that develop in the auditory brain stem after noise exposure (for further discussion of molecular changes relevant to tinnitus, see Tan et al. 2007; Knipper et al. 2009). Less well understood is the relative importance of different processes that underlie shifts in the balance of excitation and inhibition. Loss of some neural populations may simply arise as a result of transneuronal degeneration triggered by injury to hair cells and auditory nerve fibers. However, it is also possible that the changes may be compensatory adjustments to weakened input from the periphery (homeostatic plasticity), as suggested by Schaette and Kempter (2006, 2008). Changes in the balance of excitation and inhibition may also involve additional processes, such as activity-dependent plasticity or excitotoxicity. Overactivation of neurons during noise exposure might result in excess release of glutamate, which could lead to activation of NMDA receptors. The latter effect could lead to induction of LTP or long-term depression (LTD) and/or induction of excitotoxic injury, particularly to inhibitory interneurons whose small size might make them more vulnerable.

4.2 Plasticity in the Intrinsic Membrane Properties of Neurons

Lidocaine is an agent whose primary mechanism of action is to block fast voltage-dependent sodium (Na^+) channels (Onizuka et al. 2004). The fact that intravenously administered lidocaine is often found to be effective in reducing tinnitus (Reyes et al. 2002) suggests that the intrinsic membrane properties of neurons can play an important role in tinnitus. Unfortunately, the use of lidocaine as a systemic drug is impractical for tinnitus treatment owing to its short-lived benefits and its adverse effects on heart and central nervous system function that accompany long-term use (McFadden 1982).

To date, the contribution of alterations in these intrinsic membrane properties to the emergence of tinnitus has received relatively little attention. The resting membrane potential, the threshold of spike generation, the duration of action potentials, and the time course of recovery from action potentials are all properties that are determined in large part by the types and numbers of voltage-sensitive ion channels of neurons; these, in turn, directly affect the spontaneous discharge rates of neurons. Alterations in several channels [e.g., L-type calcium channels (Singer et al. 2001), Ih channels (Hille 2001), and Kv3 family of potassium channels (Akemann and Knopfel 2006), are known to influence spontaneous rate and/or patterns of spontaneous activity. Surprisingly, little work has been focused on the effects of noise on ion-conductance properties of auditory neurons that could be tinnitus inducing.

The exception is a recent study by Dong et al. (2010a) examining the effects of noise exposure on the expression of the potassium channel KCNK15. Expression of KCNK15 in the ipsilateral and contralateral CNs was not significantly altered at any time after exposure but did show a trend toward an increase in the ipsilateral CN. In contrast, expression of this channel in the contralateral IC was decreased at 2- and 4-week survival times. This decrease is consistent with an increase in excitability because the KCNK15 channel plays a role in setting the resting membrane potential (Plant et al. 2005). However, the fact that there was an increase in KCNK15 in the ipsilateral IC at 2 and 4 weeks, despite the presence of hyperactivity indicates that KCNK15 downregulation may be only one of several factors that contribute to the hyperactive state.

5 Clinical Implications

Patients with NIT now have a fair number of treatment options, which can be grouped into several categories, including acoustic therapy, electrical stimulation therapy, transcranial magnetic stimulation, pharmacotherapy, and cognitive behavioral psychotherapy. Most of these treatment approaches have been available for many years and historically have had little connection with an understanding of the underlying pathology of NIT. Noise maskers or noise generators and hearing aids are forms of acoustic therapy that are available in most audiology clinics and are designed to cover up or steer attention away from the tinnitus sound. Counseling, cognitive behavioral therapy, and most forms of pharmacotherapy are currently used to treat the person's negative emotional reaction to the tinnitus sounds. Surgery is appropriate for certain rare forms of tinnitus that result from impingements of nerves by vascular anomalies or tumors, but for NIT surgical section of the eighth nerve is not a good option (Jannetta et al. 1986; Møller et al. 1993). The remaining options, including electrical and magnetic stimulation and physical therapy specifically customized for tinnitus patients are more experimental in nature and are available in only a small number of research institutions. Summaries of the various treatment modalities have been reviewed in several books over the past decade (Snow 2004; Langguth et al. 2007; Tyler 2008) and are not repeated here. Instead, this section focuses on how the research findings discussed in the preceding text are currently impacting some of the existing clinical approaches to tinnitus management and treatment.

5.1 Current Plasticity-Based Treatments

5.1.1 Acoustic Therapy

As discussed in Sect. 2.2, tinnitus is often a temporally changing percept that varies in its psychophysical attributes over time. There is reason to suspect that such changes might be related to shifts in the level and distribution of hyperactivity along the tonotopic axis, as has been demonstrated physiologically (see Sect. 3.2.1). These shifts, in turn, might reflect shifts in the balance of excitatory and inhibitory processes at the synaptic level, which have been well characterized anatomically and pharmacologically in the auditory brain stem and cortex (see Sect. 4.1).

But even the normal, undisturbed auditory system possesses some capacity to change the loudness of sound through the process of adaptation. This ability has been demonstrated by testing loudness matches in normal hearing subjects with steady-state ongoing sounds (Hellman et al. 1997). When tested over many minutes, the loudness matches for such stimuli were found to decline toward a plateau level over periods of up to 6 min. The degree of the decline is frequency and intensity dependent, but in general, the declines are greatest for low-level high-frequency sounds between 2 and 12 kHz. Loudness adaptation is proof of concept that the loudness can be modulated. It is possible that changes in the degree of loudness adaptation may underlie temporal plasticity of tinnitus.

Recent evidence suggests that the ability of the auditory system to change the loudness of sounds, including tinnitus, might lend itself to therapeutic manipulation. This possibility was raised by Sheldrake and Hazell (1992), who noticed that many tinnitus patients became unable to perceive their tinnitus after several years of wearing noise-maskers. Unfortunately, the study did not include patients with tinnitus who did not wear noise maskers, leaving open the possibility that the masker population may have experienced a spontaneous recovery from tinnitus. However, a more detailed study of adaptation lends support to the possibility that loudness percepts can be modulated by altering the level of background noise. Formby et al. (2003) found that normal hearing subjects became less sensitive to sound (the perceived loudness of sounds decreased) after 2 weeks of stimulation with low-level noise. In contrast, those wearing earplugs for the same period of time became more sensitive after the earplugs were removed (loudness increased). They interpreted this change as an adaptive plasticity caused by a change in central gain. This finding was taken a step further by Munro and Blount (2009), who found that the increase in loudness of sound in normal hearing subjects after use of earplugs for extended periods is associated with a decrease in middle ear reflex thresholds, suggesting that the change in loudness involves an adjustment in gain at the brain stem level. These studies raise the possibility that a similar adaptive plasticity might be achievable in tinnitus patients by prolonged use of low- or moderate-level noise.

This concept has already begun to be tested experimentally and clinically. Noreña and Eggermont (2006) presented evidence showing that tinnitus-related changes in the auditory cortex induced by loud sound exposure can be prevented by placing the animals in an enriched acoustic environment consisting of moderate level noise spectrally matched to the range of the hearing loss. Animals placed in this environment immediately after loud sound exposure did not show any tinnitus-related changes in activity (hyperactivity, synchronized activity, or reorganization) following the acoustic trauma as did the exposed group that were not placed in the enriched environment.

Sound therapy based on the research findings just described has been tested clinically with some encouraging results. Davis et al. (2008) found that a high percentage of patients who were treated with an in-ear device that presented sounds with energy concentrated in the frequency range of the patients' hearing loss experienced improvement in their tinnitus. A more recent study reported that patients with tinnitus (cause unspecified) experienced improvement in their condition with noise generators worn in the ear, provided the pitch of their tinnitus was within the spectral range of the sound produced by the device (Schaette et al. 2010).

5.2 *Tinnitus Retraining Therapy*

Tinnitus retraining therapy (TRT) takes advantage of the fact that tinnitus precepts not only consist of auditory sensations, but also have attentional and emotional components. TRT combines counseling with acoustic therapy and is structured to recondition patients through directive counseling to disassociate any negative values from their tinnitus percepts. Further habituation of the auditory signal is sought by acoustic therapy in which patients wear a noise generator set at a level that is just below the threshold of audibility of the tinnitus signal. The theory of use is based on the notion, commonly used in the treatment of pain, that aberrant signals can be downregulated by the brain by reducing contrast between the aberrant signal and the background. By increasing background noise, the degree of contrast is reduced. The brain is believed to habituate better to signals with lower contrast. Over time, the level of the noise can be reduced as the brain downregulates the gain on the aberrant signal. In other words, an abnormal conditioned reflex arc is created, and because of plasticity of the brain, any conditioned reflex can be reversed with training (Jastreboff and Hazell 2004). Numerous clinical studies have been published suggesting that this form of therapy is effective. A difficulty in interpreting the data is that it is not clear to what extent the reported improvements are due to the counseling and which are due to the use of acoustic therapy. Carefully designed clinical trials are needed to address this question, but the premise on which the theory is based seems plausible.

5.2.1 Pharmacotherapy

NMDA receptors are implicated in the induction of activity-dependent plasticity, such as LTP and LTD, which could tilt the balance of excitation and inhibition toward the side of excitation (Sect. 4.1). Numerous investigations have explored NMDA receptors as possible targets for tinnitus treatment, although most of these have been used in models of salicylate-induced tinnitus, which likely involve different mechanisms than NIT (Guitton et al. 2003, 2004; Puel 2007). Some studies suggest that caroverine may have an anti-tinnitus (tinnitolytic) effect (Denk et al. 1997; Ehrenberger 2005). Caroverine acts on α-amino-3-hydroxyl-5-methyl-4-isoxazole-propionate (AMPA) receptors at low concentrations, but at high doses acts as an uncompetitive antagonist of NMDA receptors. In a placebo-controlled blind study of 60 tinnitus patients, 63% showed improvement in their tinnitus immediately after administration of caroverine at doses that block NMDA receptors (Denk et al. 1997). No improvement was reported by patients receiving placebo. Lobarinas et al. (2006) reported a slight tinnitolytic effect of the NMDA receptor antagonist memantine, although the improvement was not statistically significant. These studies were conducted in a rat animal model in which tinnitus was generally induced by treatment with salicylate. Neramexane is a low- to moderate-affinity uncompetitive NMDA receptor antagonist. Like memantine, neramexane also acts as an antagonist of serotonin and nicotinic acetylcholine receptors.

Although this drug has several applications, its effect on tinnitus remains to be clarified, but the results of a phase II clinical trial with neramexane are suggestive of a tinnitolytic effect of this agent (Althaus 2009). Similarly, a few clinical studies suggest that another NMDA receptor antagonist, acamprosate, produces some significant benefit to tinnitus patients (Azevedo and Figueiredo 2005). A tinnitolytic effect has also been observed in a rat model of NIT treated with the NMDA receptor antagonist ifenprodil (Guitton and Dudai 2007). This effect was achieved in rats with NIT by administering the antagonist at the cochlear level. This suggests that the tinnitus was of peripheral origin, opening further speculation that some contribution to NIT may originate in the auditory nerve (see Sect. 3.1).

5.2.2 Electrical and Magnetic Stimulation

Electrical stimulation has been exploited as a means of suppressing tinnitus-producing activity. The ability of electrical stimulation of higher level auditory centers to modulate tinnitus in humans subjects (De Ridder et al. 2006, 2007; Seidman et al. 2008; Fenoy et al. 2006) or tinnitus-related hyperactivity at lower levels of the auditory pathway in animal models has yielded some promising results (Zhang and Guan 2008; Song et al. 2009). Some reports indicate that tinnitus is diminished or abolished for long periods after cessation of stimulation (Seidman et al. 2008; De Ridder et al. 2006, 2007). Other studies suggest that less invasive approaches, involving use of repetitive transcranial magnetic stimulation (rTMS), may also have suppressive effects on chronic tinnitus long after stimulation is applied (Plewnia et al. 2007; Kleinjung et al. 2007; Langguth et al. 2010). However, as yet, electrical stimulation of the brain and rTMS remain experimental approaches that benefit only a subset of patients. Improvements in these forms of therapy will depend on the ability to optimize the focus of treatment with respect to the locus of the generator sites, which are likely to vary across individuals.

6 Summary and Conclusions

It should be clear from the above discussion that research on NIT has made major headway in recent years. It is now possible to offer fairly detailed accounts of what types of changes underlie NIT, where in the auditory system they occur, and what some of the cellular and molecular processes that underlie the induction of these tinnitus-producing changes might be. Noise exposure induces hyperactivity (bursting and non-bursting activity) and increases in neural synchrony at various levels of the central auditory system from the dorsal cochlear nucleus to the auditory cortex. NIT thus seems to be a systemwide problem in the state of signaling in the auditory system as a whole, not just a single brain center. How and where exactly these changes are translated into tinnitus percepts is not clear. But it seems likely that the ability of these signals to generate tinnitus stems from their resemblance to sound-evoked activity.

There is a wide range of evidence to support the concept that hyperactivity of auditory neurons is a consequence of a shift in the balance of excitation and inhibition of auditory neurons (Fig. 8.1). A link to anatomical and molecular changes is beginning to emerge with studies showing decreases in inhibition and increases in excitation. In some cases, the decreases in inhibition involve loss of inhibitory synapses at the same levels where hyperactivity is found. There is also evidence that certain receptors can change their subunit composition after noise exposure, as exemplified by the glycine receptor. Some new evidence suggests that some increases in activity may be driven by a compensatory mechanism whereby descending pathways or the receptors for their neurotransmitters may be upregulated. Neurons at the cochlear nucleus level become hypersensitive to cholinergic and trigeminal ganglion stimulation, and these changes now appear associated and possibly linked to an overall change in the number of cholinergic and glutamatergic synapses in the cochlear nucleus, particularly in the granule cell region. Conceivably, similar upregulations may occur in cholinergic and glutamatergic pathways that terminate at higher levels, thereby contributing to hyperactivity beyond the cochlear nucleus. These changes do not preclude other participating mechanisms such as changes in the intrinsic membrane properties of neurons and LTP. There is evidence for altered expression of KCNK15 channels in the IC after noise exposure. These alterations may also contribute to higher levels of activity in the affected cell populations.

This new knowledge is beginning to have clinical impact. A new generation of translational studies that are testing new therapeutic strategies for the treatment of tinnitus has been stimulated. We are seeing more scientific approaches to the testing of potentially new tinnitolytic therapies in human subjects. These include drugs that act on specific receptors implicated in tinnitus, acoustic therapies designed to reverse the tinnitus-inducing plasticity, and electrical stimulation studies targeting pathways that can turn down the gain on neurons with aberrant activity. Some of these approaches have already been tested and shown to be helpful to some tinnitus patients and are available in an increasing number of clinics.

At the same time, the new knowledge base presents important challenges. The main problem is that multiple changes have been described and multiple mechanisms proposed, and it is not yet clear which are primary in importance and which are secondary or even just mere epiphenomena. The resolution of this issue is key to our ability to identify the appropriate treatment targets that will be needed to optimize therapies. A second challenge is an understanding of whether tinnitus reflects changes that occur independently at various levels of the auditory system, or instead, represent a change that occurs at one level, which is then relayed to other levels in a feed-forward fashion. Very little work has been done to address this issue directly. A third challenge will be to target specific cell populations or changes within those populations without affecting other populations required for normal sound perception and other vital functions. This is critical if treatment options are to be applied in real clinical settings without risk of serious side effects. Lastly, while an increasing number of patients are benefiting from the available new treatments, it remains a major challenge to optimize these treatments so that they bring benefit for a majority of patients. These challenges will no doubt be difficult owing to the complexity and interacting nature of the underlying changes that occur. But there is good reason to expect that progress in meeting these challenges will follow with improvements in technology and continued growth of interest and research effort on tinnitus.

Acknowledgments This review was supported by a grant from the National Institutes on Deafness and Other Communicative Disorders R01DC009097 to J. A. K.

References

Akemann, W., & Knopfel T. (2006). Interaction of Kv3 potassium channels and resurgent sodium current influences the rate of spontaneous firing of Purkinje neurons. *Journal of Neuroscience*, 26(17), 4602–4612.

Althaus, M. (2009). Clinical development of new drugs for the treatment of tinnitus using the example of nermexane. *Proceedings of the Tinnitus Research Initiative*, 3, 38.

Atherley, G. R., Hempstock, T. I., & Noble, W. G. (1968). Study of tinnitus induced temporarily by noise. *Journal of the Acoustical Society of America*, 44, 1503–1506.

Axelsson, A., & Sandh, A. (1985). Tinnitus in noise-induced hearing loss. *British Journal of Audiology*, 19, 271–276.

Axelsson, A., & Barrenas, M. L. (1992). Tinnitus in noise-induced hearing loss. In A. L. Dancer, D. Henderson, R. J. Salvi, & R. P. Hamernik (Eds.), *Noise-induced hearing loss* (pp. 269–276). Boston: Mosby.

Azevedo, A., & Figueiredo, R. (2005). Tinnitus treatment with acamprosate: Double-blind study. *Brazilian Journal of Otorhinolaryngology*, 71, 618–623.

Bartnik, G., Hawley, M., Rogowski, M., Raj-Koziak, D., Fabijanska, A., & Formby, C. (2009). [Distortion product otoacoustic emission levels and input/output-growth functions in normal-hearing individuals with tinnitus and/or hyperacusis]. *Otolaryngology Poland*, 63(2),171–181.

Bauer, C. A., Turner, J. G., Caspary, D. M., Myers, K. S., & Brozoski, T. J. (2008). Tinnitus and inferior colliculus activity in chinchillas related to three distinct patterns of cochlear trauma. *Journal of Neuroscience Research*, 86, 2564–2578.

Berliner, K. I., Shelton, C., Hitselberger, W. E., & Luxford, W. M. (1992). Acoustic tumors: Effect of surgical removal on tinnitus. *American Journal of Otology*, 13, 13–17.

Bilak, M., Kim, J., Potashner, S. J., Bohne, B. A., & Morest, D. K. (1997). New growth of axons in the cochlear nucleus of adult chinchillas after acoustic trauma. *Experimental Neurology*, 147, 256–268.

Brozoski, T. J., Bauer, C. A., & Caspary, D. M. (2002). Elevated fusiform cell activity in the dorsal cochlear nucleus of chinchillas with psychophysical evidence of tinnitus. *Journal of Neuroscience*, 22, 2383–2390.

Burns, E. M. (1984). A comparison of variability among measurements of subjective tinnitus and objective stimuli. *Audiology*, 23, 426–440.

Cahani, M., Paul, G., & Shahar, A. (1983). Tinnitus pitch and acoustic trauma. *Audiology*, 22, 357–363.

Chang, H., Chen, K., Kaltenbach, J. A., Zhang, J., & Godfrey, D. A. (2002). Effects of acoustic trauma on dorsal cochlear nucleus neuron activity in slices. *Hearing Research*, 164, 59–68.

Chen, G-D., & Jastreboff, P. J. (1995). Salicylate-induced abnormal activity in the inferior colliculus of rats. *Hearing Research*, 82,158–178.

Chen, K., Waller, H. J., & Godfrey, D. A. (1998). Effects of endogenous acetylcholine on spontaneous activity in rat dorsal cochlear nucleus slices. *Brain Research*, 783, 219–226.

Chen, K., Waller, H. J., Godfrey, T. G., & Godfrey, D. A. (1999). Glutamatergic transmission of neuronal responses to carbachol in rat dorsal cochlear nucleus slices. *Neuroscience*, 90, 1043–1049.

Chermak, G. D., & Dengerink, J. E. (1987). Characteristics of temporary noise-induced tinnitus in male and female subjects. *Scandinavian Audiology*, 16(2), 67–73.

Cransac, H., Cottet-Emard, J. M., Hellström, S., & Peyrin, L. (1998). Specific sound-induced noradrenergic and serotonergic activation in central auditory structures. *Hearing Research*, 118, 151–156.

Davis, P. B., Wilde, R. A., Steed, L. G., & Hanley, P. J. (2008). Treatment of tinnitus with a customized acoustic neural stimulus: A controlled clinical study. *Ear Nose and Throat Journal*, 87, 330–339

Denk, D. M., Heinzl, H., Franz, P., & Ehrenberger, K. (1997). Caroverine in tinnitus treatment. A placebo-controlled blind study. *Acta Oto-Laryngologica*, 117, 825–830.

De Ridder, D., De Mulder, G., Verstraeten, E., Van der Kelen, K., Sunaert, S., Smits, M., Kovacs, S., Verlooy, J., Van de Heyning, P., & Moller, A. R. (2006). Primary and secondary auditory cortex stimulation for intractable tinnitus. *ORL Journal of Otorhinolaryngology and Related Specialties*, 68, 48–54.

De Ridder, D., De Mulder, G., Menovsky, T., Sunaert, S., & Kovacs, S. (2007). Electrical stimulation of auditory and somatosensory cortices for treatment of tinnitus and pain. *Progress in Brain Research*, 166, 377–388.

Dobie, R. A., Hoberg, K. E., & Rees, T. S. (1986). Electrical tinnitus suppression: A double-blind crossover study. *Otolaryngology, Head and Neck Surgery*, 95, 319–323.

Dong, S., Mulders, W. H., Rodger, J., & Robertson, D. (2009). Changes in neuronal activity and gene expression in guinea-pig auditory brainstem after unilateral partial hearing loss. *Neuroscience*, 159, 1164–1174.

Dong, S., Mulders, W. H., Rodger, J., Woo, S., & Robertson, D. (2010a). Acoustic trauma evokes hyperactivity and changes in gene expression in guinea-pig auditory brainstem. *European Journal of Neuroscience*, 31(9), 1616–1628.

Dong, S., Rodger, J., Mulders, W. H., & Robertson, D. (2010b). Tonotopic changes in GABA receptor expression in guinea pig inferior colliculus after partial unilateral hearing loss. *Brain Research*, 1342, 24–32.

Ebert, U., & Ostwald, J. (1992). Serotonin modulates auditory information processing in the cochlear nucleus of the rat. *Neuroscience Letters*, 145, 51–54.

Eggermont, J. J. (1995). Representation of a voice onset time continuum in primary auditory cortex of the cat. *Journal of the Acoustical Society of America*, 98, 911–920.

Eggermont, J. J. (2006). Cortical tonotopic map reorganization and its implications for treatment of tinnitus. *Acta Oto-Laryngologica*, 556(Supplementum), 9–12.

Eggermont, J. J., & Komiya, H. (2000). Moderate noise trauma in juvenile cats results in profound cortical topographic map changes in adulthood. *Hearing Research*, 142(1–2), 89–101.

Ehrenberger, K. (2005). Topical administration of caroverine in somatic tinnitus treatment: proof-of-concept study. *International Tinnitus Journal*, 11, 34–37.

Feldmann, H. (1971). Homolateral and contralateral masking of tinnitus by noise-bands and by pure tones. *Audiology*, 10, 138–144.

Fenoy, A. J., Severson, M. A., Volkov, I. O., Brugge, J. F., & Howard, M. A. 3 rd (2006). Hearing suppression induced by electrical stimulation of human auditory cortex. *Brain Research*, 1118, 75–83.

Finlayson, P. G., & Kaltenbach, J. A. (2009). Alterations in the spontaneous discharge patterns of single units in the dorsal cochlear nucleus following intense sound exposure. *Hearing Research*, 56, 104–117.

Formby, C., Sherlock, L. P., & Gold, S. L. (2003). Adaptive plasticity of loudness induced by chronic attenuation and enhancement of the acoustic background. *Journal of the Acoustical Society of America*, 114, 55–58.

Fujino, K., & Oertel, D. (2003). Bidirectional synaptic plasticity in the cerebellum-like mammalian dorsal cochlear nucleus. *Proceedings of the National Academy of Sciences of the USA*, 100, 265–270.

Gardner, G. (1984). Neurologic surgery and tinnitus. *Journal of Laryngology and Otology Supplement*, 9, 311–318.

Godfrey, D. A., Beranek, K. L., Carlson, L., Parli, J. A., Dunn, J. D., & Ross, C. D. (1990). Contribution of centrifugal innervation to choline acetyltransferase activity in the cat cochlear nucleus. *Hearing Research*, 49, 259–279.

Godfrey, D. A., Godfrey, T. G., Mikesell, N. L., Waller, H. J., Yao, W., Chen, K., & Kaltenbach, J. A. (1997). Chemistry of granular and closely related regions of the cochlear nucleus. In J. Syka (Ed.), *Acoustical signal processing in the central auditory system* (pp. 139–153). New York: Plenum Press.

Goldstein, B. A., Shulman, A., Lenhardt, M. L., Richards, D. G., Madsen, A. G., & Guinta, R. (2001). Long-term inhibition of tinnitus by UltraQuiet therapy: Preliminary report. *International Tinnitus Journal*, 7, 122–127.

Guitton, M. J., & Dudai, Y. (2007). Blockade of cochlear NMDA receptors prevents long-term tinnitus during a brief consolidation window after acoustic trauma. *Neural Plasticity*, 2007, 80904.

Guitton, M. J., Caston, J., Ruel, J., Johnson, R. M., Pujol, R., & Puel. J. L. (2003). Salicylate induces tinnitus through activation of cochlear NMDA receptors. *Journal of Neuroscience*, 23, 3944–3952.

Guitton, M. J., Wang, J., & Puel, J. L. (2004). New pharmacological strategies to restore hearing and treat tinnitus. *Acta Oto-Laryngologica*, 124, 411–415.

Harnett, M., Bernier, B., Ahn, K., & Morikawa, H. (2009). Bursting-time-dependent plasticity of NMDA receptor-mediated transmission in midbrain dopamine neurons. *Neuron*, 62, 826–838.

Hazell, J. W., & Wood, S. (1981). Tinnitus masking-a significant contribution to tinnitus management. *British Journal of Audiology*, 15, 223–230.

Heffner, H. E., & Harrington, I. A. (2002). Tinnitus in hamsters following exposure to intense sound. *Hearing Research*, 170, 83–95.

Hellman, R., Miśkiewicz, A., & Scharf, B. (1997). Loudness adaptation and excitation patterns: Effects of frequency and level. *Journal of the Acoustical Society of America*, 101, 2176–2185.

Hille, B. (2001). *Ionic channels of excitable membranes*. Sutherland, MA: Sinauer.

House, J. W., & Brackman, D. E. (1981). Tinnitus: Surgical treatment. In D. Evered & G. Lawrenson (Eds.), *CIBA Foundation Symposium 85: Tinnitus* (pp. 204–212). London: Pitman.

Illing, R. B., Kraus, K. S., & Meidinger, M. A. (2005). Reconnecting neuronal networks in the auditory brainstem following unilateral deafening. *Hearing Research*, 206, 185–199.

Irvine, D. R. F., Rajan, R., & Smith, S. (2003). Effect of restricted cochlear lesions in adult cats on the frequency organization of the inferior colliculus. *Journal of Comparative Neurology*, 467, 354–374.

Izquierdo, M. A., Gutiérrez-Conde, P. M., Merchán, M. A., & Malmierca, M. S. (2008). Non-plastic reorganization of frequency coding in the inferior colliculus of the rat following noise-induced hearing loss. *Neuroscience*, 154(1), 355–369.

Jackson, P. (1985). A comparison of the effects of eighth nerve section with lidocaine on tinnitus. *Journal of Laryngology and Otology*, 99, 663–666.

Jannetta, P. J., Moller, M. B., Moller, A. R., & Sekhar, L. N. (1986). Neurosurgical treatment of vertigo by microvascular decompression of the eighth cranial nerve. *Clinical Neurosurgery*, 33, 645–665.

Jastreboff, P. J., & Hazell, J. (2004). *Tinnitus retraining therapy*. Cambridge, UK: Cambridge University Press.

Jin, Y. M., Godfrey, D. A., Wang, J., & Kaltenbach, J. A. (2006). Effects of intense tone exposure on choline acetyltransferase activity in the hamster cochlear nucleus. *Hearing Research*, 216–217, 168–175.

Kaltenbach, J. A., & Afman, C. E. (2000). Hyperactivity in the dorsal cochlear nucleus after intense sound exposure and its resemblance to tone-evoked activity: A physiological model for tinnitus. *Hearing Research*, 140, 165–172.

Kaltenbach, J. A., & Zhang, J. (2007). Intense sound-induced plasticity in the dorsal cochlear nucleus of rats: Evidence for cholinergic receptor upregulation. *Hearing Research*, 226, 232–243.

Kaltenbach, J. A., Godfrey, D. A., Neumann, J. B., McCaslin, D. L., Afman, C. E., & Zhang, J. (1998). Changes in spontaneous neural activity in the dorsal cochlear nucleus following exposure to intense sound: Relation to threshold shift. *Hearing Research*, 124, 78–84.

Kaltenbach, J. A., Zhang, J., & Afman, C. E. (2000). Plasticity of spontaneous neural activity in the dorsal cochlear nucleus after intense sound exposure. *Hearing Research*, 147, 282–292.

Kaltenbach, J. A., Zacharek, M. A., Zhang, J. S., & Frederick, S. (2004). Activity in the dorsal cochlear nucleus of hamsters previously tested for tinnitus following intense tone exposure. *Neuroscience Letters*, 355, 121–125.

Kim, J. J., Gross, J., Morest, D. K., & Potashner, S. J. (2004a). Quantitative study of degeneration and new growth of axons and synaptic endings in the chinchilla cochlear nucleus after acoustic overstimulation. *Journal of Neuroscience Research*, 77, 829–842.

Kim, J. J., Gross, J., Potashner, S. J., & Morest, D. K. (2004b). Fine structure of long-term changes in the cochlear nucleus after acoustic overstimulation: Chronic degeneration and new growth of synaptic endings. *Journal of Neuroscience Research*, 77, 817–822.

Kim, J. J., Gross, J., Potashner, S. J., & Morest, D. K. (2004c). Fine structure of degeneration in the cochlear nucleus of the chinchilla after acoustic overstimulation. *Journal of Neuroscience Research*, 77, 798–816.

Kleinjung, T., Steffens, T., Londero, A., & Langguth, B. (2007). Transcranial magnetic stimulation (TMS) for treatment of chronic tinnitus: Clinical effects. *Progress in Brain Research*, 166, 359–367.

Klepper, A., & Herbert, H. (1991). Distribution and origin of noradrenergic and serotonergic fibers in the cochlear nucleus and inferior colliculus of the rat. *Brain Research*, 557(1–2), 190–201.

Knipper, M., Zimmermann, U., & Müller, M. (2009). Molecular aspects of tinnitus. *Hearing Research*, 266, 60–69.

Komiya, H., & Eggermont, J. J. (2000). Spontaneous firing activity of cortical neurons in adult cats with reorganized tonotopic map following pure-tone trauma *Acta Oto-Laryngologica*, 120, 750–756.

König, O., Schaette, R., Kempter, R., & Gross, M. (2006). Course of hearing loss and occurrence of tinnitus. *Hearing Research*, 221(1–2), 59–64.

Kowalska, S., & Sulkowski, W. (2001). Tinnitus in noise-induced hearing impairment. *Medical Practice*, 52(5), 305–313.

Langguth, B., Hajak, G., Kleinjung, T., Cacace, A., & Møller, A. R. (2007). *Tinnitus: Pathophysiology and Treatment.* Progress in Brain Research, Vol. 166. Amsterdam: Elsevier.

Langguth, B., Kleinjung, T., Landgrebe, M., de Ridder, D., & Hajak, G. (2010). rTMS for the treatment of tinnitus: The role of neuronavigation for coil positioning. *Neurophysiology Clinics*, 40(1), 45–58.

Lanting, C. P., de Kleine, E., & van Dijk, P. (2009). Neural activity underlying tinnitus generation: results from PET and fMRI. *Hearing Research*, 255(1–2), 1–13.

Liberman, M. C., & Kiang, N. Y-S. (1978). Acoustic trauma in cats. Cochlear pathology and auditory-nerve activity. *Acta Oto-Laryngologica*, 258(Supplementum), 1–63.

Liberman, M. C., & Dodds, L. W. (1984). Single-neuron labeling and chronic cochlear pathology II Stereocilia damage and alterations of spontaneous discharge rates *Hearing Research*, 416, 43–53.

Lobarinas, E., Yang, G., Sun, W., Ding, D., Mirza, N., Dalby-Brown, W., Hilczmayer, E., Fitzgerald, S., Zhang, L., & Salvi, R. (2006). Salicylate and quinine-induced tinnitus and effects of memantine. *Acta Oto-Laryngologica*, 556(Supplementum), 13–19.

Loeb, M., & Smith, R. P. (1967). Relation of induced tinnitus to physical characteristics of the inducing stimuli. *Journal of the Acoustical Society of America*, 43, 453–455.

Lonsbury-Martin, B. L., & Martin, G. K. (1981). Effects of moderately intense sound on auditory sensitivity in rhesus monkeys: Behavioral and neural observations. *Journal of Neurophysiology* 46, 563–586.

McFadden, D. (1982). Tinnitus: Facts, theories and treatments. *Report of Working Group 89. Committee on Hearing, Bioacoustics and Biomechanics.* Washington, DC: National Academy Press.

Meidinger, M. A., Hildebrandt-Schoenfeld, H., & Illing, R. B. (2006). Cochlear damage induces GAP-43 expression in cholinergic synapses of the cochlear nucleus in the adult rat: A light and electron microscopic study. *European Journal of Neuroscience*, 23, 3187–3199.

Meikle, M. B., & Greist, S. E. (1991). Computer data analysis: Tinnitus Data Registry. In A. Shulman (Ed.), *Tinnitus diagnosis and treatment* (pp. 416–430). Malvern, PA: Lea & Febiger.

Meleca, R. J., Kaltenbach, J. A., & Falzarano, P. R. (1997). Changes in the tonotopic map of the dorsal cochlear nucleus in hamsters with hair cell loss and radial nerve bundle degeneration. *Brain Research*, 750, 201–213.

Møller, M. B., Møller, A. R., Jannetta, P. J., & Jho, H. D. (1993). Vascular decompression surgery for severe tinnitus: Selection criteria and results. *Laryngoscope*, 103, 421–427.

Mulders, W. H., & Robertson, D. (2009). Hyperactivity in the auditory midbrain after acoustic trauma: Dependence on cochlear activity. *Neuroscience*, 164, 733–746.

Muly, S. M., Gross, J. S., Morest, D. K., & Potashner, S. J. (2002). Synaptophysin in the cochlear nucleus following acoustic trauma. *Experimental Neurolology*, 177, 202–221.

Muly, S. M., Gross, J. S., & Potashner, S. J. (2004). Noise trauma alters D-[^{3}H]aspartate release and AMPA binding in chinchilla cochlear nucleus. *Journal of Neuroscience Research*, 75(4), 585–596.

Munro, K. J., & Blount, J. (2009). Adaptive plasticity in brainstem of adult listeners following earplug-induced deprivation. *Journal of the Acoustical Society of America*, 126, 568–571.

Nicolas-Puel, C., Akbaraly, T., Lloyd, R., Berr, C., Uziel, A., Rebillard, G., & Puel, J. L. (2006). Characteristics of tinnitus in a population of 555 patients: Specificities of tinnitus induced by noise trauma. *International Tinnitus Journal*, 12, 64–70.

Noreña, A. J., & Eggermont, J. J. (2006). Enriched acoustic environment after noise trauma abolishes neural signs of tinnitus. *NeuroReport*, 17, 559–563.

Noreña, A., Micheyl, C., Chéry-Croze, S., & Collet, L. (2002). Psychoacoustic characterization of the tinnitus spectrum: Implications for the underlying mechanisms of tinnitus. *Audiology and Neurootology*, 7, 358–369.

Ochi, K., & Eggermont, J. J. (1997). Effects of quinine on neural activity in cat primary auditory cortex. *Hearing Research*, 105, 105–118.

Onizuka, S., Kasaba, T., Hamakawa, T., Ibusuki, S., & Takasaki, M. (2004). Lidocaine increases intracellular sodium concentration through voltage-dependent sodium channels in an identified lymnaea neuron. *Anesthesiology*, 101(1), 110–120.

Penner, M. J. (1983). Variability in matches to subjective tinnitus. *Journal of Speech and Hearing Research*, 26, 263–267.

Plant, L. D., Rajan, S., & Goldstein, S. A. (2005). K2P channels and their protein partners. *Current Opinions in Neurobiology*, 15(3), 326–333.

Plewnia, C., Reimold, M., Najib, A., Reischl, G., Plontke, S. K., & Gerloff, C. (2007). Moderate therapeutic efficacy of positron emission tomography-navigated repetitive transcranial magnetic stimulation for chronic tinnitus: A randomised, controlled pilot study. *Journal of Neurology, Neurosurgery and Psychiatry*, 78, 152–156.

Puel, J. L. (2007). Cochlear NMDA receptor blockade prevents salicylate-induced tinnitus. *B-ENT*, 7, 19–22.

Rajan, R., & Irvine, D. R. (1998). Absence of plasticity of the frequency map in dorsal cochlear nucleus of adult cats after unilateral partial cochlear lesions. *Journal of Comparative Neurology*, 399, 35–46.

Rajan, R., Irvine, D. R., Wise, L. Z., & Heil, P. (1993). Effect of unilateral partial cochlear lesions in adult cats on the representation of lesioned and unlesioned cochleas in primary auditory cortex. *Journal of Comparative Neurology*, 338, 17–49.

Reyes, S. A., Salvi, R. J., Burkard, R. F., Coad, M. L., Wack, D. S., Galantowicz, P. J., & Lockwood, A. H. (2002). Brain imaging of the effects of lidocaine on tinnitus. *Hearing Research*, 171(1–2), 43–50.

Roberts, L. E. (2007). Residual inhibition. *Progress in Brain Research*, 166, 487–495.

Roberts, L. E. (2011). Neural synchrony and neural plasticity in tinnitus. In A. R. Møller, B. Langguth, R. De Ridder, & T. Kleinjung (Eds.), *Textbook of tinnitus* (103–112). New York: Humana Press.

Roberts, L. E., Moffat, G., & Bosnyak, D. J. (2006). Residual inhibition functions in relation to tinnitus spectra and auditory threshold shift. *Acta Oto-Laryngologica*, 556(Supplementum), 27–33.

Roberts, L. E., Eggermont, J. J., Caspary, D. M., Shore, S. E., Melcher, J. R., & Kaltenbach, J. A. (2010). Ringing ears: The neuroscience of Tinnitus V. *Neuroscience*, 30, 14972–14979.

Robertson, D., & Irvine, D. R. (1989). Plasticity of frequency organization in auditory cortex of guinea pigs with partial unilateral deafness. *Journal of Comparative Neurology*, 282(3), 456–471.

Robinson, S. (2007). Antidepressants for treatment of tinnitus. *Progress in Brain Research*, 166, 263–271.

Schaette, R., & Kempter, R. (2006). Development of tinnitus-related neuronal hyperactivity through homeostatic plasticity after hearing loss: A computational model. *European Journal of Neuroscience*, 23(11), 3124–3138.

Schaette, R., & Kempter, R. (2008). Development of hyperactivity after hearing loss in a computational model of the dorsal cochlear nucleus depends on neuron response type. *Hearing Research*, 240(1–2), 57–72.

Schaette, R., König, O., Hornig, D., Gross, M., & Kempter, R. (2010). Acoustic stimulation treatments against tinnitus could be most effective when tinnitus pitch is within the stimulated frequency range. *Hearing Research*, 269(1–2), 95–101.

Seidman, M. D., Ridder, D. D., Elisevich, K., Bowyer, S. M., Darrat, I., Dria, J., Stach, B., Jiang, Q., Tepley, N., Ewing, J., Seidman, M., & Zhang, J. (2008). Direct electrical stimulation of Heschl's gyrus for tinnitus treatment. *Laryngoscope*, 118, 491–500.

Seki, S., & Eggermont, J. J. (2003). Changes in spontaneous firing rate and neural synchrony in cat primary auditory cortex after localized tone-induced hearing loss. *Hearing Research*, 180, 28–38.

Sheldrake, J. B., & Hazell, J. (1992). Maskers versus hearing aids in the prosthetic management of tinnitus. In J-M. Aran & R. Dauman (Eds.), *Proceedings of the IVth International Tinnitus Seminar* (pp. 395–399). Amsterdam: Kugler Ghedini.

Shi, Y., Burchiel, K. J., Anderson, V. C., & Martin, W. H. (2009). Deep brain stimulation effects in patients with tinnitus. *Otolaryngology, Head and Neck Surgery*, 141, 285–287.

Shiomi, Y., Tsuji, J., Naito, Y., Fujiki, N., & Yamamoto, N. (1997). Characteristics of DPOAE audiogram in tinnitus patients. *Hearing Research*, 108, 83–88.

Shore, S. E., Koehler, S., Oldakowski, M., Hughes, L. F., & Syed, S. (2008). Dorsal cochlear nucleus responses to somatosensory stimulation are enhanced after noise-induced hearing loss. *European Journal of Neuroscience*, 27, 155–168.

Silverstein, H. (1976). Transmeatal labyrinthectomy with and without cochleovestibular neurectomy. *Laryngoscope*, 86, 1777–1791.

Singer, J. H., Mirotznik, R. R., & Feller, M. B. (2001). Potentiation of L-type calcium channels reveals nonsynaptic mechanisms that correlate spontaneous activity in the developing mammalian retina. *Journal of Neuroscience*, 21(21), 8514–8522.

Smith, R. L. (1977). Short-term adaptation in single auditory nerve fibers: Some poststimulatory effects. *Journal of Neurophysiology*, 40, 1098–1111.

Snow, J. B. (2004). *Tinnitus: Theory and management*. Ontario, Canada: BC Decker.

Song, H. Y., Tong, Z., Wang, Y. M., Qian, S. J., Guo, R. X., & Shi, J. R. (2009). Effect of electrical stimulation of the primary auditory cortex on the spontaneous activities of the external nucleus of the inferior colliculus in a rat model of tinnitus induced by salicylate acid *Sheng Li Xue Bao*, 61, 121–126.

Tan, J., Rüttiger, L., Panford-Walsh, R., Singer, W., Schulze, H., Kilian, S. B., Hadjab, S., Zimmermann, U., Köpschall, I., Rohbock, K., & Knipper, M. (2007). Tinnitus behavior and hearing function correlate with the reciprocal expression patterns of BDNF and Arg3.1/arc in auditory neurons following acoustic trauma. *Neuroscience*, 145, 715–726.

Terry, A. M., Jones, D. M., Davis, B. R., & Slater, R. (1983). Parametric studies of tinnitus masking and residual inhibition. *British Journal of Audiology*, 17, 245–256.

Thomas, M. J., Watabe, A. M., Moody, T. D., Makhinson, M., & O'Dell, T. J. (1998). Postsynaptic complex spike bursting enables the induction of LTP by theta frequency synaptic stimulation. *Journal of Neuroscience*, 18, 7118–7126.

Thompson, A. M., & Thompson, G. C. (2001). Serotonin projection patterns to the cochlear nucleus. *Brain Research*, 907, 195–207.

Thompson, A. M., Moore, K. R., & Thompson, G. C. (1995). Distribution and origin of serotoninergic afferents to guinea pig cochlear nucleus. *Journal of Comparative Neurology*, 351, 104–116.

Turner, J. G., Brozoski, T. J., Bauer, C. A., Parrish, J. L., Myers, K., Hughes, L. F., & Caspary, D. M. (2006). Gap detection deficits in rats with tinnitus: a potential novel screening tool. *Behavioral Neuroscience*, 120, 188–195.

Tyler, R. S. (2008). *The consumer handbook on tinnitus*. Sedona: Auricle Ink Publishers.

Tzounopoulos, T. (2008). Mechanisms of synaptic plasticity in the dorsal cochlear nucleus: Plasticity-induced changes that could underlie tinnitus. *American Journal of Audiology*, 17, S170–175.

Wang, H., Brozoski, T. J., Turner, J. G., Ling, L., Parrish, J. L., Hughes, L. F., & Caspary, D. M. (2009). Plasticity at glycinergic synapses in dorsal cochlear nucleus of rats with behavioral evidence of tinnitus. *Neuroscience*, 164(2), 747–759.

Watanabe, S., Hoffman, D. A., Migliore, M., & Johnston, D. (2002). Dendritic K^+ channels contribute to spike-timing dependent long-term potentiation in hippocampal pyramidal neurons. *Proceedings of the National Academy of Sciences of the USA*, 99, 8366–8371.

Wilson, R. H., & Carhart, R. (1971). Forward and backward masking: interactions and additivity. *Journal of the Acoustical Society of America*, 49, 1254.

Zacharek, M. A., Kaltenbach, J. A., Mathog, T. A., & Zhang, J. S. (2002). Effects of cochlear ablation on noise induced hyperactivity in the hamster dorsal cochlear nucleus: Implications on the origin of noise-induced tinnitus. *Hearing Research*, 172, 137–144.

Zeng, C., Nannapaneni, N., Zhou, J., Hughes, L. F., & Shore, S. (2009). Cochlear damage changes the distribution of vesicular glutamate transporters associated with auditory and nonauditory inputs to the cochlear nucleus. *Journal of Neuroscience*, 29, 4210–4217.

Zhang, J., & Guan, Z. (2008). Modulatory effects of somatosensory electrical stimulation on neural activity of the dorsal cochlear nucleus of hamsters. *Journal of Neuroscience Research*, 86, 1178–1187.

Part III
Susceptibility and Factors Contributing to NIHL

Chapter 9
Genes That Influence Susceptibility to Noise-Induced Hearing Loss

Tzy-Wen Gong and Margaret I. Lomax

1 Introduction

The ability to detect sound is a genetic trait essential for survival and subject to evolutionary selection. With industrial operations and the use of power tools and equipment, noise impinges on many aspects of our daily life and we are subject to frequent bouts of excessive noise. Although exposure to less intense noise can lead to temporary loss of auditory response, the cumulative effect of such noise exposures over time may also cause permanent hearing loss. Even with the availability of protective devices, a substantial portion of the population with prolonged exposure to loud noise still experiences progressive, irreversible hearing loss (see Rabinowitz, Chap. 2). Tests that can predict which individuals have more noise sensitivity would allow them to make more knowledgeable choices in decisions affecting their degree of noise exposure, would enable them to take additional protective measures to minimize the impact of noise exposure (for a review of hearing protective devices, see Casali, Chap. 12), and would enable them to seek monitoring and treatment more proactively and frequently.

Our current understanding of the genetic influences on susceptibility to noise-induced hearing loss (NIHL) in humans is still rather limited, at least partially due to lack of timely diagnosis, repetitive noise exposure, and the confounding effects of age-related hearing loss (AHL). One study supporting the notion of genetic contribution to NIHL is the Finnish Twin Cohort involving 573 same-sexed pairs of 131 monozygotic and 442 dizygotic twins 31–88 years of age (Heinonen-Guzejev et al. 2005). Noise sensitivity of individuals, assessed through a questionnaire and categorized into four classes – high, quite high, quite low, and low – was more similar between

T.-W. Gong (✉)
Department of Otolaryngology-Head/Neck Surgery, Kresge Hearing Research Institute, The University of Michigan, 1150 W. Medical Center Drive, Ann Arbor, MI 48109-5648, USA
e-mail: tzywen@umich.edu

C.G. Le Prell et al. (eds.), *Noise-Induced Hearing Loss: Scientific Advances*, Springer Handbook of Auditory Research 40, DOI 10.1007/978-1-4419-9523-0_9,

monozygotic twins than between dizygotic twins. Quantitative genetic modeling also indicated significant inheritance within families. There are also some successes in humans using a candidate gene approach. These candidate genes include hereditary deafness genes and those involved in the molecular mechanisms critical for development of the auditory system and maintenance of its function. On the other hand, animal models provide invaluable tools in gene identification. In fact, mice provided the first evidence for a genetic influence on NIHL as well as recent identification of genes rendering cochlea more susceptible to noise trauma (reviewed in Ohlemiller 2006, 2008).

Inbred strains of mice, produced by successive generations of brother–sister mating, represent animal models with limited genetic variability within each strain, although considerable genetic variability occurs between strains. Many inbred mouse strains have been used by the scientific community to facilitate identification of human disease genes and our understanding of disease pathology, including deafness. For updates on mouse inbred strains and deafness models, see http://hearingimpairment.jax.org/models.html. Moreover, inbred mouse strains often harbor subtle changes in genes [single nucleotide polymorphisms (SNPs)] that produce interesting phenotypes. At least one such polymorphism has been linked to susceptibility to noise overstimulation in several different inbred mouse strains.

Advances in gene mapping in the 1990s culminated in the complete sequences for human and mouse genomes by 2000. Coupling these technical advances with the identification of large human families with inherited deafness has led to the identification of many hereditary deafness genes. As of 2010, approximately 60 human autosomal dominant (DFNA) and more than 80 autosomal recessive (DFNB) nonsyndromic deafness genes have been mapped to specific chromosomal locations (loci) and many genes underlying deafness have been identified (for updates, see the Hereditary Hearing Loss Homepage, http://hereditaryhearingloss.org).

In addition to the monogenic inheritance observed in many deaf families, deafness and susceptibility to noise can be influenced by multiple genes, each of which contributes to, or modifies, the final phenotype. Several approaches have been taken to facilitate identification of such interacting genes. Quantitative trait locus (QTL) analysis estimates the contribution of one or more genes to a measureable characteristic that varies in degree and can be used to define chromosomal regions harboring gene(s) of interests. This approach is often used in mouse studies where large numbers of progeny can be generated by crossing two or more strains. QTL analysis has been used in mouse studies to identify genes underlying progressive AHL (Erway et al. 1993; Drayton and Noben-Trauth 2006) and NIHL (Erway et al. 1996; Davis et al. 2001; White et al. 2009). In humans, QTL analysis may be less feasible because of a more limited number of subjects.

Attention has recently focused on the relationship between human genetic variations, such as the slight sequence differences or SNP that occur in genes, and specific diseases (Altshuler et al. 2008). Through the large-scale HapMap project, most of the commonly occurring SNPs have been identified and mapped (McVean et al. 2005). The HapMap project also demonstrated that blocks of closely linked SNPs, such as haplotypes, are coinherited; therefore, one of the SNPs in a region may be used to represent all the other SNPs.

This chapter provides an overview of NIHL-susceptibility genes identified to date. The proteins they encode are categorized based on corresponding function or molecular pathways affected by noise. Genes for these proteins constitute the candidates in searching for noise-susceptible genes in several human populations. Genetic studies that led to these gene identifications are described. This is followed by a discussion of recent developments in, and use of, human gene-association studies in gene identification and the impact of such studies in human populations for identifying the genetic basis of noise susceptibility and the challenges that remain.

2 Proteins Involved in Cochlear Development and Maintenance of Function

Studies in animal models enable in-depth characterization of the physiological and morphological impact of acoustic injury. Details of anatomical changes following noise have been described previously (Saunders et al. 1985; Henderson et al. 2006) and by Hu (Chap. 5). In brief, intense noise may induce mechanical damage, such as rupture of stereocilia tip links, rupture of cell–cell junctions, pillar cells buckling, and detachment of the organ of Corti from the basilar membrane. Intense noise also causes glutamate excitotoxicity, reduces cochlear blood flow, disturbs ionic balance and metabolic homeostasis, and enhances oxidative stress (Cheng et al. 2005; Henderson et al. 2006; Ohlemiller 2006; Le Prell et al. 2007). Important events following the noise exposure are removal of damaged hair cells (Hu et al. 2000, 2006) as well as pillar cells, intermediate cells of the stria, and type II and IV fibrocytes (Hirose and Liberman 2003). Loss of hair cells is irreversible and may lead to retraction of afferent nerves and permanent hearing loss. The following sections review genes that influence many of these processes. Genes with significant association are summarized in Table 9.1.

2.1 Structural Integrity of Stereocilia

In the cochlea, sensory hair cells are among those cells most susceptible to noise trauma (Frolenkov et al. 2004). The hair bundle on the apical surface of inner and outer hair cells (OHCs) consists of precisely organized actin-rich stereocilia. Noise may affect insertion of stereocilia of the OHCs into the tectorial membrane and cause mechanical damage to stereocilia. Although disconnected stereocilia of the OHCs may be reattached to the tectorial membrane during a window of time (Nordmann et al. 2000), more severe damage can be irreversible and lead to death of hair cells, and thus permanent hearing loss.

Table 9.1 Summary of genes implicated in noise-induced hearing loss

Pathway	Protein	Human gene	Significant human SNPs, haplotype	Mouse models
Stereocilia structure	Cadherin 23	CDH23	rs3802711, rs1227049, rs1227049 (Yang et al. 2006a); F1888S (Schultz et al. 2005)	*Cdh*^{A753G} (Johnson et al. 2000; Noben-Trauth et al. 2003); *waltzer* (*v*) (Di Palma et al. 2001); *modifier of deafwaddler (mdfw)* (Zheng and Johnson 2001); *salsa* (Schwander et al. 2009)
	Protocadherin 15	PCDH15	rs7095441 (Konings et al. 2009b)	*Ames waltzer* (*av*) (Alagramam et al. 2001)
Ion homeostasis	Plasma membrane Ca^{2+} pump, PMCA2	ATP2B2	V586M (Schultz et al. 2005)	*deaf waddler* (*dfw*) (Street et al. 1998); *Pmca2*$^{+/-}$ mice (Kozel et al. 2002)
	Transient receptor potential vanilloid 4	TRPV4		*Trpv4* knockout (Tabuchi et al. 2005)
	Potassium channel	KCNE1	rs2070358,[a] rs1805127, rs1805128 (Van Laer et al. 2006; Pawelczyk et al. 2009)	
		KCNQ1	rs163171 (Van Laer et al. 2006; Pawelczyk et al. 2009); rs7945327,[b] rs463924,[b] rs11022922,[b] rs718579,[b] rs2283205,[b] rs231899,[b] rs2056892[b] (Van Laer et al. 2006; Pawelczyk et al. 2009)	
		KCNQ4	rs34287852[a] (Van Laer et al. 2006; Pawelczyk et al. 2009)	
	Inwardly rectifying K^+ channel	KCNJ10	rs1130183[b] (Pawelczyk et al. 2009)	
	Connexin 32	GJB1	rs1997625[b] (Pawelczyk et al. 2009)	
	Connexin 26	GJB2	rs3751385,[a] M34T[b] (Van Laer et al. 2006; Van Eyken et al. 2007; Pawelczyk et al. 2009)	
	Connexin 30.3	GJB4	rs755931 (Pawelczyk et al. 2009)	

Oxidative stress	Catalase	CAT	rs494024,[a] rs475043,[a] rs564250,[c] rs1001179,[c] rs12273124[c] (Konings et al. 2007)	
	Cu, Zn-superoxide dismutase	SOD1		*Sod1* knockout (Ohlemiller et al. 1999a; McFadden et al. 2001)
	Mn-superoxide dismutase	SOD2	IVS3-23T/G, IVS3-60T/G (Fortunato et al. 2004)	
	Paraoxonase 2	PON2	S311C (Fortunato et al. 2004)	
	Glutathione *S*-transferase, mu	GSTM1	GSTM1 null (Rabinowitz et al. 2002; Carlsson et al. 2005)	
	Glutathione peroxidase 1	GPX1		*Gpx1* knockout (Ohlemiller et al. 2000; McFadden et al. 2001)
General stress protection	Heat shock factor 1	HSF1		*Hsf1* knockout (Fairfield et al. 2005)
	Heat shock protein 70	HSP70	Haplotype GGC, GGT (Yang et al. 2006b; Konings et al. 2009a); HSP70-1 (rs1043618), HSP70-2 (rs1061581), HSP70-hom (rs2227956),[a] Haplotype GAC,[a] CGT (Yang et al. 2006b; Konings et al. 2009a)	
	Glucocorticoid receptor	GCR		Mice adrenalectomized or pharmacologically treated (Tahera et al. 2007)
Others	Glutamate-aspartate transporter, GLAST	SLC1A3		*Slc1a3* knockout (Hakuba et al. 2000)
	Myosin heavy chain	MYH14	rs667907[a], rs588035[a] (Konings et al. 2009b)	

[a] Significant in both Swedish and Polish cohorts
[b] Significantly correlated with hearing loss when noise level is considered
[c] Significant in the Polish cohort, not the Swedish cohort

The precisely organized staircase-like morphology of the hair bundle is maintained by a series of extracellular filaments between adjacent stereocilia: kinociliary links, ankle links, lateral links, and tip links. While kinociliary links, ankle links, and some lateral links are transient structures essential for normal development of stereocilia, top portions of lateral links and tip links are permanent structures critical in maintaining auditory function (Muller 2008; Petit and Richardson 2009). The tip link between the top of one shorter stereocilium and the upper region of an adjacent taller stereocilium is believed to gate transduction channels and are thus essential for normal mechanoelectrical transduction (Hudspeth 1989). Recent findings illustrate that each tip link consists of a cadherin 23 (encoded by CDH23[1]) homodimer on the upper half and a protocadherin 15 (encoded by PCDH15) homodimer on the lower half (Kazmierczak et al. 2007). Both CDH23 and PCDH15 are members of the FAT cadherin superfamily and have substantially longer extracellular domains. Whereas classical cadherins contain five repeated, conserved extracellular domains, CDH23 and PCDH15 contain 27 and 11 repeats, respectively. In addition to the tip links in the hair bundles of mature hair cells, the transient lateral links and kinociliary links in developing hair cells also contain CDH23 and PCDH15, reflecting the critical roles of these two molecules in development as well as maintenance of hair cell function (reviewed in Muller 2008; Sakaguchi et al. 2009). Any variations in protein sequence of CDH23 and PCDH15 are likely to affect development and maintenance of these structures that are essential for sound transduction from mechanical stimuli to electrical signals.

Disease-causing mutations have been found throughout the CDH23 and PCDH15 genes with no apparent sites with an unusually high frequency of mutation. Truncation mutations affecting structure and length of ectodomains are known to cause congenital syndromic deafness in Usher syndrome (CDH23 in USH1D; PCDH15 in USH1F), with auditory, vestibular, and retinal dysfunction (Ahmed et al. 2001; Bolz et al. 2001). On the other hand, missense mutations encoding proteins of more subtle point defects tend to result in deafness without other symptoms, that is, nonsyndromic deafness (such as CDH23 in DFNB12; PCDH15 in DFNB23) (Bork et al. 2001; Ahmed et al. 2003). Similar observations have been made in mouse models. For instance, null mutations in mouse *Cdh23* and *Pcdh15* cause disorganization of hair bundles and early deterioration of auditory and vestibular sensory epithelium in *waltzer* (*v; Cdh23*$^{834-835insG}$) and *Ames waltzer (av)* mice, respectively (Alagramam et al. 2001; Di Palma et al. 2001). A missense mutation of *Cdh23*^{A2210T} in *salsa* mice leads to a milder phenotype with normal development but progressive loss of stereocilia tip links, resulting in hearing loss that resembles DFNB12 patients (reviewed in Petit and Richardson 2009; Schwander et al. 2009).

[1] Based on the guidelines for human and mouse genome nomenclature, human gene symbols are in uppercase, while mouse gene symbols are italicized and use an initial capital letter, followed by lowercase letters. For instance, symbols for the human and mouse genes encoding cadherin 23 protein are CDH23 and *Cdh23*, respectively. Protein abbreviations are in uppercase.

Evidence that CDH23 and PCDH15 play a role in NIHL comes from inbred strains of mice harboring subtle differences in gene sequences. Auditory studies of common inbred mouse strains noted that at 23 months CBA mice have the best hearing and the lowest auditory brainstem response (ABR) thresholds, whereas DBA/2J and C57BL/6J (BL6) mice exhibit early or accelerated hearing loss as they age (Erway et al. 1993). Several groups have also compared CBA with BL6 and have shown that BL6 mice with early or accelerated hearing loss are more susceptible to noise and exhibit larger threshold shifts at most frequencies tested (Hultcrantz and Li 1993; Li and Borg 1993; Li et al. 1993). CBA mice also exhibit rapid recovery of auditory function in the first 3 days after noise exposure, whereas BL6 mice showed limited recovery over time (Erway et al. 1996; Davis et al. 2001; Noben-Trauth et al. 2003).

Mapping studies identified a locus on mouse chromosome (Chr) 10 responsible for the early onset of hearing loss in the BL6 as well as in nine other common inbred strains (Erway et al. 1996; Johnson et al. 1997, 2000). This locus was originally designated *Ahl*, and later renamed as *Ahl1*, for the first locus of age-related hearing loss. Genetic complementation tests show that *Ahl* and *Nihl* are allelic, that is, localized to the same chromosomal position and indistinguishable from each other (Johnson et al. 2000). Mice inheriting two copies of the mutant *Ahl* allele (*Ahl* homozygotes) have an earlier onset of hearing loss as well as greater threshold shifts after noise exposure than those with one or two copies of the wild-type *Ahl* allele. Thus, sensitivity to noise in BL6 mice also appears to be a recessive trait that cosegregates with *Ahl*. The genetic basis of these phenotypes is an A753G polymorphism on *Cdh23* exon 7 (Davis et al. 2003; Noben-Trauth et al. 2003) that leads to a higher probability of exon 7 skipping and an in-frame deletion of 143 amino acids in the C-terminal domain of CDH23. Altered adhesion or reduced stability of CDH23 may confer susceptibility to AHL and NIHL.

The first human gene association study that assessed CDH23 as a NIHL susceptibility gene was a cross-sectional epidemiology study of Chinese industrial workers (Yang et al. 2006a). SNPs in the CDH23 gene were examined in 93 workers with NIHL versus 101 workers with normal hearing. Two CDH23 SNPs showed strong correlations with NIHL. At SNP rs3802711, individuals with the T/T genotype were at higher risk than those with the C/T genotype. At the terminal position of exon 7, individuals with the G/G genotype were more susceptible to noise than those with the A/G genotype. After adjusting for age, gender, smoking, and history of noise exposure, the C/C genotype at SNP rs1227049 was also associated with a higher risk compared to the G/G genotype. This suggests that certain CDH23 genotypes may render individuals more susceptible to noise-induced hearing impairment. A more extensive list of 35 SNPs in CDH23 has been evaluated in a Polish population and its association with NIHL is underway (Sliwinska-Kowalska et al. 2008).

PCDH15 may also be a NIHL susceptibility gene. In another association study of the same candidate gene approach, 644 SNPs covering 53 candidate genes were examined in two independent groups of NIHL subjects in Sweden and in Poland (Konings et al. 2009b). SNP rs7095441 in PCDH15 was found to be associated with NIHL in both populations.

2.2 Ion Homeostasis: Pumps, Channels, and Connexins

Maintenance of ion homeostasis is critical for auditory function. Intracellular calcium (Ca^{2+}) is an important signaling molecule that regulates neurotransmitter release by the inner hair cell (IHC). Calcium is also important in maintaining CDH23–PCDH15 interactions, which are Ca^{2+} sensitive. Type 2 plasma membrane Ca^{2+}-ATPase pump (PMCA2, ATP2B2) is an important regulator of interstereociliary Ca^{2+} levels. In the mouse inner ear, PMCA2 is localized to stereocilia and the basolateral wall of both auditory and vestibular hair cells. Considering its importance in maintaining Ca^{2+} homeostasis, this calcium pump is clearly important for NIHL. Defects in PMCA2 are likely to affect CDH23 and PCDH15 interactions and subsequent mechanosensory transduction of stereocilia as well as neurotransmitter release by the IHC. Indeed, inactivating mutations in *Atp2b2* result in deafness and vestibular dysfunction in the spontaneous mouse mutant *deaf waddler* (*dfw*) and in mice with an *Atp2b2*-targeted deletion (Kozel et al. 1998; Street et al. 1998).

Phenotypic expression of the recessive deaf mutant *dfw* that harbors inactivating mutations in PMCA2 is known to be affected by the mouse genetic background (Noben-Trauth et al. 1997). Whereas *dfw/dfw* homozygotes are deaf, *+/dfw* heterozygotes exhibit either normal hearing or an accelerated early onset of hearing loss, depending on a locus on Chr 10 which was named modifier of deaf waddler (*mdfw*). Genetic and functional studies revealed that *mdfw* is allelic to *Ahl* (Zheng and Johnson 2001). Identification of *Cdh23* as a modifier gene of *mdfw* is the first example of a modifier gene in deafness and a nice demonstration of digenic inheritance in deafness. The interesting effect of modifier genes is not limited to mouse models but is also observed in humans. Heterozygotic mutation in $ATP2B2^{V586M}$ has been shown in a family to increase severity of deafness due to $CDH23^{F1888S}$ homozygosity (Schultz et al. 2005). These studies provide convincing evidence that subtle mutations in one gene can modify the effects of a second one, leading to more extreme phenotypes. They may also provide a molecular basis for the variable vulnerability in noise sensitivity observed in human populations and suggest that CDH23 and ATP2B2 are excellent candidate genes for association studies.

In addition to its critical role in CDH23–PCDH15 interaction, calcium is also important for intracellular signaling. Transient receptor potential vanilloid 4 (TRPV4) is a Ca^{2+}-permeable cation channel that is thought to function as an osmosensory and mechanosensory receptor. In the mouse cochlea, TRPV4 is expressed in the inner and OHCs and spiral ganglion neurons (Shen et al. 2006). TRPV4 is responsible for Ca^{2+} influx into OHCs induced by hypotonic stimulation. *Trpv4* knockout mice exhibit normal cochlear morphology (in the organ of Corti, spiral ganglion neurons, stria vascularis) as well as auditory function in comparison to wild type littermates up to 5 months of age. However, *Trpv4* knockout mice exhibit greater sensitivity to noise, evident in a greater permanent threshold shift after a 4 h 128 dB pure tone noise (Tabuchi et al. 2005).

The unique ionic composition of the endolymph in scala media, with its high potassium (K^+) content, is important for hair cell function. Activation of the transduction channel in the stereocilia of hair cells allows influx of K^+ ions into IHCs via the apical transduction channel and signals release of neurotransmitters to afferent neurons. K^+ is driven out into perilymph via basolateral K^+ channels, including KCNQ4, KCNN2, and KCNMA1. There are two pathways for transporting K^+ ions back to the endolymph. The lateral pathway involves supporting cells of the organ of Corti, fibrocytes of the spiral ligament, and the stria vascularis (KCNJ10 in intermediate cells; KCNE1/KCNQ1 in marginal cells). The medial pathway passes the supporting cells in the direction of the interdental cells (reviewed in Wangemann 2006; Zdebik et al. 2009). Failure to maintain the secretion and reabsorption of fluid and ionic balance in scala media results in an enlargement of the endolymphatic compartment as seen in Ménière's disease and Pendred syndrome or collapse of the compartment as seen in Jervell and Lange-Nielsen syndrome. Mutations in genes involved in K^+ recycling and ion homeostasis often cause congenital deafness in humans and mice. These critical genes identified to-date include KCNE1, KCNJ10, KCNQ1, and KCNQ4 for K^+ channel proteins, SLC12A2 for a Na, K, Cl-cotransporter, SLC26A4 for an anion channel pendrin, and GJB2, GJB3, GJB6, and GJA1 for connexins (reviewed in Wangemann 2006).

Connexins are subunits of multimeric intercellular channels called gap junctions. These channels form pores between adjacent cells, allowing small molecules and metabolites to pass from one cell to another (Kikuchi et al. 2000; Zhao et al. 2006). GJB2 (DFNB1) encodes connexin 26 (Cx26) and is also the most frequently occurring deafness gene. The most common mutation in Caucasians is the 25delG mutation that leads to truncation of Cx26 and profound hearing loss. In addition to GJB2, mutations in GJB1 (for Cx32), GJB3 (Cx31), GJB4 (Cx30.3), GJB6 (Cx30), and GJA1 (Cx43) have also been identified in nonsyndromic deafness of either recessive or dominant inheritance. Mutations in GJA1 have been identified in syndromic deafness. In examining whether GJB2 mutation renders subjects more susceptible to NIHL, no correlation was found in 702 Polish $GJB2^{35delG}$ carriers (heterozygotes) (Van Eyken et al. 2007).

In light of the importance in maintaining correct ionic balance for normal auditory function, the aforementioned deafness genes involved in maintaining K^+ homeostasis are compelling candidates for NIHL susceptibility genes. Several gene association studies examined SNPs in these genes in populations of factory workers of industries with high levels of ambient noise, such as the auto industry, aircraft industry, steel mills, or paper mills. In a case-control study genetic analysis was carried out in the 10% most sensitive subjects compared to the 10% most resistant subjects selected from 1,261 male noise-exposed Swedish industrial workers (Van Laer et al. 2006). Ten candidate genes, for example, five connexin genes (GJB1, GJB2, GJB3, GJB4, GJB6) and five K^+-channel genes (KCNE1, KCNJ10, KCNQ1, KCNQ4, SLC12A2) were analyzed for 35 SNPs. Significant differences between susceptible and resistant individuals were found for three SNPs in KCNE1 and one SNP each in KCNQ1 and KCNQ4. After analysis for genotype or haplotype frequencies, only SNP rs2070358 in KCNE1 remained statistically significant, whereas the other SNP associations were no longer significant. There was no significant difference in allele frequencies in any SNPs in connexin genes.

Another study using a similar approach was conducted recently in 702 Polish noise-exposed workers, but with an expanded panel of 99 SNPs for the same ten candidate genes involved in K^+-recycling (Pawelczyk et al. 2009). A significant association with NIHL was found in KCNE1 SNP rs2070358 and KNCQ4 SNP rs34287852, which supports the finding in the Swedish population. In addition, significant associations were found for SNPs in GJB1, GJB2, GJB4, KCNJ10, and KCNQ1, suggesting a potential role for these genes in NIHL. Further association studies in a different population or analysis of the contribution of these SNPs to protein function will be necessary to confirm their role in NIHL.

2.3 *Oxidative Stress*

Reactive oxygen species (ROS) and reactive nitrogen species (RNS) have been suggested as major culprits in mediating noise-induced cell death in the cochlea (see Le Prell and Bao, Chap. 13). ROS are formed as natural byproducts of normal aerobic metabolism. These highly reactive ROS/RNS, including superoxide anion ($O_2{\bullet}^-$), hydroxyl free radicals (•OH), hydrogen peroxide (H_2O_2), hydrochlorous acid (HClO), nitric oxide free radicals (NO•), and peroxynitrite ($ONOO{\bullet}^-$), are vital signaling molecules for normal cellular function and are typically metabolized via cellular reduction/oxidation (redox) reactions. For instance, superoxide is reduced by superoxide dismutases (SOD) to H_2O_2, which is subsequently reduced to water by catalase or peroxidases of the glutathione and thioredoxin systems. Combined effects of these enzymes in the redox systems are considered the first-line defense mechanism in controlling ROS/RNS levels and maintaining redox homeostasis, vital to the function of many cellular proteins. Under certain conditions, ROS/RNS production can increase dramatically to levels exceeding the cell's normal redox capacity, which leads to propagation of free radicals and causes significant oxidation of lipids, proteins, and DNA, and ultimately leads to damage to cell structures.

Exposure to noise has been shown to lead to overproduction of H_2O_2 in mitochondria and to redox disequilibrium and release of cytochrome *c*, which triggers the apoptotic cascade and results in cell death (Kim et al. 2006). Accumulation and propagation of free radicals and ROS/RNS over time lead to damage to a broad area in the organ of Corti region, spiral ganglion cells, and stria vascularis (Yamane et al. 1995; Ohlemiller et al. 1999b). The link between noise and overproduction of ROS/RNS is strengthened by the protective effects against NIHL of free radical scavengers, including vitamin E, ascorbic acid, D-methionine, *N*-acetyl-L-cysteine, and the coenzyme Q mimetic idebenone (reviewed in Le Prell et al. 2007; also in Le Prell and Bao, Chap. 13).

Metabolism of superoxide and derived peroxides requires the combined effect of catalase, SOD, and enzymes in the antioxidative glutathione, glutaredoxin, and thioredoxin systems. Decreased activity in any of these components may lead to compromised antioxidative capacity and increased ROS/RNS damage.

Exposure to a moderate noise level can immediately reduce catalase activity by 30% in the stria vascularis, shown in the chinchilla model, thus contributing to decreased overall antioxidant capacity (Jacono et al. 1998). Intracochlear delivery of adenovirus-carried catalase in a guinea pig model of NIHL effectively protects hearing thresholds and reduces hair cell loss (Kawamoto et al. 2004). Association studies have been carried out in humans to examine the role of catalase in susceptibility to NIHL. Fourteen SNPs covering the entire CAT gene were genotyped and compared between the 10% most sensitive and resistant subjects selected from 1,261 Swedish workers and 4,500 Polish noise-exposed industrial workers (Konings et al. 2007). There were significant associations between noise exposure levels and SNPs rs494024 and rs475043 in CAT in both populations, while three additional SNPs were also significantly associated in the Polish population.

Partial or complete elimination of cytosolic SOD (Cu, Zn-SOD, SOD1) in mice leads to a slightly elevated baseline auditory threshold and greater sensitivity to noise (McFadden et al. 1999, 2001; Ohlemiller et al. 1999b). However, overexpression of SOD1 in transgenic mice, intracochlear gene delivery, or intraperitonel injection of purified protein does not offer additional protection against noise, drug insults, or aging (Kawamoto et al. 2004; Endo et al. 2005; Keithley et al. 2005).

In comparison to SOD1, mitochondrial SOD (Mn-SOD, SOD2) may also play a significant role. *Sod2*$^{-/-}$ knockout mice die within the first 2 postnatal weeks, whereas *Sod2*$^{+/-}$ mice appear normal with no obvious pathology (Huang et al. 1999) and possess normal hearing at a young age (Le and Keithley 2007). Overexpression of SOD2 does effectively protect hearing and hair cells from aminoglycoside ototoxicity (Kawamoto et al. 2004). Therefore, SOD2, with its close proximity to superoxide generated in the mitochondrion, is likely to play a vital role in quenching H_2O_2 and oxidative stress after noise. Indeed, SOD2 polymorphism has been shown to be associated with NIHL susceptibility in an Italian study in comparing 31 individuals with good hearing to 63 individuals with NIHL. These 94 male aircraft factory workers were selected from a cohort of 252 co-workers, based on the criteria of prolonged exposure to loud noise (average 94 dB for 20 years), use of the same protection device, and no hyperlipidemia or diabetes (Fortunato et al. 2004). Logistic regression analysis showed that two intronic polymorphisms (SOD2 IVS3-23T/G, IVS3-60T/G), among five SNPs analyzed, are associated with NIHL. Because these polymorphisms reside in an intronic region, the functional consequences of these polymorphisms are not clear. In light of the small sample size and a strong confounding factor of smoking, also suggested by previous studies (Barone et al. 1987; Cruickshanks et al. 1998; Mohammadi et al. 2010), this finding needs to be confirmed in another population or cohort.

Paraoxonases (PONs) are glycosylated proteins with multienzymatic activities involved in metabolism of organophosphates and lipid derivatives. There are three linked genes (PON1-3) encoding three PON proteins with slightly different expression patterns and substrate specificities. PON1 and PON3 are expressed in the liver and secreted into the blood, where they are associated with high-density lipoprotein (HDL) to enhance the antiatherosclerotic properties of HDL and have been implicated in protection against cardiovascular diseases. PON2 is ubiquitously expressed in most cells and is thought to exert its antioxidant effects at a cellular level. The aforementioned study of Italian aircraft factory workers exposed to prolonged loud noise also analyzed associations between NIHL and PON1 or PON2 (Fortunato et al. 2004). Logistic regression analysis showed a genetic predisposition of the PON2 311C allele in NIHL (Fortunato et al. 2004). On the other hand, there was no significant association between NIHL and the two popular polymorphisms in PON1: Q192R, and M55L.

The glutathione system plays a central role in intrinsic antioxidative mechanisms. Glutathione is a glutamate–cysteine–glycine tripeptide in which the single –SH group of the cysteine residue acts as an electron donor/acceptor. Mice lacking the ubiquitous glutathione peroxidase, GPX1, have elevated hearing thresholds and are more sensitive to noise (Ohlemiller et al. 2000; McFadden et al. 2001). Glutathione peroxidases (encoded by GPX1-6) use reduced glutathione (GSH) as an electron donor to reduce H_2O_2 to water or reduce lipid hydroxyperoxide to its corresponding alcohol, and generate glutathione disulfide (GSSG). GSSG can be reduced to regenerate GSH by glutathione reductase (encoded by GSR). In addition to regeneration from GSSG, intracellular GSH can be derived from de novo synthesis. Glutathione biosynthesis requires two ATP-dependent enzymes: γ-glutamylcysteine ligase (GCL) and glutathione synthase. Further, GSH can form conjugates via actions of glutathione *S*-transferases (GSTs), or react with electrophilic compounds nonenzymatically. In fact, glutathione conjugation is an essential aspect of both xenobiotic and normal physiological metabolism. Formation of protein-S-SG mixed disulfides, glutathione conjugation, and GSSG excretion can thus reduce the GSH pool available and increase demands of de novo synthesis to replenish the antioxidative capacity (Dickinson and Forman 2002). A low level of glutathione may render cells more susceptible to oxidative stress.

Glutathione *S*-transferases (GSTs) comprise a superfamily of enzymes including several classes of cytosolic enzymes: α (encoded by GSTA1-5), μ (GSTM1-5), θ (GSTT1-2), π (GSTP1), ω (GSTO1-2), ζ (GSTZ1), and a mitochondrial enzyme κ (GSTK1). These GST enzymes vary in their substrate affinities, tissue distributions, and levels of expression (Sundberg et al. 1993). In humans, there are genetic variations in GST genes; about 50% of Caucasians carry a deletion of GSTM1 (Board 1981) and 25–40% of all humans lack the GSTT1 gene (Pemble et al. 1994). GSTA, GSTM, and GSTP are known to be expressed in the rat cochlea, while the others have not been examined (Whitlon et al. 1999).

These GST genes were among the first candidates examined in human association studies. In a study attempting to identify NIHL susceptible genes for proteins that protect against oxidative stress, 58 workers of both genders and heterogeneous ethnicity were recruited from three American factories with noisy working environments (average noise level of 87 dBA); audiometric measurements correlated with the GSTM1 and GSTT1 genotypes (Rabinowitz et al. 2002). Individuals possessing the GSTM1 gene had significantly better high-frequency otoacoustic emissions compared to GSTM1 null individuals, suggesting a protective role for GSTM1 against NIHL. This correlation persists even after accounting for age, race, sex, and years of noise exposure. However, the association of GSTM1 genotype with hearing function was not observed in the Swedish study involving workers in paper pulp mills or in a steel factory (Carlsson et al. 2005). On the other hand, the GSTT1 genotype did not correlate with hearing function in the American or the Swedish study, neither GSTP1 nor GSR SNPs in the Swedish study. Although various GST genes appear to contribute differently to protection of hearing function, it is yet to be determined if compensatory effects exist among GST isoforms.

2.4 Protective Pathways: Stress Response and Glucocorticoids

Exposure to a mild noise has long been known to protect the ear from a subsequent louder noise. This phenomenon is known as "sound conditioning" (Niu and Canlon 2002). Several mechanisms have been proposed to explain this phenomenon. In addition to the aforementioned endogenous antioxidant systems, induction of the heat shock response and activation of glucocorticoid signaling pathway are perhaps the best developed.

Heat shock transcription factor 1 (HSF1) is an important transcription factor that controls the inducible stress response, which protects cells and tissues from numerous cellular and environmental stresses (Morimoto et al. 1997). This response, often called the heat shock response, has been conserved throughout evolution, from bacteria to humans. Stressors activate HSF1, which trimerizes, translocates to the nucleus, and binds to heat shock elements in the genes for heat shock proteins (HSPs), leading to rapid induction of these important molecular chaperones. Both heat and noise stresses are known to activate the heat shock response in the cochlea and protect it from subsequent noise trauma (Yoshida et al. 1999, 2000; Sugahara et al. 2003).

Mice lacking HSF1 (*Hsf1* knockouts) are often more sensitive to stressors and the auditory system is no exception. *Hsf1*$^{-/-}$ null mice are viable (Xiao et al. 1999) and have normal auditory thresholds (Sugahara et al. 2003; Fairfield et al. 2005). However, in response to a mild noise that produces only a temporary hearing loss in wild-type littermates, *Hsf1*$^{-/-}$ null mice suffer from a greater loss in function and in number of OHCs (Fairfield et al. 2005). Heat stress, either from whole body heat shock (Yoshida et al. 1999) or local hyperthermia (Sugahara et al. 2003), has been shown to upregulate *Hsp70* expression in the cochlea, which is presumed to play a key role in protecting hair cells from subsequent noise or other stresses.

Induction of genes for HSPs is often used to assess activation of the heat shock response, and induction of HSP70 is a favored target. There is a cluster of human HSP70 genes on Chr 6p21.3. The two inducible genes, HSP70-1 and HSP70-2, differ by only eight nucleotides in the coding region, encode an identical protein, and share similar structures and functions. The third gene in the cluster, HSP70-hom, is expressed constitutively and encodes a protein with 90% sequence identity to HSP70-1/HSP70-2. These and other HSP70 family proteins are molecular chaperones that control protein folding and prevent aggregation of proteins.

Several studies have implicated HSP as susceptibility genes for NIHL. One study examined SNPs, by restriction fragment length polymorphism (RFLP), in the three linked HSP70 genes in 194 Chinese autoworkers who had been exposed to ambient factory noise for 1 year, and who had no other confounding factors, such as additional noise exposure or exposure to known toxicants (Yang et al. 2006b). Association was determined between audiometric measurements of normal hearing versus noise-impaired subjects for three informative polymorphic markers for HSP70-1 (rs1043618), HSP70-2 (rs1061581), and HSP70-hom (rs2227956). There was no significant association between hearing loss and any of the three SNPs individually. However, those more susceptible to NIHL appeared to have higher probabilities of possessing two (GGC, GGT) of the eight possible haplotypes, that is, combinations of alleles of these three closely linked genes that are inherited together. Recently, this analysis of HSP70 genes was extended to two additional independent populations: 206 male Swedish factory workers and 238 Polish workers from different industries (Konings et al. 2009a). In this study, SNP rs2227956 in HSP70-hom showed significant association with NIHL in both sample sets, while rs1043618 and rs1061581 corresponding to the other two HSP70 genes were also significant in the Swedish sample set. Analysis of the haplotypes composed of these three SNPs revealed significant associations between NIHL and haplotype GAC in both sample sets and with haplotype CGT in the Swedish sample set. In contrast, the two significant haplotypes GGC and GGT identified previously in the Chinese sample set are infrequent or absent in the Swedish and Polish sample sets, likely due to ethnic differences. Nevertheless, associations between NIHL and HSP haplotypes in three independent populations strongly support the role of HSP70 in affecting sensitivity to noise.

Mild to moderate noise activates the hypothalamic–pituitary–adrenal (HPA) axis. This phenomenon is thought to contribute, at least in part, to the protective mechanism underlying sound conditioning to modulate noise sensitivity (Tahera et al. 2007) and to reduce the impact of a subsequent loud noise. Activation of the HPA axis results in elevation of plasma glucocorticoids (GC) and increased expression of glucocorticoid receptors (GR) in spiral ganglion neurons. GR is subsequently translocated into the nucleus, and acts as a transcription factor to regulate gene expression via the glucocorticoid response element (GRE), AP-1 site (when in combination with Fos and Jun), or NFκB-binding site (if in combination with NFκB). The critical role of GC/GR in modulating noise sensitivity is supported further by absence of protective effects by sound conditioning when animals are adrenalectomized or treated with corticosterone synthesis inhibitor metyrapone and GR antagonist RU486 (Tahera et al. 2007). This GR-mediated pathway also appears to be involved in restraint-stress-induced protection (Meltser et al. 2009), as well as glucocorticoid therapy in various hearing disorders.

2.5 Other Processes

Glutamate is an excitatory neurotransmitter released from the IHCs to afferent neurons and is thus a critical component of the mechanoelectrical transduction pathway (Ruel et al. 2007). The importance of proper neurotransmitter release in normal auditory function is demonstrated by profound hearing impairment in deletion/mutation of *Bsn* for Bassoon (a scaffolding protein in anchoring ribbons at the IHC presynaptic active zone), OTOF/DFNB9 for otoferin (involved in synaptic-vesicle exocytosis), and *Slc17a8* for vesicular glutamate transporter 3 (VGLUT3, transport of glutamate into presynaptic vesicles) (Yasunaga et al. 1999; Varga et al. 2003; Khimich et al. 2005). A subtle change in SLC17A8 has also been reported recently in nonsymdromic DFNA25 patients with progressive hearing loss that resembles AHL (Ruel et al. 2008). Similarly, subtle changes in SLC17A8 as well as other proteins involved in glutamate transmission may have a profound impact on the response to noise trauma.

Once released into synaptic cleft, glutamate, if not removed in a timely fashion, can result in overstimulation and subsequent retraction of postsynaptic afferent dendrites, the phenomenon known as excitotoxicity (Pujol and Puel 1999). Administration of glutamate antagonists, for example, MK 801, kynurenate, and carbamethione, significantly reduces noise-induced threshold shift (Ruel et al. 2007). *SLC1A3* encodes glutamate-aspartate transporter GLAST that is highly expressed in the supporting cells surrounding IHCs. In addition, it is also expressed in fibrocytes of the limbus region, spiral ganglion neurons, and spiral ligament in the cochlea, supporting the notion that GLAST functions to remove glutamate released by the inner hair cells away from the synaptic cleft and to facilitate reuse of glutamate. Deletion of *Slc1a3* results in accumulation of glutamate in perilymph, massive, prolonged swelling of afferent dendrites, and exacerbation of auditory functional loss after noise exposure (Hakuba et al. 2000). These data suggest that GLAST forms an effective system in removing glutamate and is neuroprotective against NIHL.

Nonmuscle myosin heavy chain IIC (encoded by MYH14) is mutated in DFNA4 (Donaudy et al. 2004). Individuals with mutations in DFNA4 suffer from mild to moderate progressive sensorineural hearing loss. In early postnatal mouse cochlea, MYH14 is expressed in the cochlea in the organ of Corti, the stria vascularis, as well as in Hensen and Claudius cells, external sulcus cells, and the epithelium of the spiral prominence. In spite of its role in deafness and the critical roles of other unconventional myosins in the ear, the function of MYH14 remains undetermined. SNPs in MYH14 were among the 644 SNPs for 53 candidate genes screened in two independent sets of NIHL subjects in Sweden and in Poland. Two of these, SNPs rs667907 and rs588035, were found to be associated with susceptibility to NIHL (Konings et al. 2009b).

3 Future Perspectives in Gene Identification

Identification of individual genes requires a large number of genetically highly related study subjects to provide sufficient statistical power. This is more feasible in animal models. Two ongoing mouse studies use an unbiased approach to investigate the effect of noise, namely QTL analysis. The first examined the effect of noise on endocochlear potential in two inbred mouse strains, CBA and BL6 (Ohlemiller et al. 2010). CBA mice exhibit a reversible reduction in EP after noise exposure, while BL6 mice do not. Crosses between the two strains indicated that the CBA phenotype is dominant. Using backcrosses between these strains, the authors were able to map a major QTL to Chr 18, and two minor QTLs to chromosomes 5 and 16. Additional backcrosses will be required to narrow down the chromosomal region and to eventually identify gene(s) responsible for the phenotype. The second study compared two strains with different sensitivities to noise, CAST/Ei (resistant) and BL6 (sensitive) (White et al. 2009), using the genome-tagged library of congenic strains, each of which carries defined segments of the CAST/Ei genome introduced onto the BL6 background. Some strains were sensitive to noise, while other strains showed resistance to the initial noise damage but displayed variable recovery. Thus, there appeared to be two components of NIHL: one that influences susceptibility to the initial damage, and the second that affects the recovery mechanism. Approaches such as this in the mouse may eventually lead to specific genes that affect different aspects of NIHL in humans.

There have been no real QTL studies in humans for susceptibility to NIHL, only longitudinal studies of hearing loss in specific populations. The Framingham Heart Study was initiated in 1948 with 5,209 individuals residing in Framingham, MA. That study followed this cohort for many years and later included 5,124 offspring of the original participants, plus their spouses. Hearing examinations were conducted beginning in 1973. Because it is possible to construct family pedigrees that contained two or even three generations, the analysis of age-related hearing loss (AHL) or presbycusis in the Framingham population (DeStefano et al. 2003) is a classical linkage analysis (Altshuler et al. 2008). Several genomic regions were identified that appeared to contain loci for ARHL, but the small number of individuals in each family in the study (low statistical power) was insufficient to identify individual genes. Genetic susceptibility to NIHL has not been examined in this population.

Genome-wide association studies (GWAS) have been used successfully in the last 5 years to identify multiple genes that interact and contribute to a disease phenotype. To date, no true GWAS studies have been reported for sensitivity to NIHL in humans, mainly because of limitations in identifying sufficient number of noise-sensitive individuals and noise-resistant controls with similar environmental exposure. In large-scale GWAS, DNA from thousands of individuals diagnosed with a specific disease and DNA from nonaffected matched controls are screened using DNA chips that contain between 3,000 and 5,000 of the most informative SNPs in the human genome. It enables researchers to identify SNPs associated with specific diseases (association studies). These new tools facilitate identification of genes for complex disorders, where small changes in several genes may contribute to the final phenotype (reviewed in Altshuler et al. 2008). Replication of a finding in a second, unrelated population has become an important criterion for assessing the reliability of finding (Chanock et al. 2007). Controls usually include a similar analysis of nonaffected individuals from different ethnic groups (individuals with Western European, African, or Asian ancestry), as some SNPs show higher frequencies in specific ethnic groups and are less informative for association studies. Because of the large number of DNA samples and SNPs measured, GWAS studies have been very expensive. Nevertheless, the development of databases with known allele frequencies of SNPs (www.hapmap.org) and the availability of high throughput genotyping methods, including the DNA chips mentioned earlier, makes GWAS studies feasible.

4 Challenges in Identifying Genetic Susceptibility to NIHL

4.1 Advantages and Limitations of Animal Models

This chapter illustrates the advantages of animal models, particularly mouse models, for studies of genetic susceptibility to NIHL. A major advantage for studies of NIHL in animals is the ability to precisely deliver and monitor noise, for example, continuous exposure to noise of defined frequency spectrum and energy for 1–2 h. Given the genetic homogeneity of inbred mouse strains, there is less variability in outcome measures in mice than in other experimental models. Nevertheless, it is often difficult to compare animal studies from different laboratories due to the different noise exposure protocols used in different settings.

Another advantage of animal studies for NIHL is that researchers have access to cochlear tissues at different times after noise exposure for histological studies of morphology and pathology and molecular analysis of changes in gene expression. In contrast, researchers can only examine human cochlear structures in temporal bones obtained at autopsy. Changes seen in cochlear structures in temporal bones may reflect the cumulative impact of noise exposure over many years, the effects of other environmental insults (such as ototoxic drugs and/or chemical exposures), and any pathological age-related cellular degeneration. Although collections of temporal bones are extremely useful and have provided insights into patterns of cochlear damage, the time between noise exposure and analysis of temporal bones can be quite long.

Mouse geneticists will continue to knock out or introduce known human mutations into mouse genes that encode important cochlear proteins to assess the functional effects. In addition to determining how the mutation affects auditory thresholds, it will be important to investigate sensitivity to noise, which is a rather unexplored area of study. This large collection of mouse mutants will continue to provide opportunities to assess whether or not eliminating or modifying a particular protein makes the mouse more sensitive to NIHL.

A major limitation of noise studies is the effect of genetic background on cochlear degeneration and NIHL. While some substrains of 129 are completely resistant to noise, BL6, BALB/c, and certain substrains of 129 commonly used to generate knockout or transgenic mice carry the *Cdh23*^{ahl1} allele, a major contributor to early onset of hearing loss as well as NIHL (Davis et al. 2001). NIH investigators are encouraged to backcross their knockout mutants to BL6 so that mutants are on a uniform genetic background. The presence of the *Cdh23*^{ahl1} allele may complicate assessment of auditory function in such mutants, particularly at a later age (4 months), at which time CBA mice have been shown to acquire resistance to noise (Kujawa and Liberman 2006). Therefore, selection of a mouse strain for NIHL studies should be made with caution. When working with wild-type, not mutant, mice, it may be advisable to use F1 hybrids of two inbred strains to take advantage of hybrid vigor, which reduces the variability in outcome measures even further (Frisina et al. 2009).

4.2 Confounding Issues in Human Studies of NIHL

Genetic studies of susceptibility to NIHL in humans are limited by the difficulties in accurately assessing the cumulative noise exposure of individuals. Industrial workers who have spent many years in the same industry are ideal for these initial genetic association studies of NIHL because in many industries the actual noise exposure is carefully monitored (Henderson and Hamernik, Chap. 4). However, placement of hearing protection devices, and consistency in their use, may differ across workers within an industry. Moreover, variation in the types, dosage, patterns, and duration of noise exposure in different industries makes it difficult to assess the cumulative noise exposure of individuals, particularly when workers move among different industries. In addition to work-related noise, the exposure to recreational noise (such as hunting, loud music heard through head phones, use of power tools, and more) is difficult to measure. Clearly, studies of NIHL in individuals not in industrial settings would have to rely on self-reporting of prior noise exposure through standardized questionnaires, which might lead to underreporting.

Aging and environmental factors (drugs, pollutants) can also have confounding effects on genetic studies of NIHL in humans (Morata and Johnson, Chap. 11). As industrial workers age, it becomes difficult to distinguish NIHL from presbycusis (Bielefeld, Chap. 10). NIHL is distinguished by the characteristic notch at 4 kHz in the audiogram, but as the hearing loss increases, the notch spreads and becomes indistinguishable from age-related hearing loss. Finally, with respect to self reporting, individuals often may not be aware of or do not remember whether or not they have been exposed to ototoxic drugs during their lifetimes.

A population that might be well-suited for new investigation into sensitivity to NIHL is that of military recruits and/or actively deployed military personnel serving in combat areas. Such studies will differ significantly from studies of industrial workers in the type of noise exposure. Whereas industrial workers are typically exposed to continuous loud noise at specific frequency ranges, military personnel are often exposed to discontinuous, explosive noise during firearms training and by bomb blasts in combat. Can genetic association studies on military personnel be compared, then, with studies on industrial workers? Can large number of noise-exposed individuals be combined to generate the large numbers needed for GWAS studies? These questions remain to be answered as our understanding of the genetic basis of NIHL continues to expand.

5 Summary

The genetic studies of mouse models presented in this chapter provide an initial insight into the identification of genes that underlie susceptibility to NIHL. Identification of deafness genes in either mice or humans has contributed immensely to our understanding of the molecular basis of sound transduction in the mammalian inner ear. As illustrated in this chapter, subtle modifications of some of these genes do not cause deafness, but do render the mouse more sensitive to noise. Mouse studies have not only facilitated identification of many human deafness genes, but also have directly impacted our understanding of the disease phenotypes through morphological studies. Given the precise delivery of noise with animal models, mouse mutants will continue to be a vital part of the search for NIHL susceptible genes.

This review has also illustrated that insights obtained on gene interactions through modifying genes can inform human studies. The genetic interactions between CDH23 and ATP2B2 observed in at least one family with hereditary deafness suggest that additional examples will be uncovered in the future, as differences in phenotype exhibited by individuals of the same genotype lead to identification of additional modifier genes. Such studies are likely to be influenced by our understanding of the genetic interactions in mice.

Human families with hereditary deafness, particularly those with syndromes involving deafness such as Usher syndrome, have provided rare opportunities for identifying proteins in complexes and complex protein interactions which must be intact for normal auditory function. Many of these cases involve identification of proteins involved in maintaining correct stereocilia architecture and function. Clearly, subtle sequence changes of any one of the proteins in these complexes might lead to sensitivity to noise.

To date, no genetic test exists for susceptibility to NIHL. Basic genetic studies in mice will continue to inform the discussion of genetic susceptibility to NIHL in humans. Understanding the molecular mechanisms underlying these mutations in mice should increase our understanding of susceptibility to NIHL in humans. Currently, it is not feasible or practical to sequence all known deafness genes in a single individual for all the variations in these genes that might make the individual susceptible or resistant to noise. However, with the rapid advances in whole genome sequencing and the anticipated decrease in costs, it may 1 day be possible to examine an individual's entire genome at a more cost-effective way, and to understand the cumulative effect of changes in multiple genes on susceptibility to NIHL.

References

Ahmed, Z. M., Riazuddin, S., Bernstein, S. L., Ahmed, Z., Khan, S., Griffith, A. J., Morell, R. J., Friedman, T. B., & Wilcox, E. R. (2001). Mutations of the protocadherin gene PCDH15 cause Usher syndrome type 1F. *American Journal of Human Genetics*, 69, 25–34.

Ahmed, Z. M., Riazuddin, S., Ahmad, J., Bernstein, S. L., Guo, Y., Sabar, M. F., Sieving, P., Griffith, A. J., Friedman, T. B., Belyantseva, I. A., & Wilcox, E. R. (2003). PCDH15 is expressed in the neurosensory epithelium of the eye and ear and mutant alleles are responsible for both USH1F and DFNB23. *Human Molecular Genetics*, 12, 3215–3223.

Alagramam, K. N., Murcia, C. L., Kwon, H. Y., Pawlowski, K. S., Wright, C. G., & Woychik, R. P. (2001). The mouse Ames waltzer hearing-loss mutant is caused by mutation of Pcdh15, a novel protocadherin gene. *Nature Genetics*, 27, 99–102.

Altshuler, D., Daly, M. J., & Lander, E. S. (2008). Genetic mapping in human disease. *Science*, 322, 881–888.

Barone, J. A., Peters, J. M., Garabrant, D. H., Bernstein, L., & Krebsbach, R. (1987). Smoking as a risk factor in noise-induced hearing loss. *Journal of Occupational Medicine*, 29, 741–745.

Board, P. G. (1981). Gene deletion and partial deficiency of the glutathione *S*-transferase (ligandin) system in man. *FEBS Letters*, 135, 12–14.

Bolz, H., von Brederlow, B., Ramirez, A., Bryda, E. C., Kutsche, K., Nothwang, H. G., Seeliger, M., del, C. S. C. M., Vila, M. C., Molina, O. P., Gal, A., & Kubisch, C. (2001). Mutation of CDH23, encoding a new member of the cadherin gene family, causes Usher syndrome type 1D. *Nature Genetics*, 27, 108–112.

Bork, J. M., Peters, L. M., Riazuddin, S., Bernstein, S. L., Ahmed, Z. M., Ness, S. L., Polomeno, R., Ramesh, A., Schloss, M., Srisailpathy, C. R., Wayne, S., Bellman, S., Desmukh, D., Ahmed, Z., Khan, S. N., Kaloustian, V. M., Li, X. C., Lalwani, A., Riazuddin, S., Bitner-Glindzicz, M., Nance, W. E., Liu, X. Z., Wistow, G., Smith, R. J., Griffith, A. J., Wilcox, E. R., Friedman, T. B., & Morell, R. J. (2001). Usher syndrome 1D and nonsyndromic autosomal recessive deafness DFNB12 are caused by allelic mutations of the novel cadherin-like gene CDH23. *American Journal of Human Genetics*, 68, 26–37.

Carlsson, P. I., Van Laer, L., Borg, E., Bondeson, M. L., Thys, M., Fransen, E., & Van Camp, G. (2005). The influence of genetic variation in oxidative stress genes on human noise susceptibility. *Hearing Research*, 202, 87–96.

Chanock, S. J., Manolio, T., Boehnke, M., Boerwinkle, E., Hunter, D. J., Thomas, G., Hirschhorn, J. N., Abecasis, G., Altshuler, D., Bailey-Wilson, J. E., Brooks, L. D., Cardon, L. R., Daly, M., Donnelly, P., Fraumeni, J. F., Jr., Freimer, N. B., Gerhard, D. S., Gunter, C., Guttmacher, A. E., Guyer, M. S., Harris, E. L., Hoh, J., Hoover, R., Kong, C. A., Merikangas, K. R., Morton, C. C., Palmer, L. J., Phimister, E. G., Rice, J. P., Roberts, J., Rotimi, C., Tucker, M. A., Vogan, K. J., Wacholder, S., Wijsman, E. M., Winn, D. M., & Collins, F. S. (2007). Replicating genotype-phenotype associations. *Nature*, 447, 655–660.

Cheng, A. G., Cunningham, L. L., & Rubel, E. W. (2005). Mechanisms of hair cell death and protection. *Current Opinion in Otolaryngology Head Neck Surgery*, 13, 343–348.

Cruickshanks, K. J., Klein, R., Klein, B. E., Wiley, T. L., Nondahl, D. M., & Tweed, T. S. (1998). Cigarette smoking and hearing loss: The epidemiology of hearing loss study. *JAMA*, 279, 1715–1719.

Davis, R. R., Newlander, J. K., Ling, X., Cortopassi, G. A., Krieg, E. F., & Erway, L. C. (2001). Genetic basis for susceptibility to noise-induced hearing loss in mice. *Hearing Research*, 155, 82–90.

Davis, R. R., Kozel, P., & Erway, L. C. (2003). Genetic influences in individual susceptibility to noise: A review. *Noise Health*, 5, 19–28.

DeStefano, A. L., Gates, G. A., Heard-Costa, N., Myers, R. H., & Baldwin, C. T. (2003). Genomewide linkage analysis to presbycusis in the Framingham Heart Study. *Archives of Otolaryngology Head Neck Surgery*, 129, 285–289.

Dickinson, D. A., & Forman, H. J. (2002). Cellular glutathione and thiols metabolism. *Biochemical Pharmacology*, 64, 1019–1026.

Di Palma, F., Holme, R. H., Bryda, E. C., Belyantseva, I. A., Pellegrino, R., Kachar, B., Steel, K. P., & Noben-Trauth, K. (2001). Mutations in Cdh23, encoding a new type of cadherin, cause stereocilia disorganization in waltzer, the mouse model for Usher syndrome type 1D. *Nature Genetics*, 27, 103–107.

Donaudy, F., Snoeckx, R., Pfister, M., Zenner, H. P., Blin, N., Di Stazio, M., Ferrara, A., Lanzara, C., Ficarella, R., Declau, F., Pusch, C. M., Nurnberg, P., Melchionda, S., Zelante, L., Ballana, E., Estivill, X., Van Camp, G., Gasparini, P., & Savoia, A. (2004). Nonmuscle myosin heavy-chain gene MYH14 is expressed in cochlea and mutated in patients affected by autosomal dominant hearing impairment (DFNA4). *American Journal of Human Genetics*, 74, 770–776.

Drayton, M., & Noben-Trauth, K. (2006). Mapping quantitative trait loci for hearing loss in Black Swiss mice. *Hearing Research*, 212, 128–139.

Endo, T., Nakagawa, T., Iguchi, F., Kita, T., Okano, T., Sha, S. H., Schacht, J., Shiga, A., Kim, T. S., & Ito, J. (2005). Elevation of superoxide dismutase increases acoustic trauma from noise exposure. *Free Radical Biology and Medicine*, 38, 492–498.

Erway, L. C., Willott, J. F., Archer, J. R., & Harrison, D. E. (1993). Genetics of age-related hearing loss in mice: I. Inbred and F1 hybrid strains. *Hearing Research*, 65, 125–132.

Erway, L. C., Shiau, Y. W., Davis, R. R., & Krieg, E. F. (1996). Genetics of age-related hearing loss in mice. III. Susceptibility of inbred and F1 hybrid strains to noise-induced hearing loss. *Hearing Research*, 93, 181–187.

Fairfield, D. A., Lomax, M. I., Dootz, G. A., Chen, S., Galecki, A. T., Benjamin, I. J., Dolan, D. F., & Altschuler, R. A. (2005). Heat shock factor 1-deficient mice exhibit decreased recovery of hearing following noise overstimulation. *Journal of Neuroscience Research*, 81, 589–596.

Fortunato, G., Marciano, E., Zarrilli, F., Mazzaccara, C., Intrieri, M., Calcagno, G., Vitale, D. F., La Manna, P., Saulino, C., Marcelli, V., & Sacchetti, L. (2004). Paraoxonase and superoxide dismutase gene polymorphisms and noise-induced hearing loss. *Clinical Chemistry*, 50, 2012–2018.

Frisina, R. D., Singh, A., Bak, M., Bozorg, S., Seth, R., & Zhu, X. (2009). F1 (CBAxC57) mice show superior hearing in old age relative to their parental strains: Hybrid vigor or a new animal model for "Golden Ears"? *Neurobiology of Aging*. Epub October 28, 2009. PMID: 19879021; PMCID: PMC2891213

Frolenkov, G. I., Belyantseva, I. A., Friedman, T. B., & Griffith, A. J. (2004). Genetic insights into the morphogenesis of inner ear hair cells. *Nature Review Genetics*, 5, 489–498.

Hakuba, N., Koga, K., Gyo, K., Usami, S. I., & Tanaka, K. (2000). Exacerbation of noise-induced hearing loss in mice lacking the glutamate transporter GLAST. *Journal of Neuroscience*, 20, 8750–8753.

Heinonen-Guzejev, M., Vuorinen, H. S., Mussalo-Rauhamaa, H., Heikkila, K., Koskenvuo, M., & Kaprio, J. (2005). Genetic component of noise sensitivity. *Twin Research in Human Genetics*, 8, 245–249.

Henderson, D., Bielefeld, E. C., Harris, K. C., & Hu, B. H. (2006). The role of oxidative stress in noise-induced hearing loss. *Ear and Hearing*, 27, 1–19.

Hirose, K., & Liberman, M. C. (2003). Lateral wall histopathology and endocochlear potential in the noise-damaged mouse cochlea. *Journal of the Association of Research in Otolaryngology*, 4, 339–352.

Hu, B. H., Guo, W., Wang, P. Y., Henderson, D., & Jiang, S. C. (2000). Intense noise-induced apoptosis in hair cells of guinea pig cochleae. *Acta Oto-Laryngologica*, 120, 19–24.

Hu, B. H., Henderson, D., & Nicotera, T. M. (2006). Extremely rapid induction of outer hair cell apoptosis in the chinchilla cochlea following exposure to impulse noise. *Hearing Research*, 211, 16–25.

Huang, T. T., Carlson, E. J., Raineri, I., Gillespie, A. M., Kozy, H., & Epstein, C. J. (1999). The use of transgenic and mutant mice to study oxygen free radical metabolism. *Annals of the New York Academy of Sciences*, 893, 95–112.

Hudspeth, A. J. (1989). Mechanoelectrical transduction by hair cells of the bullfrog's sacculus. *Progress in Brain Research*, 80, 129–135; discussion 127–128.

Hultcrantz, M., & Li, H. S. (1993). Inner ear morphology in CBA/Ca and C57BL/6J mice in relationship to noise, age and phenotype. *European Archives of Otorhinolaryngology*, 250, 257–264.

Jacono, A. A., Hu, B., Kopke, R. D., Henderson, D., Van De Water, T. R., & Steinman, H. M. (1998). Changes in cochlear antioxidant enzyme activity after sound conditioning and noise exposure in the chinchilla. *Hearing Research*, 117, 31–38.

Johnson, K. R., Erway, L. C., Cook, S. A., Willott, J. F., & Zheng, Q. Y. (1997). A major gene affecting age-related hearing loss in C57BL/6J mice. *Hearing Research*, 114, 83–92.

Johnson, K. R., Zheng, Q. Y., & Erway, L. C. (2000). A major gene affecting age-related hearing loss is common to at least ten inbred strains of mice. *Genomics*, 70, 171–180.

Kawamoto, K., Sha, S. H., Minoda, R., Izumikawa, M., Kuriyama, H., Schacht, J., & Raphael, Y. (2004). Antioxidant gene therapy can protect hearing and hair cells from ototoxicity. *Molecular Therapy*, 9, 173–181.

Kazmierczak, P., Sakaguchi, H., Tokita, J., Wilson-Kubalek, E. M., Milligan, R. A., Muller, U., & Kachar, B. (2007). Cadherin 23 and protocadherin 15 interact to form tip-link filaments in sensory hair cells. *Nature*, 449, 87–91.

Keithley, E. M., Canto, C., Zheng, Q. Y., Wang, X., Fischel-Ghodsian, N., & Johnson, K. R. (2005). Cu/Zn superoxide dismutase and age-related hearing loss. *Hearing Research*, 209, 76–85.

Khimich, D., Nouvian, R., Pujol, R., Tom Dieck, S., Egner, A., Gundelfinger, E. D., & Moser, T. (2005). Hair cell synaptic ribbons are essential for synchronous auditory signalling. *Nature*, 434, 889–894.

Kikuchi, T., Adams, J. C., Miyabe, Y., So, E., & Kobayashi, T. (2000). Potassium ion recycling pathway via gap junction systems in the mammalian cochlea and its interruption in hereditary nonsyndromic deafness. *Medical Electron Microscroscopy*, 33, 51–56.

Kim, R., Emi, M., & Tanabe, K. (2006). Role of mitochondria as the gardens of cell death. *Cancer Chemotherapy and Pharmacology*, 57, 545–553.

Konings, A., Van Laer, L., Pawelczyk, M., Carlsson, P. I., Bondeson, M. L., Rajkowska, E., Dudarewicz, A., Vandevelde, A., Fransen, E., Huyghe, J., Borg, E., Sliwinska-Kowalska, M., & Van Camp, G. (2007). Association between variations in CAT and noise-induced hearing loss in two independent noise-exposed populations. *Human Molecular Genetics*, 16, 1872–1883.

Konings, A., Van Laer, L., Michel, S., Pawelczyk, M., Carlsson, P. I., Bondeson, M. L., Rajkowska, E., Dudarewicz, A., Vandevelde, A., Fransen, E., Huyghe, J., Borg, E., Sliwinska-Kowalska, M., & Van Camp, G. (2009a). Variations in HSP70 genes associated with noise-induced hearing loss in two independent populations. *European Journal of Human Genetics*, 17, 329–335.

Konings, A., Van Laer, L., Wiktorek-Smagur, A., Rajkowska, E., Pawelczyk, M., Carlsson, P. I., Bondeson, M. L., Dudarewicz, A., Vandevelde, A., Fransen, E., Huyghe, J., Borg, E., Sliwinska-Kowalska, M., & Van Camp, G. (2009b). Candidate gene association study for noise-induced hearing loss in two independent noise-exposed populations. *Annals of Human Genetics*, 73, 215–224.

Kozel, P. J., Friedman, R. A., Erway, L. C., Yamoah, E. N., Liu, L. H., Riddle, T., Duffy, J. J., Doetschman, T., Miller, M. L., Cardell, E. L., & Shull, G. E. (1998). Balance and hearing deficits in mice with a null mutation in the gene encoding plasma membrane Ca^{2+}-ATPase isoform 2. *Journal of Biological Chemistry*, 273, 18693–18696.

Kozel, P. J., Davis, R. R., Krieg, E. F., Shull, G. E., & Erway, L. C. (2002). Deficiency in plasma membrane calcium ATPase isoform 2 increases susceptibility to noise-induced hearing loss in mice. *Hearing Research*, 164, 231–239.

Kujawa, S. G., & Liberman, M. C. (2006). Acceleration of age-related hearing loss by early noise exposure: Evidence of a misspent youth. *Journal of Neuroscience*, 26, 2115–2123.

Le, T., & Keithley, E. M. (2007). Effects of antioxidants on the aging inner ear. *Hearing Research*, 226, 194–202.

Le Prell, C. G., Yamashita, D., Minami, S. B., Yamasoba, T., & Miller, J. M. (2007). Mechanisms of noise-induced hearing loss indicate multiple methods of prevention. *Hearing Research*, 226, 22–43.

Li, H. S., & Borg, E. (1993). Auditory degeneration after acoustic trauma in two genotypes of mice. *Hearing Research*, 68, 19–27.

Li, H. S., Hultcrantz, M., & Borg, E. (1993). Influence of age on noise-induced permanent threshold shifts in CBA/Ca and C57BL/6J mice. *Audiology*, 32, 195–204.

McFadden, S. L., Ding, D., Reaume, A. G., Flood, D. G., & Salvi, R. J. (1999). Age-related cochlear hair cell loss is enhanced in mice lacking copper/zinc superoxide dismutase. *Neurobiology of Aging*, 20, 1–8.

McFadden, S. L., Ohlemiller, K. K., Ding, D., Shero, M., & Salvi, R. J. (2001). The influence of superoxide dismutase and glutathione peroxidase deficiencies on noise-induced hearing loss in mice. *Noise Health*, 3, 49–64.

McVean, G., Spencer, C. C., & Chaix, R. (2005). Perspectives on human genetic variation from the HapMap Project. *PLoS Genetics*, 1, e54.

Meltser, I., Tahera, Y., & Canlon, B. (2009). Glucocorticoid receptor and mitogen-activated protein kinase activity after restraint stress and acoustic trauma. *Journal of Neurotrauma*, 26, 1835–1845.

Mohammadi, S., Mazhari, M. M., Mehrparvar, A. H., & Attarchi, M. S. (2010). Cigarette smoking and occupational noise-induced hearing loss. *Eur J Public Health*, 20, 452–455.

Morimoto, R. I., Kline, M. P., Bimston, D. N., & Cotto, J. J. (1997). The heat-shock response: Regulation and function of heat-shock proteins and molecular chaperones. *Essays in Biochemistry*, 32, 17–29.

Muller, U. (2008). Cadherins and mechanotransduction by hair cells. *Current Opinion in Cell Biology*, 20, 557–566.

Niu, X., & Canlon, B. (2002). Protective mechanisms of sound conditioning. *Advances in Otorhinolaryngology*, 59, 96–105.

Noben-Trauth, K., Zheng, Q. Y., Johnson, K. R., & Nishina, P. M. (1997). *mdfw*: A deafness susceptibility locus that interacts with deaf waddler (dfw). *Genomics*, 44, 266–272.

Noben-Trauth, K., Zheng, Q. Y., & Johnson, K. R. (2003). Association of cadherin 23 with polygenic inheritance and genetic modification of sensorineural hearing loss. *Nature Genetics*, 35, 21–23.

Nordmann, A. S., Bohne, B. A., & Harding, G. W. (2000). Histopathological differences between temporary and permanent threshold shift. *Hearing Research*, 139, 13–30.

Ohlemiller, K. K. (2006). Contributions of mouse models to understanding of age- and noise-related hearing loss. *Brain Research*, 1091, 89–102.

Ohlemiller, K. K. (2008). Recent findings and emerging questions in cochlear noise injury. *Hearing Research*, 245, 5–17.

Ohlemiller, K. K., Wright, J. S., & Dugan, L. L. (1999a). Early elevation of cochlear reactive oxygen species following noise exposure. *Audiology Neurootology*, 4, 229–236.

Ohlemiller, K. K., McFadden, S. L., Ding, D. L., Flood, D. G., Reaume, A. G., Hoffman, E. K., Scott, R. W., Wright, J. S., Putcha, G. V., & Salvi, R. J. (1999b). Targeted deletion of the cytosolic Cu/Zn-superoxide dismutase gene (Sod1) increases susceptibility to noise-induced hearing loss. *Audiology Neurootology*, 4, 237–246.

Ohlemiller, K. K., McFadden, S. L., Ding, D. L., Lear, P. M., & Ho, Y. S. (2000). Targeted mutation of the gene for cellular glutathione peroxidase (Gpx1) increases noise-induced hearing loss in mice. *Journal of the Association of Research in Otolaryngology*, 1, 243–254.

Ohlemiller, K. K., Rosen, A. D., & Gagnon, P. M. (2010). A major effect QTL on chromosome 18 for noise injury to the mouse cochlear lateral wall. *Hearing Research*, 260, 47–53.

Pawelczyk, M., Van Laer, L., Fransen, E., Rajkowska, E., Konings, A., Carlsson, P. I., Borg, E., Van Camp, G., & Sliwinska-Kowalska, M. (2009). Analysis of gene polymorphisms associated with K ion circulation in the inner ear of patients susceptible and resistant to noise-induced hearing loss. *Annals of Human Genetics*, 73, 411–421.

Pemble, S., Schroeder, K. R., Spencer, S. R., Meyer, D. J., Hallier, E., Bolt, H. M., Ketterer, B., & Taylor, J. B. (1994). Human glutathione S-transferase theta (GSTT1): cDNA cloning and the characterization of a genetic polymorphism. *Biochemical Journal*, 300 (Pt 1), 271–276.

Petit, C., & Richardson, G. P. (2009). Linking genes underlying deafness to hair-bundle development and function. *Nature Neuroscience*, 12, 703–710.

Pujol, R., & Puel, J. L. (1999). Excitotoxicity, synaptic repair, and functional recovery in the mammalian cochlea: A review of recent findings. *Annals of the New York Academy of Sciences*, 884, 249–254.

Rabinowitz, P. M., Pierce Wise, J., Sr., Hur Mobo, B., Antonucci, P. G., Powell, C., & Slade, M. (2002). Antioxidant status and hearing function in noise-exposed workers. *Hearing Research*, 173, 164–171.

Ruel, J., Wang, J., Rebillard, G., Eybalin, M., Lloyd, R., Pujol, R., & Puel, J. L. (2007). Physiology, pharmacology and plasticity at the inner hair cell synaptic complex. *Hearing Research*, 227, 19–27.

Ruel, J., Emery, S., Nouvian, R., Bersot, T., Amilhon, B., Van Rybroek, J. M., Rebillard, G., Lenoir, M., Eybalin, M., Delprat, B., Sivakumaran, T. A., Giros, B., El Mestikawy, S., Moser, T., Smith, R. J., Lesperance, M. M., & Puel, J. L. (2008). Impairment of SLC17A8 encoding vesicular glutamate transporter-3, VGLUT3, underlies nonsyndromic deafness DFNA25 and inner hair cell dysfunction in null mice. *American Journal of Human Genetics*, 83, 278–292.

Sakaguchi, H., Tokita, J., Muller, U., & Kachar, B. (2009). Tip links in hair cells: Molecular composition and role in hearing loss. *Current Opinion in Otolaryngology Head Neck Surgery*, 17, 388–393.

Saunders, J. C., Dear, S. P., & Schneider, M. E. (1985). The anatomical consequences of acoustic injury: A review and tutorial. *Journal of the Acoustical Society of America*, 78, 833–860.

Schultz, J. M., Yang, Y., Caride, A. J., Filoteo, A. G., Penheiter, A. R., Lagziel, A., Morell, R. J., Mohiddin, S. A., Fananapazir, L., Madeo, A. C., Penniston, J. T., & Griffith, A. J. (2005). Modification of human hearing loss by plasma-membrane calcium pump PMCA2. *New England Journal of Medicine*, 352, 1557–1564.

Schwander, M., Xiong, W., Tokita, J., Lelli, A., Elledge, H. M., Kazmierczak, P., Sczaniecka, A., Kolatkar, A., Wiltshire, T., Kuhn, P., Holt, J. R., Kachar, B., Tarantino, L., & Muller, U. (2009). A mouse model for nonsyndromic deafness (DFNB12) links hearing loss to defects in tip links of mechanosensory hair cells. *Proceedings of the National Academy of Sciences of the, USA*, 106, 5252–5257.

Shen, J., Harada, N., Kubo, N., Liu, B., Mizuno, A., Suzuki, M., & Yamashita, T. (2006). Functional expression of transient receptor potential vanilloid 4 in the mouse cochlea. *NeuroReport*, 17, 135–139.

Sliwinska-Kowalska, M., Noben-Trauth, K., Pawelczyk, M., & Kowalski, T. J. (2008). Single nucleotide polymorphisms in the cadherin 23 (CDH23) gene in Polish workers exposed to industrial noise. *American Journal of Human Biology*, 20, 481–483.

Street, V. A., McKee-Johnson, J. W., Fonseca, R. C., Tempel, B. L., & Noben-Trauth, K. (1998). Mutations in a plasma membrane Ca^{2+-}ATPase gene cause deafness in deafwaddler mice. *Nature Genetics*, 19, 390–394.

Sugahara, K., Inouye, S., Izu, H., Katoh, Y., Katsuki, K., Takemoto, T., Shimogori, H., Yamashita, H., & Nakai, A. (2003). Heat shock transcription factor HSF1 is required for survival of sensory hair cells against acoustic overexposure. *Hearing Research*, 182, 88–96.

Sundberg, A. G., Nilsson, R., Appelkvist, E. L., & Dallner, G. (1993). Immunohistochemical localization of alpha and pi class glutathione transferases in normal human tissues. *Pharmacology and Toxicology*, 72, 321–331.

Tabuchi, K., Suzuki, M., Mizuno, A., & Hara, A. (2005). Hearing impairment in TRPV4 knockout mice. *Neuroscience Letters*, 382, 304–308.

Tahera, Y., Meltser, I., Johansson, P., Salman, H., & Canlon, B. (2007). Sound conditioning protects hearing by activating the hypothalamic-pituitary-adrenal axis. *Neurobiology of Disease*, 25, 189–197.

Van Eyken, E., Van Laer, L., Fransen, E., Topsakal, V., Hendrickx, J. J., Demeester, K., Van de Heyning, P., Maki-Torkko, E., Hannula, S., Sorri, M., Jensen, M., Parving, A., Bille, M., Baur, M., Pfister, M., Bonaconsa, A., Mazzoli, M., Orzan, E., Espeso, A., Stephens, D., Verbruggen, K., Huyghe, J., Dhooge, I., Huygen, P., Kremer, H., Cremers, C., Kunst, S., Manninen, M., Pyykko, I., Rajkowska, E., Pawelczyk, M., Sliwinska-Kowalska, M., Steffens, M., Wienker, T., & Van Camp, G. (2007). The contribution of GJB2 (Connexin 26) 35delG to age-related hearing impairment and noise-induced hearing loss. *Otology Neurotology*, 28, 970–975.

Van Laer, L., Carlsson, P. I., Ottschytsch, N., Bondeson, M. L., Konings, A., Vandevelde, A., Dieltjens, N., Fransen, E., Snyders, D., Borg, E., Raes, A., & Van Camp, G. (2006). The contribution of genes involved in potassium-recycling in the inner ear to noise-induced hearing loss. *Human Mutation*, 27, 786–795.

Varga, R., Kelley, P. M., Keats, B. J., Starr, A., Leal, S. M., Cohn, E., & Kimberling, W. J. (2003). Non-syndromic recessive auditory neuropathy is the result of mutations in the otoferlin (OTOF) gene. *Journal of Medical Genetics*, 40, 45–50.

Wangemann, P. (2006). Supporting sensory transduction: Cochlear fluid homeostasis and the endocochlear potential. *Journal of Physiology (London)*, 576, 11–21.

White, C. H., Ohmen, J. D., Sheth, S., Zebboudj, A. F., McHugh, R. K., Hoffman, L. F., Lusis, A. J., Davis, R. C., & Friedman, R. A. (2009). Genome-wide screening for genetic loci associated with noise-induced hearing loss. *Mammalian Genome*, 20, 207–213.

Whitlon, D. S., Wright, L. S., Nelson, S. A., Szakaly, R., & Siegel, F. L. (1999). Maturation of cochlear glutathione-*S*-transferases correlates with the end of the sensitive period for ototoxicity. *Hearing Research*, 137, 43–50.

Xiao, X., Zuo, X., Davis, A. A., McMillan, D. R., Curry, B. B., Richardson, J. A., & Benjamin, I. J. (1999). HSF1 is required for extra-embryonic development, postnatal growth and protection during inflammatory responses in mice. *EMBO Journal*, 18, 5943–5952.

Yamane, H., Nakai, Y., Takayama, M., Konishi, K., Iguchi, H., Nakagawa, T., Shibata, S., Kato, A., Sunami, K., & Kawakatsu, C. (1995). The emergence of free radicals after acoustic trauma and strial blood flow. *Acta Oto-Laryngologica*, 519(Supplement), 87–92.

Yang, M., Tan, H., Zheng, J. R., Wang, F., Jiang, C., He, M., Chen, Y., & Wu, T. (2006a). [Association of cadherin CDH23 gene polymorphisms with noise induced hearing loss in Chinese workers]. *Wei Sheng Yan Jiu*, 35, 19–22.

Yang, M., Tan, H., Yang, Q., Wang, F., Yao, H., Wei, Q., Tanguay, R. M., & Wu, T. (2006b). Association of hsp70 polymorphisms with risk of noise-induced hearing loss in Chinese automobile workers. *Cell Stress and Chaperones*, 11, 233–239.

Yasunaga, S., Grati, M., Cohen-Salmon, M., El-Amraoui, A., Mustapha, M., Salem, N., El-Zir, E., Loiselet, J., & Petit, C. (1999). A mutation in OTOF, encoding otoferlin, a FER-1–like protein, causes DFNB9, a nonsyndromic form of deafness. *Nature Genetics*, 21, 363–369.

Yoshida, N., Kristiansen, A., & Liberman, M. C. (1999). Heat stress and protection from permanent acoustic injury in mice. *Journal of Neuroscience*, 19, 10116–10124.

Yoshida, N., Hequembourg, S. J., Atencio, C. A., Rosowski, J. J., & Liberman, M. C. (2000). Acoustic injury in mice: 129/SvEv is exceptionally resistant to noise-induced hearing loss. *Hearing Research*, 141, 97–106.

Zdebik, A. A., Wangemann, P., & Jentsch, T. J. (2009). Potassium ion movement in the inner ear: Insights from genetic disease and mouse models. *Physiology (Bethesda)*, 24, 307–316.

Zhao, H. B., Kikuchi, T., Ngezahayo, A., & White, T. W. (2006). Gap junctions and cochlear homeostasis. *Journal of Membrane Biology*, 209, 177–186.

Zheng, Q. Y., & Johnson, K. R. (2001). Hearing loss associated with the modifier of deaf waddler (mdfw) locus corresponds with age-related hearing loss in 12 inbred strains of mice. *Hearing Research*, 154, 45–53.

Chapter 10
Effects of Early Noise Exposure on Subsequent Age-Related Changes in Hearing

Eric C. Bielefeld

1 Introduction

Estimates of the prevalence of hearing loss in the adult population of the United States have suggested that 28 million (NIDCD 1989) to 29 million (Agrawal et al. 2008) Americans had hearing loss in the speech frequency range (defined by the investigators as 0.5, 1, 2, and 4 kHz). The number rises to 55 million Americans when high-frequency hearing loss (defined as 3, 4, and 6 kHz) is also considered (Agrawal et al. 2008). Estimates suggest that as much as 40% of the population has hearing impairment (defined as 25 dB or greater threshold average of 0.5, 1, 2, and 4 kHz in at least one ear) by age 65 (Gates et al. 1990; Cruickshanks et al. 1998). By age 85, the percentage of people living with hearing loss (defined as above) is estimated at 60–80% (Gates et al. 1990; Desai et al. 2001). Thus, within the elderly segment of our society, hearing loss is among the most prevalent handicapping conditions. Age-related hearing loss (ARHL) continues to be an expanding health care problem for the aging U.S. population, one that has a major influence on the fields of medicine, audiology, and hearing science. Although numerous factors can contribute to hearing loss (pathology to the conductive auditory system, ototoxic medications, ear trauma, ear disease, etc.), much of the hearing loss affecting the adult population can be attributed to two factors: noise and aging. Although the exact contribution of each of these to an individual's hearing loss is impossible to determine, in most adults with hearing loss one or both of these factors is present, and their relative contributions have been widely discussed.

E.C. Bielefeld (✉)
Department of Speech and Hearing Science, The Ohio State University,
1070 Carmack Road, Columbus, OH 43220, USA
e-mail: bielefeld.6@osu.edu

C.G. Le Prell et al. (eds.), *Noise-Induced Hearing Loss: Scientific Advances*,
Springer Handbook of Auditory Research 40, DOI 10.1007/978-1-4419-9523-0_10,

2 Involvement of Noise in the Trajectory of ARHL

Rabinowitz (Chap. 2) provides a detailed discussion of current prevalence estimates for hazardous noise exposure and noise-induced hearing loss (NIHL). When assessing NIHL in an adult patient, the challenge for the clinician is to determine the relative contributions of NIHL, ARHL, and any other ototoxic insults to the patient's total hearing loss. The International Organization for Standardization (ISO) created a standard on "Acoustics – Determination of occupational noise exposure and estimation of noise-induced hearing impairment" in 1990 that includes audiometric assessments of a population of listeners with documented noise exposures. The standard is number 1999 (ISO-1999 1990). Data from the ISO-1999 Database A provide an index of the levels of threshold shift in the speech frequencies that can be attributed to the aging process, without contributions from noise or any form of auditory pathology. The population measured in Database A was screened for any history of noise, occupational or recreational, as well as any history of other auditory diseases or pathological insults. Database B from ISO-1999 demonstrates the increased hearing loss that occurs across a population that includes those with lifelong contributions of noise and auditory pathologies to their hearing losses. The population measured in this database was not screened for noise or ototoxic history. Database A provides evidence that screening of the population does not result in a population that is free of hearing loss. Thus, it can be concluded that ARHL does occur independently of noise or any other ototoxic exposures. But Database B demonstrates that a significant component of hearing loss in the adult population can be attributed to factors other than strictly aging, and that those factors result in a greater disparity in hearing loss between men and women ages 30–60 years. Recent evidence also suggests that the data in Database B may not be directly relevant to younger generations of aging people. Zhan et al. (2010) longitudinally examined participants in their previous epidemiology of hearing loss study (Cruickshanks et al. 1998) from Beaver Dam, Wisconsin. They included in their analyses hearing losses in the generation of offspring from their study participants. They found that the younger generation exhibited lower prevalence of hearing loss (average thresholds at 0.5, 1, 2, and 4 kHz) than participants of equivalent age from an earlier generation. For example, participants ages 65–69 years born in 1940–1944 had a lower prevalence of hearing loss than 65–69-year-olds born between 1925 and 1929 (Zhan et al. 2010). The finding suggests several important ideas about NIHL and ARHL. First, ARHL is indeed modifiable based on history of noise and ototoxicity (consistent with the comparison of ISO-1999 Database A and Database B). Second, the younger generation appears to be living a lifestyle that is more conducive to the retention of hearing into old age (though the study was restricted to primarily non-Hispanic Caucasian, Midwestern Americans). This finding suggests that hearing conservation awareness in occupational and nonoccupational settings has been effective for the "baby boomer" generation. Whether the effectiveness of hearing conservation continues into current younger generations will not be clear for decades. Finally, the data from Zhan et al. (2010) suggest that the ISO-1999 databases will need to be updated regularly to reflect generational differences in ARHL, and that Database B at least will possibly change significantly from one generation to the next.

An issue that comparison of ISO-1999 Database A to Database B cannot address is whether the ARHL that is demonstrated in Database A is the result of true age-induced hearing loss (hearing loss caused directly by the aging process) or if it is the product of a lifetime of accumulated subclinical noise or other ototoxic exposures. Studies comparing hearing losses in the aging populations of industrialized societies versus nonindustrialized societies are consistent with the findings of ISO-1999 that ARHL exists even without noise history, but that noise history will exacerbate hearing loss across an aging population. Rosen et al. (1962) studied the Mabaans in the Sudan, a population of people exposed to very low noise doses throughout their lifespans. They demonstrated significantly less hearing loss when compared with age-matched groups from an industrialized society (Rosen et al. 1962). Analysis of the data introduced the possibility that, although a pure ARHL did appear to exist, ARHL measured from populations in industrial societies was heavily influenced by noise and other exogenous factors. A lingering question at the conclusion of the study was whether the key influencing factor in the reduced ARHL was inherent resistance to ARHL in the Mabaan population rather than the reduced exposure to noise (for review of genes that influence hearing loss, see Gong and Lomax, Chap. 9).

To demonstrate that the population differences in ARHL were not the result of inherent characteristics of the population or the culture that resulted in differential susceptibilities to ARHL, Goycoolea et al. (1986) examined the population of Easter Island, dividing the subjects into groups that had never left the nonindustrialized (and non-noisy) island, those who had left for 1–5 years, and those who had left the island for more than 5 years. The group that had spent the most time in industrialized societies had the greatest hearing loss. The group that had left the island for 1–5 years had less hearing loss than the group that was off for more than 5 years, but had more hearing loss than the group that had never left the island. The group that had never left had the lowest amount of hearing loss among those tested (Goycoolea et al. 1986). Because the tested groups were all part of the same population native to the island, genetic sources of variability underlying inherent susceptibility to ARHL were assumed to be consistent across groups. Therefore, the differential degree of hearing loss was attributed to the time spent in the industrialized societies. This study, like that of Rosen et al. (1962) and the comparison of ISO-1999 Database A to Database B, also provides evidence for an inherent hearing loss that results directly from the aging process and that is independent of noise or other ototoxic factors.

3 Types of Interactions Between Noise and Aging in Medical–Legal Evaluation of Hearing Loss

Although it seems likely that accumulated noise exposures influence ARHL and contribute to an aged individual's hearing loss, the key question for medical–legal evaluation of hearing loss is whether NIHL during youth or adulthood interacts with ARHL and alters its trajectory. One of the great challenges in medical–legal evaluation of NIHL is the task of evaluating an individual's audiogram and parsing out the contribution of noise to the patient's hearing loss. This requires determination of the relative contribution of other (non-noise) factors, of which age is the most common and typically most significant (Dobie 1993). In their most basic forms, NIHL and ARHL would be expected to follow one of three interactive relationships: additive, synergistic, or antagonistic. As reviewed in the upcoming sections describing human and animal studies of noise and aging, there is evidence for all three forms of interaction.

1. Additive: The additive interaction simply takes the contribution from one factor and adds it to the contribution from the second factor. An assumption underlying this approach is that the two contributing factors (noise and aging) are operating independently of one another. If a patient presents with a 40-dB threshold shift at 3 kHz, and 20 dB of that hearing loss is estimated to have been the result of noise exposure, the assumption is then that the patient would have had a 20-dB hearing loss at 3 kHz from aging alone had the noise exposure never occurred.
2. Synergistic: The second form of interaction is synergistic, in which two variables combined create a greater effect than the addition of the two variables independently. In this case, if a patient presented with 40 dB of hearing loss at 3 kHz and 20 dB of the loss could be attributed directly to noise-induced permanent threshold shift (PTS), the assumption would be that the 20 dB NIHL led to a greater ARHL than would otherwise have occurred. The 20 dB of ARHL that was calculated using the additive method thus becomes too high of an attribution of hearing loss due to aging. The assumption in this case is that, if the noise exposure had not taken place, the patient's ARHL would be less than 20 dB, and therefore the patient's total hearing loss would be less than 20 dB at 3 kHz. The synergistic interaction means that the 20 dB NIHL caused a portion of the 20-dB ARHL.
3. Antagonistic: Finally, the last general category of interaction for consideration is antagonistic. In an antagonistic interaction, the two variables combine for less of an effect than would be predicted by adding the two variables' independent contributions. In this case, if a patient presented with 40 dB of hearing loss at 3 kHz and 20 dB of the loss could be attributed directly to noise-induced PTS, the assumption would be that the 20 dB NIHL led to a reduced ARHL compared to that which would have occurred without the noise. The 20 dB of ARHL that is calculated in an additive method thus becomes too low of an attribution to aging. The assumption in this case is that, if the noise exposure had not taken place, the patient's ARHL would be greater than 20 dB, and therefore the patient's total hearing loss would be more than 20 dB at the 3-kHz frequency in the absence of any noise insult.

The challenge with determining the nature of the noise–age interaction in the human is that so few patients had controlled noise conditions, and had noise exposures that were confined to early in their lives, before the potential onset of any ARHL. Thus, for most patients, NIHL and ARHL happened concurrently, and there is no way to accurately attribute part of the hearing loss to noise and part of the hearing loss to aging. This issue is the key concern with allocation of hearing loss in medical–legal environments. Experiments using controlled noise exposure at specific times during the lifespan can be performed in animal models, and those experiments are discussed in the text that follows.

ISO-1999 advocates an additive formula of summing a noise contribution to an ARHL for use in medical–legal evaluation of hearing loss. The formula includes a correction factor for high-level thresholds shifts. The ARHL is a fixed amount, based on Database A or B, and depends on the individual patient's age and gender. The potential weakness with this approach lies in the required assumption that the noise exposure does not alter the trajectory of the ARHL. If there is a synergistic interaction between noise and aging, then an additive approach is underestimating the contribution of NIHL to any individual patient's hearing loss. If there is an antagonistic interaction between noise and aging, then the additive approach is overestimating the contribution of NIHL.

4 Age-Related Hearing Loss in Humans

The complexity of human ARHL is a significant part of what makes the interaction of NIHL and ARHL challenging to define. Human ARHL can be classified into different categories based on the underlying cochlear pathology in a system developed by Harold F. Schuknecht, MD. The system was based on audiograms and postmortem temporal bone examinations of patients with suspected ARHL. The system, first developed in the 1960s (Schuknecht 1964) and refined in multiple publications as late as the early 1990s (Schuknecht and Gacek 1993), still works as an effective classification system for ARHL. What is most interesting about the classification system is that, while most patients have some degree of overlap in their underlying pathologies, each of the categories is capable of generating a sloping, high-frequency audiogram consistent with ISO-1999 Database A or B. Thus, it is not possible to look at an audiogram and conclude into which category of pathology an individual patient is best classified. Currently, there are six pathological categories for peripheral ARHL: Sensory, Neural, Metabolic, Cochlear Conductive, Indeterminate, and Mixed. Sensory ARHL is caused primarily by missing outer hair cells (OHCs). Functionally, OHC loss results in decreased otoacoustic emissions amplitudes (Ohlms et al. 1991) that frequently accompany a sharply sloping hearing loss on the audiogram (Schuknecht 1964). Neural ARHL is associated with degeneration of the afferent auditory nerve fibers, often leading to impaired word discrimination ability (Schuknecht 1964; Schuknecht and Gacek 1993). Metabolic ARHL has an underlying pathology of degeneration of the stria vascularis, leading to loss of the endocochlear potential (EP) (Schulte and Schmiedt 1992). Originally thought to be associated with a flat hearing loss, newer evidence has shown that loss of the EP has more profound effects on the base of the cochlea and can indeed produce a hearing loss that is more severe in the high frequencies (Schmiedt et al. 2002; Mills and Schmiedt 2004). Cochlear Conductive ARHL is an evenly sloping hearing loss that Schuknecht (1964) hypothesized was the result of changes in the stiffness properties of the basilar membrane. The cochleae of those patients with the evenly sloping hearing loss did not show damage to OHCs, spiral ganglion cells (SGCs), or stria vascularis. Although the underlying pathology of altered cochlear mechanics has not yet been proven with postmortem temporal bone examinations, the hypothesis is logical, and the category remains a part of the classification scheme for ARHL. Indeterminate ARHL is a category that included several temporal bone cases that showed hearing loss consistent with Sensory or Metabolic ARHL, but that showed no underlying pathology consistent with Sensory, Neural, or Metabolic ARHL. Schuknecht and Gacek (1993) speculated that the underlying pathology might be dysfunction of the cochlear cells, a pathology that could yield audiometry consistent with Sensory ARHL, but would not be detectable with light microscopic examination of the temporal bones. Schuknecht and Gacek (1993) found this "Indeterminate" pathology pattern in 5 of 21 temporal bones they examined, suggesting that Indeterminate ARHL could affect nearly 25% of ARHL patients. The sixth ARHL category is Mixed, and is characterized by cochleae with any combination of the above five categories. Finally, in addition to the variety of cochlear pathologies that can result in ARHL, there is Central ARHL, in which changes occur in the central auditory system. Typically assessed and diagnosed in humans using central auditory behavioral and physiologic tests, functional deficits independent of cochlear hearing loss have been detected in numerous aging populations (Otto and McCandless 1982; Welsh et al. 1985; Stach et al. 1985), leading to the development of Central ARHL as a unique category of ARHL. The extent to which Central ARHL can occur as the primary pathology underlying ARHL, or if it is likely to occur as a secondary degeneration following peripheral damage, is currently unclear and likely to vary between individuals (Stach et al. 1990).

5 Human Studies of Noise and Aging Interactions

Although ISO has advocated for an additive relationship between noise and aging, it appears that this is because there is not definitive evidence to the contrary. Without definitive evidence of a synergistic or antagonistic interaction, an additive interaction is the most fair to the two parties (plaintiff and defendant) in a medical–legal case involving NIHL. However, the human studies to be reviewed in this section, and the animal studies to be reviewed in the next section, indicate that there are complex interactions between noise exposure and later ARHL.

Gates et al. (2000) examined the progression of ARHL in human subjects from the Framingham Heart Study that displayed audiometric notches (encompassing the 3–6 kHz frequency range) that were attributed to history of noise exposure. The investigators classified their subjects into three groups: those with no significant noise notch (less than 15 dB threshold shift in the 3–6-kHz range), those with a small noise notch (15–34 dB threshold shift in the 3–6-kHz range), and those with large noise notches (35 dB or greater threshold shift in the 3–6-kHz range). As the subjects aged over a 15-year period, the data indicated two key findings.

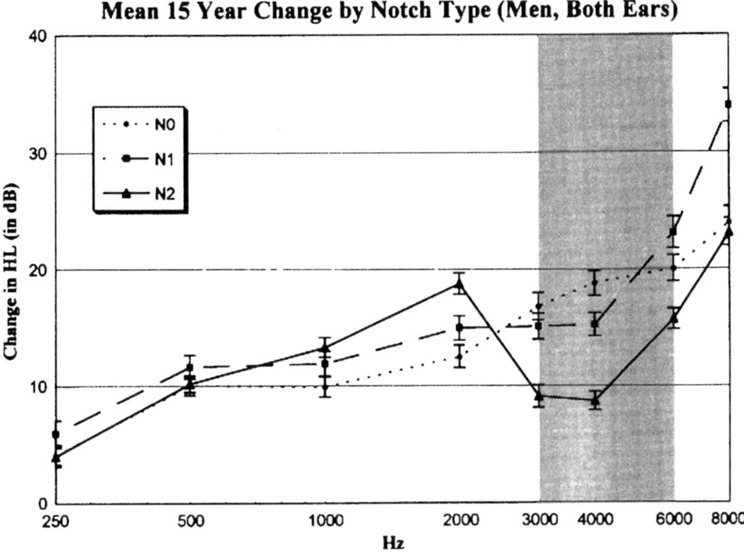

Fig. 10.1 Age-related threshold shift in patients with noise notches (3–6 kHz) on the audiogram. N0 group showed no noise notch (<15 dB notch depth). N1 group had shallow notch (15–35 dB notch depth). N2 had large notch (<35 dB notch depth). Notch depth was measured at study onset, and the data presented here are the changes in thresholds across the audiometric test range. In the notch frequencies of 4–6 kHz (in the shaded region), the subjects with the large noise notches (N2) showed less age-related threshold shift than the groups with little (N1) or no noise notch (N0). But the group with large notches (N2) did show greater age-related threshold shift at the adjacent frequency of 2 kHz (Reprinted from Gates et al. (2000), with permission from Elsevier)

First, in the frequencies within the noise notch (4–6 kHz), ARHL was smaller in the group with the large noise notch than in the group with no notch. At 2 kHz, the opposite was true; ARHL was worse in the subjects with the large audiometric notches compared with subjects without noise notches (Fig. 10.1) (Gates et al. 2000). Thus, within the noise notch, there was an antagonistic interaction between noise and aging. Having the NIHL led to a reduction in the amount of threshold shift from aging. At the frequency adjacent to the noise notch, 2 kHz, the interaction between noise and aging was synergistic. Having the noise notch at 3–6 kHz exacerbated the ARHL at 2 kHz. Thus, the general conclusion from the study is that the interaction between noise and aging is not only complex, but also frequency dependent. Although study of a population of recruited subjects in South Carolina found no relationship between history of NIHL and the trajectory of ARHL (Lee et al. 2005), the findings of the Gates et al. (2000) study were replicated in a subject population in Gothenburg, Sweden. Subjects with a history of noise exposure had more ARHL in the frequencies adjacent to the noise notch as they aged from age 70 to 75 years (Rosenhall 2003).

6 ARHL in Animal Models

Studying ARHL in the human is an inherently complex process, due to the long lifespan, inherent genetic and phenotypic differences between populations, and the long list of exogenous factors that influence the aging process and could influence the aging process for the auditory system. For those reasons, numerous animal models have been used in the study of ARHL. The advantages to these models are the shorter lifespans, control over many of the exogenous influences that could alter the aging trajectory, and the ability to perform anatomical and histological evaluations at any point during the course of the aging process.

6.1 Sensory ARHL

Mouse models of ARHL have been used frequently and have provided unique insights into the aging trajectory of the cochlea, as well as considerable insight into the genetics underlying ARHL (see reviews by Ohlemiller 2006, 2008). The CBA/Ca mouse strain and C57BL/6 mouse strain have been employed frequently as models of ARHL due to OHC loss (Schuknecht's Sensory ARHL). The CBA/Ca mouse has an approximately 30-month average lifespan, and is fairly resistant to ARHL up to 18 months of age (Hunter and Willott 1987). After 18 months, hearing declines progressively beginning in the high frequencies and moving to the low frequencies (Li and Borg 1991). The cochlear pathology underlying the ARHL in the CBA/Ca mouse is progressive loss of OHCs and inner hair cells (IHCs) (Li and Hultcrantz 1994; Spongr et al. 1997). The C57BL/6 mouse is a model of accelerated ARHL. The C57BL/6 mouse demonstrates a much more rapid onset of hearing loss and more rapid decline than the CBA/Ca mouse (Li and Borg 1991), as well as an increased susceptibility to NIHL (Li et al. 1993; Ohlemiller et al. 2000). Like the CBA/Ca mouse, the underlying pathology of the ARHL in the C57BL/6 is early stereocilia damage and OHC degeneration, which is rapidly followed by massive OHC and IHC loss. By 1 year of age, OHCs and IHCs can be completely absent (Li and Hultcrantz 1994; Spongr et al. 1997).

As is the case with NIHL (reviews by Henderson et al. 2006; Le Prell et al. 2007) and hearing loss from ototoxic drug exposure (reviews by Rybak et al. 2007; Rizzo and Hirose 2007), age-related OHC loss in animal models of aging appears to mediated by oxidative stress (Someya et al. 2009, 2010) and caspase-mediated apoptosis (Hu et al. 2008). Unlike noise and drug ototoxicity, there is no specific triggering event (noise exposure, drug administration, etc.) that allows for analysis of the cell death pathways in a discrete interval following the insult to the cochlea. Thus, it is difficult to take a specific snapshot of the cochlea during the age-related cochlear degeneration phase(s) to obtain a clear indication of the dominant mechanisms in age-related cochlear damage and loss of OHCs and IHCs. The mechanisms are likely to be very complex, and will be the subject of ongoing investigations in the future.

6.2 Metabolic ARHL

The Mongolian gerbil (*Meriones unguiculatus*) develops ARHL that is consistent with a pattern of Metabolic ARHL. By 36 months of age, the Mongolian gerbil shows a 15–35 dB threshold shift (Mills et al. 1990), with large variability in individual animals' age-related threshold shifts (Mills et al. 1990). The hearing loss is associated with underlying damage to the stria vascularis (Gratton and Schulte 1995; Thomopoulos et al. 1997) and spiral ligament fibrocytes (Spicer and Schulte 2002). The damage to stria vascularis is associated with a reduction in mitochondrial ATP production in the strial marginal cells (Spicer and Schulte 2005), leading to a loss of $Na,^{+}$ K^{+}-ATPase activity and a decrease in the EP (Schulte and Schmiedt 1992; Gratton et al. 1997). The loss of EPs is thought to be the causative factor in the Mongolian gerbil's loss of hearing sensitivity with age. Thus, the Mongolian gerbil is used as a model of Metabolic ARHL. Numerous mouse models are also being employed as models of Metabolic ARHL, including Tyrp1B^{-lt}, BALB/cJ, CBA/CaJ, NOD.NON-H2^{nbl}/LtJ, and C57BL/6-Tyr^{c-2J} (see review by Ohlemiller 2009). Each has shown depressed EP in a portion of their populations as they age.

6.3 Central ARHL

Animal models have also been used to observe age-related central auditory nervous system changes. One of the hallmarks of the damaged auditory system is the appearance of plastic changes in the organization of the central auditory structures after peripheral deafferentation, whether the peripheral deafferentation comes from noise or from aging (for detailed review of noise-induced neural plasticity, see Kaltenbach, Chap. 8). It is unknown in many of the animal models studied how much (if any) of the central changes that have been observed occur as primary age-related degeneration, and how much is secondary to peripheral damage. For the purposes of this discussion, rat models of ARHL are briefly reviewed as examples of aging of the central auditory nervous system. Changes in γ-aminobutyric acid (GABA) distribution in the inferior colliculus and auditory cortex have been identified and implicated in potential Central ARHL. GABA is a major inhibitory neurotransmitter involved in maintaining frequency processing in the auditory cortex (Chang et al. 2005) and has a major role in the plastic changes in the auditory cortex that occur from aging or peripheral deafferentation (see review by Caspary et al. 2008). Overall GABA levels and GABA release were decreased in aged Fischer 344 rats' inferior colliculi (Caspary et al. 1990). In the auditory cortex, glutamic acid decarboxylase, a GABA synthetic enzyme, was found to be lower in older Brown/Norway rats, suggesting an overall reduction in GABA in the auditory cortex with advancing age. Similar reductions were not found in nearby, nonauditory cortices (Ling et al. 2005). The findings of age-related GABA loss indicate possible changes in cortical auditory physiology and processing in aging animals.

7 Animal Studies of Interactions of Noise and Aging

In an attempt to assess the validity of the additive rule recommended by ISO for attributing hearing loss to noise or aging, Mills et al. (1996, 1997) conducted a series of studies of Mongolian gerbils that were exposed to noise and then allowed to age. The first such study utilized a long-term exposure to 85 dBA noise that lasted from age 8 months to age 34 months. Because the exposure extended so deep in the animals' lifespans, there was no effective way to determine what portion of the threshold was age related and what portion was noise induced. The two damage processes overlapped, obscuring what the PTS induced by the long-term noise exposure would have been if it had been independent of the aging process (Mills et al. 1997). The investigators therefore undertook a second study in which the noise exposure was a short-duration (1 h), monaural, tonal exposure (113 dB SPL) delivered to the animals at age 18 months. The animals were assessed for PTS 6 weeks after the exposure and then allowed to age in quiet until age 36 months. Preexposure thresholds were generally equivalent in the exposed and unexposed ears (Fig. 10.2a), and the ears that were exposed to the noise demonstrated significantly worse thresholds (PTS) at 6 weeks postexposure (Fig. 10.2b). At 36 months of age, the noise + aging ears had significantly less hearing loss than predicted by an ISO additive method (see Fig. 10.2c) (Mills et al. 1997). The results showed that the quiet aged ears' hearing losses "caught up" to the noise + aging ears' thresholds (although they did not catch up completely; the noise + aging animals had higher thresholds at some frequencies at 36 months compared to the quiet aged animals). Thus, for the Mongolian gerbil, the noise–aging interaction appears to be antagonistic, that is, NIHL at a younger age led to a reduction in the hearing loss that was attributed to aging. The findings suggest that, from a cochlear pathology perspective, as the noise damage and aging damage produced a largely antagonistic interaction (with the exception being at 4 kHz, where the age–noise interaction was closer to additive; see Fig. 10.2b, c), some of the damage created by the noise would have been created by the aging process. Because the cochlear structures were already damaged/disordered by the noise, any additional effects of aging on those structures had limited additional impact on hearing.

In addition to the findings in the Mongolian gerbil, the mouse model has been tested to assess the interactive effect of noise and aging. Kujawa and Liberman (2006) assessed the effects of noise exposure on the aging trajectory of the CBA/CaJ mouse. The mice were exposed to a short-duration (2-h) noise at various ages from 4 to 124 weeks of age. The animals were then allowed to age into a time span when ARHL was predicted to occur. Based on ABR threshold shifts, the investigators found a synergistic interaction in which the early noise exposure exacerbated the severity of the subsequent ARHL. The effect was greatest for those animals exposed to the noise at a younger age, but it was still present in those exposed as late as 32 weeks of age. Although the synergistic interaction is an intriguing finding on its own, the findings were complicated further by the fact that the noise + aging group did not demonstrate an increased degree of threshold shift in distortion product otoacoustic emissions (DPOAEs), even though they demonstrated exacerbated ARHL as measured by the ABR. This divergence with respect to measurement

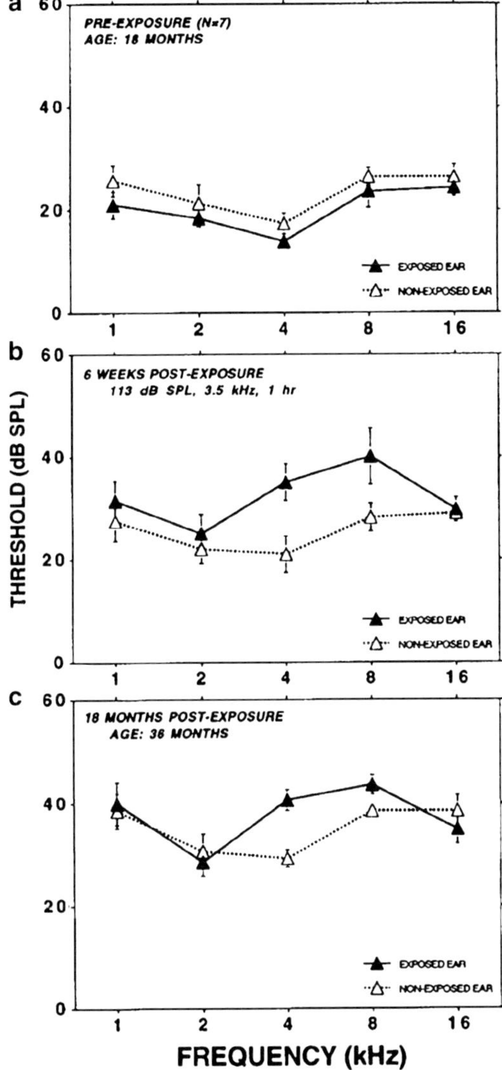

Fig. 10.2 Auditory thresholds in Mongolian gerbils. (**a**) Thresholds at 18 months of age, before any noise exposure. (**b**) Thresholds 6 weeks after a hazardous noise exposure of a 3.5 kHz pure tone at 113 dB SPL for 1 h. Note the higher thresholds in the ears exposed to the noise (*dark triangles*), suggesting significant threshold shift. (**c**) Thresholds for the noise-exposed and unexposed ears at 36 months of age. Note that the unexposed ears' thresholds (*open triangles*) are much closer to the noise-exposed ears' thresholds (*dark triangles*) than they are in (**b**) (Reprinted from Mills et al. (1997), with permission from the Acoustical Society of America)

technique implied that the cochlear pathology shifted from primarily OHC loss in the quiet-aged mice (Li and Hultcrantz 1994; Spongr et al. 1997) to a different pathology in the noise + aging mice. Anatomical evaluations showed an increased age-related loss of SGCs in the cochleae with noise exposure history. The selective SGC loss did not occur in the aging cochleae without noise exposure history, nor did it occur in animals with noise exposure but no ARHL. The finding of primary SGC loss prompted the investigators to propose that the noise + aging mice are models of Neural ARHL (Kujawa and Liberman 2006). This hypothesis has since been modified (Kujawa and Liberman 2009) to reflect that the SGC loss may be the direct result of noise exposure. The synapses are lost in the acute phase shortly after the noise, but the cell bodies of the SGCs degenerate over a longer period of weeks or months. This degenerative process occurs even without PTS or hair cell loss. Thus, the possibility exists that much Neural ARHL may in fact be cumulative noise-induced SGC damage. For patients who report no noise exposure history, the Neural ARHL with which they present clinically may actually be evidence of accumulated subclinical noise exposures.

The findings in the CBA/CaJ mouse studies introduce the intriguing possibility that the relationship between NIHL and ARHL is more complex than that which can be assessed using the pure tone audiogram, and that the notion of additive, synergistic, or antagonistic interactions between noise and aging as assessed on the pure tone audiogram may not be the extent of the complexity of the issue. Neural ARHL often includes substantial difficulties with word discrimination (Schuknecht 1964). If indeed there is a population of patients with Neural ARHL from the combined insults of noise and aging, then that is a patient population that would be expected to struggle to derive the required rehabilitative benefit from amplification or cochlear implants.

To summarize the Mongolian gerbil and CBA/CaJ mouse studies, the Mongolian gerbil appears to display an antagonistic interaction. The history of NIHL leads to a reduction in the hearing loss at 36 months that can be attributed to aging, which could reflect the notion that noise and aging lead to disability/death of the same sets of structures. If noise has already disabled a set of cochlear structures, then aging has less impact on total auditory function because the structures it would target have already ceased to contribute to auditory sensitivity. The CBA/CaJ mouse, conversely, shows a synergistic interaction between noise and aging, in which a prior NIHL leads to more severe ARHL later in the lifespan. Further complicating the interaction is the finding that NIHL at a younger age may not only exacerbate the ARHL, but either change the pathology underlying it from OHC pathology to SGC pathology, or add SGC pathology to the age-related OHC pathology. Either of these patterns of cochlear damage could manifest in a decreased ability to perceive and interpret complex sounds.

8 Possible Mechanisms of NIHL and ARHL Interactions

A factor that remains unclear in both animal and human studies of noise–age interactions is the interaction(s) of pathologies underlying the NIHL and ARHL. With NIHL, the dominant underlying pathology is damage/death of the OHCs (Henderson et al. 2006; also see Hu, Chap. 5). In addition to the effects on the OHCs, noise exposure causes transient glutamate excitotoxicity at the junctions between the IHCs and the type I afferent auditory nerve fibers of the spiral ganglion (Puel et al. 1998). This glutamate excitotoxicity has been proposed as a possible cause of long-term, age-related degeneration of the SGCs (Kujawa and Liberman 2006). The Mongolian gerbil (Mills et al. 1996, 1997) and CBA/CaJ mouse (Kujawa and Liberman 2006) displayed very different interactions of noise and aging. It is possible that the nature of the differing interactions was due to the differing nature of pathologies underlying the ARHL. In the Mongolian gerbil, the primary degeneration occurs in stria vascularis, leading to a loss of the EP and loss of cochlear amplification. It is possible that damaging/killing OHCs with noise, robbing the ear of cochlear amplification, is a pathology that overlaps with the aging process. Because the cochlear amplification is already lost due to OHC death, loss of the EP with aging has little effect on hearing. Thus, there is an antagonistic interaction in which NIHL leads to reduced ARHL. The situation in the CBA/CaJ mouse is very different in that noise exposure induces significant SGC pathology in addition to OHC death, or in the absence of OHC death. The SGC pathology may be a direct consequence of the noise, a consequence of the interaction of noise and age, or a combination of both direct and interactive pathologies (Kujawa and Liberman 2006, 2009). Thus, the data from studies of animal models of the noise-age interaction suggest that multiple variables could be considered when hypothesizing about the nature and mechanisms of noise–age interactions in the human.

In addition to noise-aging interactions in the cochlea, it seems possible that interactions would take place in the central auditory system as well. Noise exposure at a relatively young age could alter the trajectory of the aging of the central auditory system. Just as there are age-related changes to the central system (see earlier), cochlear damage from noise is causes plastic changes in the central auditory system (for detailed review, see Kaltenbach, Chap. 8), including in the cochlear nucleus (Kaltenbach et al. 1992), inferior colliculus (Wang et al. 1996), and the auditory cortex (Salvi et al. 2000). Combined peripheral deafferentation from noise and aging could remove inputs to the central auditory system, leading to an altered pattern of plastic reorganization that is unique to the noise–aging interaction. Coupled with direct effects of aging on the central auditory system, these plastic changes could manifest in changes in processing of auditory stimuli, including complex stimuli such as speech.

9 Summary

There is growing evidence that the interaction between noise exposure before onset of ARHL and the subsequent ARHL is a complex one that may vary across individuals or populations. Of particular consideration with the human aging population is how the interactions of noise and aging may differ across subsets of the ARHL population. Patients with NIHL comprise a heterogeneous population with respect to the timing and magnitude of the noise exposures. As can be seen from ISO-1999, as well as numerous investigations on the topic (Rosen et al. 1962; Hinchcliffe 1964; Goycoolea et al. 1986), a cumulative history of occupational and recreational noise is a risk factor for more severe ARHL, even if the noise exposures do not induce discrete NIHLs as individual events. In addition to this population of cumulative noise exposure history, a segment of the population is exposed to occupational noise regularly over a period of decades. Another segment gets high-level impulse or impact noise (130–170 dB pSPL) capable of inducing acoustic trauma. It seems likely that the different patterns of NIHL across the aging patient population are likely to interact with ARHL in different ways. Further complicating the relationship between noise and aging is the heterogeneity of the ARHL population. As described in Sect. 4, at least five different discrete pathologies are believed to affect the aging cochlea alone or in combination (Schuknecht 1964; Schuknect and Gacek 1993). In addition, the human population, with its wide variety of noise and ototoxicity histories, may be dominated by the patients who demonstrate Mixed ARHL, in which they demonstrate a blend of multiple pathologies with a broad spectrum of resulting hearing losses (Allen and Eddins 2010). With the disparate findings about the noise–age interaction in humans with ARHL and in animal models of ARHL, it is reasonable to suggest that the interactions of noise and aging in the human population may vary with the severity of the NIHL, the age at which the NIHL is acquired, the presence or absence of a mechanical trauma as part of the NIHL, and the particular cochlear pathology (or pathologies) underlying the ARHL. Further study on these variables in a wide variety of animal models of ARHL seems necessary before more comprehensive conclusions can be drawn about the nature of the noise–age interaction in any particular clinical patient's hearing loss profile.

References

Agrawal, Y., Platz, E. A., & Niparko, J. K. (2008). Prevalence of hearing loss and differences by demographic characteristics among US adults: Data from the National Health and Nutrition Examination Survey, 1999–2004. *Archives of Internal Medicine*, 168(14), 1522–1530.

Allen, P. D, & Eddins, D. A. (2010) Presbycusis phenotypes form a heterogeneous continuum when ordered by degree and configuration of hearing loss. *Hearing Research*, 264(1–2), 10–20.

Caspary, D. M., Raza, A., Lawhorn Armour, B. A., Pippin, J., & Arneric, S. P. (1990). Immunocytochemical and neurochemical evidence for age-related loss of GABA in the inferior colliculus: Implications for neural presbycusis. *Journal of Neuroscience*, 10(7), 2363–2372.

Caspary, D. M., Ling, L. L., Turner, J. G., & Hughes, L. F. (2008). Inhibitory neurotransmission, plasticity and aging in the mammalian central auditory system. *The Journal of Experimental Biology*, 211(11), 1781–1791.

Chang, E. F., Bao, S., Imaizumi, K., Schreiner, C. E., & Merzenich, M. M. (2005). Development of spectral and temporal response selectivity in the auditory cortex. *Proceedings of the National Academy of Sciences of the USA*, 102(45), 16460–16465.

Cruickshanks, K. J., Wiley, T. L., Tweed, T. S., Klein, B. E., Klein, R., Mares-Perlman, J. A., & Nondahl, D. M. (1998). Prevalence of hearing loss in older adults in Beaver Dam, Wisconsin. The Epidemiology of Hearing Loss Study. *American Journal of Epidemiology*, 148(9), 879–886.

Desai, M., Pratt, L. A., Lentzner, H., & Robinson, K. N. (2001). Trends in vision and hearing among older Americans. *Aging Trends*, Mar(2), 1–8.

Dobie, R. (1993). Medical-legal evaluation of hearing loss (2nd edition). San Diego: Singular.

Gates, G. A., Cooper, J. C., Jr, Kannel, W. B., & Miller, N. J. (1990). Hearing in the elderly: The Framingham cohort, 1983–1985. Part I. Basic audiometric test results. *Ear and Hearing*, 11(4), 247–256.

Gates, G. A., Schmid, P., Kujawa, S. G., Nam, B., & D'Agostino, R. (2000). Longitudinal threshold changes in older men with audiometric notches. *Hearing Research*, 141(1–2), 220–228.

Goycoolea, M. V., Goycoolea, H. G., Farfan, C. R., Rodriguez, L. G., Martinez, G. C., & Vidal, R. (1986). Effect of life in industrialized societies on hearing in natives of Easter Island. *Laryngoscope*, 96, 1391–1396.

Gratton, M. A., & Schulte, B. A. (1995). Alterations in microvasculature are associated with atrophy of the stria vascularis in quiet-aged gerbils. *Hearing Research*, 82, 44–52.

Gratton, M. A., Smyth, B. J., Lam, C. F., Boettcher, F. A., & Schmiedt, R. A. (1997). Decline in the endocochlear potential corresponds to decreased Na,K-ATPase activity in the lateral wall of quiet-aged gerbils. *Hearing Research*, 108(1–2), 9–16.

Henderson, D., Bielefeld, E. C., Harris, K. C., & Hu, B. H. (2006). The role of oxidative stress in noise-induced hearing loss. *Ear and Hearing*, 27, 1–19.

Hinchcliffe, R. (1964). Hearing levels of elderly in Jamaica. *Annals of Otology, Rhinology, and Laryngology*, 73, 1012–1019.

Hu, B. H., Yang, W. P., Bielefeld, E. C., Li, M., Chen, G. D., & Henderson, D. (2008). Apoptotic outer hair cell death in the cochleae of aging Fischer 344/NHsd rats. *Hearing Research*, 241, 26–33.

Hunter, K. P., & Willott, J. F. (1987). Aging and the auditory brainstem response in mice with severe or minimal presbycusis. *Hearing Research*, 30(2–3), 207–218.

ISO1999. (1990). Acoustics—Determination of occupational noise exposure and estimation of noise-induced hearing impairment. International Organization for Standardization. http://www.iso.org/iso/catalogue_detail.htm?csnumber=6759.

Kaltenbach, J. A., Czaja, J. M., & Kaplan, C. R. (1992). Changes in the tonotopic map of the dorsal cochlear nucleus following induction of cochlear lesions by exposure to intense sound. *Hearing Research*, 59(2), 213–223.

Kujawa, S. G., & Liberman, M. C. (2006). Acceleration of age-related hearing loss by early noise exposure: Evidence of a misspent youth. *Journal of Neuroscience*, 26(7), 2115–2123.

Kujawa, S. G., & Liberman, M. C. (2009). Adding insult to injury: Cochlear nerve degeneration after "temporary" noise-induced hearing loss. *Journal of Neuroscience*, 29(45), 14077–14085.

Le Prell, C. G., Yamashita, D., Minami, S. B., Yamasoba, T., & Miller, J. M. (2007). Mechanisms of noise-induced hearing loss indicate multiple methods of prevention. *Hearing Research*, 226(1–2), 22–43.

Lee, F. S., Matthews, L. J., Dubno, J. R., & Mills, J. H. (2005). Longitudinal study of pure-tone thresholds in older persons. *Ear and Hearing*, 26, 1–11.

Li, H. S., & Borg, E. (1991). Age-related loss of auditory sensitivity in two mouse genotypes. *Acta Oto-Laryngologica*, 111(5), 827–834.

Li, H. S., & Hultcrantz, M. (1994). Age-related degeneration of the organ of Corti in two genotypes of mice. *Journal for Oto-rhino-laryngology and Its Related Species*, 56(2), 61–67.

Li, H. S., Hultcrantz, M, & Borg, E. (1993). Influence of age on noise-induced permanent threshold shifts in CBA/Ca and C57BL/6J mice. *Audiology*, 32(3), 195–204.

Ling, L. L., Hughes, L. F., & Caspary, D. M. (2005). Age-related loss of the GABA synthetic enzyme glutamic acid decarboxylase in rat primary auditory cortex. *Neuroscience*, 132(4), 1103–1113.

Mills, D. M., & Schmiedt, R. A. (2004). Metabolic presbycusis: Differential changes in auditory brainstem and otoacoustic emission responses with chronic furosemide application in the gerbil. *Journal of the Association for Research in Otolaryngology*, 2004, 5(1), 1–10.

Mills, J. H., Schmiedt, R. A., & Kulish, L. F. (1990). Age-related changes in auditory potentials of Mongolian gerbil. *Hearing Research*, 46(3), 201–210.

Mills, J. H., Lee, F. S., Dubno, J. R., & Boettcher, F. A. (1996). Interactions between age-related and noise-induced hearing loss. In A. Axelsson, H. Borchgrevink, R. P. Hamernik, P. A. Hellstron, D. Henderson, & R. J. Salvi (Eds.), *Scientific basis of noise-induced hearing loss*. New York: Thieme.

Mills, J. H., Boettcher, F. A., & Dubno, J. R. (1997). Interaction of noise-induced permanent threshold shift and age-related threshold shift. *Journal of the Acoustical Society of America*, 101(3), 1681–1686.

National Institute on Deafness and Other Communication Disorders. (1989). Research Plan. Bethesda, MD: National Institutes of Health.
Ohlemiller, K. K. (2006). Contributions of mouse models to understanding of age- and noise-related hearing loss. *Brain Research*, 1091(1), 89–102.
Ohlemiller, K. K. (2008). Recent findings and emerging questions in cochlear noise injury. *Hearing Research*, 245(1–2), 5–17.
Ohlemiller, K. K. (2009). Mechanisms and genes in human strial presbycusis from animal models. *Brain Research*, 1277, 70–83.
Ohlemiller, K. K., Wright, J. S., & Heidbreder A. F. (2000). Vulnerability to noise-induced hearing loss in 'middle-aged' and young adult mice: A dose-response approach in CBA, C57BL, and BALB inbred strains. *Hearing Research*, 147, 239–247.
Ohlms, L. A., Lonsbury-Martin, B. L., & Martin, G. K. (1991). Acoustic-distortion products: Separation of sensory from neural dysfunction in sensorineural hearing loss in human beings and rabbits. *Otoloaryngology Head and Neck Surgery*, 104(2), 159–174.
Otto, W. C., & McCandless, G. A. (1982). Aging and auditory site of lesion. *Ear and Hearing*, 3(3), 110–117.
Puel, J. L., Ruel, J., Gervais d'Aldin, C., & Pujol, R. (1998) Excitotoxicity and repair of cochlear synapses after noise-trauma induced hearing loss. *NeuroReport*, 9, 2109–2114.
Rizzo, M. D., & Hirose, K. (2007). Aminoglycoside ototoxicity. *Current Opinions in Otolaryngology Head and Neck Surgery*, 15(5), 352–357.
Rosen, S., Bergman, M., Plester, D., El-Mofty, A., & Hamad Satti, M. (1962). Presbycusis study of a relatively noise-free population in the Sudan. *Transactions of the American Otological Society*, 50, 135–152.
Rosenhall U. (2003). The influence of ageing on noise-induced hearing loss. *Noise and Health*, 5(20), 47–53.
Rybak, L. P., Whitworth, C. A., Mukherjea, D., & Ramkumar, V. (2007). Mechanisms of cisplatin-induced ototoxicity and prevention. *Hearing Research*, 226(1–2), 157–167.
Salvi, R. J., Wang, J., & Ding, D. (2000). Auditory plasticity and hyperactivity following cochlear damage. *Hearing Research*, 147(1–2), 261–274.
Schmiedt, R. A., Lang, H., Okamura, H. O., & Schulte, B. A. (2002). Effects of furosemide applied chronically to the round window: A model of metabolic presbyacusis. *Journal of Neuroscience*, 22(21), 9643–9650.
Schuknecht, H. F. (1964). Further observations on the pathology of presbycusis. *Archives of Otolaryngology*, 80, 369–382.
Schuknecht, H. F., & Gacek, M. R. (1993). Cochlear pathology in presbycusis. *Annals of Otology, Rhinology, and Laryngology*, 102(1 Pt 2), 1–16.
Schulte, B. A., & Schmiedt, R. A. (1992). Lateral wall Na,K-ATPase and endocochlear potentials decline with age in quiet-reared gerbils. *Hearing Research*, 61(1–2), 35–46.
Someya, S., Xu, J., Kondo, K., Ding, D., Salvi, R. J., Yamasoba, T., Rabinovitch, P. S., Weindruch, R., Leeuwenburgh, C., Tanokura, M., & Prolla, T. A. (2009). Age-related hearing loss in C57BL/6J mice is mediated by Bak-dependent mitochondrial apoptosis. *Proceedings of the National Academy of Sciences of the USA*, 106(46), 19432–19437.
Someya, S., Tanokura, M., Weindruch, R., Prolla, T. A., & Yamasoba, T. (2010). Effects of caloric restriction on age-related hearing loss in rodents and rhesus monkeys. *Current Aging Science*, 3(1), 20–25.
Spicer, S. S., & Schulte, B. A. (2002) Spiral ligament pathology in quiet-aged gerbils. *Hearing Research*, 172, 172–185.
Spicer, S. S., & Schulte, B. A. (2005) Pathologic changes of presbycusis begin in secondary processes and spread to primary processes of strial marginal cells. *Hearing Research*, 205, 225–240.
Spongr, V. P., Flood, D. G., Frisina, R. D., & Salvi, R. J. (1997) Quantitative measures of hair cell loss in CBA and C57BL/6 mice throughout their life spans. *Journal of the Acoustical Society of America*, 101, 3546–3553.

Stach, B. A., Jerger, J., & Fleming, K. A. (1985).Central presbyacusis: A longitudinal study. *Ear and Hearing*, 6(6), 304–306.

Stach, B. A., Spretnjak, M. L., & Jerger, J. (1990). The prevalence of central presbyacusis in a clinical population. *Journal of the American Academy of Audiology*, 1(2), 109–115.

Thomopoulos, G. N., Spicer, S. S., Gratton, M. A., & Schulte, B. A. (1997) Age-related thickening of basement membrane in stria vascularis capillaries. *Hearing Research*, 111, 31–41.

Wang, J., Salvi, R. J., & Powers, N. (1996). Plasticity of response properties of inferior colliculus neurons following acute cochlear damage. *Journal of Neurophysiology*, 75(1), 171–183.

Welsh, L. W., Welsh, J. J., & Healy, M. P. (1985). Central presbyacusis. *Laryngoscope*, 95(2), 128–136.

Zhan, W., Cruickshanks, K. J., Klein, B. E., Klein, R., Huang, G. H., Pankow, J. S., Gangnon, R. E., & Tweed, T. S. (2010). Generational differences in the prevalence of hearing impairment in older adults. *American Journal of Epidemiology*, 171(2), 260–266.

Chapter 11
Effects of Exposure to Chemicals on Noise-Induced Hearing Loss

Thais C. Morata and Ann-Christin Johnson

1 Introduction

Several factors have been studied in an effort to explain why the prevalence and degree of noise-induced hearing loss (NIHL) can vary so much within a group and among groups. Some of the factors studied to date include variations in exposure (see Henderson and Hamernik, Chap. 4), age (see Rabinowitz, Chap. 2; Bielefeld, Chap. 10), gender, genetics (see Gong and Lomax, Chap. 9), race, and general health indicators, such as blood pressure and use of certain medications (Toppila et al. 2000). The focus of the present chapter is the interaction of ototoxic industrial chemicals with noise, which results in increased hearing loss.

Hearing loss can occur after ingestion of certain drugs due to their effects on the peripheral auditory system or central nervous system. The mechanisms of action of ototoxic substances may involve the entire organ, specific cells within the organ, components of specific cells, or individual biochemical pathways. Drugs and other substances that alter hearing or equilibrium by acting primarily at the level of the brain stem or the central auditory pathways are considered to be neurotoxic and not strictly ototoxic (Hawkins 1976).

The ototoxicity of therapeutic drugs has been recognized since the nineteenth century. Schacht and Hawkins (2006) reviewed initial reports that associated the intake of certain drugs such as quinine and acetylsalicylic acid with temporary hearing loss as well as dizziness and tinnitus. In the 1940s, permanent damage to the cochlea was reported in several patients treated with the newly discovered drug

---

Disclaimer: The findings and conclusions in this chapter are those of the authors and do not necessarily represent the views of the National Institute for Occupational Safety and Health.

T.C. Morata (✉)
National Institute for Occupational Safety and Health,
4676 Columbia Parkway, Cincinnati, OH 45226-1998, USA
e-mail: tmorata@cdc.gov

C.G. Le Prell et al. (eds.), *Noise-Induced Hearing Loss: Scientific Advances*,
Springer Handbook of Auditory Research 40, DOI 10.1007/978-1-4419-9523-0_11,

for treatment of tuberculosis, the aminoglycoside antibiotic streptomycin (Hinshaw and Feldman 1945). Today there are many well known ototoxic drugs used in clinical situations. Most of them (antibiotics, chemotherapeutics, diuretics, and antimalaria drugs) are used despite these negative side effects to treat other serious, sometimes life-threatening conditions. In the developed nations, and in some developing ones, the prescription of these drugs will trigger "ototoxicity monitoring" of patients to allow early detection of auditory effects and, when necessary, audiologic interventions to address the hearing impairment (AAA 2009).

In contrast, only in the past 20 years has the ototoxicity of chemicals found in the environment from contaminants in air, food or water, and in the workplace become a concern for researchers, toxicologists, audiologists, and other healthcare professionals. Initial reports described the ototoxicity of environmental chemicals after acute intoxications or poisonings, and these reports included observations that hearing loss was more common and sometimes more severe in work settings where chemical exposures occurred (Barregård and Axelsson 1984). Since then, considerable progress toward understanding the effects of certain environmental and occupational chemicals on the auditory system and their interactions with noise has been made (Fechter et al. 1987; Morata 1989; Lataye et al. 2000). Today, ototoxic properties have been identified for multiple classes of industrial chemicals, including solvents, metals, asphyxiants, pesticides, and polychlorinated biphenyls (PCBs). The rest of this chapter reviews the ototoxicity of these compounds and their interactions with noise.

Ototoxicants are of interest in the work environment, not only because of their actions on the hearing system of humans, but also because they may interact with each other and with noise when exposure is combined (simultaneously or sequentially). It is well known that the effects of many drugs or agents, when given concurrently, cannot necessarily be predicted on the basis of their individual effects. In such instances, the damage incurred by agents acting together may exceed the simple summation of the damage each agent produces alone (Prosen and Stebbins 1980; Humes 1984). This synergistic effect is separate from, and perhaps more dangerous than, simple additive effects, as these synergistic effects are difficult to predict. Because noise is the most common exposure that causes hearing loss in humans, special attention has been given to the combined exposure to noise and agents with ototoxic effects on the auditory system.

Solvents and carbon monoxide are the environmental/occupational chemicals most extensively studied to date because of their ubiquitous industrial use. These are chemicals that are widely used in several industrial sectors. Studies conducted with animal subjects have shown that some solvents can reach the inner ear through the blood stream even before they are metabolized. Solvents were found in the endolymph and perilymph, and these solvents not only caused damage to some inner ear structures, but also impaired auditory function (Campo et al. 1999). The onset, site, mechanism, and extent of ototoxic damage of these toxicants vary according to risk factors that include type of chemical, level and duration of chemical exposure, interactions between chemicals or noise, noise exposure level, and duration. Dose–response properties have not been precisely identified, but it appears that risk increases with increasing exposure, as is the case with ototoxic drugs such as cisplatin (used in chemotherapy) and aminoglycoside antibiotics (Laurell and Jungelius 1990; Halsey et al. 2005).

Ototoxic drugs often cause a high-frequency hearing loss whereas the hearing loss caused by occupational exposure to chemicals can be very similar to a hearing loss caused by excessive noise. Because noise exposure is so common in modern societies, this might explain the delay in recognizing the risk to hearing that these chemicals can pose.

Pure-tone audiometry, the standard clinical test used to determine a person's hearing sensitivity at specific frequencies, offers little information as to the relative health of inner (IHCs) and outer hair cells (OHCs), and the neural population. In other words, pure-tone audiometry does not provide information on the cause of the hearing loss. Other hearing tests such as word recognition, auditory reflex, and otoacoustic emission tests can help identify the site of damage. This information may help to differentiate the effects of chemicals from the effects of noise, as chemicals can affect more central portions of the auditory system (Ödkvist et al. 1987; Möller et al. 1989). In the presence of central deficits not only will sounds be perceived as less loud, but they may also be perceived as distorted.

2 Auditory Effects of Chemicals in the Work Environment

During the past two decades, scientific investigations have yielded new insights into the ototoxicity properties of a growing number of environmental and occupational chemicals (for detailed reviews, see Campo et al. 2009; Johnson and Morata 2010). This chapter summarizes the three classes of chemicals – solvents, asphyxiants, and metals – in which studies have included noise exposure as a cofactor. In addition, key details of the critical studies in animals and humans of auditory effects of chemicals for their corresponding No Observed Adverse Effect Level (NOAELs) and Lowest Observed Adverse Effect Level (LOAELs) are presented in Tables 11.1 and 11.2.

2.1 Solvents

There is robust evidence that exposure to solvents such as toluene, styrene, and xylene produces cochlear lesions (for details see Table 11.1) (Sullivan et al. 1988; Johnson and Canlon 1994; Campo et al. 2001). Clinical and occupational studies have also linked exposures to a variety of solvents (e.g., styrene, solvent mixtures, and jet fuels) with disorders in the central auditory pathway (Laukli and Hansen 1995; Fuente et al. 2006; Johnson et al. 2006).

Organic solvent ototoxicity was suggested in the 1960s (Lehnhardt 1965), but was not clearly demonstrated until the 1980s (for a review of the body of evidence, see Johnson and Morata 2010). Barregård and Axelsson (1984) reviewed five early occupational studies and four case reports that showed that the incidence of sensorineural hearing loss was higher than expected in noise-exposed workers who

Table 11.1 Critical studies in animals of auditory effects of chemicals with corresponding No Observed Adverse Effect Level (NOAELs) and Lowest Observed Adverse Effect Level LOAELs for selected substances (in ppm if not otherwise stated). Studies with combined noise exposure are also included

NOAEL	LOAEL	Noise level (N)	Exposure regimen	Species	Method	Endpoint(s), auditory effects only	Reference
Styrene (STY)							
STY alone							
–	200 mg/kg bw ~250 ppm	–	Gavage: 5 day/week, 3 weeks	Rats	ABR, ME	OHC loss at 200 mg/kg bw (lowest dose tested) Hearing loss at 300 mg/kg bw	Chen et al. (2007)
–	300	–	Inhalation: 6 h/day, 5 day/week, 4 weeks	Young rats, forced to be active	ABR, ME	OHC loss at 300 ppm (lowest dose tested) Hearing loss at 500 ppm	Lataye et al. (2005)
300	600	–	Inhalation: 12 h/day, 5 day/week, 4 weeks	Rats	ABR, ME	Hearing loss and OHC loss	Mäkitie et al. (2002)
STY combined with noise (N)							
–	400	85 dB Leq8h, OBN at 8 Hz, (86.2 dB SPL)	Inhalation and N: 6 h/day, 5 day/week, 4 weeks	Young rats, active	ABR, ME	OHC loss (synergism, only dose tested) Hearing loss not different from N	Lataye et al. (2005)
300	600	100–105 dB SPL	Inhalation and N: 12 h/day, 5 day/week, 4 weeks	Rats	ABR, ME	Hearing loss and OHC loss at 600 ppm (synergism) Hearing loss not different from N at 300 ppm	Mäkitie et al. (2003)

Toluene (TOL)							
TOL alone							
–	1,000	–	Inhalation: 6 h/day, 5 day/week, 4 weeks	Rats	ABR, ME	OHC loss (lowest dose tested) No hearing loss	Campo et al. (1997), Loquet et al. (1999)
700	1,000	–	Inhalation: 14 h/day, 7 day/week, 16 weeks	Young rats	ABR, BA, CAR	Hearing loss	Pryor et al. (1984)
TOL combined with noise (N)							
500	–	90 dB SPL, steady WBN 4-20 kHz (~87 dB Leq8h)	Inhalation: 6 h/day, 5 day/week, 90 days N: 4 h/day, 5 day/week, 90 days	Rats	ABR DPOAE	Hearing loss not different from N (highest dose tested)	Lund and Kristiansen (2008)
500	1,000	96 dB SPL (~90 dB Leq8h)	Inhalation: 6 h/day, 10 days N: 2 h/day, 10 days (after TOL exposure)	Young rats	ABR	Hearing loss at 1,000 ppm (synergism) Hearing loss not different from N at 500 ppm	Brandt-Lassen et al. (2000)

(continued)

Table 11.1 (continued)

NOAEL	LOAEL	Noise level (N)	Exposure regimen	Species	Method	Endpoint(s), auditory effects only	Reference
Xylene (XYL)							
p-XYL alone, mixed XYL (10% *o*-, 80% *m*-, and 10% *p*-XYL, EBZ content not given) and mixture (*o*-, *m*-, and *p*-XYL and EBZ)							
450 *p*-XYL	900 *p*-XYL	–	Inhalation: 6 h/day, 6 day/week, 13 weeks	Rats	ABR, ME	OHC loss (*o*- and *m*-xylene not ototoxic)	Gagnaire et al. (2001)
	250 Mixture (~50 *p*-XYL +50 EBZ)	–	Inhalation: 6 h/day, 5 day/week, 13 weeks	Rats	ABR, ME	OHC loss at 250 ppm mixture (synergism, lowest dose tested) Hearing loss at 1,000 ppm mixture (synergism)	Gagnaire et al. (2007a)
XYL combined with noise (N)							
No data							
Ethylbenzene (EBZ)							
EBZ alone and in mixture (EBZ and *o*-, *m*-, and *p*-XYL)							
–	200	–	Inhalation: 6 h/day, 5 day/week, 13 weeks	Rats	ABR, ME	OHC loss (lowest dose tested) Hearing loss at 400 ppm	Gagnaire et al. (2007b)
300	400	–	Inhalation: 6 h/day, 5 days	Rats	CAP, ME	Hearing loss and OHC loss	Cappaert et al. (2000)

EBZ combined with noise (N)							
–	300	95 or 105 dB SPL broadband 1.5–12.5 kHz	Inhalation: 6 h/day, 5 days N: 8 h/day, 5 days	Rats	CAP, DPOAE, ME,	OHC loss (when combined with 105 dB), (synergism, lowest dose tested) Hearing loss not different from N at 400 ppm	Cappaert et al. (2001)
Trichloroethylene (TCE)							
TCE alone							
–	2,000	–	Inhalation: 12 h/day, 7 day/week, 3 weeks	Rats	ABR	Hearing loss (mid-frequency, lowest dose tested)	Rebert et al. (1991)
1,600	2,400 1,707 (BMC)	–	Inhalation: 6 h/day, 5 day/week, 13 weeks	Rats	RMA at 16 kHz only	Hearing loss (mid-frequency) BMC: 15-dB increase in hearing threshold	Crofton and Zhao (1997)
2,400	3,200 1,418 (BMC)	–	Inhalation: 6 h/day, 5 day/week, 4 weeks	Rats	RMA at 16 kHz only	Hearing loss (mid-frequency) BMC: 15-dB increase in hearing threshold	Crofton and Zhao (1997)

(continued)

Table 11.1 (continued)

NOAEL	LOAEL	Noise level (N)	Exposure regimen	Species	Method	Endpoint(s), auditory effects only	Reference
TCE combined with noise (N)							
–	3,000	95 dB SPL	Inhalation: 18 h/day, 5 day/week, 3 weeks	Rats	RMA	Hearing loss (mid-frequency, only dose tested). Generally additive effect. Synergism at 4 kHz	Muijser et al. (2000)
Solvent mixtures							
Jet fuel (JP-8) alone							
–	1,000 mg/m^3	–	Inhalation: 4 h/day, 1 or 5 days	Rats	CAP, DPOAE, ME	Hearing loss (decrease in DPOAE amplitude) after repeated exposure. No OHC loss. Only dose tested	Fechter et al. (2007)
Jet fuel (JP-8) combined with noise (N)							
–	1,000 mg/m^3	97, 102 or 105 dB OBN 8 kHz	Inhalation: 4 h/day, 1 or 5 days N: 4 h (105 dB): 1 h/day, 5 day (102 dB) or 4 h/day, 5 days (97 dB)	Rats	CAP, DPOAE, ME	Hearing loss (decrease in DPOAE amplitude) and OHC loss greater than by N; OHC loss only after repeated exposure. Only dose tested	Fechter et al. (2007)

Lead (Pb) (blood level)							
Pb alone							
35 μg/dl	55 μg/dl	–	In diet: prenatal to ~10 years of age	Monkeys	ABR	Auditory effects (prolongations of ABR latencies)	Lilienthal and Winneke (1996)
35–40 μg/dl	–	–	In milk: birth to 1 or 2 years of age	Monkeys	ABR, DPOAE, Tympanometry	No auditory effects	Lasky et al. (2001)
Pb combined with noise (N)							
No data							
Carbon monoxide (CO)							
CO alone							
1,500		–	3.5–9.5 h	Rats	CAP, CM	No auditory effects	Chen and Fechter (1999)
CO combined with noise (N)							
300	500	110 dB Lin, or 115 dB Lin, 4 kHz OBN	Inhalation: 9.5 h N: 8 h	Rats	CAP, CM	Potentiation of NIHL (synergism)	Chen et al. (1999)
300	500	100 dB Lin, 13.6 kHz OBN	Inhalation: 9.5 h	Rats	CAP	Potentiation of NIHL increased linearly as CO increased between 500 and 1,500 ppm (synergism)	Fechter et al. (2000)
	BMCs: 194, 320		N: 8 h			BMCs: Increase in auditory threshold equivalent to 10% of the effect by N at 194 ppm and 5-dB potentiation of NIHL at 320 ppm	

(continued)

Table 11.1 (continued)

NOAEL	LOAEL	Noise level (N)	Exposure regimen	Species	Method	Endpoint(s), auditory effects only	Reference
300	500	84 dB SPL Leq8h, impulsive	Inhalation: 8 h/day, 10 days N: 6 h/day, 10 days	Rats	ABR, DPOAE	Potentiation of NIHL (synergism)	Lund and Kristiansen (2008)
Hydrogen cyanide (HCN)							
HCN alone							
50	–	–	Inhalation: 3.5 h	Rats	CAP, ME	No hearing loss or OHC loss	Fechter et al. (2002)
HCN combined with noise (N)							
10 ppm	30 ppm	100 dB Lin, OBN BMCs: 2–16 ppm	Inhalation: 3.5 h N: 2 h	Rats	CAP, ME	Potentiation of NIHL and OHC loss BMCs (lower bounds): Impaired auditory threshold 10% above the effect by N at 2 ppm, 5-dB potentiation of NIHL at 9 ppm	Fechter et al. (2002)
Acrylonitrile (ACN)							
ACN alone							
50 mg/kg	–	–	s.c. injection: 1, 2 or 5 days	Rats	CAP	No hearing loss (transient loss of auditory threshold sensitivity immediately after exposure). Only dose tested	Fechter et al. (2003, 2004)

ACN combined with noise (N)							
–	50 mg/kg	108 dB OBN	s.c. injection: 1 or 2 days N: 8 h/day, 1 days	Rats	CAP	Potentiation of NIHL (permanent loss of auditory threshold sensitivity). Only dose tested	Fechter et al. (2003)
–	50 mg/kg	105 dB Lin OBN	s.c. injection: 1 or 5 days N: 4 h/day, 1 or 5 days	Rats	CAP	Potentiation of NIHL (permanent loss of auditory threshold sensitivity). Only dose tested	Fechter et al. (2004)
–	50 mg/kg	95 or 97 dB OBN	s.c. injection: 5 days N: 4 h/day, 5 days	Rats	CAP, DPOAE, ME	Hearing loss (permanent threshold shifts, decrease in DPOAE amplitudes) and OHC loss. No effect of ACN or N alone. Only dose tested	Pouyatos et al. (2005b)

ABR auditory brain stem response, *ASR* acoustic startle response, *BA* behavioral audiometry, *CAP* compound action potential, *CAR* conditioned avoidance response, *CM* cochlear microphonics, *DPOAE* distortion product otoacoustic emissions, *i.p.* intraperitoneal, *LOAEL* lowest observed adverse effect level, *MAEP* middle latency auditory evoked potentials, *ME* morphological examination, *NIHL* noise-induced hearing loss, *NOAEL* no observed adverse effect level, *N* noise, *OBN* octave band noise, *OHC* outer hair cell, *PTA* pure-tone audiometry, *RMA* reflex modification audiometry, *s.c.* subcutaneous, *SPL* sound pressure level, *SP* summating potential, *ppm* parts per million

Table 11.2 Critical studies on auditory effects in humans (exposure levels in ppm if not stated otherwise). Hearing loss means changes measured with pure-tone audiometry. Auditory dysfunction means changes measured in the central auditory system with evoked potential testing or other central tests

Current exposure level, mean ± SD, (range)	Exposure duration	Noise (N)	Size of study group	Results and comments	Reference
Styrene (STY)					
3.5 (0.05–22) (STY)	17 (1–39) years (STY)	≤84 dBA (STY)	65 STY	Hearing loss and auditory dysfunction (speech) in STY and STY + N compared to N and controls. Biological marker for STY (urinary mandelic acid) associated with hearing loss	Morata et al. (2002), Johnson et al. (2006)
2.8 (0.007–12) (STY + N)	15 (2–37) years (STY + N)	89 dB (STY + N)	89 STY + N		
Average work–life exposure: 18 (STY)		86 dBA (N)	78 N		
14 (STY + N)			81 controls		
~5 (0.7–14) (estimated from urinary mandelic acid + phenylglyoxylic acid)	7 ± 6.2 years	73 dBA	32 STY 60 controls	Hearing loss in STY compared to the age-matched controls	Mascagni et al. (2007)
8 (0.1–93)	9.4 ± 8.9 years	≤85 dB	44 STY	Hearing loss in high-frequency range in STY subgroup ($n = 54$) exposed >16 ppm for ≥5 years. The effect correlated to STY in air and to biological marker of STY (urinary mandelic acid)	Morioka et al. (1999)
<50 for 87/93 workers			49 STY in mixture including TOL 33 controls		
Average work–life exposure:		80 dBA (STY)	194 STY	Hearing loss in STY and in STY + N compared to N and controls. Average work-life exposure to STY correlated to hearing loss	Śliwińska-Kowalska et al. (2003)
14 ± 9.3 (STY)		89 dBA (STY + N)	55 STY + N		
8 ± 6 (STY + N)		89 dBA (N)	66 N 157 controls		

22 (3.7–46)	5.4 years	69–76 dBA (STY) 82–86 dBA (N)	19 STY 18 N 11 controls	Hearing loss in high-frequency range in STY exposed compared to N and controls	Morioka et al. (2000)
Toluene (TOL)					
26 ± 20 Life–time weighted exposure: 45 ± 17		81–82 dBA	192 TOL No controls	No hearing loss. TOL levels or duration not associated with hearing loss. NOAEL estimated to 50 ppm by authors	Schäper et al. (2003)
(9–37)	12 (2–24 years) (TOL + N) 6 (3–15 years) (N)	88–98 dBA TOL + N and N	50 TOL + N 50 N 40 controls Workers with nl hearing.	>60% of TOL + N or N had no response in TEOAE vs. 27% of controls 49% of TOL + N had no contralateral inhibition in TEOAE vs. 17% of N and 7% of controls	Bernardi (2000)
34 (2–89) (estimated from TOL in blood)	21.4 (4–30) years	Not given	49 TOL 59 controls	Auditory dysfunction of ABR	Vrca et al. (1996)
(0.04–244) in mixture + N ≤50 for 109/124 workers (mixture incl. ethyl acetate and ethanol)	Duration: 7.7 (1–25) years	71–93 dBA	124 TOL (in mix) + N No controls	Hearing loss in 49%. Biological marker (urinary hippuric acid) for TOL correlated with hearing loss	Morata et al. (1997)
97	12–14 years	Not given	40 TOL 40 controls	Auditory dysfunction of ABR shown in TOL exposed (only workers with normal PTA included)	Abbate et al. (1993)
Carbon disulfide (CS_2)					
3–8 (ventilation improved 14 years ago)	2–7 years up to >20 years	Not given	25 CS_2 (2–7 years) 34 CS_2 (>20 years) 40 controls	Auditory dysfunction of ABR shown in CS_2 exposed >20 years	Hirata et al. (1992)

(continued)

Table 11.2 (continued)

Current exposure level, mean ± SD, (range)	Exposure duration	Noise (N)	Size of study group	Results and comments	Reference
1.6–20.1		80–91 dBA (CS_2 + N) 83–90 dBA (N)	131 CS_2 + N 105 N 110 controls	Higher prevalence (68%) of hearing loss in exposed than in N and controls. Greatly increased risk for exposures > 14.6 ppm	Chang et al. (2003)
Lead (blood levels, μg/dl)					
Median: 30 (12–59)		Not given	22 Pb 14 controls	Auditory dysfunction (P300) in Pb Blood Pb level correlated with auditory dysfunction.	Araki et al. (1992)
28 ± 8 (4–62) Lifetime weighted average: 39 ± 12 (4–66) Lifetime integrated blood Pb index: 719 μg–year/dl	17 (0.2–26) years	Not given	359 Pb No controls	Pb exposure interfered with ABRs in a dose-dependent manner. Current and lifetime weighted average blood Pb level associated with the ABR wave I latency while the lifetime index was associated with wave III latency	Bleecker et al. (2003)
57		86 dBA Leq	220 Pb 119 controls	Hearing loss correlated to high and long-term Pb exposure index (duration of employment and ambient Pb concentration). No correlation to N alone or to the interaction between N and short- or long-term Pb exposure	Wu et al. (2000)

7 μg/dL (blood)	Mean 8.5 years	77–85.5	248 exposed to metals & noise 11 controls	Lead was the only metal in blood found significantly correlated with hearing loss for most tested sound frequencies ($p<0.05$ to $p<0.0001$). After adjustment for age and noise level, lead >7 μg/dL was significantly associated with hearing loss at the sound frequencies of 3,000–8,000 Hz with odds ratios raging from 3.06 to 6.26 ($p<0.05$–$p<0.005$)	Hwang et al. (2009)
Carbon monoxide (CO)					
(16–35)		85–90 dBA CO+N 90–91 dBA N	2 CO 2 CO+N 3 N 21 controls	Hearing loss in CO+N at 8 kHz	Lacerda (2007)
Hydrogen cyanide (HCN), acrylonitrile (ACN), 3,3′-iminodipropionitrile (IDPN)					
No data					

ABR auditory brain stem response, NOAEL no observed adverse effect level, *N* noise, *OR* odds ratio, *PTA* pure-tone audiometry, *SD* standard deviation, *TEOAE* transient evoked otoacoustic emissions, ppm parts per million, *nl* normal

were also exposed to solvents. An ototraumatic interaction between noise and organic solvents was suggested, and biological plausibility was discussed. Because organic solvents are known for their neurotoxic effects in both the central and the peripheral nervous systems, it was argued that solvents might injure the sensory cells and peripheral endings in the cochlea. It was further hypothesized that, because solvent-related effects have been detected in the brain, central auditory disorders could also be expected.

2.1.1 Biological Basis for the Solvents Auditory Effects. Studies with Experimental Animals

The aromatic solvents of the alkylbenzene family (e.g., toluene, ethylbenzene, and xylene) have been found to affect the auditory system. The relative ototoxicity varies among the aromatic solvents. Styrene has been shown to be more ototoxic than toluene, and, of the xylene isomers, only *p*-xylene is ototoxic (for details see Table 11.1) (Gagnaire and Langlais 2005; Maguin et al. 2006). Toluene, styrene, ethylbenzene, and thrichloroethylene have been found to interact synergistically with noise (Johnson and Morata 2010). Aliphatic solvents, including *n*-hexane and *n*-heptane, as well as carbon disulfide, are neurotoxic substances that can affect the auditory system.

In animal experiments as well as in human studies, the ototoxic effect after inhalation of organic solvents has been established using many different methods. In animal studies, the most sensitive method to discover the ototoxic effect of solvent is morphological studies detecting the loss of hair cells in the cochlea as shown after toluene and styrene exposure in rats (for details see Table 11.1) (Johnson and Canlon 1994; Lataye et al. 2000, 2005; Mäkitie et al. 2002).

The OHCs are electromotile, that is, these cells change their length in response to sound stimulation (Ashmore 1987; Dallos et al. 1991). Electromotility and the resulting hair cell active process depends on calcium concentration within the hair cells. Thus, OHCs may be vulnerable to ototoxic agents that interfere with intracellular calcium regulation. In vitro studies with isolated OHCs exposed to toluene have shown dysmorphia and impaired regulation of intracellular levels of free calcium (Liu and Fechter 1997). Changes occurred rapidly with in vitro exposure to 100 μM toluene, a level predicted to accumulate in the brains of humans if exposed to 80–100 ppm of toluene in air (Liu and Fechter 1997).

Gagnaire and Langlais (2005) studied 21 different aromatic solvents and found that only 8 (toluene, *p*-xylene, ethylbenzene, *n*-propylbenzene, styrene, α-methylstyrene, *trans*-β-methylstyrene, and allylbenzene) caused loss of hair cells. The degree of hair cell loss differed after exposure to those solvents. The degree of ototoxicity was not clearly related to the octanol/water partition of the solvent, but correlations between some structural properties and ototoxicity were observed (Gagnaire and Langlais 2005).

Studies comparing auditory effects of the three isomers of xylene (*ortho*-, *meta*-, and *para*-xylene) have shown that only *p*-xylene is ototoxic (Gagnaire et al. 2001).

Both *o*- and *m*-xylene induce liver enzymes and they are therefore eliminated from the body of rats faster than *p*-xylene. Given slower clearance, *p*-xylene reaches a higher level in the blood and also gives rise to more potentially toxic intermediates than the other two isomers, which could explain why only *p*-xylene is ototoxic (Maguin et al. 2006). However, Gagnaire et al. (2007b) showed that when a higher dosage was used to obtain the same blood and brain levels with *m*-xylene as with a known ototoxic dose of *p*-xylene, there was still no ototoxic effect observed after exposure to *m*-xylene. Taken together, the differences in metabolic rates probably do not explain the different ototoxic potentials of the xylene isomers (Gagnaire et al. 2007b).

Laboratory investigations appear to identify a common pattern of cochlear dysfunction and injury after solvent exposure. This pattern, produced by toluene, styrene, xylenes, and trichloroethylene, involves impairment of OHCs that normally encode middle frequency tones and are located in the middle turns of the cochlea (for details see Table 11.1) (Crofton and Zhao 1993; Crofton et al. 1994; Campo et al. 2001). This tonotopicity of the cochlear damage is different from that induced by aminoglycoside antibiotics, which affect mainly the high-frequency area of the cochlea. The pattern of damage is probably due to the intoxication route taken by the solvents to reach the organ of Corti, as shown for styrene (Campo et al. 2001; Lataye et al. 2001). In these studies, as well as in a study by Chen et al. (2007), it was shown that styrene reaches the hair cells in the cochlea from the blood via the stria vascularis and through the supporting cells. This explains why the third row of OHCs is affected first, that is, this row of OHCs is closer to the supporting cells.

The ototoxic effects of styrene exposure continue even after chemical exposure has ended (Campo et al. 2001). The intoxication route of the solvents likely explains why the ototoxic effect of styrene progresses beyond the cessation of styrene exposures to 700 ppm and above. Organ exposure continues some time after cessation of contaminated air exposure because of the time taken to clear the chemicals from the tissues (Loquet et al. 2000; Campo et al. 2001; see Hu, Chap. 5 for more information on the apoptotic process of cell death and its duration).

In rats, levels of solvents were measured in the blood, brain, auditory nerves, organ of Corti, and in cerebrospinal fluid and the perilymph from the inner ear after exposure to either toluene or styrene for 1 day. In this study solvents were detectable in the tissues but not in the fluids, indicating that toluene and styrene are transported through the tissues of the organ of Corti rather than through the fluids of the inner ear (Campo et al. 2001).

Chen et al. (2007) also measured the concentration of styrene in different regions in the cochlea and found a higher solvent concentration in the tissues of the middle region with lower levels in the apex and the basal turn, explaining the higher vulnerability in the middle frequency region. The reason for the higher concentration in the middle region is not fully understood.

Trichloroethylene has been shown via electrophysiological testing and cochlear histopathology to impair IHC and spiral ganglion cell function. Loss of spiral ganglion cells was significant in the middle turn of the cochlea, but not in the basal turn. The data confirm that the behaviourally measured loss in auditory function is a consequence of cochlear impairment, with the spiral ganglion cells being a prominent target of this solvent (Fechter et al. 1998).

Effects of toluene exposure on the central auditory pathways in rats have been further investigated in two recent studies. In these experiments, toluene inhibited the auditory efferent system by modifying the response of the protective acoustic reflexes from the efferent system originating from the olivary complex in the brain stem. In these experiments, toluene acted in the same way as other known cholinergic receptor antagonists (Campo et al. 2007; Lataye et al. 2007). Maguin et al. (2009) showed that toluene acts also on the regulation of acetylcholine release in muscles by blocking the voltage-gated Ca^{2+} channels involved in the protective middle ear reflex exhibited by the stapedius muscle. This reflex is also mediated by efferent motorneurons emanating from the olive complex in the brain stem. These studies (Campo et al. 2007; Lataye et al. 2007; Maguin et al. 2009) provide interesting insight into the mechanism of the interaction between solvents and noise. One hypothesis is that solvents reduce the protective middle ear reflex as well as the efferent reflex, thus making noise more damaging to the inner ear in the presence of solvent exposure.

Several experimental studies have shown that noise exposure produces reactive oxygen species (ROS) in the inner ear (Yamane et al. 1995; Henderson et al. 2006; Le Prell et al. 2007; for detailed discussion see Le Prell and Bao, Chap. 13). Accumulating evidence links ROS to cochlear damage for both ototoxicants and noise trauma (Evans and Halliwell 1999; Kopke et al. 1999; Fechter et al. 2002), which may also explain the interaction between noise and oxidizing chemical agents such as solvents and asphyxiants. It has been shown that combinations of non-damaging noise and oxidizing chemical agents lead to oxidative stress that causes the death of hair cells in the inner ear. Recent evidence for apoptotic cell death came from activated caspase pathways observed in the OHCs after styrene exposure in rats (Fechter et al. 2000, 2002, 2003; Pouyatos et al. 2005a,b; Chen et al. 2007).

Solvent-induced hearing loss is species dependent. Rats are sensitive to solvents, whereas guinea pigs and chinchillas have been unaffected in studies to date. Davis et al. (2002) reported no effects in the chinchilla auditory system after toluene exposure alone or in combination with noise. The authors argued that the chinchilla liver detoxified the toluene. Hepatic microsomes from chinchillas, rats, and humans were tested for the ability to metabolize toluene to the more water-soluble, and less toxic, compound benzyl alcohol. Chinchillas had higher levels and activities of liver cytochrome P450 (CYP) enzymes than both rats and humans, suggesting that a more effective metabolism of toluene was possible in chinchillas. Similar observations were reported by Lataye et al. (2003), who measured toluene and styrene levels after exposures in rat and guinea pig models. Lataye et al. (2003) found that styrene concentration in rat blood samples was four times higher than in guinea pig blood samples. The authors concluded that interspecies differences in susceptibility may be explained by (1) different amounts of solvent transported by blood and corresponding differences in the amount of solvent reaching the organ of Corti, (2) differences in metabolism, (3) differences in glutathione (considered as an endogenous defense against ROS in the inner ear) within the sensory epithelium, and (4) morphological differences of the lateral wall membranes of the OHCs of the cochlea (Lataye et al. 2003). Consistent with these predictions, when Gagnaire et al. (2007b) investigated the difference in blood and brain levels of *p*-xylene between guinea pigs and rats, the blood level of *p*-xylene in guinea pigs was only half of that in the rat and the level in the brain reached only about 20–30% of that in rats. Rats also had four times slower elimination rate than guinea pigs. Thus, toxicokinetic factors likely contribute to the species difference between rats and guinea pigs and between rats and chinchillas (Davis et al. 2002; Lataye et al. 2003; Gagnaire et al. 2007a,b). Solvent metabolism in humans is closer to that of rats than to that of guinea pigs (Lataye et al. 2003).

2.1.2 Evidence from Studies on Effects of Occupational Exposures

Human data are consistent with the evidence from animal studies that toluene, styrene, and solvent mixtures are ototoxic (for details see Table 11.2) (Johnson and Morata 2010). Solvent-induced hearing losses are often moderate to severe, as is also the case with noise-induced hearing loss (NIHL). The audiometric high-frequency "notch" common in NIHL is often present after long-term chemical exposures, although some reports indicate that a wider range of audiometric frequencies are affected when compared to the range of frequencies affected by noise.

The prevalence of solvent-induced hearing loss (identified through pure-tone audiometry) varies across studies. This is often also the case in studies with NIHL, a finding that has been explained by the wide range of possible exposure scenarios and the influence of modifying or confounding factors as mentioned earlier herein. Different definitions of hearing loss have been used, and the criteria used to define hearing loss have a critical impact on the prevalence of screening referral rates (for recent discussion, see Le Prell et al. 2011). Other important limitations to human studies include often insufficient and/or unreliable exposure history data (for both chemicals and noise), as well as a lack of comparability between study and control groups regarding solvent and noise exposures.

In several studies, central auditory tests have been performed to complement pure-tone audiometry outcomes in workers exposed to solvents or metals (e.g., Ödkvist et al. 1992; Fuente et al. 2006; Johnson et al. 2006; for details see Table 11.2). The results from these studies indicate that solvents can also affect the central auditory system and impair sound discrimination.

Although less well investigated, trichloroethylene and carbon disulphide have also been associated with negative auditory effects in humans. Observed auditory effects of *n*-hexane have been interpreted as a sign of its well known central nervous system toxicity. There are no published studies on the potential for ototoxicity of xylenes, ethylbenzene, chlorobenzene, or *n*-heptane in humans, even though xylene, ethylbenzene and chlorobenzene are common components in solvent mixtures. This represents an area of considerable importance for future investigations. Moreover, it is important to assess the risks associated with solvent mixtures. Such studies are critically important; however, formal risk assessment criteria will be challenging to define because of the many different chemical mixtures used across different workplaces. Exposure to solvent mixtures containing ototoxic solvents (e.g., styrene, toluene, xylenes, ethylbenzene, trichloroethylene, *n*-hexane, jet fuels, and white spirit) clearly may cause auditory effects (Johnson and Morata 2010).

2.2 *Asphyxiants*

Asphyxiants are vapors or gases that can cause unconsciousness or death by suffocation. They act by interfering with oxygen delivery or utilization. Asphyxiation, or suffocation, occurs when the blood does not deliver enough oxygen to the body. Some chemical asphyxiants, such as carbon monoxide (CO), reduce the blood's ability to carry oxygen. Others, such as cyanide, interfere with the body's utilization of oxygen. The chemical asphyxiants that have been studied for their ototoxicity include CO, hydrogen cyanide, acrylonitrile, and 3,3′-iminodipropio-nitrile.

2.2.1 Carbon Monoxide

The auditory effects of carbon monoxide (CO) in combination with noise have been examined in numerous animal experiments. The majority of studies were performed on rats. This species demonstrates a much higher resistance to acute CO intoxication than humans. In rats, a lethal dose for a 30-min exposure is 5,000 ppm; in humans, the lethal dose is 1,500 ppm (Rao and Fechter 2000a). Experiments in rats show that CO does not alter auditory function even when used at up to lethal doses (Chen and Fechter 1999). However, when delivered alone or in combination with toluene, CO exposure may potentiate NIHL (Lund and Kristiansen 2004).

CO can potentiate NIHL at noise exposure conditions that have limited effects on auditory function alone (Young et al. 1987; Fechter et al. 1988). Under intermittent noise exposure with long quiet periods, CO exposure may produce unexpectedly large, permanent threshold shifts (Chen et al. 1999; Rao and Fechter 2000b). Surprisingly, the data did not validate the anticipated relationship between the percentage of time that noise is present (noise duty cycle) and increasing hearing loss. Instead, the mildest noise duty cycle (noise exposure interrupted with quiet breaks) produced maximal hearing loss when CO was also present. Otherwise, when CO was absent, hearing loss was reduced because of the quiet breaks.

Auditory function was compared in rats that had been exposed 4 weeks earlier to CO alone, noise alone, CO in combination with noise, or air in a chamber (no noise, no added CO). The compound action potential threshold evoked by pure-tone stimuli was used as a measure of auditory sensitivity. Potentiation of NIHL by CO increased linearly as CO concentration increased (Chen et al. 1999; Fechter et al. 2000).

Benchmark dose software (BMDS) from the U.S. Environmental Protection Agency (EPA) fits mathematical models to dose–response data, with the end result being selection of a "benchmark dose" that meets a preselected "benchmark response." BMDS was used to determine a benchmark concentration of CO that produced either an increase in auditory threshold equivalent to 10% of the effect of noise alone or a 5-dB potentiation of NIHL (Fechter et al. 2000). Without exposure to CO, the noise effect was not significant, but in the combined exposure scenarios, responses were significantly poorer.

Rao and Fechter (2000a) explored the ability of phenyl-*N*-*tert*-butylnitrone (PBN), a spin trap agent that forms adducts with free radicals, to protect against the combined effects of noise and CO on auditory function in rats. Intraperitoneal injection of PBN both pre- and postexposure to CO and noise protected against the permanent hearing loss. Protection did not occur when PBN was given only postexposure (Rao and Fechter 2000a). Thus, although these results help to establish a role for oxidative stress in the interaction of CO and noise, PBN does not offer an effective therapeutic treatment strategy. Different antioxidant agents clearly vary with respect to post-noise treatment outcomes (for review, see Le Prell and Bao, Chap. 13), and it is possible that other antioxidant agents might provide an effective post-CO rescue strategy. Additional studies are required to confirm the potential for protection against CO–noise interactions.

Several reports have documented that hearing loss is one of the outcomes associated with acute CO poisoning in humans. Unlike the findings in animal studies, noise exposure was not a necessary factor for the auditory problems to occur. In an early investigation, Lumio (1948) examined 700 patients suspected of having CO poisoning. With the exception of 6% of the study population who worked in factories, most of the participants were from occupations in which workers were exposed to CO from gas generators used in automobiles. Occupational noise exposure and health indicators that could contribute to hearing loss were accounted for through a general medical examination and interview. Among those who were diagnosed with chronic CO poisoning, 78% had hearing loss. The hearing disorders were evident in the extended high-frequency region of the audiogram.

A more recent study involved an examination of a database containing workers' charts collected by the Quebec National Public Health Institute between 1983 and 1996 (Lacerda et al. 2005; Lacerda 2007; Leroux et al. 2008). The effect of CO with noise exposure below 90 dBA was not significant at any frequency. Workers who were exposed to CO and to noise levels above 90 dBA displayed significantly poorer hearing thresholds at high frequencies (3,000–6,000 Hz) than workers without CO exposure, but with equivalent noise exposure. The magnitude of the shift in hearing thresholds was influenced by the number of noise exposure years.

Studies conducted by the same team in Canada also examined if a combined, nonoccupational exposure to noise and CO could affect the hearing thresholds of workers with occupational noise exposure. Information was available on their occupational noise exposure levels, but estimated for their nonoccupational exposures. Results indicated a significant interaction between audiometric results of a specific test frequency and nonoccupational CO exposure, years of occupational noise exposure, and current occupational noise exposure level. Nonoccupational noise exposure had a marginal effect on hearing thresholds when compared to nonoccupational noise and CO exposure (which had a larger effect on hearing thresholds). However, these effects were only observed in the group with at least 15 years of occupational noise exposure associated with concurrent nonoccupational exposure to CO and noise (Lacerda 2007; Leroux et al. 2008).

Further, the effects of occupational exposure to low concentrations of CO and noise on hearing status of a small subsample of workers ($n = 28$) were also explored. The environmental CO levels ranged between 16 and 35 ppm and the biological CO (carboxyhaemoglobin) levels ranged from 2% to 3%. The audiometric data indicated that combined CO and noise exposure had an effect on hearing at 8,000 Hz, as measured via both pure-tone audiometry and distortion product otoacoustic emissions, but this was based on only two individuals (Lacerda 2007). Although conclusions drawn from small case studies are necessarily limited, the data clearly indicate this is an important area for future research.

Ahn et al. (2005) conducted a nested case-control study in a cohort of male iron and steel workers exposed to low concentrations of CO. The study group comprised 770 cases and 2,574 incidence density age-matched controls. Quantitative CO and noise exposure data were available from a job-exposure matrix. The odds ratio for hearing loss (4 kHz threshold ≥35 dB) was 2.5 (95% CI 1.2–5.0) for exposure levels greater than 20 ppm of CO, after controlling for noise exposure level, body mass index, smoking, hypertension, and diabetes.

2.2.2 Hydrogen Cyanide

Cyanides are chemical compounds that contain a cyano functional group, CN^-. The cyanide ion has a single negative charge and consists of a carbon that is triply bonded to a nitrogen atom. Cyanide is often used as a shorthand term for hydrogen cyanide. It is used in tempering steel, dyeing, explosives, engraving, and the production of acrylic resin plastic and other organic chemical products. Hydrogen cyanide is contained in the exhaust of vehicles, in tobacco smoke, and in the smoke of burning nitrogen-containing plastics.

In an investigation on the auditory effects of hydrogen cyanide by Fechter et al. (2002), rats were exposed to hydrogen cyanide alone for 3.5 h or in combination with 2 h of octave band noise exposure. Additional groups received noise exposure alone (which did cause an auditory effect) and no treatment other than placement in a quiet inhalation chamber with clean air. Hydrogen cyanide alone did not cause significant hearing loss or hair cell loss. The combined exposure to noise and hydrogen cyanide caused a cyanide dose–dependent compound action potential (CAP) threshold impairment and OHC that exceeded the noise exposure alone. At 30 ppm, the potentiation of NIHL achieved statistical significance. A risk assessment analysis was conducted for the auditory threshold data using the BMDS software described earlier. A continuous model showed that the data could be described by a linear function. For a benchmark response corresponding to a 5-dB increase in the auditory threshold above the effect of noise alone, the lower bound of the 95% CI for the benchmark dose was 9 ppm. The benchmark dose that impaired the auditory threshold 10% above the effect of noise alone had a lower bound of 2 ppm. The lower bound of the hydrogen cyanide dose that produced a one standard deviation elevation in NIHLwas 16 ppm (Fechter et al. 2002). Auditory effects of hydrogen cyanide in humans have not been studied.

2.2.3 Acrylonitrile

A nitrile is any organic compound that has a carbon atom and a nitrogen atom triply bonded together, that is, a –C≡N functional group. The prefix cyano- is used in chemical nomenclature to indicate the presence of a nitrile group in a molecule. Acrylonitrile is used in the production of other chemicals such as plastics, synthetic rubber, and acrylic fibers, and is 1 of the 50 most commonly produced industrial chemicals.

Four animal experiments on the effects of acrylonitrile on the auditory system have been conducted (Fechter et al. 2003, 2004; Pouyatos et al. 2005b, 2007). Acrylonitrile potentiates NIHL as a consequence of oxidative stress. The metabolism of acrylonitrile involves conjugation with glutathione, resulting in rapid and pronounced depletion of this antioxidant in many organs including brain, liver, and kidney. It also results in cyanide formation via a secondary oxidative pathway. The studies indicate that the OHCs are the main target of toxicity.

Acrylonitrile alone elevated auditory thresholds temporarily in rats. No effects were seen after 3 weeks. Acrylonitrile in combination with noise increased auditory threshold impairment relative to rats receiving noise only when measured 3 weeks after exposure (Fechter et al. 2003). Combined exposure for 5 days to acrylonitrile and moderate noise caused permanent hearing loss and OHC loss in rats. Individually, neither acrylonitrile nor noise caused these effects (Pouyatos et al. 2005b).

Rats treated daily with phenyl-*N*-*tert*-butylnitrone (PBN, spin-trap agent that sequesters ROS) before and again after acrylonitrile and noise treatment for 5 consecutive days showed approximately the same auditory impairment as did rats receiving noise only. Thus, PBN blocked the potentiation of NIHL (Fechter et al. 2004). L-*N*-acetylcysteine (antioxidant, pro-glutathione drug) treatment of rats also decreased auditory loss and hair cell loss resulting from combined exposure to acrylonitrile and moderate noise (Pouyatos et al. 2007).

None of the studies conducted to date have shown that acrylonitrile exposure alone damages the auditory system of the rat; however, acrylonitrile *does* potentiate NIHL at noise levels that are realistic in terms of human exposure. However, the acrylonitrile exposure route used in the animal studies (subcutaneous injection) differs from that experienced by workers, and the doses of acrylonitrile in animal studies were greater. Because the widespread use of acrylonitrile in industry occurs in settings where noise exposure is also present, the identified synergistic mechanism may be important for occupational health. However, to date, auditory effects of acrylonitrile in humans have not been studied.

3 Lead

Metals that have been studied for their ototoxicity include lead, mercury, and organotins. Lead and mercury may affect both the cochlea (Rice and Gilbert 1992; Rice 1997) and the central auditory pathways (Discalzi et al. 1993; Otto and Fox 1993; Lasky et al. 1995). Lead is the only metal that has been investigated with respect to coexposure with noise.

Several experiments have been conducted regarding the effects of lead exposure on the auditory system. In guinea pigs, lead exposure induced dysfunction of the vestibulocochlear nerve, but it did not induce dysfunction of the organ of Corti and the stria vascularis (Yamamura et al. 1989). In contrast, cochlear effects by lead were reported in studies with monkeys (Lasky et al. 1995; Rice 1997).

Researchers have conducted occupational studies of the effects of lead exposure on the auditory system (Discalzi et al. 1993; Farahat et al. 1997; Wu et al. 2000; Hwang et al. 2009). They have not always reported noise levels, particularly because most studies examined the effects of lead on the central auditory system, which has not been historically considered to be affected by noise exposure (although see Kaltenbach, Chap. 8 for discussion of noise-induced plasticity in the central nervous system). However, because of the nature of the work performed, it is likely that the studied workers were also exposed to noise.

Studies conducted with lead-exposed workers consistently report an association between lead exposure and central auditory effects. Chronic lead exposure impaired conduction in the auditory nerve and the auditory pathway in the lower brain stem. Blood lead concentrations significantly correlated with abnormalities in the recorded evoked potentials (Araki et al. 1992; Bleecker et al. 2003).

One study that tested for statistical interaction between lead and noise showed no significant interaction (Wu et al. 2000). More recently, in a study with workers from a steel plant, Hwang et al. (2009) reported that lead was the only metal in blood that significantly correlated with hearing loss for most tested sound frequencies, after adjusting for age and noise level.

Studies on chemical exposures outside the work environment are largely beyond the scope of the present chapter, given challenges in obtaining accurate measurements outside of workplace monitoring. However, in the case of lead, it is worth noting that several studies conducted with children have shown ototoxic effects (Schwartz and Otto 1987, 1991; Osman et al. 1999). Paradoxically, these effects have not been seen in cases of extreme plumbism, either in adults or in children (Buchanan et al. 1999).

4 Summary

The precise conditions, such as the specific concentration or period of time, one would need to be exposed to the studied chemicals to suffer an effect have not been identified for most of the chemicals studied and described in this chapter. The dose–response lowest observed adverse effect level (LOAEL) and no-observed adverse effect level (NOAEL) have been identified in animal experiments for a few substances.

Researchers have demonstrated that by adding other stressors such as impact noise or CO, or by ensuring that subjects are active during chemical exposure, the lowest level of solvent exposure needed to elicit an auditory effect is reduced (Lataye et al. 2005; Lund & Kristiansen, 2008). Moreover, there is a difference in the lowest chemical level necessary to cause an effect in humans compared to that measured in experimental animals. When compared, the levels necessary to produce an auditory effect are lower (posing a greater risk) in humans than in animals.

Increased vulnerability of humans may occur because humans are generally exposed to solvents in combination with a multitude of other factors (several exposures, physical demands, etc.) whereas animal experiments typically involve isolated solvent exposures. Another complication in determining the concentration needed for a hearing loss to occur in humans exists because individuals often do not know the specific chemical concentration to which they have been exposed, and because many factors can interact in causing an effect. Unfortunately, cases of hearing loss have been observed after chemical exposures that were within permissible limits.

The evidence that interactions in which adverse effects are greater than the sum of the individual effects have been reported between noise and chemicals raises serious concern. It has been noted that as one adds stressors, the LOAEL and NOAEL of the other agent can be lowered. For example, a single exposure to a particular chemical in quiet may not elicit a toxic response, yet the same exposure in the presence of high-level noise can create a hearing loss (when either alone would not).

Another challenge in this area is that the number of chemicals studied to date is very small, particularly when one considers the enormous number of existing industrial chemicals and the thousands of new ones placed on the market every year. It is therefore of crucial importance to understand not only the specific mechanisms by which individual chemicals affect the auditory system, but also general mechanisms of damage common to multiple chemicals.

Cell damage can occur via several different mechanisms and result in an auditory disorder; however, there are common features shared between damaging mechanisms resulting from a physical agent (such as noise) and some of the ototoxic chemicals. Damage to the OHCs can be driven by the formation of free radicals, including reactive oxygen species and reactive nitrogen species. The generation of free radicals has been associated with cellular injury in different organ systems. It is a basic mechanism of toxicity, and is part of at least one mechanism underlying NIHL, as explained in further detail by Le Prell and Bao (Chap. 13). Other chemicals such as metals and pesticides may affect both the cochlea and the central auditory pathways, depending on the substance.

When specific ototoxicity information is not available on a particular chemical, individuals concerned about the potential risk factors should look for information on the agent's general toxicity, as well as toxicity related to damage to the kidneys and nerves (nephrotoxicity and neurotoxicity, respectively). Information on whether a chemical produces reactive free radicals could also give some clues about that agent's potential ototoxicity. Glutathione is an important cellular antioxidant that limits cell damage by reactive oxygen species (for detailed discussion see Le Prell and Bao, Chap. 13). Evidence is available, for instance, indicating that ototoxicity due to noise plus CO or hydrogen cyanide exposure is mediated by free radicals. For this reason, information on certain chemicals being associated with free radicals or glutathione depletion could also help in the decision to examine a chemical for potential ototoxicity.

Although the focus of this chapter is the interaction of occupational chemicals and noise, it is important to remember that exposures to these chemicals can occur outside the work environment. Nonoccupational exposure can result from any activities that involve solvents, paint, polyurethanes, paint thinners, degreasers, and fuels.

In conclusion, the chemicals described in this chapter have been associated with negative auditory effects in animals. These chemicals are substances with diverse chemical structures, implying multiple targets for injury within the auditory system and multiple possible underlying mechanisms. This complexity represents an obstacle in identifying the chemical structural features necessary for a chemical to be ototoxic.

Another challenge to comprehensive understanding is that different species respond differently to the same chemical, perhaps because of metabolic differences or other species-specific differences. On the other hand, this has offered some clues as to the toxic action (Fechter 1989; Davis et al. 2002; Lataye et al. 2003), such as the role of reactive oxygen species and glutathione.

Because noise is often present in the occupational arena, there is a need to incorporate noise exposure in the investigations of ototoxicity of industrial chemicals. This adds to the complexity of the problem. Little is known about combined chemical exposures, and even less is known about mechanisms for interaction between a chemical and physical agent, in this case noise, which makes prediction of the outcome challenging.

To date, the existing human studies have been designed to generate or test hypotheses regarding general toxicity, instead of examining dose–response relationships. Limitations of the studies to date, such as incorrect study design, insufficient characterization of the exposure levels of chemicals and noise, lack of details on if and how other risk factors were accounted for, and so forth, preclude the use of their results in estimating dose–response relationships. Thus, NOAELs and LOAELs for the chemicals cannot yet be documented for the chemicals covered in the present chapter.

Styrene, toluene, lead, and CO are the substances that have been most extensively studied to date, owing to their relevance to occupational health and evidence of their general toxicity or neurotoxicity. Studies conducted with experimental animals provide the most robust evidence regarding mechanisms and dose–effect relationships between agents and effects on the auditory function or physiology. The human studies confirm the relevance of the animal studies findings to occupational health.

Acknowledgments This chapter is dedicated to the memory of Dr. Derek E. Dunn.

References

American Academy of Audiology (AAA). (2009). *American Academy of Audiology Position Statement and Clinical Practice Guidelines. Ototoxicity Monitoring*. Reston, VA: American Academy of Audiology.

Abbate, C., Giorgianni, C., Munao, F., & Brecciaroli, R. (1993). Neurotoxicity induced by exposure to toluene. An electrophysiologic study. *International Archives of Occupational and Environmental Health*, 64(6), 389–392.

Ahn, Y. S., Morata, T. C., Stayner, L. T., & Smith, R. (2005). Hearing loss among iron and steel workers exposed to low levels of carbon monoxide and noise. *Abstract of the Ninth International Symposium on Neurobehavioral Methods and Effects in Occupational and Environmental Health*. Gyeongju, Korea, September 26–29, 2005.

Araki, S., Murata, K., Yokoyama, K., & Uchida, E. (1992). Auditory event-related potential (P300) in relation to peripheral nerve conduction in workers exposed to lead, zinc, and copper: Effects of lead on cognitive function and central nervous system. *American Journal of Industrial Medicine*, 21(4), 539–547.

Ashmore, J. F. (1987). A fast motile response in guinea pig outer hair cells: The cellular basis of the cochhear amplifier. *Journal of Physiology*, 388(1–2), 323–347.

Barregård, L., & Axelsson, A. (1984). Is there an ototraumatic interaction between noise and solvents? *Scandinavian Audiology*, 13(3), 151–155.

Bernardi, A. P. A. (2000). *Workers exposed to noise and toluene: Study of otoacoustic emissions and contraletral suppression*. São Paulo, Brazil: Faculdade de Saúde Pública da Universidade de São Paulo (Master's degree dissertation in Portuguese).

Bleecker, M. L., Ford, D. P., Lindgren, K. N., Scheetz, K., & Tiburzi, M. J. (2003). Association of chronic and current measures of lead exposure with different components of brainstem auditory evoked potentials. *Neurotoxicology*, 24(4–5), 625–631.

Brandt-Lassen, R., Lund, S. P., & Jepsen, G. B. (2000). Rats exposed to toluene and noise may develop loss of auditory sensitivity due to synergistic interaction. *Noise and Health*, 3(9), 33–44.

Buchanan, L. H., Counter, S. A., Ortega, F., & Laurell, G. (1999). Distortion product oto-acoustic emissions in Andean children and adults with chronic lead intoxication. *Acta Oto-Laryngologica*, 119(6), 652–658.

Campo, P., Lataye, R., Cossec, B., & Placidi, V. (1997). Toluene-induced hearing loss: A mid-frequency location of the cochlear lesions. *Neurotoxicology and Teratology*, 19(2), 129–140.

Campo, P., Loquet, G., Blachère, V., & Roure, M. (1999). Toluene and styrene: Intoxication route in the rat cochlea. *Neurotoxicology and Teratology*, 21(4), 427–434.

Campo, P., Lataye, R., Loquet, G., & Bonnet, P. (2001). Styrene-induced hearing loss: A membrane insult. *Hearing Research*, 154(1–2), 170–180.

Campo, P., Maguin, K., & Lataye, R. (2007). Effects of aromatic solvents on acoustic reflexes mediated by central auditory pathways. *Toxicological Sciences*, 99(2), 582–590.

Campo, P., Maguin, K., Gabriel, S., Möller, A., Nies, E., Gomez, M. D. S., & Toppila, E. (2009). European Agency for Safety and Health at Work. Combined exposure to noise and ototoxic substances (60 pp.). Luxembourg: Office for Official Publications of the European Communities.

Cappaert, N. L., Klis, S. F., Baretta, A. B., Muijser, H., & Smoorenburg, G. F. (2000). Ethyl benzene-induced ototoxicity in rats: A dose-dependent mid-frequency hearing loss. *Journal of the Association for Research in Otolaryngology*, 1(3), 292–299.

Cappaert, N. L., Klis, S. F., Muijser, H., Kulig, B. M., & Smoorenburg, G. F. (2001). Simultaneous exposure to ethyl benzene and noise: Synergistic effects on outer hair cells. *Hearing Research*, 162(1–2), 67–79.

Chang, S. J., Shih, T. S., Chou, T. C., Chen, C. J., Chang, H. Y., & Sung, F. C. (2003). Hearing loss in workers exposed to carbon disulfide and noise. *Environmental Health Perspectives*, 111(13), 1620–1624.

Chen, G. D., & Fechter, L. D. (1999). Potentiation of octave-band noise induced auditory impairment by carbon monoxide. *Hearing Research*, 132(1–2), 149–159.

Chen, G. D., McWilliams, M. L., & Fechter, L. D. (1999). Intermittent noise-induced hearing loss and the influence of carbon monoxide. *Hearing Research*, 138(1–2), 181–191.

Chen, G. D., Chi, L. H., Kostyniak, P. J., & Henderson, D. (2007). Styrene induced alterations in biomarkers of exposure and effects in the cochlea: Mechanisms of hearing loss. *Toxicological Sciences*, 98(1), 167–177.

Crofton, K. M., & Zhao, X. (1993). Mid-frequency hearing loss in rats following inhalation exposure to trichloroethylene: Evidence from reflex modification audiometry. *Neurotoxicology and Teratology*, 15(6), 413–423.

Crofton, K. M., & Zhao, X. (1997). The ototoxicity of trichloroethylene: Extrapolation and relevance of high-concentration, short-duration animal exposure data. *Fundamental and Applied Toxicology*, 38(1), 101–106.

Crofton, K. M., Lassiter, T. L., & Rebert, C. S. (1994). Solvent-induced ototoxicity in rats: An atypical selective mid-frequency hearing deficit. *Hearing Research*, 80(1), 25–30.

Dallos, P., Evans, B. N., & Hallworth, R. (1991). Nature of the motor element in electrokinetic shape changes of cochlear outer hair cells. *Nature*, 350(6314), 155–157.

Davis, R. R., Murphy, W. J., Snawder, J. E., Striley, C. A., Henderson, D., Khan, A., --Krieg, E.F. (2002). Susceptibility to the ototoxic properties of toluene is species specific. *Hearing Research*, 166(1–2), 24–32.

Discalzi, G., Fabbro. D., Meliga, F., Mocellini, A., & Capellaro, F. (1993). Effects of occupational exposure to mercury and lead on brainstem auditory evoked potentials. *International Journal of Psychophysiology*, 14(1), 21–25.

Evans, P., & Halliwell, B. (1999). Free radicals and hearing. Cause, consequence, and criteria. *Annals of the New York Academy of Sciences*, 884, 19–40.

Farahat, T. M., Abdel-Rasoul, G. M., El-Assy, A. R., Kandil, S. H., & Kabil, M. K. (1997). Hearing thresholds of workers in a printing facility. *Environmental Research*, 73(2), 189–192.

Fechter, L. D. (1989). A mechanistic basis for interactions between noise and chemical exposure. *Archives of Complex Environmental Studies*, 1(1), 23–28.

Fechter, L. D., Thorne, P. R., & Nuttall A. L. (1987). Effects of carbon monoxide on cochlear electrophysiology and blood flow. *Hearing Research*, 27(1), 37–45.

Fechter, L. D., Young, J. S., & Carlisle, L. (1988). Potentiation of noise induced threshold shifts and hair cell loss by carbon monoxide. *Hearing Research*, 34(1), 39–47.

Fechter, L. D., Liu, Y, Herr, D. W., & Crofton, K. M. (1998). Trichloroethylene ototoxicity: Evidence for a cochlear origin. *Toxicology Sciences*, 42(1), 28–35.

Fechter, L. D., Chen, G. D., Rao, D., & Larabee, J. (2000). Predicting exposure conditions that facilitate the potentiation of noise-induced hearing loss by carbon monoxide. *Toxicological Sciences*, 58(2), 315–323.

Fechter, L. D., Chen, G. D., & Johnson, D. L. (2002). Potentiation of noise-induced hearing loss by low concentrations of hydrogen cyanide in rats. *Toxicological Sciences*, 66(1), 131–138.

Fechter, L. D., Klis, S. F., Shirwany, N. A., Moore, T. G., & Rao, D. B. (2003). Acrylonitrile produces transient cochlear function loss and potentiates permanent noise-induced hearing loss. *Toxicological Sciences*, 75(1), 117–123.

Fechter, L. D., Gearhart, C., & Shirwany, N. A. (2004). Acrylonitrile potentiates noise-induced hearing loss in rat. *Journal of the Association for Research in Otolaryngology*, 5(1), 90–98.

Fechter, L. D., Gearhart, C., Fulton, S., Campbell, J., Fisher, J., Na, K., Cocker, D., Nelson-Miller, A., Moon, P., & Pouyatos, B. (2007). JP-8 jet fuel can promote auditory impairment resulting from subsequent noise exposure in rats. *Toxicological Sciences*, 98(2), 510–525.

Fuente, A., McPherson, B., Munoz, V., & Pablo Espina, J. (2006). Assessment of central auditory processing in a group of workers exposed to solvents. *Acta Oto-Laryngologica*, 126(11), 1188–1194.

Gagnaire, F., & Langlais, C. (2005). Relative ototoxicity of 21 aromatic solvents. *Archives of Toxicology*, 79(6), 346–354.

Gagnaire, F., Marignac, B., Langlais, C., & Bonnet, P. (2001). Ototoxicity in rats exposed to ortho-, meta- and para-xylene vapours for 13 weeks. *Pharmacology and Toxicology*, 89(1), 6–14.

Gagnaire, F., Langlais, C., Grossmann, S., & Wild, P. (2007a). Ototoxicity in rats exposed to ethylbenzene and to two technical xylene vapours for 13 weeks. *Archives of Toxicology*, 81(2), 127–143

Gagnaire, F., Marignac, B., Blachere, V., Grossmann, S., & Langlais, C. (2007b). The role of toxicokinetics in xylene-induced ototoxicity in the rat and guinea pig. *Toxicology*, 231(2–3), 147–158.

Halsey, K., Skjönsberg, A., Ulfendahl, M., & Dolan, D. F. (2005). Efferent-mediated adaptation of the DPOAE as a predictor of aminoglycoside toxicity. *Hearing Research*, 201(1–2), 99–108.

Hawkins, J. E. (1976). Drug ototoxicity. In W. D, Keidel & W. D. Neff (Eds.), *Handbook of sensory physiology* (Vol. V/3, pp. 707–748). Heidelberg: Springer-Verlag.

Henderson, D., Bielefeld, E. C., Harris, K. C., & Hu, B. H. (2006). The role of oxidative stress in noise-induced hearing loss. *Ear and Hearing*, 27(1), 1–19.

Hinshaw, H. C., & Feldman, W. H. (1945). Streptomycin in treatment of clinical tuberculosis: A preliminary report. *Proceedings of Staff Meeting, Mayo Clinic*, 20, 313.

Hirata, M., Ogawa, Y., Okayama, A., & Goto, S. (1992). A cross-sectional study on the brainstem auditory evoked potential among workers exposed to carbon disulfide. *International Archives of Occupational and Environmental Health*, 64(5), 321–324.

Humes, L. E. (1984). Noise-induced hearing loss as influenced by other agents and by some physical characteristics of the individual. *Journal of the Acoustical Society of America*, 76(5), 1318–1329.

Hwang, Y. H., Chiang, H. Y., Yen-Jean, M. C., & Wang, J. D. (2009). The association between low levels of lead in blood and occupational noise-induced hearing loss in steel workers. *The Science of the Total Environ*, 408(1), 43–9.

Johnson, A. C., & Canlon, B. (1994). Progressive hair cell loss induced by toluene exposure. *Hearing Research*, 75(1–2), 201–208.

Johnson, A. C., & Morata, T. C. (2010). *Occupational exposure to chemicals and hearing impairment*. The Nordic Expert Group for Criteria Documentation of Health Risks of Chemicals, Nordic Expert Group. *Arbete och Hälsa*, 44(4), 1–177.

Johnson, A. C., Morata, T. C., Lindblad, A. C., Nylén, P. R., Svensson, E. B., Krieg, E., Aksentijevic, A., & Prasher, D. (2006). Audiological findings in workers exposed to styrene alone or in concert with noise. *Noise and Health*, 8(3), 45–57.

Kopke, R., Allen, K. A., Henderson, D., Hoffer, M., Frenz, D., & Van de Water, T. (1999). A radical demise. Toxins and trauma share common pathways in hair cell death. *Annals of the New York Academy Sciences*, 884, 171–191.

Lacerda, A. B. M. (2007). *Effets de l'exposition chronique au monoxyde de carbone et au bruit sur l'audition*. Montréal, Canada: Faculté des études supérieures de l'Université de Montréal, (Doctoral thesis in French).

Lacerda, A., Leroux, T., & Gagne, J. P. (2005). Noise and carbon monoxide exposure increases hearing loss in workers. In *Proceedings of the 149th meeting of the Acoustical Society of America*, Vancouver, Canada, May 16–20.

Lasky, R. E., Maier, M. M., Snodgrass, E. B., Hecox, K. E., & Laughlin, N. K. (1995). The effects of lead on otoacoustic emissions and auditory evoked potentials in monkeys. *Neurotoxicology and Teratology*, 17(6), 633–644.

Lasky, R. E., Luck, M. L., Torre, P, 3 rd & Laughlin, N. (2001). The effects of early lead exposure on auditory function in rhesus monkeys. *Neurotoxicology and Teratology*, 23(6), 639–649.

Lataye, R., Campo, P., & Loquet, G. (2000). Combined effects of noise and styrene exposure on hearing function in the rat. *Hearing Research*, 139(1–2), 86–96.

Lataye, R., Campo, P., Barthelemy, C., Loquet, G., & Bonnet, P. (2001). Cochlear pathology induced by styrene. *Neurotoxicology and Teratology*, 23(1), 71–79.

Lataye, R., Campo, P., Pouyatos, B., Cossec, B., Blachere, V., & Morel, G. (2003). Solvent ototoxicity in the rat and guinea pig. *Neurotoxicology and Teratology*, 25(1), 39–50.

Lataye, R., Campo, P., Loquet, G., & Morel, G. (2005). Combined effects of noise and styrene on hearing: Comparison between active and sedentary rats. *Noise and Health*, 7(27), 49–64.

Lataye, R., Maguin, K., & Campo, P. (2007). Increase in cochlear microphonic potential after toluene administration. *Hearing Research*, 230(1–2), 34–42.

Laukli, E., & Hansen, P. W. (1995). An audiometric test battery for the evaluation of occupational exposure to industrial solvents. *Acta Oto-Laryngologica*, 115(2), 162–164.

Laurell, G., & Jungelius, U. (1990). High-dose cisplatin treatment: Hearing loss and plasma concentrations. *Laryngoscope*, 100(7), 724–734.

Le Prell, C. G., Yamashita, D., Minami, S. B., Yamasoba, T., & Miller, J. M. (2007). Mechanisms of noise-induced hearing loss indicate multiple methods of prevention. *Hearing Research*, 226(1–2), 22–43.

Le Prell, C. G., Hensley, B. N., Campbell, K. C. M., Hall, J. W. III, & Guire, K. (2011). Evidence of hearing loss in a "normally-hearing" college-student population. *International Journal of Audiology*, 50(Supplement 1), S21–31.

Lehnhardt, E. (1965). [Occupational injuries to the ear]. *Archiv für Ohren-, Nasen- und Kehlkopfheilkund, vereinigt mit Zeitschrift für Hals-, Nasen- und Ohrenheilkunde*, 185, 1–242 (in German).

Leroux, T., Lacerda, A., & Gagne, J. P. (2008). Auditory effects of chronic exposure to carbon monoxide and noise among workers. In: *Proceedings of the 9th International Congress on Noise as a Public Health Problem (ICBEN)*, Foxwood, Connecticut, July 21–25, 2008.

Lilienthal, H., & Winneke, G. (1996). Lead effects on the brain stem auditory evoked potential in monkeys during and after the treatment phase. *Neurotoxicology and Teratology*, 18(1), 17–32.

Liu, Y., & Fechter, L. D. (1997). Toluene disrupts outer hair cell morphometry and intracellular calcium homeostasis in cochlear cells of guinea pigs. *Toxicology and Applied Pharmacology*, 142(2), 270–277.

Loquet, G., Campo, P., & Lataye, R. (1999). Comparison of toluene-induced and styrene-induced hearing losses. *Neurotoxicology and Teratology*, 21(4), 689–697.

Loquet, G., Campo, P., Lataye, R., Cossec, B., & Bonnet, P. (2000). Combined effects of exposure to styrene and ethanol on the auditory function in the rat. *Hearing Research*, 148(1–2), 173–180.

Lumio, J. S. (1948). Hearing deficiencies caused by carbon monoxide (generator gas). *Acta Oto Laryngologica*, 71(Supplement), 1–112.

Lund, S. P., & Kristiansen, G. B. (2004). Studies on the auditory effects of combined exposures to noise, toluene, and carbon monoxide. *Noise and industrial chemicals: Interaction effects on hearing and balance* (pp. 56–76). NoiseChem. Key Action 4: Environmental and Health 2001–2004, Final Report.

Lund, S. P., & Kristiansen, G. B. (2008). Hazards to hearing from combined exposure to toluene and noise in rats. *International Journal for Occupational Medicine and Environmental Health*, 21(1), 47–57.

Maguin, K., Lataye, R., Campo, P., Cossec, B., Burgart, M., & Waniusiow, D. (2006). Ototoxicity of the three xylene isomers in the rat. *Neurotoxicology and Teratology*, 28(6), 648–656.

Maguin, K., Campo, P., & Parietti-Winkler, C. (2009). Toluene can perturb the neuronal voltage-dependent Ca^{2+} channels involved in the middle-ear reflex. *Toxicological Sciences*, 107(2), 473–481.

Mäkitie, A., Pirvola, U., Pyykkö, I., Sakakibara, H., Riihimäki, V., & Ylikoski, J. (2002). Functional and morphological effects of styrene on the auditory system of the rat. *Archives of Toxicology*, 76(1), 40–47.

Mäkitie, A.A., Pirvola, U., Pyykkö, I., Sakakibara, H., Riihimäki, V., & Ylikoski, J. (2003). The ototoxic interaction of styrene and noise. *Hearing Research*, 179(1–2), 9–20.

Mascagni, P., Formenti, C., Pettazzoni, M., Feltrin, G., & Toffoletto, F. (2007). [Hearing function and solvent exposure: Study of a worker population exposed to styrene]. *Giornale Italiano di Medicina de Lavoro ed Ergonomia*, 29(3 Supplement), 277–279 (in Italian with English abstract).

Möller, C., Ödkvist, L. M., Thell, J., Larsby, B., Hyden, D., Bergholtz, L. M., & Tham, R. (1989). Otoneurological findings in psycho-organic syndrome caused by industrial solvent exposure. *Acta Oto-Laryngologica*, 107(1), 5–12.

Morata, T. C. (1989). Study of the effects of simultaneous exposure to noise and carbon disulfide on workers' hearing. *Scandinavian Audiology*, 18(1), 53–58.

Morata, T.C., Fiorini, A.C., Fischer, F.M., Colacioppo, S., Wallingford, K.M., Krieg, E.F., Dunn, D.E., Gozzoli, L., Padrão, M.A., & Cesar, C.L. (1997). Toluene-induced hearing loss among rotogravure printing workers. *Scandinavian Journal of Work Environment and Health*, 23(4), 289–98.

Morata, T. C., Johnson, A. C., Nylén, P., Svensson, E. B., Cheng, J., Krieg, E. F., Lindblad, A. C., Ernstgård, L., & Franks, J. (2002). Audiometric findings in workers exposed to low levels of styrene and noise. *Journal of Occupational and Environmental Medicine*, 44(9), 806–814.

Morioka, I., Kuroda, M., Miyashita, K., & Takeda, S. (1999). Evaluation of organic solvent ototoxicity by the upper limit of hearing. *Archives of Environmental Health*, 54(5), 341–346.

Morioka, I., Miyai, N., Yamamoto, H., & Miyashita, K. (2000). Evaluation of combined effect of organic solvents and noise by the upper limit of hearing. *Industrial Health*, 38(2), 252–257.

Muijser, H., Lammers, J. H., & Kullig, B. M. (2000). Effects of exposure to trichloroethylene and noise on hearing in rats. *Noise and Health*, 2(1), 57–66.

Osman, K., Pawlas, K., Schutz, A., Gazdzik, M., Sokal, J. A., & Vahter, M. (1999). Lead exposure and hearing effects in children in Katowice, Poland. *Environmental Research*, 80(1), 1–8.

Otto, D. A., & Fox, D. A. (1993). Auditory and visual dysfunction following lead exposure. *Neurotoxicology*, 14(2–3), 191–207.

Pouyatos, B., Campo, P., & Lataye, R. (2005a). Influence of age on noise- and styrene-induced hearing loss in the Long-Evans rat. *Environmental Toxicology and Pharmacology*, 19(3), 561–570.

Pouyatos, B., Gearhart, C. A., & Fechter, L. D. (2005b). Acrylonitrile potentiates hearing loss and cochlear damage induced by moderate noise exposure in rats. *Toxicology and Applied Pharmacology*, 204(1), 46–56.

Pouyatos, B., Gearhart, C., Nelson-Miller, A., Fulton, S., & Fechter, L. (2007). Oxidative stress pathways in the potentiation of noise-induced hearing loss by acrylonitrile. *Hearing Research*, 224(1–2), 61–74.

Prosen, C. A., & Stebbins, W. C. (1980). Ototoxicity. In P. S. Spencer, & H. H. Schaumburg (Eds.), *Experimental and clinical neurotoxicology* (pp. 62–76). Baltimore: Williams & Wilkins.

Pryor, G. T., Rebert, C. S., Dickinson, J., & Feeney, E. M. (1984). Factors affecting toluene-induced ototoxicity in rats. *Neurobehavioral Toxicology and Teratology*, 6(3), 223–238.

Rao, D., & Fechter, L. D. (2000a). Protective effects of phenyl-*N-tert*-butylnitrone on the potentiation of noise-induced hearing loss by carbon monoxide. *Toxicology and Applied Pharmacology*, 167(2), 125–131.

Rao, D. B., & Fechter, L. D. (2000b). Increased noise severity limits potentiation of noise induced hearing loss by carbon monoxide. *Hearing Research*, 150(1–2), 206–214.

Rebert, C. S., Day, V. L., Matteucci, M. J., & Pryor, G. T. (1991). Sensory-evoked potentials in rats chronically exposed to trichloroethylene: Predominant auditory dysfunction. *Neurotoxicology and Teratology*, 13(1), 83–90.

Rice, D. C. (1997). Effects of lifetime lead exposure in monkeys on detection of pure tones. *Fundamental and Applied Toxicology*, 36(2), 112–118.

Rice, D. C., & Gilbert, S. G. (1992). Exposure to methyl mercury from birth to adulthood impairs high-frequency hearing in monkeys. *Toxicology and Applied Pharmacology*, 115(1), 6–10.

Schacht, J., & Hawkins, J. E. (2006). Sketches of otohistory. Part 11. Ototoxicity: Drug-induced hearing loss. *Audiology and Neurootology*, 11(1), 1–6.

Schwartz, J., & Otto, D. (1987). Blood lead, hearing thresholds, and neurobehavioral development in children and youth. *Archives of Environmental Health*, 42(3), 153–160.

Schwartz, J., & Otto, D. (1991). Lead and minor hearing impairment. *Archives of Environmental Health*, 46(5), 300–305.

Schäper, M., Demes, P., Zupanic, M., Blaszkewicz, M., & Seeber, A. (2003). Occupational toluene exposure and auditory function: Results from a follow-up study. *Annals of Occupational Hygiene*, 47(6), 493–502.

Śliwińska-Kowalska, M., Zamyslowska-Szmytke, E., Szymczak, W., Kotylo, P., Fiszer, M., Wesolowski, W., & Pawlaczyk-Luszcynska, M. (2003). Ototoxic effects of occupational exposure to styrene and co-exposure to styrene and noise. *Journal of Occupational and Environmental Medicine*, 45(1), 15–24.

Sullivan, M. J., Rarey, K. E., & Conolly, R. B. (1988). Ototoxicity of toluene in rats. *Neurotoxicology and Teratology*, 10(6), 525–530.

Toppila, E., Pyykkö, I., Starck, J., Kaksonen, R., & Ishizaki, H. (2000). Individual risk factors in the development of noise-induced hearing loss. *Noise Health*, 2(8), 59–70.

Vrca, A., Karacic, V., Bozicevic, D., Bozikov, V., & Malinar, M. (1996). Brainstem auditory evoked potentials in individuals exposed to long-term low concentrations of toluene. *American Journal of Industrial Medicine*, 30(1), 62–66.

Wu, T. N., Shen, C. Y., Lai, J. S., Goo, C. F., Ko, K. N., Chi, H. Y., Chang, P. Y., & Liou, S. H. (2000). Effects of lead and noise exposures on hearing ability. *Archives of Environmental Health*, 55(2), 109–114.

Yamamura, K., Terayama, K., Yamamoto, N., Kohyama, A., & Kishi, R. (1989). Effects of acute lead acetate exposure on adult guinea pigs: Electrophysiological study of the inner ear. *Fundamental and Applied Toxicology*, 13(3), 509–515.

Yamane, H., Nakai, Y., Takayama, M., Iguchi, H., Nakagawa, T., & Kojima, A. (1995). Appearance of free radicals in the guinea pig inner ear after noise-induced acoustic trauma. *European Archives of Oto-rhino-laryngol*ogy, 252(8), 504–508.

Young, J. S., Upchurch, M. B., Kaufman, M. J., & Fechter, L. D. (1987). Carbon monoxide exposure potentiates high-frequency auditory threshold shifts induced by noise. *Hearing Research*, 26(1), 37–43.

Ödkvist, L. M., Arlinger, S. D., Edling, C., Larsby, B., & Bergholtz, L. M. (1987). Audiological and vestibulo-oculomotor findings in workers exposed to solvents and jet fuel. *Scandinavian Audiology*, 16(1), 75–81.

Ödkvist, L. M., Möller, C., & Thuomas, K. A. (1992). Otoneurologic disturbances caused by solvent pollution. *Otolaryngology Head and Neck Surgery*, 106(6), 687–692.

Part IV
Protection and Repair

Chapter 12
Hearing Protection Devices: Regulation, Current Trends, and Emerging Technologies

John G. Casali

1 Introduction

1.1 Brief History and Applications of Hearing Protection Devices

1.1.1 The First Hearing Protectors

Hearing protection devices (HPDs), used to guard the human ear against incurring hearing loss due to noise, have been in existence at least since the early 1900s even though their use in United States (U.S.) workplaces was not regulated by law until 1971. In fact, in 1911, the famous band leader John Phillip Sousa complained to his friend and fellow skeet trapshooter J. A. R. Elliott that shooting traps "took a toll on his ears and was beginning to affect his livelihood [as a musician]." Elliott, being an inventor, then developed and patented (in eight countries, no less), the "Elliott Perfect Ear Protector," and it became a commercial success (Baldwin 2004). After using the "Elliott Protector," which was among the first commercially available hearing protectors, Sousa wrote in a letter to Elliott on January 20, 1913: "I consider your invention to lessen the shock of loud noises or overwhelming vibrations of sound of great comfort. The Elliott Perfect Ear Protector is a great success in affording protection from concussions to a sensitive ear. As a shock absorber it is invaluable" (Baldwin 2004). Unfortunately, U.S. industrial workers did not experience common use of effective hearing protection technology until many years later, even though simple cotton plugs were known to be used in some workplaces before the turn of the nineteenth century (e.g., Barr 1896). (Even today, some individuals incorrectly assume that cotton suffices as a hearing protector.) The lack of protection in early mechanized industries,

---

J.G. Casali (✉)
Department of Industrial and Systems Engineering, Auditory Systems Laboratory, Virginia Tech, Blacksburg, VA 24061, USA
e-mail: jcasali@vt.edu

C.G. Le Prell et al. (eds.), *Noise-Induced Hearing Loss: Scientific Advances*, Springer Handbook of Auditory Research 40, DOI 10.1007/978-1-4419-9523-0_12,

coupled with high noise exposures, resulted in hearing loss and related problems, such as tinnitus, that manifested in workers and that, very tragically, were often viewed as an accepted consequence of the occupation. As such, the terms "blacksmith's deafness" and "boilermaker's ear" were coined (Fosboke 1831; Holt 1882; Berger 2003a).

1.1.2 Hearing Protection in the Military and Industry, Including Regulations

Through the first half of the twentieth century, HPDs were not commonly used in U.S. workplaces or for most leisure-time exposures; however, the U.S. military has recognized their importance at least since World War II for protection against the effects of noise-emitting ordnance as well as loud machinery such as tanks and aircraft. (For a detailed review, see Grantham, Chap. 3.) In fact, one of the earliest regulations on hearing conservation was U.S. Air Force regulation 160-3 of 1948 (Department of the Air Force 1948), which specifically called for the use of HPDs. However, in U.S. workplaces, though a few industrial hearing conservation programs appeared in the 1940s and 1950s (Berger 2003a), hearing protection was not promulgated into law until May of 1971, with the "Occupational Noise Exposure Standard" of the Occupational Safety and Health Act (OSHA). The OSHA Noise Standard was the first legal requirement, based on exposure levels, for hearing protection in general industry (OSHA 1971a), and a similar law was promulgated for construction work (OSHA 1971b), these occupational settings being where the great majority of U.S. citizens were, and continue to be associated with the greatest risk for hearing loss due to noise exposure. Many employers reacted to this first widely applicable OSHA regulation simply by providing hearing protectors, instead of placing an emphasis on implementation of more permanent engineering noise controls and "buy-quiet" strategies for replacing machinery.

Later, in 1983, the legal advent of the OSHA Hearing Conservation Amendment (OSHA 1983) for General Industry immediately resulted in the use of HPDs to proliferate in U.S. industrial workplaces because this Amendment required a choice of HPDs to be *supplied* to any worker exposed to above an 85 dBA time-weighted-average (TWA), or 50% noise dose, for an 8-h workday, with the measurement taken on the "slow" scale and using a 5-dB exchange rate between exposure dBA level and time of exposure. Other industries, including airline, truck, and bus carriers; railroads; and oil and gas well drilling have separate, and generally less comprehensive, noise and hearing conservation regulations than those of OSHA (1971a, 1983) for general industry, and unfortunately, as of the date of this chapter, there has never been an analogue to the OSHA Hearing Conservation Amendment of 1983 adopted into law for the construction industry, in which noise levels can be quite hazardous (Casali and Lancaster 2008), although one was drafted and proposed earlier in this decade. Finally, in the mining industry, hearing protection has been addressed in its regulation, first under the Federal Coal Mine Health and Safety Act of 1969 and later under the Federal Mine Safety and Health Amendments Act of 1977. In 1999, the Mine Safety and Health Administration (MSHA) issued a more comprehensive noise regulation that governed all forms of mining (MSHA 1999). The major point about all of these historical milestones in U.S. federal regulatory development is that hearing protectors were not really addressed in U.S. occupational safety and health law until the late 1960s and early 1970s, depending upon the industry involved.

1.1.3 Hearing Protection Devices Outside of Work: Protection and Annoyance Reduction

Though actual usage data are elusive, there is a general indication that HPDs are becoming more popular in nonoccupational usage among the general public as awareness for noise-induced hearing loss increases. Non-workplace usage of HPDs is needed for protection of hearing against noises produced by *certain* power tools (especially some pneumatic, hydraulic, and gasoline-powered devices), lawn care equipment, recreational vehicles, target shooting and hunting, spectator events, amplified music, and many other sources. In fact, some of these activities, for example, recreational firearm use (Nondahl et al. 2000) and attendance at motorsport events such as monster truck races (Casali 1990), pose exposure levels that can result in permanent hearing loss and that equal or exceed the levels experienced by workers in many industries. Therefore, HPD usage is important in life outside of work, to reduce the energy to the ear from a plethora of sources.

HPDs are beneficial beyond the realm of hearing loss prevention, however, and are sometimes applied in noisy environments that may pose no real threat to hearing, but that are disturbing nonetheless. For example, HPDs are used for reduction of noise annoyance in settings such as the passenger cabins of commercial aircraft, in subways or buses, and for aid in sleeping in noisy environments pervaded by such annoyances as traffic noise or snoring (although HPD use while sleeping is *not recommended* where the audibility of acoustic signals, such as smoke alarms or building enunciators, is important). Hearing protection features are now incorporated into other products worn on the ears, such as headphones for music rendition or headsets for aircraft cockpit communications, to provide improved audio signal fidelity and speech intelligibility. In fact, an effectively attenuating headphone/headset design can help improve the signal/speech-to-noise ratio at the ear such that the audibility of earphone output content is improved, while at the same time providing protection from ambient noise hazards (Robinson and Casali 2003). However, in using such devices, the user must be aware that due to the device's reduction of outside sound levels as well as to the device's earphone acoustical output, *external* signals, such as traffic or sirens in the vicinity of a jogger or auditory warnings in an aircraft cockpit, could be missed because they are attenuated or masked, respectively. Of course, these effects of headphones/headsets on the situational awareness of the wearer are very dependent upon the design of the HPD (such as the frequency spectrum of attenuation produced by passive or active noise cancellation features), and further, the operational effects are also very situation- and user-specific (Casali and Gerges 2006).

Location: SOURCE -------------------> PATH ------------------> RECEIVER

Countermeasures: Engineering Controls Engineering Controls Administrative Controls *Hearing Protection*

Fig. 12.1 Systems approach to reduction of noise exposures

2 Hearing Protection Versus (or as Supplement to) Engineering Noise Control

2.1 Systems Approach to Noise Abatement

A straightforward systems approach to noise abatement is often advocated, wherein efforts to reduce or eliminate noise exposures are concentrated in three primary locations, as shown in Fig. 12.1. In this approach, HPDs are, from a systems perspective, the last line of defense in the chain (Berger 2003b; Casali 2006; Gerges and Casali 2007). This is because an HPD's protective success depends on human intervention and behaviors, and not simply on an industrial employer to supply the correct protectors, or on a consumer to purchase them. Moreover, HPD success depends heavily on the need for the user to fit and use them properly. Therefore, HPDs are an *active* countermeasure implemented *at the receiver who must apply them properly*, as opposed to a *passive* countermeasure such as engineering controls that are implemented *at the noise source* or *in its propagation path*. In other words, the HPD, being an active countermeasure akin to an automobile seat belt, depends on the human to use it, and to do so properly to obtain protection. In contrast, engineered noise control, being a passive countermeasure akin to crashworthy structural design in an automobile, does not depend upon the human for successful performance. The focus of noise reduction efforts should certainly be on the development and purchase of quieter consumer products or industrial machinery, and on reduction of noise via engineering means at the source or in its path; however, it is true that in many cases hearing protection supplants engineering and administrative controls (which time-limit workers' exposures) when such controls are not practical, available, or economical, or when an employer or product manufacturer simply does not place engineering control at high priority. It is, in large part, for these reasons that the use of HPDs has proliferated, especially in industrial and military situations (Gerges and Casali 2007). In some cases, HPDs constitute the only countermeasure because it is difficult, if not infeasible, to adequately "engineer-out" noises, such as those from certain military weapons, excavation equipment, explosives, or aircraft. Further, in some situations, engineering and administrative controls may have already been implemented, but are insufficient at abating the noise to acceptable (or even OSHA-legal) levels; in these instances, hearing protection may be used to "make up the difference" needed in the noise reduction effort.

2.2 Hearing Protection Device Performance Limitations

There is no question that if noise exposures persist as sufficiently high, even after noise control efforts, the wearing of HPDs is the only means of protecting the user's hearing against the injurious effects of noise. Permanent noise-induced hearing loss most typically comprises an insidious neural injury that is progressive with continued exposures, and this neural loss is manifested in the hair cells of the cochlea of the inner ear; however, it can also comprise immediate acoustic trauma if the elastic limits of the tympanum, ossicular chain of the middle ear, or cochlear structures are exceeded by a powerful acoustic insult, such as an explosion. (These anatomical and physiological manifestations of hearing loss are covered in Hu, Chap. 5.) For the great majority of noises to which people are exposed, *if* HPDs are properly selected, fitted, and worn, they are indeed *effective* at preventing these noise-induced hearing losses. In industry, where 90% of the noise exposures are at TWA levels less than or equal to 95 dBA, all that is really needed (at least for OSHA 1983 Hearing Conservation Amendment compliance) is 10 dB of actual in situ performance, a performance level that the majority of HPDs, when properly fit and worn, can readily provide (Berger 2003b). However, HPDs are not a panacea, and if *not* properly selected and worn, they may provide negligible attenuation, and thus can be expected to be ineffective, as demonstrated by Park and Casali (1991) and many others. Further, owing to the acoustic pathways that flank or bypass an HPD, including air leaks around the HPD's seal, HPD material transmission, HPD vibration, and HPD-flanking via bone conduction, rare but extremely high noise levels may exceed the attenuation capabilities of even the highest-attenuation, well-fit, conventional passive HPD to the point that its protective effectiveness is insufficient. For example, in 8-h daily TWA exposures that exceed about 105 dBA, and especially those with dominant low-frequency content below about 500 Hz, *double* passive hearing protection (i.e., an earmuff worn over a well-fit earplug) is advisable (Berger 2003b). In view of the fact that workers often complain about wearing even one earmuff or pair of earplugs, convincing them of the necessity of wearing both devices during the course of a workday can indeed be a challenge.

In even more severe noise environments, such as those present on aircraft carrier decks during military flight operations, the prevailing noise exposures can be extremely high, with short-term levels ranging from 146 to 153 dBA at 50 ft. from various military jet aircraft on afterburner power (McKinley 2001). Such levels overtax even the capabilities of double passive HPDs, so very specialized HPDs are thus required, and these are discussed later herein under emerging technologies. The point is, although HPDs certainly can be effective in the great majority of industrial and most other noise exposures, conventional HPDs simply cannot physically provide the amount of protection needed in certain situations. Thus, engineering controls or administrative (limiting time of exposure) countermeasures, or both, are essential.

3 Hearing Protection Devices: By Regulation and by Voluntary Use

As stated earlier, by far the majority of HPDs used in the U.S. are those worn by industrial workers, with military personnel comprising the second largest user group. Both of these groups are governed by the aforementioned regulations. Keeping in mind that *regulation* is a major part of the reason that HPD use has proliferated, it is instrumental to review, for example, just how the law that affects the majority group (i.e., U.S. workers) regulates hearing protection in relation to other noise controls.

3.1 OSHA General Industry and Construction Regulations Circa 1971

3.1.1 OSHA: General Industry Noise

The OSHA (1971a) Noise Standard for General Industry (29 CFR 1910.95(a)) specifies, "Protection against the effects of noise exposure shall be required when the sound levels exceed those shown in Table G-16 when measured on the A-scale of a standard sound level meter at slow response." (This table specifies the 90 dBA TWA "criterion level" for an 8-h exposure, including the 5-dB trading relationship between increased noise exposures and allowable exposure durations per day, e.g., 95 dBA TWA allowed for 4 h, 100 dBA TWA allowed for 2 h, and similar level/time fractional variants, along with a not-to-exceed 140 dB peak sound pressure level for impulsive or impact noise). Furthermore, 29 CFR 1910.95(b)(1) states, "When employees are subjected to sound exceeding those listed in Table G-16, feasible administrative or engineering controls shall be utilized. If such controls fail to reduce sound levels within the levels of Table G-16, personal protective equipment shall be provided and used to reduce sound levels within the levels of the table." Thus, this *chronologically first* OSHA regulation, via the words "shall be utilized," *required* feasible administrative or engineering controls to be utilized in priority over hearing protection. However, history has since demonstrated that a significant weakness in terminology was the word *feasible*, which was, unfortunately, not specifically defined in terms of technical, economical, or other feasibility criteria. Therefore, this wording left room for industries to claim infeasibility of engineering noise controls, sometimes when it was not justifiable.

Nonetheless, only in the case where engineering or administrative controls failed to reduce noise to within the limits of Table G-16, were hearing protectors to be relied upon under OSHA (1971a). Thus, it is important to recognize that even in the earliest of OSHA's general industry regulations, *HPDs were regulated as an addition to, and not a replacement for, administrative or engineering noise controls.* However, the fact remains that in practice in many industrial plants, HPDs are relied upon as the *first* line of defense against noise hazards to workers' hearing, which is not in accordance with the letter of the OSHA law.

3.1.2 OSHA: Construction Noise

The law for construction work, 29 CFR 1926.52 (OSHA 1971b), incorporates a Table D-2 for exposure limits that duplicates Table G-16 in the general industry regulation (OSHA 1971a), and it also includes the same statement regarding reliance on administrative and engineering controls in priority to hearing protectors. However, the construction regulation, in 29 CFR 1926.101 (OSHA 1971c), also added the statements (a) "Whenever it is not feasible to reduce the noise levels or duration of exposures to those specified in Table D-2, Permissible Noise Exposures, in 1926.52, ear protective devices shall be provided and used. (b) Ear protective devices inserted in the ear shall be fitted or determined individually by competent persons. (c) Plain cotton is not an acceptable protective device." While the Construction Standard's subparts (b) and (c) may have been a slight improvement over OSHA's General Industry standard in 1971 with regard to specificity of hearing protection, the Construction Standard still remains much weaker overall since it has never been updated with the addition of a Hearing Conservation Amendment, as was the General Industry Standard in 1983, discussed next.

3.2 OSHA General Industry: Hearing Conservation Amendment Circa 1983

The Hearing Conservation Amendment (OSHA 1983) significantly improved the original OSHA (1971a) Noise Standard by specifying the requirements for a multifaceted hearing conservation program when daily TWA noise exposures exceed 85 dBA (equivalent to a 50% noise dose). Reliance on engineering and administrative noise controls remained as described in the preceding text in the 1971 General Industry Standard (OSHA 1971a). However, in addition to other facets of a hearing conservation program (including noise monitoring, employee notification, audiometric testing, worker training, access to information and training materials, and exposure and audiometric recordkeeping), the Amendment provided much more specificity on the application of hearing protection. Perhaps most significant was the addition, at 29 CFR 1910.95(i), of the statement (1) that "Employers shall make hearing protectors available to all employees exposed to an 8-h time-weighted average of 85 dB or greater at no cost to the employees. Hearing protectors shall be replaced as necessary." Furthermore, the Amendment added (2) that "Employers shall ensure that hearing protectors are worn: (1) By an employee who is required by paragraph 1910.95(b)(1) of this section to wear personal protective equipment, [i.e., mandatory HPD use at exposures equal or greater than 90 dBA TWA], and (2) By any employee who is exposed to an 8-h TWA of 85 dB or greater, and who: has not had a baseline audiogram, or who has experienced a standard threshold shift [as defined by OSHA]." (Author's additions in brackets.) In addition, under paragraph (3), employers were required to provide a "variety of suitable hearing protectors" for the employee to select from, (4) training in the use and care of all hearing protectors provided, and (5) to ensure that proper initial fitting and supervision in the correct use of all hearing protectors. Finally, the OSHA (1983) Amendment, via part (j), specified computational procedures for evaluating the HPDs for adequacy of protection in specific noise exposures, with the requirement that the protected exposure levels be brought to less than or equal to 90 dBA TWA, or to less than or equal to 85 dBA TWA if the worker has experienced a standard threshold shift. The reader is referred to OSHA (1983) and Casali (2006) for more details on computing HPD adequacy.

It is important to note that the OSHA (1983) Hearing Conservation Amendment was a very significant regulatory development that greatly impacted the requirements for hearing protection, and thus the numbers of HPDs that were supplied in occupational settings dramatically increased as a result. Although engineering or administrative controls were still required for TWA exposures above 90 dBA, the Amendment brought, at no cost to the worker, a selection of HPDs to each one who was exposed to 85 dBA TWA or above. The 5 dBA difference between the 90 dBA OSHA "criterion" level imposed as a result of OSHA (1971a), and the 85 dBA OSHA "action" level imposed as a result of the 1983 OSHA Hearing Conservation Amendment, comprised an exposure window where in thousands of theretofore *unprotected-by-law* noise-exposed workers were thereafter required to be supplied with a selection of suitable hearing protection. In this sense, the new 85 dBA TWA action level was a major step forward in protecting workers against the hazards of noise exposures; however, the fact that HPDs were required by the OSHA Hearing Conservation Amendment to be provided at that exposure level should not be taken as an indication that HPDs are preferable to engineering noise controls that could be implemented and beneficial at that 85 dBA TWA level (rather than at the *required* 90 dBA TWA level), and that do not require human intervention on a daily basis.

3.3 Hearing Protection Devices in Nonoccupational Settings

3.3.1 Home and Recreational Use of Hearing Protection

In stark contrast to the use of hearing protection in most occupational settings in the U.S. which as discussed in the preceding text, is mandated by federal statute, hearing protection use in recreational and home settings is generally up to the individual's discretion. In most cases, this discretion will be exercised, or not, based on the individual's cognizance of what constitutes an unsafe noise or not, and his or her own tendencies in risk-taking behavior. Although there are programs to educate civilians about noise-related dangers and the importance of HPD use, public awareness about the hazardous effects of noise exposure is indeed low and programs to improve this awareness are needed (WHO 1997; ASHA 2009). However, in addition to the conventional passive hearing protectors that have been available for consumer purchase for decades, there have been recent improvements such as HPDs that are styled and sized specifically for children, HPDs designed for spectator events (e.g., earmuffs that incorporate radios for use at sporting events), HPDs with signal pass-through circuitry (e.g., electronic earmuffs for hunters), lightweight active noise cancellation HPDs for reducing low-frequency noise in aircraft cabins, uniform-attenuation earplugs for musicians and concert attendees, and other innovations to be discussed later herein and reviewed in detail elsewhere (Casali 2010a, b). Thus, a broad variety of attractive HPD and related products are available for consumer use.

3.3.2 Local Ordinances on Hearing Protection

Also in contrast to the long-standing federal OSHA laws for occupational exposures, if noise in the community and recreational settings is governed at all, it is usually by local ordinances that typically relate more to noise annoyance than to hearing hazard risks (Casali 2006). However, there are situations in which local ordinances may require the use of HPDs, such as in certain venues where recreational exposures are loud due to amplified music or gaming arcades, and in such venues there should be warning signs that stipulate that hearing protection is required upon entry. Nonetheless, such ordinances to protect the health of the civilian are in the minority, as governance of such exposures is hit-or-miss and often arises only as a result of public complaint or civil litigation for premises liability. Again, such exposures beg the question as to why the noise is simply not controlled to within safe limits by engineering means or by simply "turning the volume down," rather than by warning the attendee to wear hearing protection and depending upon the attendee to have such protection at hand.

4 Hearing Protection Device Attenuation Data and Labeling Regulations

4.1 Labeled Versus In-field Attenuation Performance

The labeling of hearing protector performance has been the subject of much debate for well over two decades, with much of the attention given to the lack of correspondence between on-package, Environmental Protection Agency (EPA)-required attenuation data and the actual protection achieved by users in the field (Berger and Casali 1997; Casali and Robinson 2003). For OSHA (1983) and other applications, the protective adequacy of an HPD for a given noise exposure is determined by subtracting, in a prescribed manner, the attenuation data required by the EPA to be included on protector packaging from the TWA noise exposure for the affected worker (see OSHA 1983; Appendix B: Methods for Estimating the Adequacy of Hearing Protector Attenuation). These attenuation data must be obtained from psychophysical real-ear-attenuation-at-threshold (REAT) tests at nine 1/3 octave bands with centers from 125 to 8,000 Hz that are performed on human listeners, and the signed, arithmetic difference between the thresholds with the HPD on and without it constitutes the attenuation at a given frequency. Spectral attenuation statistics (means and standard deviations) and the broadband single number Noise Reduction Rating (NRR), which is computed therefrom, are provided, and either the spectral data or the NRR can be applied for estimating HPD adequacy for a given exposure, per Appendix B of OSHA (1983). These labeled ratings are the primary means by which end-users compare different HPDs on a common basis and make determinations of whether adequate protection and OSHA compliance will be attained for a given noise environment. Therefore, the accuracy and validity of the ratings are of high importance.

4.2 *Current EPA-Required Hearing Protector Labeling and Cited Test Standards*

The labeling of hearing protectors is controlled by the EPA via federal law per 40 CFR Part 211, Subpart B, which was promulgated in September, 1979 (EPA 1979) and remains in effect as of this writing. This section applies to "any device or material, capable of being worn on the head or in the ear canal, that is sold wholly or in part on the basis of its ability to reduce unwanted sound that enters the user's ears (40 CFR Part 211, Subpart B). Unfortunately, the currently-prevailing law references an outdated and now-superseded ANSI standard (ANSI S3.19-1974) for obtaining the real-ear attenuation of threshold data on which the EPA primary HPD label, inclusive of the NRR, is based. The data appearing on HPD packaging are obtained under optimal laboratory conditions with properly fitted protectors, and with trained, well-practiced human subjects. In no way does the "experimenter-fit" protocol and other aspects of the currently required (by the EPA) test procedure (ANSI S3.19-1974) represent the conditions under which HPDs are selected, fit, and used in the workplace, and this has been demonstrated in numerous research studies (e.g., Park and Casali 1991; Berger and Casali 1997; Berger et al. 1998). Therefore, the attenuation data used in the octave band or NRR formulae discussed earlier herein are inflated and cannot be assumed as representative of the protection that will actually be achieved in the field. As evidence, Fig. 12.2 presents the results of a review by Berger (2003b) of research studies in which manufacturers' on-package NRRs (in the background) were compared against NRRs computed from actual subjects taken with their HPDs from field settings (in the foreground). Clearly, the differences between laboratory and field estimates of HPD attenuation are large, and more so for earplugs than earmuffs, and the hearing conservationist or HPD consumer must take this into account when selecting protectors.

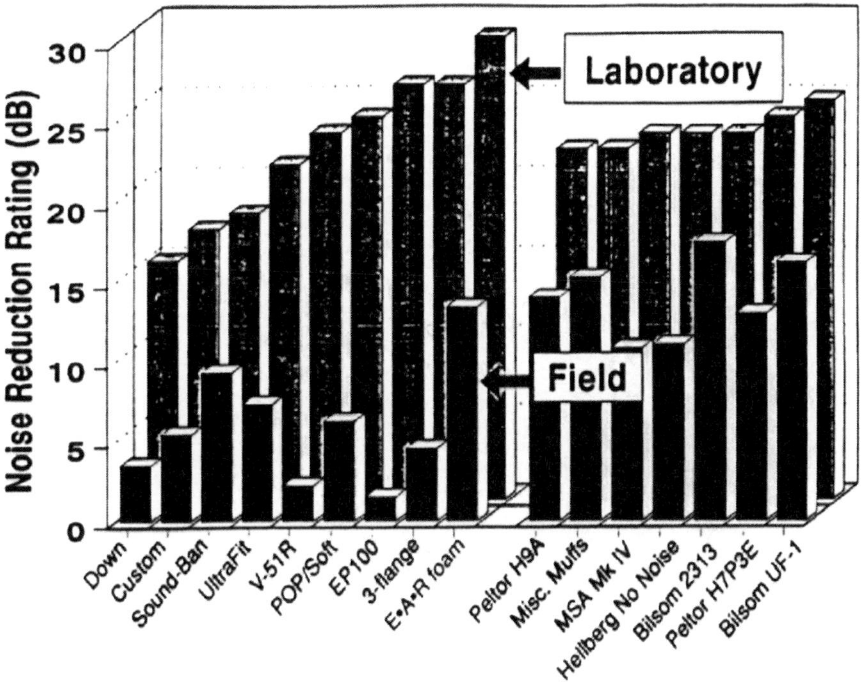

Fig. 12.2 Comparison of hearing protection device NRRs by device type: manufacturers' laboratory data in background versus real-world "field" data in foreground (Adapted with permission from Berger (2003b), Fig. 10.18)

4.3 Proposed EPA Rule on Hearing Protector Labeling and its Cited Test Standards

ANSI Working Group S12/WG11 developed a new testing standard, ANSI S12.6-1997 (R2008), which contains both a "Method A" provision for experimenter-supervised fitting of the HPD, and a "Method B" provision for subject (not experimenter) fitting of the HPD using relatively naive (not trained) subjects. Further, ANSI S12.6-1997(R2008) is much improved over the current ANSI S3.19-1974 testing standard in its experimental controls and human factors aspects of test protocol. Yet in spite of the fact that the Method B (Subject-Fit) testing protocol of ANSI S12.6-1997(R2008) has been demonstrated in scientific field experiments to yield attenuation data that are more representative of those achievable under workplace conditions wherein a quality hearing conservation program is operated (e.g., Berger et al. 1998), as of this writing Method A (experimenter-supervised fit of the HPD) has been selected by the EPA in its recently proposed revised regulation, or "Rule," for HPD testing and labeling. In this regard, the EPA has given notice (see EPA 2009: EPA Docket OAR-2003-0024) in public workshops and presentations that there is a plan to revise the 1979 labeling Rule, including replacement of ANSI

S3.19-1974 with the current ANSI standard (ANSI S12.6) to obtain the passive attenuation data along with a new means of broadband rating, for which a range of values, rather than a single value as is the current NRR, will likely be used (ANSI 2008). Further, the proposed Rule includes elements of another testing standard (ANSI S12.42) to enable physical, microphone-based testing in real ears and acoustical test fixtures (ANSI 2010); this will enable comprehensive testing of active noise cancellation and active/passive level-dependent protectors, as well as certain other HPD types that are currently not amenable to the 1979 EPA regulation for labeling, and thus that cannot currently be marketed as hearing protectors under the letter of the regulation. In addition, elements of ANSI S12.68 are included to prescribe methods for estimating protected exposure levels under HPDs (ANSI 2007). *At press time for this chapter, the aforementioned details of the EPA proposed labeling Rule had yet to be finalized; however, updates may be found on the EPA docket at* www.regulations.gov, *docket number: EPA-HQ-OAR-2003-0,024.*

5 Hearing Protection Devices: Brief Overview of Technologies and Effects on Audibility

5.1 Types of Conventional Passive Hearing Protection Devices

So-called *conventional* hearing protectors constitute the vast majority of HPDs, and these devices achieve attenuation of noise strictly by static, passive means without the use of dynamic, mechanical elements, such as valves or reactive ports, or by electronic circuitry, such as active noise cancellation or electronically modulated signal pass-through technology. The attenuation capabilities of conventional passive HPDs are afforded by a combination of acoustical factors, including the airborne sound transmission loss imposed by the construction materials; reflection characteristics of the HPD against incident sound waves; quality and integrity of seal against the ear canal walls, or against the outer ear or its surrounding tissue to prevent air leakage; ability of the HPD to engage the ear canal walls and dampen canal vibrations; ability of the HPD to reduce the bone conduction flanking pathway (only via certain full helmet designs); and the resonance frequency characteristics and acoustical impedances of the HPD. As discussed earlier herein, when properly selected for the situation and user, and correctly worn, these *conventional* HPDs, either singly or in combination (i.e., an earmuff worn over an earplug) yield adequate protection in nearly all industrial and recreational environments, and in many, but not all, military exposures.

Conventional HPDs are divided into four general types, with each type defined by the means of interface of the HPD to the ear or head. *Earplugs* consist of vinyl, silicone, spun fiberglass, cotton/wax combinations, and closed-cell foam products that are inserted into the ear canal to form a noise-blocking seal. Proper fit to the user's ears and training in insertion procedures are critical to the success of earplugs. A related but different category of HPD is the *semi-insert* or *ear canal cap* which consists of earplug-like pods that are positioned at the rim of the ear canal or in the concha bowl of the outer ear (pinna), and held in place by a lightweight headband, which on some devices can be positioned under the chin, behind the head, or over the head. The headband of an ear canal cap is often useful for positioning the device around the neck for storage when the user moves out of the noise. *Earmuffs* consist of earcups, usually of a rigid plastic material with an absorptive liner, that completely enclose the outer ear and seal around it with foam- or fluid-filled cushions. A headband connects the earcups, and as with ear canal caps, on some models of earmuffs the headband is adjustable so that it can be worn over the head, behind the neck, or under the chin, depending upon the presence of other headgear, such as a welder's mask. *Helmets* that enclose a large portion of the head are usually designed to provide impact protection, but when they contain integrated earmuff cups or a suitable liner material that seals around the ears, they also afford hearing protection (Berger and Casali 1997). Furthermore, for extreme noises that substantially transmit sound through bone conduction to the neural ear, helmets that cover the temporal and mandibular areas, as well as the cranium, can provide additional protection against bone-conducted noise (Gerges and Casali 2007).

In general terms, as a group, earplugs provide better attenuation than earmuffs below about 500 Hz and equivalent or greater protection above 2,000 Hz. At intermediate frequencies, earmuffs often have the advantage in attenuation, but this is not a hard and fast rule (Gerges and Casali 2007). Earmuffs are generally more easily fit by the user than either earplugs or canal caps, and depending on the temperature and humidity of the environment, the earmuff can be uncomfortable (in hot or high humidity environment) or a welcome ear insulator (in a cold environment). Semi-inserts generally offer less attenuation and lower comfort than earplugs or earmuffs, but because they are readily storable around the neck, they are convenient for those workers who frequently move in and out of noise. A comprehensive review of conventional HPDs and their applications may be found in Gerges and Casali (2007) as well as in Berger (2003b).

Although conventional HPDs offer adequate protection for most noise exposures, a potential disadvantage that is due to the very nature of the attenuation that they provide, i.e., via static, passive means, is the concomitant, deleterious effect on hearing quality and auditory performance that sometimes arises, depending on the user's hearing ability as well as the noise and signal conditions. For more detailed information on the effects of HPDs on speech communication and signal audibility, the reader is referred to several book chapters and reports (Suter 1989; Robinson and Casali 2003; Casali and Gerges 2006); however, a brief review is provided next.

5.2 Effects of Conventional Hearing Protection Devices on Signal and Speech Audibility

5.2.1 The Dilemma of Over- Versus Under-Protection

Users may reject hearing protection if it compromises their hearing to an extent wherein sounds no longer appear natural, signals cannot be detected, or speech cannot be understood. In some cases, too much attenuation may be provided by an HPD for a particular noise situation, with the concomitant effect that the user's hearing is unnecessarily degraded. In lay terms, this is commonly (but somewhat confusingly) referred to as *"overprotection."* The safety professional often faces a dilemma in selecting HPDs for the workforce that provide adequate attenuation for the noise threat at hand, but also that do not provide so much attenuation that the worker cannot hear important signals and speech communications. This is the dilemma of *underprotection* versus *overprotection*. To emphasize its magnitude in a legal sense, the view of the injured worker, acting as a worker's compensation or civil tort plaintiff, is sometimes as follows: "The hearing protector provided inadequate noise attenuation for defending my ears against the damaging effects of noise, so I lost my hearing over time," or, "The hearing protector provided more attenuation than needed for the noise that I was exposed to at work, and therefore was the primary cause of the accident when I was prevented from hearing the forklift's backup alarm and was run over." Although these are extreme statements, they indeed have some validity in certain circumstances if an HPD is not properly matched to a worker's needs, the noise exposure, and any hearing-critical requirements inherent in a job. In civil court, these arguments give rise to theories on which a legal foundation for potential recovery of damages may be based. For example, from a product liability perspective, these claimed "failings" of the HPD would typically fall under the legal premise of defective design or availability of superior alternative design features, and/or breach of warranty. The threat of litigation should be of concern for both HPD manufacturers as well as employers who purchase and prescribe the HPDs for workers. All of the above safety, ethical, and legal issues speak to the need for proper matching of HPDs to workers and job requirements, improved HPD designs to benefit audibility, and perhaps moreover, the need to emphasize engineering noise controls in priority over HPDs.

5.3 Technology Augmentations for Improved Audibility and Auditory Perception

5.3.1 Level-Dependent (Amplitude-Sensitive) Hearing Protectors

Overall, the research evidence on normal hearers generally suggests that conventional passive HPDs have little or no degrading effect on the wearer's understanding of external speech and signals in ambient noise levels above about 80 dBA, and may even yield some improvements with a crossover between disadvantage to advantage between 80 and 90 dBA (see reviews in Berger and Casali 1997; Casali and Gerges 2006). However, conventional HPDs often do cause increased misunderstanding and poorer detection, compared to unprotected conditions, in lower sound levels, where HPDs are not typically needed for hearing defense anyway, but may be applied for reduction of annoyance. In intermittent noise, HPDs may be worn during quiet periods so that when a loud noise occurs, the wearer will be protected. However, during those quiet periods, conventional passive HPDs typically reduce hearing acuity.

In certain of these cases, the technology enhancements that are incorporated into *level-dependent* (also called amplitude-sensitive) augmented HPDs have potential benefit. These include *passive* devices that provide minimal or moderate attenuation in sound levels up to about 110 dB, and sharply, instantaneously increased attenuation at higher levels, such as the AEARO Arc™ and Combat Arms™ earplugs, as reviewed in Casali (2010a). Also, the level-dependent category includes electronically modulated sound transmission HPDs that provide *amplification* within a passband of external sounds during quiet but rapid signal compression (sharply, rapidly decreased gain) as incident noise levels increase; these devices are extensively reviewed in Casali (2010b). Published research experiments on both passive and electronic level-dependent devices are scant; however, a few very recent studies have been reported. In those, certain commercial and military electronic earmuff-based versions of level-dependent devices have *not* been associated with a demonstrated improvement in signal audibility over conventional HPDs using vehicle backup alarms as signals, and have resulted in poorer audibility than that achieved with passive level-dependent HPDs (the aforementioned Arc™ or Combat Arms™ earplugs). For example: using military enemy camp noise as the signal for detection distances (Casali et al. 2009), using backup alarms as the signal to be localized in azimuth (Alali and Casali 2011 and Fig. 12.5), and using gunshots as the signal to be localized in azimuth (Casali and Keady 2010). In the latter two experiments, certain prototype level-dependent earplugs (Etymotic Research EB 1 and EB 15 BlastPlgs™), which incorporate unity gain in low noise levels, rapid compression circuitry in high noise levels, and minimal electrical hum/static (i.e., low noise floor), did show an advantage over the electronic level-dependent earmuffs and near equivalence to the passive, level-dependent earplugs; an example of these data appears in Fig. 12.3 from the Casali and Keady (2010) gunshot localization experiment. However, as can be gleaned from Fig. 12.3, gunshot localization accuracy under none of the augmented hearing protectors was equivalent to that achieved by the open (unoccluded) ear.

Electronically modulated sound transmission HPDs have also been suggested as a potential benefit to individuals suffering from hearing loss, in an effort to improve their "fitness for duty" in certain noisy environments. However, the benefits of such devices to hearing-impaired persons have not been empirically demonstrated through scientific research. Nonetheless, these individuals are often in need of auditory accommodation, as noise- and age-induced hearing losses generally occur in the high–frequency regions first, and for those so impaired, the effects of conventional HPDs on speech perception and signal detection are not clear cut. Due to their already elevated thresholds for mid–to–high frequency speech phonemes and most warning signals being further raised by the hearing protector, hearing-impaired individuals are usually disadvantaged in their hearing by conventional HPDs. Though there is not consensus across studies, comprehensive reviews (e.g., Suter 1989) have concluded that sufficiently hearing-impaired individuals will usually experience additional reductions in communication abilities with conventional HPDs worn in noise. Because hearing-impaired individuals are thus auditorially compromised by both the noise in their workplaces and the use of passive, conventional HPDs, more research needs to be devoted to the development of hearing-assistive technologies that can concurrently provide hearing protection in noisy environments.

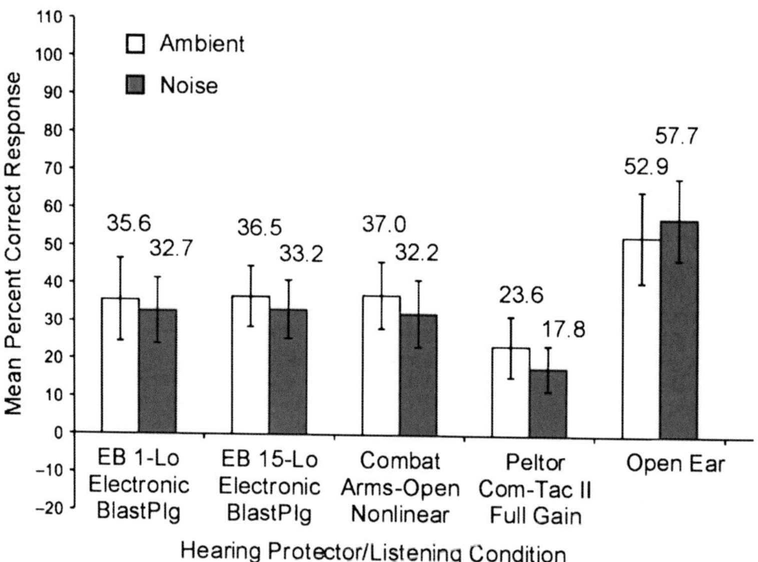

Fig. 12.3 Localization of gunshots in azimuth as measured by mean percent correct response to the actual direction, as a function of hearing protector/listening condition, in 40 dBA rural ambient vs. 82 dBA military truck idle noise. EB electronic earplugs were in low gain setting, Combat Arms™ earplug was in open (level-dependent) position, and Peltor earmuff was in full gain setting. Brackets represent 95% confidence interval based on *t*-distribution (Adapted with permission from Casali and Keady (2010), Fig. 23, with modifications by the author)

5.3.2 Augmentations Related to Spectrally-Nonlinear Passive Attenuation of Conventional Protectors

It is important to recognize that conventional passive HPDs cannot differentiate and selectively pass speech or nonverbal signal (or speech) energy versus noise energy at a given frequency. Therefore, conventional HPDs do not improve the speech/noise ratio in a given frequency band, which is the most important factor for achieving reliable signal detection or speech intelligibility. As shown in Fig. 12.4, the conventional earplugs (denoted by the labels fiberglass, premolded, and foam) attenuate high-frequency sound substantially more than low-frequency sound; therefore, they attenuate the power of high-frequency consonant sounds that are important for word discrimination and also attenuate the frequencies that are dominant in many warning signals more than they attenuate the lower frequencies. This nonlinear attenuation profile, which generally increases with frequency for most conventional earplugs and nearly all conventional earmuffs, allows more low-frequency than high-frequency noise through the protector, thus enabling an associated upward spread of masking to occur if the penetrating noise levels are high enough (Robinson and Casali 2003).

Certain augmented HPD technologies help to overcome the weaknesses of conventional HPDs as to low-frequency attenuation in particular; these include a variety

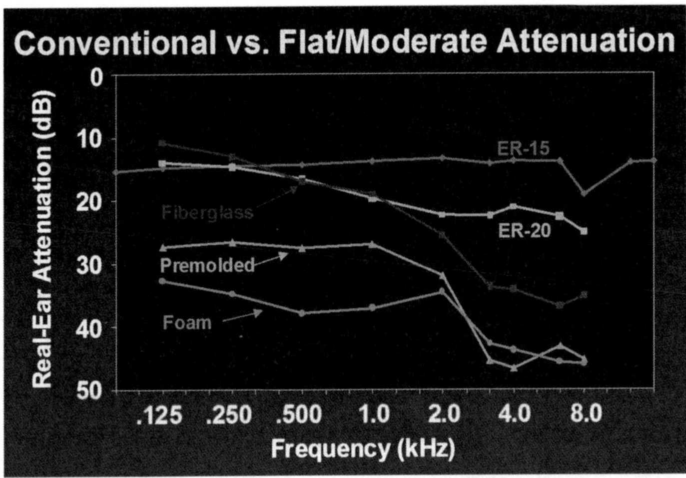

Fig. 12.4 Spectral attenuation obtained with real-ear attenuation at threshold (REAT) procedures for three conventional passive earplugs (premolded, user-molded foam, and spun fiberglass) and two uniform (*flat*) attenuation, custom-molded earplugs (ER-15, ER-20) (Courtesy of E. H. Berger, AEARO-3M, and used with permission)

of *active noise reduction* (*ANR*) devices that, through capture of the offending noise and electronic phase-cancellation of it at the ear via superposition through feedback and feed-forward control loops, bolster the low-frequency attenuation of passive HPDs below about 1,000 Hz. A review of both analog and digital examples of ANR technology, with relevant performance-based studies, appears in Casali (2010b). ANR is especially beneficial to, and thus implemented in earmuffs, which are generally weakest in low-frequency attenuation and which also provide space for the electronics of ANR circuitry to be packaged in/on the muff; however, ANR has also appeared in earplug designs in the past decade (McKinley 2001; Casali 2010b). Concomitant benefits of properly designed ANR-based HPDs include the reduction of upward spread of masking of low-frequency noise into the speech and warning signal bandwidths, as well as reduction of noise annoyance in certain environments that are dominated by low frequencies, such as jet aircraft cockpits and passenger cabins (Casali and Robinson 2003).

The tendency of conventional HPDs to exhibit a sloping, nonlinear attenuation profile versus frequency creates an imbalance from the listener's perspective because the relative amplitudes of different frequencies are heard differently than they would be without the HPD and thus, broadband acoustic signals are heard as spectrally different from normal; in other words, they sound more *bassy* (Berger 2003b; Casali 2010a). Thus, the spectral quality of a sound is altered, and sound interpretation, which is important in certain jobs that rely on aural inspection (e.g., machining, mining, engine troubleshooting) and leisure activities (e.g., performing or listening to music), may suffer as a result. This is one of the reasons that *uniform* (*also called "flat"*) *attenuation* HPDs have been developed as an augmentation technology,

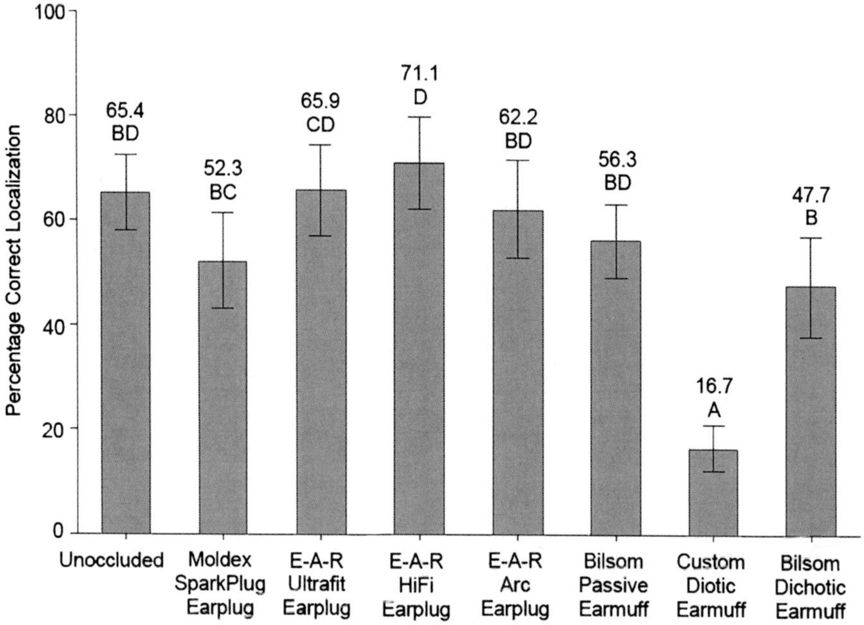

Fig. 12.5 Localization of vehicular backup alarm in azimuth as measured by mean percent correct response to the actual direction, as a function of hearing protector/listening condition, in 90 dBA pink masking noise. Electronically modulated sound transmission earmuffs (diotic and dichotic) were in full gain setting and Arc™ earplug was in open (level-dependent) position. Brackets represent 95% confidence intervals and means with different letters are significantly different at $p \leq 0.05$ (Adapted with permission from Alali and Casali (2011), Fig. 10)

and the attenuation curves for two such devices are depicted as the ER-15 and ER-20 in Fig. 12.4. These devices have proven to be more popular with musicians than conventional HPDs because the flat attenuation products do not disrupt relative perceptions of the loudness of various pitches (Casali and Robinson 2003).

Although there is minimal auditory performance-related research data on flat attenuation earplugs, in one experiment these products were found to provide a small but statistically significant advantage over certain conventional HPDs (e.g., foam earplugs) and electronically-modulated sound transmission earmuffs in a masked auditory localization task, namely that of determining the directional approach of a vehicle backup alarm in 90 dBA pink noise (Alali and Casali 2011). The mean percent correct localization data for each HPD in this experiment, compared to the open (unoccluded) ear, appear in Fig. 12.5.

5.3.3 Auditory Perceptual Issues from Covering the Pinnae

Because some of the high-frequency binaural cues (especially above about 4,000 Hz) that depend on the pinnae are altered by HPDs, judgments of sound direction and distance may be compromised. Earmuffs, which completely obscure the pinnae, may interfere with localization in the vertical plane and also tend to cause horizontal plane errors in both contralateral (left–right) and ipsilateral (front–back) judgments (Suter 1989). Earplugs may result in some ipsilateral judgment errors, but generally cause fewer localization errors than muffs because they do not completely destroy the pinna's cueing. In an effort to compensate for the lost pinnae-derived cues for sound localization that are typically destroyed with application of an earmuff, *dichotic electronically modulated sound transmission* HPDs have been developed and marketed for many applications, including military situation awareness needs, industrial environs, and even game hunting.

Electronically modulated sound transmission HPDs, covered at length in Casali (2010b), have at least one external microphone on each earmuff cup, which transmits a specified passband of the noise incident upon each microphone to a small loudspeaker under the earmuff cup. Binaural cues, at least to some degree, are thus maintained with these HPDs, assuming their between-ear gain controls are properly balanced, their microphones are sufficiently directional, and their passband includes frequencies *outside* the range that cannot be typically localized, that is, they need to transduce and pass-through frequencies below about 1,500 Hz for interaural phase timing cues and above about 3,000 Hz for interaural intensity cues (e.g., Hartmann 1999). Nonetheless, the aforementioned experiment, which employed an industrial/construction dichotic sound transmission earmuff in the azimuthal localization task of determining the approach direction of a vehicular backup alarm, demonstrated no advantage in localization with the electronic dichotic muff over a conventional earmuff or earplug, and in fact, worse localization occurred with such a muff design than with flat attenuation and certain conventional earplugs (see Fig. 12.5 and Alali and Casali 2011). Similar results occurred with a military dichotic sound transmission earmuff in the azimuthal localization task of determining the direction of a gunshot (Casali and Keady 2010). In this applied field experiment, the Peltor Com Tac II electronic earmuff generally displayed poorer localization performance in response to gunshots than two Etymotic electronic earplugs, a Combat Arms™ passive level dependent earplug, and the open ear (see Fig. 12.3).

5.4 Other Hearing Protection Device Augmentation Technologies to Date

The above-mentioned HPD augmentations of uniform attenuation, ANR, electronically modulated sound transmission, and level-dependent or amplitude-sensitive attenuation all have their place in certain applications for HPDs. The augmentations discussed herein are not an exhaustive list, and other devices that fall under different categories are currently available or in final prototyped stages. For more information on these technologies, the reader is referred to two parallel review papers by Casali, one concerning passive technologies (Casali 2010a) and the other concerning powered electronic technologies (Casali 2010b). It is important to recognize that the major goals of all aforementioned augmentation features are to foster the use of hearing protection by producing HPDs that are more acceptable to the user population; amenable to the occupational, military, or recreational environment; and can be tailored to the noise exposure. These goals are generally aimed at affording better hearing perception under a protected state, which in some cases may indeed result in a safer worker, soldier, or other user. However, although these goals are noble, as evidenced in this chapter they are not always realized in practice with actual products.

6 Emerging HPD Technologies and a View Toward the Future

To overcome some of the limitations of HPDs discussed earlier, several recent innovations in HPD technology either have been recently developed and prototyped, and in some cases, they are now commercially available. These emerging technologies continue to be refined, which is important so that the demands created by noise exposures and the need for situational awareness on the part of the wearer can be met. Note that in the brief overview that follows, examples of each technology that are known to the author are provided; however, by no means does this connote that these particular examples are being advocated or promoted, nor is the listing intended to be exhaustive.

6.1 Passive Adjustable-Attenuation Hearing Protectors

To help "tailor" the attenuation of an HPD to a particular noise problem (i.e., in lieu of selecting *different* HPDs for different exposures), earplug designs have recently been developed that allow the user some level of control over the amount of attenuation achieved (Casali 2010a). These devices incorporate a leakage path that is adjustable via the setting of a valve that obstructs a tunnel or "vent" cut through the body of the plug (e.g., a Dutch earplug, the Variphone™) or via selection from a choice of available filters or dampers that are inserted into the vent (e.g., Canadian devices including the Sonomax SonoCustom™and the Custom Protect Ear dB Blocker™).

6.2 Verifiable-Attenuation Hearing Protectors

Toward the objective of establishing a quality fit of an HPD to a user, and furthermore in keeping with the OSHA (1983) Hearing Conservation Amendment's requirement, per 29 CFR 1910.95 (i)(5), that "the employer shall insure proper initial fitting and supervise the correct use of all hearing protectors," several systems have been developed that essentially verify the attenuation attained on a given user.

For example, the original SonoPass™ system, now called the AEARO-3M E•A•RFit™, is sold as a system with a probe tube microphone test apparatus that verifies the amount of attenuation achieved via microphone-in-real-ear, noise reduction measurement techniques on each user as they are fit with the product. Another similar system is the Sperian VeriPRO™. It is important to recognize that all of these verifiable-attenuation HPDs basically measure attenuation by placing a microphone or probe tube through a duct that runs lengthwise through an earplug, and then taking a noise reduction measurement in a sound field. "Verified attenuation" is thus established *for that ducted earplug for that particular fitting on the user.* Thereafter, for actual use in the field, the ducted earplug is replaced by a solid (i.e., nonducted) earplug of the same type/model, or alternatively, a noise-blocking insert is used to occlude the duct. It also must be recognized that the attenuation obtained by these systems is *not* recognized by OSHA as a means for determining the adequacy of the hearing protector for a given noise exposure, nor is it a replacement for the EPA-required label of HPD attenuation data (though in the future, by additional legislative policy amendments, it is *possible* that it could be recognized for either or both of those applications).

6.3 Tactical Communications and Protection Systems (TCAPS)

In the past few years, several devices have been developed that have multiple objectives, and depending upon the particular product, these objectives have included some of all of the following: hearing protection from continuous noise, hearing protection from impulsive noise (particularly gunfire), measurement of protected noise exposure (i.e., at the ear under the HPD), enhancement of hearing for ambient sounds and uttered speech, and two-way communications capabilities (Casali 2010b). These products are typically designed with military or law enforcement applications in mind, giving rise to the moniker "TCAPS." All of these products incorporate sound transmission circuitry to transduce ambient sounds via a microphone on the outside of the HPD, and then those sounds are bandpass-filtered to an amplified earphone inside the HPD. Using elements of rapid-response automatic gain control with high pass-through gain capability, these devices can serve as assistive listening devices for military and other applications, to aid in threat detection, sound localization, and hearing of low-level speech, and then when gunfire occurs, the amplification rapidly decays (if designed properly), causing the device to quickly revert to a passive hearing protector.

TCAPS devices typically have more sophisticated and powerful pass-through filtering/gain circuits than do the industrial versions of sound transmission earmuffs discussed earlier herein. Further, these systems include elements that provide two-way communications capabilities, with some versions incorporating microphones covertly located under the HPD within the ear canal, that is, an "ear canal microphone," to pick up the wearer's voice by bone/tissue conduction. Still another feature of at least one device is a means for transducing the noise level under the HPD, determining cumulative noise exposure from that, and then using these data to modulate the system pass-through gain. Examples of products in the category of TCAPS include the Communications and Enhancement Protection System™ (CEPS) from Communications & Ear Protection, Inc., the QuietPro™ by NACRE AS (Norwegian), and the Silynx QuietOPS™, among others. Owing to their recent development, some of these devices have not undergone experimentation to determine their operational performance effectiveness; however, some of the products have already been deployed for use in military combat settings. Further discussion of situational awareness enhancement HPDs, and experimentation on a subset of them, can be found in Casali et al. (2009) and Casali (2010b).

6.4 Multicomponent Hearing Protection Systems for Extreme Noise

Although their application is highly specialized, multicomponent HPD systems have recently been developed and tested for use in noise environments that greatly exceed the attenuation capabilities of even double passive protectors–earmuffs worn over earplugs. Probably the most prominent of these environments is the aforementioned aircraft carrier deck during flight operations, where sound pressure levels up to the mid-150 dBA range occur in the vicinity of flight deck personnel (McKinley 2001). However, certain large-caliber weapons and explosive blasting can also produce exceedingly high exposures. In response to these extreme threats to hearing, specialized HPDs have been developed, and for the aircraft carrier deck application, these have included multiple components of staged hearing protection elements. Such HPDs provide both high passive attenuation through very deep-insertion, custom-molded earplugs, coupled with active noise reduction in the in-canal sound field *under* the earplug, all covered with a tightly fitted earmuff with highly compliant cushions (McKinley 2001). Other development efforts have applied full-head-coverage helmets with circumaural, active noise cancellation earcups inside, all worn over deeply fitted passive earplugs.

6.5 Composite Materials and Nanostructures in Hearing Protectors

A few HPDs have been developed using a composite combination of materials, typically in sandwich- or concentric-type construction, to reap the benefits of the impedance-mismatching (and the resultant attenuation benefit) that occurs with materials that differ in density, elasticity, reflectivity, and other physical parameters. Furthermore, there is some recent work in exploiting nanomaterials, particularly those with controllable cellular structures, for use in earplug and earmuff materials, with an eye toward "tuning" the attenuation of such devices by cellular modification, even dynamically, for specific noise threats. The use of composite materials and nanostructures in hearing protection is at present in its infancy, and deserves further research and development attention because of its strong potential.

7 Summary and Recommendations

Based on the technology and applied research overview herein, several conclusions and recommendations are warranted, as follows.

1. Advancements in hearing protection have been significant in the past three decades, especially in regard to devices that incorporate better comfort features, hearing-assistive and situational-awareness systems, uniform spectral attenuation, verifiable attenuation for the individual user, and active noise cancellation. Although some of these advancements are aimed at specific applications, as a whole they render modern HPDs as more useful and beneficial, and there is currently a wide assortment of available types and models that enable appropriate selection for most noise exposure situations and user requirements.
2. It is important to recognize that although OSHA federal law for General Industry (29 CFR 1910.95; OSHA 1983) requires hearing protection to be supplied to all workers exposed to an 8-h TWA of 85 dBA or above, and that usage becomes mandatory at a TWA of 90 dBA (or at 85 dBA for a worker who has exhibited a standard threshold shift), *the language of the law puts engineering controls and administrative controls at priority over hearing protection.* That is, when noise reaches a TWA of 90 dBA, feasible engineering or administrative controls are *required.* At these levels, hearing protection is indicated when the other controls *fail* to reduce the noise exposures to within a TWA of 90 dBA, and too often the argument in justification of the *failure* of legally higher-priority engineering controls is one of technical and/or economic *infeasibility*. When this argument arises, very stringent criteria for demonstrating infeasibility should be met, because engineering noise control offers significant long-term advantages over personal hearing protection. Furthermore, it should be obvious that if workplace noise levels are limited by engineering controls, "buy quiet" programs, or via other means to below the OSHA action level of 85 dBA TWA, then HPDs are completely obviated from an OSHA legal standpoint, even though the devices may still be desirable for reduction of noise annoyance or to ensure that the noise exposure hazard is fully mitigated.
3. In the recreational and home consumer setting, an emphasis on engineering of quieter products, such as power tools, toys, and recreational vehicles, will be of significant benefit because there will be less need to rely on HPDs. In some instances, although the use of HPDs in nonoccupational settings may pose a safer situation for the user from a hearing loss prevention perspective, it may also pose an unsafe situation if the user's ability to hear hazards in the vicinity, such as approaching vehicles or warning alarms, is compromised.

4. Given the fact that hearing protectors will always remain an item of personal protective equipment that is dependent upon human behavior to establish protection, the weak link is, and will always be, the human factor. It is certainly important that hearing protection manufacturers continue to emphasize the development of HPDs that are comfortable, easily sized and fit to the user, straightforward to match to the hearing-critical needs of a particular job or situation (inclusive of situation awareness and communications enhancements where needed), and in cooperation with the EPA, to provide attenuation performance labeling that reflects the protective performance of the products *as they are actually used.*
5. Certain technologies used in HPDs, such as active noise cancellation and amplitude-sensitive (or level-dependent) attenuation, are not amenable to the currently in-force EPA (1979) labeling regulation which cites ANSI S3.19-1974 as the attenuation test method. Thus, any final version of the recently proposed EPA labeling Rule (EPA 2009) must include the additional testing standards cited in that proposed Rule, so that beneficial HPD augmentation technologies can be tested, labeled, and sold as hearing protection devices.
6. Via advances in computational power and miniaturization of components, noise exposure measurement can now be obtained under earplugs and earmuffs for provision of a *protected* noise dose for a given HPD wearer. Because such a protected dosimetry measurement is an "end-of-pipe" metric that incorporates *all* noise energy present at the ear (inclusive of both incident noise as well as earphone output), further refinement and testing of these systems is warranted, and when the technology is proven, OSHA regulations should be revised to allow protected dosimetry measurements to suffice for workplace noise monitoring on individual workers.
7. In view that many occupational, nonoccupational, and military situations require HPD users to be vigilant to and cognizant of signals and communications in their vicinity, more effort is needed in the development of HPD systems that do not degrade, and in fact improve the users' situational awareness for certain acoustical stimuli in their environments.
8. Organized programs to heighten public education and awareness of the hazards of noise, the causes and implications of noise-induced hearing loss, and the causes and implications of noise-induced annoyance need to be put at higher priority by federal agencies including the National Institutes of Health and the EPA. These efforts should include information regarding the selection and use of hearing protection; however, it is also of great importance to educate the public about reducing their noise exposures by buying quieter products, making intelligent decisions about not frequenting high noise exposure events, and being vigilant to and active in public policy decision-making about community noise zoning issues and consumer product noise emissions.

Finally, returning to the systems approach to noise abatement shown in Fig. 12.1, it is important to reiterate that *hearing protection is a noise countermeasure that is implementable only at the very end of the noise propagation chain*, that is, at the receiver's ear. In the great majority of noise exposure situations, the priority should be to reduce or eliminate noise at its source or in its path through engineering controls, and not to rely on hearing protection at the receiver to curb the noise just before it enters the ears. Hearing protection, though effective when selected and applied properly, is not a panacea for combating the risks posed by noise, and its effectiveness will always be dependent on human behavior. It should thus not be viewed as a replacement for noise control engineering. However, in those minority cases where noise control engineering's afforded reduction is simply insufficient, or it is *truly* economically or technically infeasible (as perhaps with a personally shouldered, high-caliber weapon), hearing protection devices necessarily become the primary countermeasure.

References

Alali, K., & Casali, J. G. (2011). Vehicle backup alarm localization (or not): Effects of passive and electronic hearing protectors, ambient noise level, and backup alarm spectral content. *Noise and Health Journal*, 13(51), 99–112.

ANSI (1974). Method for the Measurement of Real-Ear Protection of Hearing Protectors and Physical Attenuation of Earmuffs. Standard S3.19-1974. New York: American National Standards Institute, Inc.

ANSI (2007). Methods of Estimating Effective A-Weighted Sound Pressure Levels When Hearing Protectors are Worn. Standard S12.68-2007. New York: American National Standards Institute, Inc.

ANSI (2008). Methods for Measuring the Real-Ear Attenuation of Hearing Protectors. Standard S12.6-1997(R2008). New York: American National Standards Institute, Inc.

ANSI (2010). Methods for the Measurement of Insertion Loss of Hearing Protection Devices in Continuous or Impulsive Noise Using Microphone-In-Real-Ear or Acoustic Test Fixture Procedures. S12.42-1995(R2010). New York: American National Standards Institute, Inc.

ASHA (2009). Noise and hearing loss – Noise is difficult to define! Retrieved from http://www.asha.org/public/hearing/disorders/noise.htm.

Baldwin, D. (2004) J.A.R Elliott, premier trapshooter. *Trap and Field*, July, 104. (Also : http://www.traphof.org/roadtoyesterday/elliott-JAR.htm).

Barr, T. (1896). *Manual of diseases of the ear.* Glasgow, Scotland: James Maclehose and Sons.

Berger, E. H. (2003a). Noise control and hearing conservation: Why do it? In E. H. Berger, L. H. Royster, J. D. Royster, D. P. Driscoll, & M. Layne (Eds.), *The noise manual*, (Revised 5th ed., pp. 2–17)., Fairfax, VA: American Industrial Hygiene Association. 2–

Berger, E. H. (2003b). Hearing protection devices. In E. H. Berger, L. H. Royster, J. D. Royster, D. P. Driscoll, & M. Layne (Eds.), *The noise manual*, (Revised 5th ed., pp. 379–454). Fairfax, VA: American Industrial Hygiene Association. 9–

Berger, E. H., & Casali, J. G. (1997). Hearing protection devices. In M. Crocker (Ed.) *Encyclopedia of acoustics*, (pp. 967–981). New York: John Wiley & Sons. 7–

Berger, E. H., Franks, J. R., Behar, A., Casali, J. G., Dixon-Ernst, C., Kieper, R. W., Merry, C. J., Mozo, B. T., Nixon, C. W., Ohlin, D., Royster, J. D., & Royster, L. H. (1998). Development of a new standard laboratory protocol for estimating the field attenuation of hearing protection devices, Part III: The validity of using subject-fit data. *Journal of the Acoustical Society of America*, *103*(2), 665–672.

Casali, J. G. (1990). Listening closely to the noise of monster trucks. *Roanoke Times and World News*, March 5, 1990, p. A11.

Casali, J. G. (2006). Sound and noise. In G. Salvendy (Ed.) *Handbook of human factors*, (3 rd ed., 612–642). New York: John Wiley & Sons.

Casali, J. G. (2010a). Passive augmentations in hearing protection technology circa 2010 including flat-attenuation, passive level-dependent, passive wave resonance, passive adjustable attenuation, and adjustable-fit devices: Review of design, testing, and research. *International Journal of Acoustics and Vibration*, 15(4), 187–195.

Casali, J. G. (2010b). Powered electronic augmentations in hearing protection technology circa 2010 including Active Noise Reduction, electronically-modulated sound transmission, and tactical communications devices: Review of design, testing, and research. *International Journal of Acoustics and Vibration*, 15, 168–186.

Casali, J. G., & Robinson, G. S. (2003). Augmented hearing protection devices: Active noise reduction, level-dependent, sound transmission, uniform attenuation, and adjustable devices—technology overview and performance testing issues. *EPA Docket OAR-2003–0024*, Washington, DC: U.S. Environmental Protection Agency Workshop on Hearing Protection Devices, March 27–28. Also at: www.regulations.gov, reference Docket OAR-2003–0024.

Casali, J. G., & Gerges, S. (2006). Protection and enhancement of hearing in noise. In R. C. Williges (Ed.), *Reviews of human factors and ergonomics*, (Vol. 2, pp. 195–240). Santa Monica, CA: Human Factors and Ergonomics Society. 5–

Casali, J. G., & Lancaster, J. A. (2008). *Quantification and solutions to impediments to speech communication and signal detection in the construction industry.* Blacksburg, VA: Virginia Tech, Dept. of Industrial and Systems Engineering, Technical Report 200803, (Audio Lab 4/15/08–3–HP), April 15.

Casali, J. G., & Keady, J. P. (2010). *In-field localization of gunshots in azimuth under two prototype etymotic and two production Aearo-Peltor hearing protection enhancement devices intended for military ground soldier applications.* Blacksburg, VA: Virginia Tech ISE Department Technical Report 201003, IRDL-Audio Lab Number 6/30/10–3–HP), June 30.

Casali, J. G., Ahroon, W. A., & Lancaster, J. A. (2009). A field investigation of hearing protection and hearing enhancement in one device: For soldiers whose ears and lives depend upon it. *Noise and Health Journal*, 11(42), 69–90.

Department of the Air Force (1948). *Precautionary measures against noise hazards.* AFR 160–3, Washington, DC: U.S. Air Force.

EPA. (1979). Noise labeling requirements for hearing protectors. U.S. Environmental Protection Agency, 40 CFR 211, *Federal Register*, *44*(190), 56130–56147.

EPA. (2009). *EPA Docket OAR-2003–0024.* U.S. Environmental Protection Agency workshop on hearing protection devices. Washington, DC, March 27–28. Retrieved from www.regulations.gov, reference Docket OAR-2003–0024.

Fosboke, J. (1831). Practical observations on the pathology and treatment of deafness, No. II. *Lancet VI*, 645–648.

Gerges, S., & Casali, J. G. (2007). Ear protectors. In M. Crocker (Ed.), *Handbook of noise and vibration control*, (pp. 364–376). New York: John Wiley & Sons. 4–

Hartmann, W. M. (1999). How we localize sound. *Physics Today.* 52(11), 24–29.

Holt, E. E. (1882). Boiler-maker's deafness and hearing in noise. *Transactions of American Otology Society*, 3, 34–44.

McKinley, R. (2001). Future aircraft carrier noise. *Proceedings of the International Military Noise Conference*, Baltimore, MD, April 24–26.

MSHA (1999). Health Standards for Occupational Noise Exposure; Final Rule. Mine Safety and Health Administration. 30 CFR Part 62, 64. *Federal Register.*

Nondahl, D. M., Cruickshanks, K. J., Wiley, T. L., Klein, R., Klein, B. E. K., & Tweed, T.S. (2000). Recreational firearm use and hearing loss. *Archives of Family Medicine*, 9, April, 352–357.

OSHA (1971a). Occupational Noise Exposure (General Industry). Occupational Safety and Health Administration. 29 CFR 1910.95, *Federal Register.*

OSHA (1971b). Occupational Noise Exposure (Construction Industry). Occupational Safety and Health Administration, 29 CFR 1926.52. *Federal Register.*

OSHA (1971c). Hearing Protection (Construction Industry). Occupational Safety and Health Administration, 29 CFR 1926.101. *Federal Register.*

OSHA (1983). Occupational Noise Exposure; Hearing Conservation Amendment; Final Rule. Occupational Safety and Health Administration, 29 CFR 1910.95. *Federal Register*.

Park, M. Y., & Casali, J. G. (1991). A controlled investigation of in-field attenuation performance of selected insert, earmuff, and canal cap hearing protectors. *Human Factors*, 33(6), 693–714.

Robinson, G. S., & Casali, J. G. (2003). Speech communications and signal detection in noise. In E. H. Berger, L. H. Royster, J. D. Royster, D. P. Driscoll, & M. Layne (Eds.), *The noise manual* (Revised 5th ed., pp. 567–600). Fairfax, VA: American Industrial Hygiene Association. 7–

Suter, A. H. (1989). *The effects of hearing protectors on speech communication and the perception of warning signals* (AMCMS Code 611102.74A0011), Aberdeen Proving Ground, MD: U.S. Army Human Engineering Laboratory.

WHO (1997). *Prevention of noise-induced hearing*, Report of a consultation at the World Health Organization, Geneva, Switzerland, October 28–30.

Chapter 13
Prevention of Noise-Induced Hearing Loss: Potential Therapeutic Agents

Colleen G. Le Prell and Jianxin Bao

1 Introduction

Noise-induced hearing loss (NIHL) is a significant clinical, social, and economic issue. The development of novel therapeutic agents to reduce NIHL would increase the potential protection of workers exposed to occupational noise, as well as noise-exposed military populations. Adolescents and young adults may perhaps benefit as well. A better understanding of oxidative stress during and after noise, and activation of other mechanisms of cellular and molecular events that lead to cell death subsequent to noise insult, has advanced the potential to identify and develop novel therapeutic agents. Identification of oxidative stress and improved knowledge as to how cells die has been particularly significant for the development of novel therapeutic agents to reduce NIHL. Widespread clinical acceptance of any novel therapeutic will be driven by demonstration that the agent reduces permanent noise-induced threshold shift (PTS) in randomized, placebo-controlled, prospective human clinical trials. Identification and access to populations that develop PTS despite the use of traditional hearing protection devices (which are ethically required in such studies) is challenging. Moreover, such studies are necessarily slow, requiring years of data collection from each individual subject, given that NIHL is generally slow to develop. Given these and other obstacles, many groups are turning to temporary threshold shift (TTS) models for initial human proof of concept testing. Thus, the relationship between PTS and TTS, and the clinical relevance of TTS deficits, are reviewed first.

C.G. Le Prell (✉)
Department of Speech, Language, and Hearing Science, University of Florida, 101 S. Newell Road, Gainesville, FL 32610, USA
e-mail: colleeng@phhp.ufl.edu.

C.G. Le Prell et al. (eds.), *Noise-Induced Hearing Loss: Scientific Advances*, Springer Handbook of Auditory Research 40, DOI 10.1007/978-1-4419-9523-0_13,

2 Relationship of Permanent Threshold Shift and Temporary Threshold Shift

The histopathological correlates of PTS and TTS have been well described (for recent examples, see Wang et al. 2002 and Hu, Chap. 5). TTS is the reversible hearing loss that occurs immediately post-noise and recovers over a period of several hours or days postexposure. PTS is the permanent hearing loss that fails to resolve with additional post-noise recovery time. An example of the time course of recovery subsequent to robust TTS, resolving to a more moderate PTS, is shown in Fig. 13.1.

A variety of data are consistent with a *functional* continuum through which PTS deficits can be predicted based on measured TTS, once TTS is sufficiently robust to result in PTS (e.g., see Henderson et al. 1991). Indeed, data from more than 900 chinchillas indicate TTS measured 24 h post-trauma correlates well with PTS (Hamernik et al. 2002). In contrast, other data have revealed a *morphological* continuum. The elegant use of survival-fixation techniques showed PTS was best predicted by hair cell loss and neural degeneration, not TTS deficits (Nordmann et al. 2000). These relationships did not necessarily hold with smaller PTS deficits, however. It is possible to observe PTS of 40–50 dB in the absence of hair cell loss;

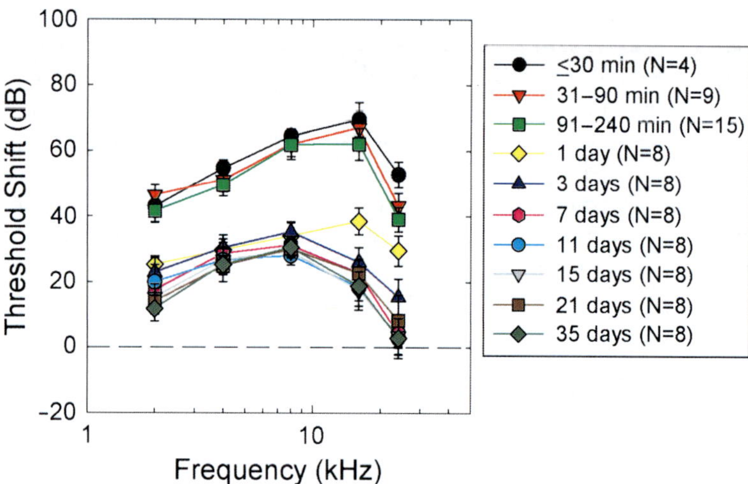

Fig. 13.1 Guinea pigs were exposed to a 114 dB SPL OBN centered at 4 kHz for 4 h. Robust temporary threshold shifts (TTS) were observed at all post-noise test times on the day of exposure (extending up to 4 h post-noise). Significant recovery occurred by 24 h (1 day) post-noise, and there was additional statistically significant recovery from day 1 to day 3 post-noise. There was a small but statistically reliable improvement from day 3 to day 7, with no additional statistically significant recovery after day 7 post-noise. These data suggest that the TTS recovery process was complete by 7 days post-noise, and remaining deficits were permanent threshold shifts (PTS). PTS varied with frequency, with the largest deficits at a frequency approximately 0.5 octaves higher than the range of frequencies encompassed by the OBN

hair cell stereocilia deficits provide one possible morphological correlate for PTS observed in the absence of hair cell loss (Wang et al. 2002).

Unlike functional and morphological changes that seem to fall along a continuum with increasing noise levels, the *molecular* response to TTS and PTS-inducing sounds appears to be quite different. The *Bcl-2* family of genes includes both proapoptotic genes (*Bax*, *Bak*) and antiapoptotic genes (*Bcl-2*, *Bcl-xl*) (for review, see Danial 2007, 2009). After PTS-inducing noise, the *Bcl-2*-associated death promoter (BAD) translocates from the mitochondria to the nucleus in outer hair cells (Vicente-Torres and Schacht 2006), resulting in *Bak* (proapoptotic) gene expression in the outer hair cells (Yamashita et al. 2008). In contrast, after TTS-inducing noise, the *Bcl-xl* antiapoptotic gene is expressed in outer hair cells (Yamashita et al. 2008). These data indicate that as noise levels increase, at some point they reach a level that causes an upregulation of *Bcl-xl*, presumably to protect the cell from damage; but as the level of sound exceeds the ability of the cell to survive, noise induces an upregulation of *Bak* to kill the cell purposefully via apoptotic pathways. As reviewed in the text that follows, oxidative stress is almost certainly an element of both PTS and TTS, and some antioxidants reduce both PTS and TTS.

2.1 Clinical Relevance of TTS

One issue of particular clinical relevance is the issue of whether repeat TTS will ultimately present as a PTS. Indirect evidence from real-world populations supports the possibility that repeat TTS exposures can ultimately result in PTS at the exposure frequency. TTS and PTS were detected at the same frequencies in two populations of motorcycle riders, one tested for TTS after a 60-min 80-mph motorcycle run and the other tested for PTS given a long-term riding history (McCombe et al. 1995). These data confirm and provide a real-world extension of Mills et al. (1981), who carefully measured TTS in noise-exposed human subjects, and compared these data to PTS data from workers already exposed to long-term noise. They described a compelling relationship between asymptotic threshold shift (ATS) and PTS. ATS is a stable, plateau-level of hearing change that is achieved after 8–12 h of noise exposure, depending on exposure level and frequency. ATS requires multiple threshold measurements during the course of the noise exposure as ATS is achieved only once TTS is no longer increasing as a function of continued exposure. An ATS could be considered a TTS as long as complete recovery is observed post-noise, but it is also possible that an ATS would not fully resolve, and that some PTS may result from the exposure. Most, if not all, of the literature on otoprotection is based on TTS at short-post noise intervals, or PTS at longer post-noise intervals, and it is rarely determined if the measured TTS was a stable ATS or not. Thus, this chapter focuses on TTS and PTS without respect to ATS.

Although the possibility of PTS after multiple repeat TTS-inducing noise exposures suggests an important opportunity to reduce PTS by preventing those repeat TTS insults, other new data suggest the possibility that even a single robust TTS may have far greater clinical relevance with respect to an eventual PTS than had been previously assumed. Noise exposure that resulted in TTS, but not PTS, produced an accelerated hearing loss as a result of progressive neural degeneration with age (Kujawa and Liberman 2006). Subjects in those studies were mice, and they were exposed to an 8–16-kHz octave band noise (OBN) at 100 dB SPL for 2 h; this exposure produces robust TTS, with 40 dB deficits measured 1 day post-noise (Kujawa and Liberman 2009). Kujawa and Liberman (2009) recently revisited the issue of neural degeneration after TTS. They observed rapid, extensive loss of synaptic contacts between hair cells and nerve fibers 24 h post-noise, during the period of TTS, as well as progressive long-term neural degeneration, subsequent to recovery from the TTS threshold deficits. Neural loss occurred even though the hair cell population was intact and normal threshold function had returned. These data suggest TTS has the potential to be more harmful than previously believed; however, there are multiple important questions to be resolved. First, except in the case of unanticipated impulse noise exposures, 40-dB TTS deficits are likely relatively rare in most normal human populations. It is therefore critical that the long-term neural sequella of smaller (less than 40 dB) TTS deficits be determined to better estimate potential risk for most human populations. Second, although these outcomes are robust in mouse models, and there is no a priori reason to assume these outcomes will not be observed in other mammalian species, it is important that long-term post-TTS outcomes be measured in other mammalian species. In the meantime, these data clearly support the possibility that reducing TTS could potentially be clinically beneficial over the long term.

In the following sections, the evidence suggesting antioxidants may be useful in reducing PTS and TTS is reviewed. Prevention of PTS is a goal with clear and compelling clinical relevance, and many studies have evaluated the potential for prevention of PTS. Prevention of TTS has been examined to a lesser extent, but, with these recent studies on long-term post-TTS outcomes (Kujawa and Liberman 2006, 2009), there will no doubt be increased interest in the potential to reduce TTS deficits. Human clinical trials have evaluated the potential to reduce TTS in human subjects exposed to discotheque noise (Kramer et al. 2006) as well as military personnel exposed to weapons noise (Attias et al. 2003), with mixed success across agents and investigations, as reviewed later in this chapter. In the next section, evidence that oxidative stress contributes to NIHL is first reviewed.

3 Oxidative Stress and NIHL

When sound waves arrive at the tympanic membrane, the middle ear ossicles are put in motion, resulting in pressure concentrated at the oval window, where these bones contact the cochlea. This pressure displaces fluid inside the cochlea in a "wave-like" pattern, following the pressure waves at the ear drum. This fluid motion displaces the basilar membrane. The outer hair cells (OHCs), essential for normal hearing thresholds, sit between the basilar membrane and the tectorial membrane. With basilar membrane motion, a shearing force is exerted on the OHCs. Until a decade ago, it was generally thought that most, if not all, NIHL occurred when this shearing force caused direct mechanical destruction of hair cells and supporting structures, with perhaps some contribution of reduced blood flow (for review, see Le Prell et al. 2007b). Because NIHL was assumed to result from direct mechanical destruction, mechanical devices (ear plugs, ear muffs) that reduce sound coming into the ear were assumed to be the only strategies for reducing NIHL. Impulse noise and other very loud sounds clearly do cause mechanical damage (Wang et al. 2002; Ohlemiller 2008); however, cell damage after most other noise insults can be largely a byproduct of oxidative stress (for reviews, see Henderson et al. 2006; Le Prell et al. 2007b; Abi-Hachem et al. 2010). The timeline for free radical formation in the inner ear is well characterized, with immediate noise-induced free radical production, well documented increases during the first 1–2 h postexposure (see Ohlemiller et al. 1999b; Yamashita et al. 2004), and maximum immunocytochemical labeling of oxidative stress byproducts at 7–10 days post-noise (Yamashita et al. 2004).

Understanding free radical production inside hair cells and other cells in the inner ear requires a basic understanding of normal cell biology. In brief, nutrients (glucose) and oxygen enter the cells, and mitochondria convert the available nutrients and oxygen to a usable form of energy: ATP. The mitochondrial transport system, which produces most ATP, consumes some 85% of all oxygen used by cells, and, under normal physiological conditions, approximately 1–5% of the oxygen consumed by the mitochondria is converted to superoxide (O_2^-), hydrogen peroxide (H_2O_2), or other free radical species (for review see Chow et al. 1999). This process produces waste in the form of carbon dioxide (CO_2) and water (H_2O), and there is a steady stream of "leaked" electrons during the process of converting the starting products into ATP plus waste products. Leaked electrons are a major source of superoxide radical production. Whereas the carbon dioxide and water waste products are safely excreted, the leaked electrons – the free radicals – must be neutralized. Endogenous defense against toxic free radical accumulation is mediated by a group of antioxidant molecules, including superoxide dismutases (SODs); catalase; glutathione (GSH); and glutathione peroxidase (GPx), an enzyme that catalyzes the formation of GSH. This is not an exhaustive list; for example, malate dehydrogenase (MDH) and lactate dehydrogenous (LDH) are antioxidant enzymes that are upregulated in perilymph immediately after noise insult, with increases lasting 1 month or longer (Juhn and Ward 1979).

4 Endogenous Antioxidant Production

4.1 Superoxide Dismutase

Superoxide dismutases (SODs) are enzymes that catalyze the destruction of superoxide into the less toxic oxygen (O_2) and hydrogen peroxide (H_2O_2) radicals. There are three major SOD families, each with different metal cofactors.

Cu-Zn-SOD binds both copper and zinc, Fe-SOD and Mn-SOD bind either iron or manganese, and Ni-SOD binds nickel. A total of three human SODs have been identified. SOD1 (Cu-Zn-SOD) is found in cytoplasm, SOD2 (Mn-SOD) is found in mitochondria, and SOD3 (Cu-Zn-SOD) is found in extracellular areas. Knockout of the SOD1 enzyme in mice leaves them more vulnerable to noise insult than mice that produce SOD1 (Ohlemiller et al. 1999a; for review, see McFadden et al. 2001). Similarly, increasing SOD1 levels before noise insult leaves guinea pigs less vulnerable to noise exposure (Cassandro et al. 2003). Observations of decreased SOD production in basal regions of the cochlea (compared to more apical regions, see Ying and Balaban 2009) has been suggested to underlie the increased vulnerability of basal hair cells to noise, drugs, and other environmental stressors described by Sha et al. (2001). Genetic variation in human SOD1 (Liu et al. 2010) and SOD2 (Fortunato et al. 2004; Chang et al. 2009) has recently been linked to vulnerability to NIHL in humans as well. These data are reviewed in detail by Gong and Lomax (Chap. 9).

4.2 *Catalase*

Catalase is a common enzyme found in nearly all living organisms exposed to oxygen, and it works closely with SOD to neutralize free radicals. After SOD converts superoxide to hydrogen peroxide, catalase spurs the decomposition of hydrogen peroxide into water and oxygen. Catalase is highly efficient; one molecule of catalase can convert millions of molecules of hydrogen peroxide to water and oxygen per second. Decreased catalase levels after noise exposure have been interpreted as reflecting the consumption of catalase as a consequence of free radical neutralization; catalase levels in noise-exposed textile workers were lower than catalase levels in control subjects (Yildirim et al. 2007). As with SOD1, genetic variation in human catalase polymorphisms has recently been linked to vulnerability to NIHL in humans (Konings et al. 2007).

4.3 *Glutathione*

Glutathione (GSH) is another key endogenous antioxidant system, and it is found in the inner ear (Usami et al. 1996). GSH is produced endogenously from cysteine, glutamic acid, and glycine, and it protects cells against electron leaks by donating a hydrogen atom to stabilize free radicals. In its reduced state, GSH is ready to donate electrons. Glutathione peroxidase (GPx) catalyzes (stimulates) reduction of free radicals by GSH. Electron donation leaves GSH in an oxidized (reactive) state. Two reactive GSH molecules bind to each other to form glutathione disulfide (GSSG). Glutathione reductase then reduces GSSG back to GSH. The antioxidant activity of GSH has been attributed specifically to cysteine, and methionine is an amino acid that serves as a precursor to cysteine. Enzymes involved in GSH production also require minerals, such as iron, magnesium, copper, selenium, and manganese. Selenium, for example, activates the formation of GSH. Knockout of the GPx1 enzyme in mice significantly impairs endogenous defense against NIHL; mice that lack GPx1 are more vulnerable to noise insult than mice that produce GPx1 (Ohlemiller et al. 2000).

Reduced GSH production, after treatment with L-buthionine-(*S*, *R*)-sulfoximine (BSO) to inhibit endogenous GSH synthesis, rendered the ear more vulnerable to noise insult (Yamasoba et al. 1998). In contrast, upregulation of GSH with 2-oxothiazolidine-4-carboxylate (OTC), a pro-cysteine drug that promotes rapid restoration of GSH, reduced noise trauma (Yamasoba et al. 1998). With the discovery that cell death after noise exposure (and resulting NIHL) was largely a byproduct of oxidative stress, and that endogenous antioxidant status could be used to influence cell survival and hearing outcomes post-noise, it became possible to define antioxidant interventions that attenuate NIHL in animals, and many of these strategies have proven effective (for reviews, see Henderson et al. 2006; Le Prell et al. 2007b). Although many pharmacologic strategies for preventing NIHL require multiple years of safety testing and dose development, there are a small number of therapeutic strategies that have the potential for rapid translation to military and other noise-exposed human populations. Multiple groups have pursued GSH-based therapeutic strategies in recent years, and the current status of three glutathione treatment strategies are described below. Dietary antioxidants are then described as they also act as potent free radical scavengers.

5 Free Radical Scavengers for Prevention of NIHL

5.1 N-Acetylcysteine

N-Acetylcysteine (NAC) is the agent for which the most data on protection against NIHL are currently available (see Table 13.1; all tables summarize antioxidant reduction in PTS at the frequency at which the greatest PTS was measured in control subjects in each individual study). Studies have commonly shown 20- to 25-dB reductions in PTS when NAC is delivered as a single pre-noise treatment (Ohinata et al. 2003), a multidose pre-noise treatment (Lorito et al. 2008), or a multidose post-noise treatment (Coleman et al. 2007a), with one study reporting better than 30-dB protection (Bielefeld et al. 2007; but see also Tamir et al. 2010, who reported less than 10-dB protection). Earlier post-noise intervention has generally been more effective than later post-noise intervention, with at least a 50% reduction in protection as treatment delays increase to 4 h post-noise or longer (Coleman et al. 2007a; Lorito et al. 2008). As doses decrease (from a typical 325 mg/kg dose down to 100 mg/kg or even 50 mg/kg), protection decreases (Bielefeld et al. 2007). Protection is also significantly reduced when NAC is delivered via oral gavage instead of injections (Bielefeld et al. 2007).

In contrast to the above positive outcomes, NAC did not provide protection against a longer-lasting (8 h/day × 5 days) insult (Hamernik et al. 2008), and some

Table 13.1 *N*-Acetylcysteine for protection against NIHL

	Species	NAC dose	Noise	Threshold shift at frequencies with greatest PTS	Protection (reduction in PTS)
Kopke et al. (2000)	Chinchilla	325 mg/kg i.p. plus 50 mg/kg salicylate; 1 h pre- and 1 h post-, then 2×/day × 2 days	OBN centered at 4 kHz, 105 dB SPL × 6 h	PTS reduced from ~40 dB at 2 and 4 kHz to ~10 dB	30 dB
		325 mg/kg i.p. plus 50 mg/kg salicylate; 1 h post-, then 2×/day × 2 days		PTS reduced from ~40 dB at 2 and 4 kHz to ~10 dB at 2 kHz and ~20 dB at 4 kHz	20–30 dB
Kopke et al. (2001)	Chinchilla	325 mg/kg i.p. plus 25 mg/kg salicylate; 1 h pre- and 1 h post-, then 2×/day × 2 days	OBN centered at 4 kHz, 105 dB SPL × 6 h	PTS reduced from ~45 dB at 6 kHz to ~22 dB	23 dB
		325 mg/kg i.p. plus 50 mg/kg salicylate; as above		PTS reduced from ~45 dB at 6 kHz to ~15 dB	30 dB
		325 mg/kg i.p. plus 75 mg/kg salicylate; as above		PTS reduced from ~45 dB at 6 kHz to ~25 dB	20 dB
Ohinata et al. (2003)	Guinea pig	500 mg/kg i.p. 30-min pre-noise	OBN centered at 4 kHz, 115 dB SPL × 4 h	PTS reduced from ~40 dB at 4 and 8 kHz to ~25 dB at 4 and ~15 dB at 8 kHz	15–25 dB
Duan et al. (2004)	Rat	350 mg/kg i.p. 1 h pre-, then immed, and 3 h post ("3× dose")	50 impulses × 160 dB pSPL	PTS reduced from ~20 dB at 4 kHz to ~10 dB	10 dB
		350 mg/kg i.p. 1 day and 1 h pre-, then immed, 3 h, and 1 day post ("5× dose")		PTS *increased* from ~20 dB at 4 kHz to ~40 dB	NA, thresholds worse in treated animals
Kopke et al. (2005)	Chinchilla	325 mg/kg i.p., 2×/day × 2 days pre-, 1 h pre and 1 h post-noise, then 2×/day × 2 days post-noise	150 impulses × 155-dB pSPL	PTS reduced from ~40 dB at 6 kHz to ~12 dB	28 dB
Bielefeld et al. (2005)	Chinchilla	325 mg/kg i.p., 1 h pre-noise on 4 days of noise	OBN centered at 4 kHz, 100 dB SPL × 6 h/day × 4 days	PTS reduced from ~30 dB at 4 kHz to ~18 dB	12 dB

Kramer et al. (2006)	Human	900 mg effervescent tablet in 8 oz. water 30 min before nightclub entry	L_{avg} = 98.1 dBA × 2 h (L_{avg} range = 92.5–102.8 dBA)	14 dB TTS at 4 kHz not reduced by treatment	NA, no PTS
Coleman et al. (2007a)	Chinchilla	325 mg/kg i.p., 1 h post then 2×/day × 2 days	OBN centered at 4 kHz, 105 dB SPL × 6 h	PTS reduced from ~43 dB at 6 kHz to ~21 dB	22 dB
		325 mg/kg i.p., 4 h post-, then 2×/day × 2 days		PTS reduced from ~43 dB at 6 kHz to ~25 dB	18 dB
		325 mg/kg i.p., 12 h post-, then 2×/day × 2 days		PTS reduced from ~43 dB at 6 kHz to ~37 dB	6 dB; N.S.
Bielefeld et al. (2007)	Chinchilla	325 mg/kg i.p., 2×/day × 2 days pre-, 1 h pre- and 1 h post-noise, then 2×/day × 2 days post ("expt 1")	117–130 dB pSPL impacts, 98 dBA Gaussian noise × 2 h (L_{eq} = 105 dBA)	PTS reduced from ~65 dB at 4, 6, and 8 kHz to ~10–12 dB at 4, 6, and 8 kHz	53–55 dB
		325 mg/kg i.p., 2×/day × 2 days pre-, 1 h pre- and 1 h post-, then 2×/day × 2 days post-noise ("expt 2")	150 impulses × 155-dB pSPL	PTS reduced from ~35–40 dB at 4 and 6 kHz to ~5–10 dB at 4 and 6 kHz	30 dB
		100 mg/kg i.p., 2×/day × 2 days pre-, 1 h pre- and 1 h post-, then 2×/day × 2 days post-noise ("expt 2")		PTS reduced from ~35–40 dB at 4 and 6 kHz to ~15 dB at 4 and 6 kHz	20–25 dB
		50 mg/kg i.p., 2×/day × 2 days pre-, 1 h pre- and 1 h post-, then 2×/day × 2 days post-noise ("expt 2")		PTS reduced from ~35–40 dB at 4 and 6 kHz to ~15 dB at 4 and 6 kHz	20–25 dB
		325 mg/kg p.o., 2×/day × 2 days pre-, 1 h pre- and 1 h post-, then 2×/day × 2 days post-noise ("expt 3")	OBN centered at 4 kHz, 105 dB SPL × 6 h	PTS reduced from ~42 dB at 4 and 6 kHz to ~32 dB at 6 and 8 kHz	10 dB

(continued)

Table 13.1 (continued)

	Species	NAC dose	Noise	Threshold shift at frequencies with greatest PTS	Protection (reduction in PTS)
Choi et al. (2008)	Chinchilla	100 mg/kg plus 50 mg/kg 4-OHPBN i.p., 4 h post-, then 2×/day × 2 days post-	OBN centered at 4 kHz, 105 dB SPL × 6 h	PTS reduced from ~38 dB at 6 kHz in controls, to ~17 dB at 6 kHz w/50 mg/kg 4-OHPBN dB, to ~7 dB at 6 kHz with 4-OHPBN plus NAC	10 dB relative to: OHPBN control; 31 dB relative to no drug control
Lorito et al. (2008)	Rat	375 mg/kg i.p., 2×/day × 2 days starting 1 h pre-noise	OBN centered at 8 kHz, 105 dB SPL × 4 h	PTS reduced from ~25 dB at 16 kHz to ~4 dB	21 dB
		375 mg/kg i.p., immediately pre-noise		PTS reduced from ~25 dB at 16 kHz to ~5 dB	20 dB
		375 mg/kg i.p., 24 h post-noise		PTS reduced from ~25 dB at 16 kHz to ~14 dB	11 dB; N.S.
Hamernik et al. (2008)	Chinchilla	325 mg/kg i.p., 2×/day × 2 days pre-, 1 h pre- and 1 h post-, then 2×/day × 2 days post-	119–132 dB pSPL impacts superimposed on 97 dB A Gaussian noise × 8 h/day × 5 days (L_{eq} = 105 dBA)	PTS *increased* from ~57 dB at 4 kHz to ~63 dB	NA, thresholds worse in treated animals
Fetoni et al. (2009b)	Guinea pig	500 mg/kg i.p., immed post-, then 2×/day × 2 days post-	6 kHz pure-tone, 120 dB SPL × 30 min	PTS reduced from ~55 dB at 8 kHz to ~40 dB	15 dB
Tamir et al. (2010)	Mouse (Sabra)	325 mg/kg i.p., 2×/day × 2 days pre-, 1 h pre- and 1 h post-, then 2×/day × 2 days post-	Broadband noise, 113 dB SPL × 3.5 h	PTS reduced from ~24 dB (click) to ~17 dB	7 dB; N.S.
Lin et al. (2010)	Human	1,200 mg/day × 12 days	average daily 8 h time-weighted average, range: 88.4–89.4 dB	Average TTS at 3, 4, and 6 kHz reduced from 2.77 dB to 2.45 dB	NA, No PTS

NAC, *N*-acetylcysteine; *OBN*, octave band noise; *4-OHPBN*, 4-hydroxyphenyl-*N-tert*-butyl nitrone; *PTS*, permanent threshold shift; *SPL*, sound pressure level; *TTS*, temporary threshold shift; N.S., not statistically significant

NAC dose paradigms can in fact potentiate NIHL (Duan et al. 2004). One possibility is that bioavailability of NAC quickly decreases over the short term, making pre-noise dosing paradigms inadequate for longer (8-h) exposures. One strategy for potentially improving protection is to use combinations of agents that act on multiple targets. For example, salicylate is a potent scavenger of hydroxyl radicals. When NAC was combined with salicylate (Kopke et al. 2000, 2001), or 4-hydroxyphenyl-*N-tert*-butyl nitrone (4-OHPBN, which traps hydroxyl radicals, superoxide anions, and other free radicals, see Choi et al. 2008), threshold deficits were reduced by approximately 30 dB at the frequencies at which the largest noise insult was measured. Unfortunately, neither of those studies included a NAC alone control condition, although comparisons with other studies (summarized in Table 13.1) suggest that the effects of NAC combined with other antioxidants were perhaps generally better than the effects of NAC alone.

Given the number of investigations that have shown some level of protection against NIHL with NAC treatment (see Table 13.1), there have been some early efforts to determine whether the protective effects measured in animal models would translate to humans. After Toppila et al. (2002) presented preliminary data showing no protection against TTS in human subjects treated with a 400-mg dose of NAC, Kramer et al. (2006) evaluated a higher (900 mg) dose of NAC but similarly found no reliable group differences suggesting no protection against TTS; in addition, they found no protection against noise-induced decreases in distortion product otoacoustic emission (DPOAE) amplitude. In a third study, a 900-mg NAC supplement delivered three times/day was evaluated in 566 U.S. Marine recruits exposed to impulse noise (see Lynch and Kil 2005; Kopke et al. 2007). Outcomes from that study have not yet appeared in the peer-reviewed literature as of the time this chapter was prepared. Most recently, in a prospective double-blind, crossover study, 53 male workers exposed to 88–89 dB daily occupational noise were randomly assigned to receive either NAC (1,200 mg/day, 14 days) or placebo in random order during the two arms of the cross-over trial. In other words, all subjects received both treatments, with treatment order randomized across subjects. The NAC-induced reduction in TTS was small, with 2.5-dB shift-related TTS in workers during NAC treatment, and 2.8-dB shift-related TTS in workers during the placebo control, although the difference was statistically significant (Lin et al. 2010). Genotype data were also collected, and Lin et al. (2010) reported that pairwise comparisons revealed threshold protection was limited to the subset of workers with null genotypes in both GSTM1 and GSTT1, polymorphisms that reduce endogenous glutathione *S*-transferase enzyme activity and thus potentially reduce endogenous protection. Although NAC doses can likely be safely increased in future studies based on the FDA-approved oral dosing regimen for NAC use in cases of acute acetaminophen overdose, adverse events do occur at a high rate with the FDA-approved oral dose paradigm. Some 23% of patients experience nausea or vomiting, or both, at that dose (Heard 2010). Clearly, the acute high-dose treatments used after acetaminophen overdose would not be appropriate for chronic (daily) use.

Taken together, while the data on prevention of short-term changes in human hearing using NAC have not been encouraging, this may be a consequence of the low doses selected for initial human testing, it may be a consequence of failure to translate protection from animals to humans, or it may be the case that TTS models are simply not appropriate for showing NAC-mediated protection in human subjects. Although some studies in animal models showed small (~10 dB) reductions in TTS at early post-noise times (Kopke et al. 2005; Bielefeld et al. 2007), other animal studies have revealed minimal (Duan et al. 2004) or no (Fetoni et al. 2009b) reduction in TTS with NAC treatment. Regardless, with mixed evidence supporting the translation of NAC from animal models to human subjects, other agents have received increasing attention in recent years.

5.2 D-Methionine

Noise induces lipid peroxidation (shown by malondialdehyde and 4-hydroxynonenal formation in cochlear samples). D-Methionine reduces noise-induced lipid peroxidation, and it increases SOD and catalase (Samson et al. 2008). Most of these effects have been attributed to either enhanced production of intracellular GSH, selective increases in mitochondrial GSH, or secondary effects on SOD and catalase. Detailed reviews of the mechanisms of action of D-methionine are available (Campbell et al. 2007; Vuyyuri et al. 2008).

Much of the earliest work on D-methionine as an otoprotective agent focused on protection against hearing loss induced by the chemotherapeutic drug cisplatin (for reviews, see Campbell et al. 2007; Campbell and Le Prell 2011). Work in this area actively continues, with preliminary evidence suggesting protection against cisplatin-induced hearing loss in human patients (Campbell et al. 2009). In the smaller number of studies that have evaluated use of D-methionine to reduce NIHL, this compound has been shown to reduce PTS (see Table 13.2).

PTS has been reduced by approximately 20 dB given a pretreatment paradigm (Kopke et al. 2002), and, when Campbell et al. (2007) extended these observations to a posttreatment model, PTS was still reduced by approximately 15 dB. With a smaller number of treatments, the effect was reduced to approximately 10 dB (Samson et al. 2008). Although D-methionine does not appear to influence TTS after noise exposures lasting 4–6 h (Kopke et al. 2002; Samson et al. 2008), Cheng et al. (2008) demonstrated that D-methionine reduced TTS in guinea pigs exposed to a more moderate noise insult (broadband noise, 105 dB SPL × 10 min). In that study, TTS immediately post-noise was reduced by approximately 10 dB, and TTS 1-day post-noise was reduced by approximately 5 dB. Taken together, across species, noise insults, and drug delivery times, D-methionine has consistently reduced PTS by approximately 10–20 dB at the frequencies where the greatest damage was observed, with mixed outcomes in TTS models.

The protection provided by D-methione has been smaller than that described for some other agents when the metric used for comparisons is reduction in PTS at most vulnerable frequency (as in Tables 13.1 and 13.4). It is important to note that D-methionine has generally reduced PTS to within 5–10 dB of pre-noise baseline

Table 13.2 D-Methionine for protection against NIHL

	Species	D-Methionine dose	Noise	Threshold shift at frequencies with greatest PTS	Protection (reduction in PTS)
Kopke et al. (2002)	Chinchilla	200 mg/kg i. p., 2×/day × 2 days pre-, 1 h pre- and 1 h post-, then 2×/day × 2 days post-noise	OBN centered at 4 kHz, 105 dB SPL × 6 h	PTS reduced from ~30 dB at 8 kHz to ~10 dB	20 dB
Campbell et al. (2007)	Chinchilla	200 mg/kg i. p., 1 h post-, then 2×/day × 2 days post-	OBN centered at 4 kHz, 105 dB SPL × 6 h	PTS reduced from ~20 dB at 2 and 4 kHz to ~5 dB	15 dB
Samson et al. (2008)	Mouse (C57BL/6)	400 mg/kg i.p., 1 h pre- and 1 h post-	OBN centered at 4 kHz, 110 dB SPL × 4 h	PTS reduced from ~15 dB at 4 and 8 kHz to ~5 dB	10 dB; 8 kHz, N.S.
Cheng et al. (2008)	Guinea pig	300 mg/kg i.p. 1 h pre-noise	broadband noise, 105 dB SPL × 10 min	TTS immed post-noise reduced from ~15 dB to ~4 dB (click); TTS 1 day post-noise reduced from ~8 dB to ~2 dB	NA; no PTS in control subjects

thresholds. Although this suggests relatively complete protection, the control subjects in the D-methionine studies experienced 30 dB PTS deficits, or lower, whereas subjects in studies with other agents have commonly experienced 40–50 dB PTS deficits, or greater. Thus, additional studies are needed to determine whether D-methionine-mediated protection would be as complete if D-methionine were to be tested using a more traumatic noise model comparable to that used to evaluate other protective agents. Data such as these are critical for accurately comparing efficacy across agents.

5.3 Ebselen

After noise exposure, glutathione peroxidase (GPx) levels decrease. With the depletion of GPx, which catalyzes the reduction of free radicals by GSH, endogenous antioxidant defense is compromised. Ebselen is a selenium-containing compound that catalyzes GSSG formation from GSH even more efficiently than GPx (Wendel et al. 1984). It has potent anti-inflammatory effects [attributed to downregulation of tumor necrosis factor-alpha (TNF-α); see Tewari et al. 2009], and it reduces ischemia-reperfusion injury in models of acute lung injury (Hamacher et al. 2009). Other data show that ebselen prevents the mitochondrial membrane permeability transition (MPT) pores from forming or opening, or both (Tak and Park 2009). When opened, the MPT pores can allow glutathione and other molecules to flow out of the mitochondria, reducing the ability to neutralize ROS. There is an increasing body of evidence suggesting that ebselen reduces NIHL (see Table 13.3, which summarizes ebselen-mediated reduction in PTS at the frequency at which the greatest PTS was measured in control subjects). Most of the data to date have shown PTS to be reduced by 10–30 dB at the frequency at which the greatest deficits were observed (Pourbakht and Yamasoba 2003; Lynch et al. 2004; Lynch and Kil 2005; Kil et al. 2007), with a clear increase in protection after optimization of the dosing paradigm. Robust protection has been observed in a TTS model as well (Yamasoba et al. 2005). Most studies evaluating this compound for protection against PTS have used oral administration of the test agent, which should facilitate the translation to human trials.

5.4 Vitamin A and β-Carotene

Vitamin A deficiencies increase NIHL (Biesalski et al. 1990), suggesting the potential for protection against NIHL using vitamin A or its precursors, such as β-carotene. β-Carotene efficiently scavenges singlet oxygen, quenches peroxyl radicals, and prevents lipid peroxidation (for reviews, see Burton et al. 1985; Krinsky 1989, 1998; Schafer et al. 2002; Siems et al. 2005). β-Carotene is metabolized to retinol and retinyl esters (i.e., vitamin A) and stored in the liver. Protection against NIHL with preformed vitamin A supplements has been documented (see Table 13.4, which summarizes reduction in PTS at the frequency at which the greatest PTS was measured in control subjects for a variety of dietary nutrient agents). Both pre-

Table 13.3 Ebselen for protection against NIHL

	Species	Ebselen dose	Noise	Threshold shift at frequencies with greatest PTS	Protection (reduction in PTS)
Pourbakht and Yamasoba (2003)	Guinea pig	10 mg/kg (p.o.) 1 h pre- and 18 h post-	5 h × 125 dB SPL, 4 kHz OBN	PTS reduced from ~55 dB at 4 kHz to ~10 dB	45 dB
		30 mg/kg (p.o.) 1 h pre- and 18 h post-		PTS reduced from ~55 dB at 4 kHz to ~25 dB	30 dB
Lynch et al. (2004)	Rat	16 mg/kg (i.p.) 2×/day 1 day pre-, 1 h pre- and 1 h post-, then 2×/day × 1 day	4 h × 115 dB SPL, 4–16 kHz spectrum	PTS reduced from ~15 dB to ~5 dB (click stimuli)	10 dB
		4 mg/kg (p.o.) 2×/day 1 day pre-, 1 h pre- and 1 h post-, then 2×/day × 1 day		PTS reduced from ~55 dB at 16 kHz to ~40 dB	15 dB
		4 mg/kg (p.o.) 2×/day 1 day pre-, 1 h pre- and 1 h post-, then 2×/day × 1 day, repeated 3 weeks later	4 h × 110 dB SPL, 4–16 kHz spectrum; second exposure 3 weeks later	PTS reduced from ~40 dB at 16 kHz to ~30 dB	10 dB
Yamasoba et al. (2005)	Guinea pig	10 mg/kg (p.o.) 1 h pre-noise	3 h × 115 dB SPL, 4 kHz OBN	TTS immed post-noise reduced ~30–35 dB at 4, 8, and 16 kHz	NA; no PTS in control subjects
Lynch and Kil (2005)	Rat	4 mg/kg (p.o.) 2×/day 1 day pre-, on day of noise, and 1 day post-noise	4 h × 113 dB SPL, 8 kHz OBN	PTS reduced from ~35 dB at 16 kHz to ~25 dB	10 dB
		4 mg/kg (p.o.) 2×/day 1 day pre-, on day of noise, and 12 days post-noise		PTS reduced from ~35 dB at 16 kHz to ~15 dB	20 dB
Kil et al. (2007)	Rat	4 mg/kg (p.o.) 2×/day × 1 day pre-, 1 h pre- and 1 h post-, then 2×/day × 1 day post-noise	4 h × 113 dB SPL, 4–16 kHz spectrum	PTS reduced from ~50 dB at 32 kHz to ~30 dB	20 dB
		4 mg/kg (p.o.): 2×/day × 1 day pre-noise, 1 h pre- and 1 h post-noise, then 2×/day for 12 days post-noise		PTS reduced from ~50 dB at 32 kHz to ~20 dB	30 dB

Table 13.4 Micronutrients for protection against NIHL

	Species	Nutrient dose and form	Noise	Threshold shift at frequencies with greatest PTS	Protection (reduction in PTS)
Vitamin A					
Biesalski et al. (1990)	Guinea pig	100 IU retinyl acetate/day vs. vitamin A deficient diet; ~110 days	White noise; 90 dB A × 15 min	NA; authors report% reduction in CAP amplitude at short post-noise times	NA; Amplitude recovered to 94% of baseline (supplemented) vs. 68% (depleted)
Ahn et al. (2005)	Mouse (BALB/c)	1 mg/kg all-*trans* retinoic acid p.o., 2×/day × 2 days pre- and on the 3 days of noise	white noise (300–10,000 Hz), 122 dB SPL × 3 h/day × 3 days	PTS reduced from ~65 dB to ~25 dB (click)	40 dB
Shim et al. (2009)	Mouse (BALB/c)	1 mg/kg all-*trans* retinoic acid p.o., daily for 5 days; first dose 1 h post-noise	white noise (300–10,000 Hz), 122 dB SPL × 3 h/day × 3 days	PTS reduced from ~59 dB to ~41 dB (click)	20 dB
		1 mg/kg all-*trans* retinoic acid p.o., daily for 5 days; first dose 1 day post-ncise		PTS reduced from ~59 dB to ~40 dB (click)	20 dB
		1 mg/kg all-*trans* retinoic acid p.o., daily for 5 days; first dose 2 days post-noise		PTS reduced from ~59 dB to ~44 dB (click stimuli)	15 dB
		1 mg/kg all-*trans* retinoic acid p.o., daily for 5 days; first dose 3 days post-noise		PTS unchanged; ~59 dB vs. ~56 dB (click stimuli)	5 dB; N.S.
Vitamin C					
Branis and Burda (1988)	Guinea pig	500 mg/kg ascorbic acid, i.p., 48 h, 24 h, and 5 min pre-noise	2 h × 113–118 dB SPL, 1/3 rd OBN centered at 4 kHz	No functional tests conducted; hair cell protection at 50 days post-noise reported	NA
Derekoy et al. (2004)	Rabbit	500 mg/kg ascorbic acid, i.m., 2×/day × 2.5 days pre-noise	1 h × 100 dB SPL × 1,000 Hz	TEOAE preserved in treated subjects	NA

McFadden et al. (2005)	Guinea pig	L-2-Pascorbylolyphosphate, delivered as 5,000 mg/kg chow for 35 days, starting 1 month pre-noise	6 h × 114 dB SPL, 4 kHz OBN	PTS reduced from ~40 dB at 8 and 12 kHz to ~25 dB at 8 kHz and ~20 dB at 12 kHz	15–20 dB
Heinrich et al. (2008)	Guinea pig	25 mg/kg/day ascorbic acid in drinking water	1 h × 90-dB SPL clicks (100 μs/click × 20/s)	PTS unchanged; ~30 dB in controls and treated subjects (click stimuli)	0 dB; N.S.
		525 mg/kg/day ascorbic acid in drinking water		PTS reduced from ~30 dB to ~25 dB (click stimuli)	5 dB; N.S.
Vitamin E					
Hou et al. (2003)	Guinea pig	10 mg/kg α-tocopherol i.p. 1×/day × 3 days pre-noise, 3 days of noise, and 3 days post – noise	OBN centered at 4 kHz, 100 dB SPL × 8 h/day × 3 days	PTS reduced from ~20 dB at 4 kHz to ~10 dB	10 dB
		50 mg/kg α-tocopherol i.p. 1×/day × 3 days pre-noise, 3 days of noise, and 3 days post-noise		PTS reduced from ~20 dB at 4 kHz to ~8 dB	12 dB
Yamashita et al. (2005)	Guinea pig	50 mg/kg trolox i.p. plus 75 mg/kg salicylate s.c. 2×/day × 3 days pre-noise, day of noise, 10 days post-noise	OBN centered at 4 kHz, 120 dB SPL × 5 h	PTS reduced from ~50 dB at 16 kHz to ~20 dB	30 dB
Fetoni et al. (2008)	Guinea pig	100 mg/kg α-tocopherol acetate (1,360 IU) i.m. 1 h pre-noise plus 1×/day × 3 days post-noise	6 kHz × 120 dB SPL × 40 min	PTS reduced from ~45 dB at 8, 12, and 16 kHz to ~10 dB at 8 kHz, ~5 dB at 12 kHz, and ~0 dB at 16 kHz	35–45 dB

(continued)

Table 13.4 (continued)

	Species	Nutrient dose and form	Noise	Threshold shift at frequencies with greatest PTS	Protection (reduction in PTS)
Magnesium					
Ising et al. (1982)	Guinea pig	4.5 mmol/l Mg in drinking water, ad libitum	85 dBA with 80–100 dB pSPL impulses, 12 h/day × 8 weeks; then 95 dBAwith 80–100 dB pSPL impulses, 16 h/day × 4 weeks	PTS reduced from ~34 dB to ~17 dB (click)	17 dB
Joachims et al. (1983)	Rat	130 mmol/kg Mg content added tc ad libitum diet	2-s broadband noise-bursts, 4–16 kHz × 114 dB SPL × 16 h (average interpulse interval 6 s)	PTS reduced from 24 dB to 14 dB	10 dB
Joachims et al. (1993); Attias et al. (1994)	Human	6.7 mmol (167 mg) Mg aspartate 1×/day during training period	M16 firearm training 6 days/week × 8 weeks; ~420 shots per person, mean peak level = 164 dB A	11% of Mg-treated ears had PTS >25 dB; 25% of control ears had PTS >25 dB	NA

Scheibe et al. (2000)	Guinea pig	39 mmol $MgCl_2$/l in drinking water, ad libitum starting 2 weeks pre-noise	Single pistol shot at 187 dB peak	PTS ~5 dB or less in all groups; TTS at 30 kHz reduced from ~52 dB to ~42 dB	NA
			1,000 impulses at 150 dB peak SPL	PTS ~5 dB or less in all groups; TTS at 30 kHz reduced from ~26 dB to ~12 dB	NA
			2,280 impulses at 167 peak SPL	PTS at 7.5 kHz reduced from ~35 dB to ~10 dB; TTS at 15 kHz reduced from ~65 dB to ~48 dB	25 dB
Scheibe et al. (2002)	Guinea pig	1.14 mmol $MgSO_4$/kg s.c. (in 2 injections 2 h apart) × 3 days starting immed post-, plus 39 mmol $MgCl_2$/l water × 1 week	2,280 impulses at 167 peak SPL	PTS at 16 kHz reduced from ~53 dB to ~48 dB	5 dB; N.S
		2.28 mmol $MgSO_4$/kg s.c. (in 2 injections 2 h apart) × 3 days starting immed post-, plus 39 mmol $MgCl_2$/l water × 1 week		PTS at 16 kHz reduced from ~53 dB to ~42 dB	11 dB
		2.85 mmol $MgSO_4$/kg s.c. (in 2 injections 2 h apart) × 3 days starting immed post-, plus 39 mmol $MgCl_2$/l water × 1 week		PTS at 16 kHz reduced from ~53 dB to ~23 dB	30 dB
		2.85 mmol $MgSO_4$/kg s.c. (in 2 injections 2 h apart) × 3 days starting 1 min post-, plus 39 mmol $MgCl_2$/l water × 1 week		PTS at 16 kHz reduced from ~55 dB to ~35 dB	20 dB
		2.85 mmol $MgSO_4$/kg s.c. (in 2 injections 2 h apart) × 3 days starting 2 h post-, plus 39 mmol $MgCl_2$/l water × 1 week		PTS at 16 kHz reduced from ~55 dB to ~45 dB	10 dB; N.S.
		2.85 mmol $MgSO_4$/kg s.c. (in 2 injections 2 h apart) × 3 days starting 4 h post-, plus 39 mmol $MgCl_2$/l water × 1 week		PTS at 16 kHz reduced from ~55 dB to ~43 dB	12 dB; N.S.

(continued)

Table 13.4 (continued)

	Species	Nutrient dose and form	Noise	Threshold shift at frequencies with greatest PTS	Protection (reduction in PTS)
Attias et al. (2003)	Guinea pig	39 mmol $MgCl_2$/l in drinking water, ad libitum starting 2 weeks pre-noise	60 impulses at 167 peak SPL	NA; DPOAE threshold shift of 20 dB at 3 and 4 kHz reduced to 0 dB at 3 kHz and 5 dB at 4 kHz at 50–60 min post-noise	NA; DPOAE threshold shift reduced 15–20 dB at 50–60 min post-noise
Attias et al. (2004)	Human	122 mg Mg, delivered as Mg aspartate	90 dB SL white noise × 10 min	No PTS; TTS at 6 kHz reduced from ~20 dB to ~10 dB	NA
Sendowski et al. (2006)	Guinea pig	350 mg $MgSO_4$/kg s.c. × 3 days starting 1 h post-, plus 3.7 g $MgCl_2$/l water × 1 week	3 blank gunshots from FAMAS F1 rifle; 170 dB peak	Maximum PTS ~20 dB	10 dB; N.S.
			3 blank gunshots from FAMAS F1 rifle; 176 dB peak	Maximum PTS ~40 dB	20 dB; N.S.
Le Prell et al. (2007a)	Guinea pig	2.85 mmol/kg $MgSO_4$ s.c., 1×/day 1 h pre-noise then 1×/day × 4 days post-noise	OBN centered at 4 kHz, 120 dB SPL × 5 h	Maximum PTS ~55 dB	5 dB; N.S.
Abaamrane et al. (2009)	Guinea pig	350 mg $MgSO_4$/kg s.c. × 3 days starting 1 h post-noise, plus 3.7 g $MgCl_2$/l water × 1 week	3 blank gunshots from FA-MAS F1 rifle; 170 dB peak	PTS reduced from ~25 dB at 16 kHz to ~13 dB	12 dB
		350 mg $MgSO_4$/kg s.c. × 3 days starting 1 h post-noise, plus 3.7 g of $MgCl_2$/l water × 1 month		PTS reduced from ~25 dB at 16 kHz to ~10 dB	15 dB

Beta-carotene, vitamins C and E, and magnesium					
Le Prell et al. (2007a)	Guinea pig	2.1 mg/kg beta-carotene, p.o., 71.4 mg/kg L-threoascorbic acid, s.c., 26 mg/kg trolox, s.c.; 2.85 mmol/kg $MgSO_4$ s.c., 1 h pre–noise, then 1×/day × 4 days post-noise	OBN centered at 4 kHz, 120 dB SPL × 5 h	PTS reduced from ~55 dB at 8 kHz to ~20 dB	35 dB
Le Prell et al. (2011a)	Mouse (CBA/J)	Harlan-Teklad Diet TD. 07,215; 1 kg chow contained: 77 mg beta-carotene, 2,250 mg ascorbic acid, 863 α-tocopherol, and 2,656 mg Mg × 28 days pre-noise	8–16 kHz OBN; 113–115 dB SPL × 2 h	PTS reduced from ~50 dB at 20 kHz to ~40 dB; not statistically significant	10 dB; N.S.
		Harlan-Teklad Diet TD. 07,506; 1 kg chow contained: 224 mg beta-carotene, 3,600 mg ascorbic acid, 2,650 α-tocopherol, and 4,500 mg Mg × 28 days pre-noise		PTS reduced from ~50 dB at 20 kHz to ~35 dB	15 dB
Le Prell et al. (2011b)	Guinea pig	2.1 mg/kg beta-carotene, p.o., 71.4 mg/kg L-threoascorbic acid, s.c., 26 mg/kg trolox, s.c.; 2.85 mmol/kg $MgSO_4$ s.c., 1×/day × 1 day pre- and 4 days post-noise	OBN centered at 4 kHz, 110 dB SPL × 4 h	TTS reduced from ~65 dB at 8–32 kHz to ~55 dB; PTS reduced from ~20 dB at 8–32 kHz to ~10 dB	10 dB
Tamir et al. (2010)	Mouse (Sabra)	20 mg/kg retinoic acid, i.p.; 200 mg/kg vitamin C, s.c.; 65 mg/kg α-tocopherol, i.p.; 60 mg/kg Mg s.c; 1×/day × 1 day pre- and 5 days post-noise	Broadband noise, 113 dB SPL × 3.5 h	PTS reduced from ~24 dB (click) to ~15 dB	9 dB

(Ahn et al. 2005), and post- (Shim et al. 2009) noise treatments with vitamin A (delivered as 1 mg/kg all-*trans* retinoic acid) reduced PTS. Treatments starting 2 days pre-noise were the most effective, providing a 40-dB reduction in PTS. When treatment onset was delayed until 1–2 days post-noise, reductions in PTS were smaller (i.e., 15–20 dB). Of particular relevance to the human inner ear, increased serum levels of retinol and provitamin A carotenoids were clearly associated with a decreased prevalence of hearing impairment in a community-based epidemiological study in Japan (Michikawa et al. 2009, although the opposite effect was observed in a population of Australian subjects; see Spankovich et al. 2011).

When sufficient vitamin A stores exist, metabolism to vitamin A ceases and β-carotene circulates in plasma. Dietary supplements increase plasma levels of β-carotene in healthy human subjects (Albanes et al. 1992), and intracellular GSH increases as well, as shown in animals maintained on a β-carotene supplement (Takeda et al. 2008). Increases in intracellular GSH have also been shown in cells cultured with β-carotene (Ben-Dor et al. 2005; Imamura et al. 2006), and some evidence suggests that β-carotene preferentially accumulates in mitochondria (Mayne and Parker 1986). Although there are no published reports confirming a specific role for β-carotene in protecting the inner ear, a combination that includes β-carotene has been used to reduce NIHL in guinea pigs (Le Prell et al. 2007a) and mice (Le Prell et al. 2011a). The role of β-carotene in protecting the guinea pig inner ear is somewhat unclear, as β-carotene could not be detected in plasma samples from guinea pig subjects treated with the same dose of agents when vitamin levels were explicitly measured in a second, later, study (Le Prell et al. 2011b).

Some care does need to be taken in distinguishing vitamin A precursors (the carotenoids) and preformed vitamin A (retinol, retinoic acid). Carotenoids are not toxic in animals or humans, and are nonteratogenic even at high doses in animals (for review, see Dolk et al. 1999). Carotenoids are clearly distinguished from preformed vitamin A, as high levels of preformed vitamin A (>10,000 IU/day) have been reported to increase the risk of birth defects (Rothman et al. 1995). Recent recommendations suggest pregnant women should not consume more than 5,000 IU/day from vitamin A supplements (for review, see Dolk et al. 1999). These recommendations are in distinct contrast to those for β-carotene, which has not been implicated in teratogenesis at any dose (see Miller et al. 1998). High-level supplements are also questionable for those with a long-term history of cigarette smoking or current cigarette smokers given that high-level β-carotene supplements have been linked to an increased risk of lung cancer in high-quality, randomized, placebo-controlled trials (for brief review, see Le Prell et al. 2011b).

5.5 Vitamin C

All mammals, except for fruit bats, guinea pigs, monkeys, and humans, synthesize their own endogenous vitamin C (Chatterjee 1973; Chatterjee et al. 1975; Birney et al. 1976). Vitamin C treatment increases intracellular GSH (Harapanhalli et al. 1996; Jagetia et al. 2003; Derekoy et al. 2004) and, specifically, mitochondrial GSH (Jain et al. 1992).

GPx and SOD levels (Jagetia et al. 2003), and catalase enzyme levels (Derekoy et al. 2004), also increase with vitamin C supplements. In addition to enhancing endogenous defense against oxidative stress, vitamin C directly reduces free radicals (for review, see Evans and Halliwell 1999). Vitamin C is a "preferred" antioxidant in that when vitamin C donates electrons to quench harmful free radicals, the oxidized vitamin C byproducts (ascorbyl radical and dehydroascrobic acid) are relatively unreactive free radicals, and they can be reduced back to ascorbic acid by GSH, or via at least three different enzyme pathways (for review see Padayatty et al. 2003). Vitamin C scavenges superoxide, singlet oxygen, and hydroxyl radicals (Bendich et al. 1986; for review see Sauberlich 1994). Scavenging of oxygen radicals by vitamin C occurs in the aqueous phase; that is, in cellular cytoplasm rather than within the lipid membranes (Niki 1987a, b). Although not found in the cell membranes, vitamin C enhances recycling of α-tocopherol, which then moves into the lipid membranes, and thus vitamin C does help prevent lipid peroxidation (Sato et al. 1990; Niki 1991; Chan 1993; for review see Sauberlich 1994). Vitamin C reduces noise-induced malondialdehyde formation (Derekoy et al. 2004) and reduces PTS by up to 20 dB (McFadden et al. 2005); in addition, it reduces cisplatin-induced ototoxicity and hearing loss (Lopez-Gonzalez et al. 2000; Bertolaso et al. 2001; Weijl et al. 2004). A combination that includes vitamin C has been used to reduce NIHL in guinea pigs (Le Prell et al. 2007a) and mice (Le Prell et al. 2011a), and increased intake of vitamin C has been linked to improved hearing outcomes in a human population (Spankovich et al. 2011).

5.6 Vitamin E

Just as vitamin C is widely considered to be the most important water-soluble antioxidant, vitamin E is widely considered to be the most important fat-soluble antioxidant (Burton et al. 1985). Vitamin E is a generic term used to capture the tocopherol family; α-tocopherol is the most biologically active antioxidant (for review see Kappus and Diplock 1992). Vitamin E is lipophilic, is found in cell membranes, and it prevents lipid peroxidation by scavenging lipid peroxyl radicals (see Burton et al. 1983). Vitamin E directly regulates mitochondrial generation of superoxide and hydrogen peroxide; it may do so by preventing electron leakage or mediating the superoxide generation systems directly, or it may act by directly scavenging superoxide as it is generated (Chow et al. 1999; Chow 2001). Vitamin E is a donor antioxidant; it donates electrons to lipid peroxyl radicals resulting in formation of less toxic lipid hydroperoxide and a vitamin E radical that can then be reduced back to vitamin E by either vitamin C or by GSH (for reviews see Burton et al. 1985; Rezk et al. 2004). By reacting with and reducing peroxyl radicals, vitamin E inhibits the propagation of lipid peroxidation (for review, see Schafer et al. 2002). Vitamin E (delivered as synthetic vitamin E, Trolox, or α-tocopherol) reduces NIHL (Rabinowitz et al. 2002; Hou et al. 2003; Yamashita et al. 2005), as well as cisplatin-ototoxicity (Lopez-Gonzalez et al. 2000; Teranishi et al. 2001; Kalkanis et al. 2004). Protection is dose dependent (as shown in Table 13.4) with higher doses (i.e., 100 mg/kg) providing up to 45 dB protection against PTS. In the study by Hou et al. (2003), vitamin E doses of

50 mg/kg also reduced TTS by approximately 8 dB at the frequency where the greatest deficits were detected. A combination that includes vitamin E (in the form of Trolox, a water-soluble form of vitamin E) has been used to reduce NIHL in guinea pigs (Le Prell et al. 2007a), and mice, with mice fed a supplemented chow containing increased levels of or α-tocopherol (Le Prell et al. 2011a). Increased intake of vitamin E has also been linked to improved hearing outcomes in humans (Spankovich et al. 2011).

5.7 *Magnesium*

Magnesium supplements reduce NIHL in humans and animals (see Table 13.4). Protection is dose dependent, and treatments that begin shortly after the noise insult are more effective than those initiated at longer post-noise intervals (Scheibe et al. 2002). Treatments lasting longer than 7 days post-noise may be the most effective treatment, at least after impulse noise insult (Sendowski et al. 2006; Abaamrane et al. 2009). In contrast to the aforementioned positive outcomes, Walden et al. (2000) reported naturally occurring magnesium did not reliably correlate with NIHL in a population of male U.S. Army soldiers from a single combat unit with long-term (8–18 years) exposure to high-level weapons noise. Although normal dietary levels of magnesium did not meaningfully influence NIHL in that population, this does not preclude a reliable relationship when magnesium consumption is supplemented. Magnesium levels in the cochlear perilymph clearly vary with high-level dietary magnesium supplement (Joachims et al. 1983; Scheibe et al. 1999; Attias et al. 2003).

The effects of magnesium on NIHL have been suggested to be a consequence of effects of magnesium on blood flow. Magnesium prevents noise-induced decreases in cochlear blood flow (Haupt and Scheibe 2002). Unlike most tissues, in which increased metabolism is associated with increased blood flow, high levels of noise decrease blood flow to the inner ear (for review, see Le Prell et al. 2007b). Reduced cochlear blood flow has significant implications for metabolic homeostasis in the cochlea, as cellular metabolism clearly depends on adequate O_2 and nutrients as well as elimination of waste products (e.g., Miller et al. 1996). In addition to well-documented vasodilating properties, magnesium modulates calcium channel permeability, influx of calcium into cochlear hair cells, and glutamate release (Gunther et al. 1989; Cevette et al. 2003). Both increased activity at glutamate receptors (either after noise or during infusion of glutamate receptor agonists) and deficits in calcium homeostasis have been linked to hearing loss (for reviews see Le Prell et al. 2001, 2004, 2007b). Magnesium is also a *N*-methyl-D-aspartate (NMDA)-receptor antagonist. The fact that the NMDA-receptor antagonist MK-801 reduces the effects of noise (Duan et al. 2000; Ohinata et al. 2003), ischemia (Konig et al. 2003), and excitotoxic (Janssen 1992) or ototoxic drugs (Basile et al. 1996; Duan et al. 2000) suggests another potential protective mechanism for magnesium. Finally, magnesium is increasingly considered to directly mediate both oxidative stress and DNA repair (Wolf et al. 2007; Wolf and Trapani 2008; Wolf et al. 2008, 2009). Regardless of the specific mechanism of action, magnesium supplements clearly attenuate NIHL.

5.8 *Nutrient Combination: β-Carotene, Vitamins C and E, and Magnesium*

Synergistic protective effects of combinations of ascorbic acid and α-tocopherol, β-carotene and α-tocopherol, and β-carotene and ascorbic acid have been well described outside of the auditory system (see, e.g., Yeum et al. 2009). A synergistic interaction between several vitamins (β-carotene, and vitamins C and E) and magnesium was explicitly shown in the inner ear by Le Prell et al. (2007a). Although neither the vitamin combination nor the magnesium alone conferred protection against NIHL, the combination of the vitamins with magnesium resulted in robust (30–35 dB) reductions in NIHL in guinea pigs (see Fig. 13.2a, b), although lesser protection was recently reported by Tamir et al. (2010), who used mice as subjects. The reductions in PTS obtained with this micronutrient combination, combined with the extensive safety data for vitamins (Age-Related Eye Disease Study Research Group 2001), relatively low cost of dietary supplements, and widespread availability of dietary supplements serve to make this an appealing therapeutic intervention, and new data are now available from several additional studies.

In the first of two recent studies, protection against NIHL was approximately 15 dB in mice treated with a dietary nutrient formulation (see Fig. 13.2c, d). Protection was dose dependent (Le Prell et al. 2011a), suggesting the possibility that increasing the levels of the vitamins in the custom dietary formulation would provide increased protection. This is an empirical issue, and additional data are critical. In the second investigation, there was evidence for reductions in TTS measured 1 day post-noise (see Fig. 13.2e), as well as reduction of PTS measured 7 days post-noise (see Fig. 13.2f) after a more moderate noise exposure (Le Prell et al. 2011b). Guinea pigs in the control group had an approximately 20 dB PTS, much like chinchillas in the control group for the D-methionine studies; nutrient-treated animals had thresholds within 5–10 dB of baseline, much like D-methionine treated chinchillas described in Sect. 5.2. As noted previously, there is an urgent need for systematic comparisons within the same species, using the same noise insult, to allow direct comparisons across agents.

6 Other Therapeutics

6.1 *Pancaspase Inhibitors*

Free radical scavengers that reduce oxidative stress events intervene in early cell death events and prevent the initiation of later apoptotic cascades leading to cell death. In contrast, pancaspase inhibitors are agents that intervene in the final stages of cell death to prevent "executioner" activities of the caspases. Oxidative stress

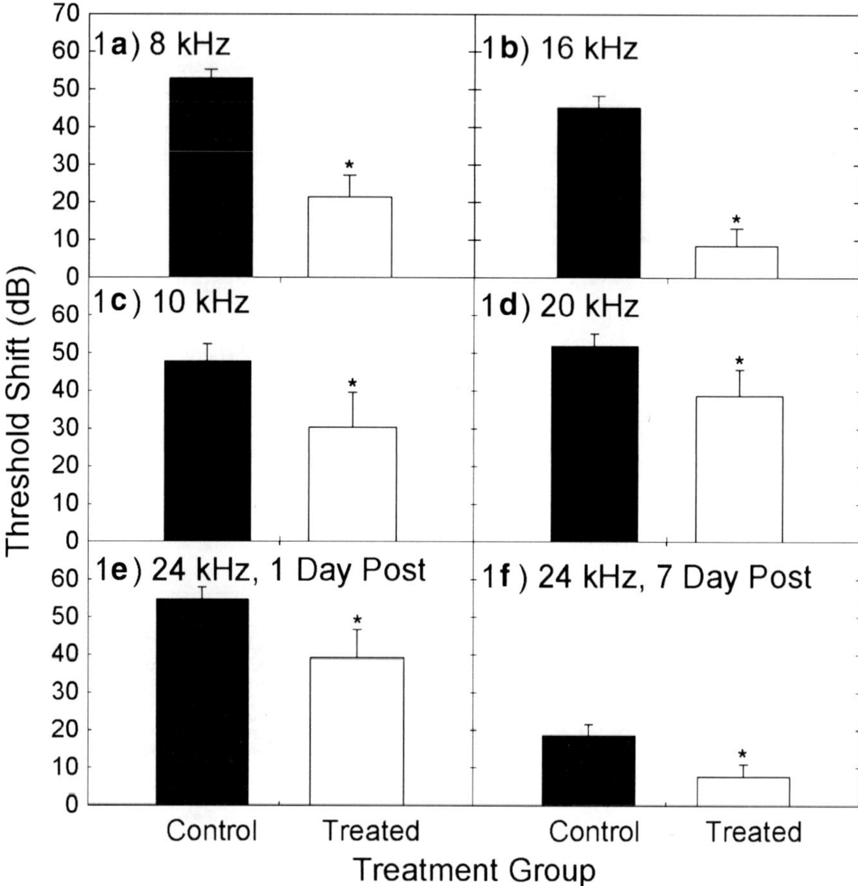

Fig. 13.2 Combination therapy, with β-carotene, vitamins C and E, and magnesium, was highly effective in preventing permanent noise-induced threshold shift (PTS) in guinea pigs exposed to OBN centered at 4 kHz for 5 h at a level of 120 dB SPL (**a**, **b**; see Le Prell et al. 2007a). Vitamin E was delivered in the form of Trolox, a water-soluble analogue of vitamin E, and all active agents were injected starting 1 h before noise insult. This combination of agents reduced PTS to a lesser extent in mice that were fed a supplemented diet containing increased levels of the active agents; mice were maintained on nutritionally complete control diet or supplemented diet for 28 days pre-noise, with noise being an 8–16 kHz OBN for 2 h at a level of 113–115 dB SPL (**c**, **d**; see Le Prell et al. 2011a). Finally, using the same dose paradigm as in Le Prell et al. (2007a) but a shorter, less intense noise insult (OBN centered at 4 kHz for 4 h at a level of 110 dB SPL), this combination of agents reduced acute threshold shift in the first 24 h post-noise (**e**) as well as 7-days post post-noise (**f**) (see Le Prell et al. 2011b). All data are means ± SEM. Threshold shift is shown at the frequencies at which control animals had the greatest threshold deficits; pattern of threshold shift across frequency was similar in treated animals and controls; that is, if control animals had the greatest PTS at 8 kHz, the treated animals did as well. Study details are summarized in Table 13.4

driven via the mitochondrial energy pathways can ultimately lead to cell death in the inner ear via translocation of apoptosome inducing factor (AIF) or endonuclease G (EndoG) from the mitochondria into the cell nucleus, or, it can lead to cell death subsequent to caspase activation (for review, see Le Prell et al. 2007b). Caspase-mediated cell death can be driven not only by these mitochondria-mediated pathways (often termed intrinsic pathways), but also by a receptor-activated (extrinsic) pathway. Caspases-1, -3, -8, and -9, are activated in the inner ear after noise exposure or other stressors, with expression observed in hair cells, spiral ganglion cells, stria vascularis and spiral ligament, and lateral wall (for reviews, see Cheng et al. 2005; Eshraghi and Van De Water 2006; Le Prell et al. 2007b). Evidence that caspases-5, -6, -7 and -10 are involved in apoptosis in the inner ear is emerging (Eshraghi and Van De Water 2006), and there is now preliminary evidence that caspase-2 is also activated by noise (Le Prell et al. 2010b). Caspase inhibitors have been shown to prevent cell death in the inner ear after aminoglycoside treatment (Nakagawa et al. 2003; Corbacella et al. 2004; Okuda et al. 2005; Wei et al. 2005), chemotherapeutics (Zhang et al. 2003), and hypoxia and neurotrophic factor (NTF) withdrawal (Cheng et al. 1999), and despite the need for phase I safety data, this class of agents has been suggested as promising for novel NIHL therapeutics (Le Prell et al. 2007b; Abi-Hachem et al. 2010). Caspase inhibitors have been proposed as novel antiapoptotic drug therapies based on animal experiments in other models as well (Canbay et al. 2004; Faubel and Edelstein 2005; Park et al. 2006; Colak et al. 2009). Conduct of a successful human clinical trial with oral dosing of a pan-caspase inhibitor in patients with chronic hepatitis C (Shiffman et al. 2010) highlights the need to evaluate these agents within a model of NIHL. The remainder of this section discusses different potential therapeutic agents that act at intermediate points in the pathways to cell death.

6.2 *Calcium Channel Blockers*

The classic evidence for noise-induced neural "excitotoxicity" is the swelling of auditory neurons observed after noise exposure (Robertson 1983; Puel et al. 1998; Yamasoba et al. 2005). The swelling has been attributed to glutamate-induced excitotoxicity, given that application of glutamate (Janssen et al. 1991) or glutamate agonists (Puel et al. 1991, 1994; Le Prell et al. 2004) results in similar neural swelling and comparable temporary functional deficits. In brief, glutamate is released from the inner hair cells (IHCs) in response to sound, and toxic concentrations of this excitatory amino acid can accumulate in the synapse between the IHCs and auditory nerve dendrites post-noise. This increases sodium and potassium ion flux across the post-synaptic membranes, drives calcium (Ca^{2+}) influx, and increases the passive entry of chlorine. The osmotic imbalance results in entry of fluid into cells, which produces the observed swelling, and ultimately, rupturing of dendritic membranes followed by neural degeneration (for review, see Le Prell et al. 2001).

In addition to the classic "excitotoxic" pathway, a second Ca^{2+}-dependent *oxidative cell death* pathway has now been well characterized in neural systems. Death of neuronal cells (in multiple biological systems) via this oxidative pathway is attributed to nitric oxide synthase (NOS) production and RNS formation after the Ca^{2+} influx (for review, see Le Prell et al. 2007b).

Fessenden and Schacht (1998) proposed a model in which Ca^{2+} entry activates neuronal NOS in auditory neurons, thereby driving NO production, which reacts with superoxide to form the highly toxic peroxynitrite radical, and ultimately results in neural degeneration. Adam-Vizi and Starkov (2010) recently reviewed the evidence that Ca^{2+} directly results in oxidative stress; they described multiple demonstrations that an overaccumulation of Ca^{2+} can drive ROS production. If ROS and RNS species significantly contribute to neural trauma, then free radical scavengers might act not only to preserve hair cells, but also to preserve auditory neurons. Consistent with this, ebselen, which is an efficient scavenger of peroxynitrite as well as a glutathione peroxidase mimic, prevented noise-induced swelling of the auditory nerve dendrites (Yamasoba et al. 2005). Vitamins E and C also effectively preserve auditory nerve populations, an effect observed after drug insult (Maruyama et al. 2007). Although neural protection after noise insult has not been directly demonstrated using combination nutrient therapy, decreased TTS at short post-noise intervals (such as that shown in Fig. 13.2e) is consistent with neural protection.

Noise exposure increases Ca^{2+} concentration not only in afferent dendrites, but also in hair cells (Fridberger and Ulfendahl 1996; Fridberger et al. 1998). The mechanisms through which changes in Ca^{2+} concentration may lead to cell death have been well reviewed (Szydlowska and Tymianski 2010). In brief, Ca^{2+} influx and accumulation can influence calpains, phospholipase A_2, caspases, calmodulin-dependent protein kinase (CaMK), endonucleases, and ion transporters, with effects including (but not limited to) mitochondrial damage cell membrane disruption, cytoskeletal breakdown, free radical production, NO production, DNA fragmentation, and swelling (see Fig. 13.3 in Szydlowska and Tymianski 2010 for excellent overview). These elevations in intracellular calcium have been specifically implicated in hair cell damage and hearing loss post-noise, as deficits induced by noise (Heinrich et al. 1999, 2005), chemotherapeutics (So et al. 2005), or H_2O_2 (Dehne et al. 2000), can be blocked by calcium channel blockers.

Because Ca^{2+} deregulation is a widely recognized contributor to neuronal injury, the potential of Ca^{2+} channel blockers to reduce NIHL has been specifically examined. There are at least five types of Ca^{2+} channels (L-, N-, P/Q-, R-, and T-type). They have different pharmacological profiles, but can be broadly classified into low- and high-voltage activated channels, with the L-type and T-type channels perhaps being the best characterized channels (for review see Kisilevsky and Zamponi 2008; Shin et al. 2008; Suzuki et al. 2010). Initial studies focused on agents that block the L-type Ca^{2+} channels, given that the voltage-dependent Ca^{2+} channels in mammalian OHCs have been generally considered to possess the properties of L-type currents (see Inagaki et al. 2008). Early results were disappointing, however. Both diltiazem and a second L-type voltage-dependent channel blocker, nimodipine, failed to prevent TTS, PTS, or noise-induced depression of DPOAE amplitude (Boettcher 1996; Ison et al. 1997; Boettcher et al. 1998). Surprisingly, diltiazem was subsequently shown to protect OHC ultrastructure, an effect that was presumably mediated by the drug-induced decrease in the amount of Ca^{2+} measured in OHCs after noise insult (Heinrich et al. 1999; Maurer et al. 1999). Since then, four L-type Ca^{2+} channel blockers, including diltiazem and nimodipine, as well as verapamil and nicardipine, were shown to reduce both OHC loss and PTS induced by noise (Uemaetomari et al. 2009). Protection was dose dependent, which may explain the various contradictory outcomes to date with L-type Ca^{2+} channel blockers.

Although L-type Ca^{2+} channels were initially the focus of auditory research investigations, the low resting potential of OHCs and their slight depolarization upon sound stimulation have recently been taken as evidence that the low voltage-activated T-type channels may contribute to Ca^{2+} regulation in OHCs (Inagaki et al. 2008). Indeed, these channels were expressed at the mRNA and protein levels, and OHC current was blocked by the T-type-specific antagonist mibefradil (Inagaki et al. 2008). Newer studies have therefore focused on two T-type Ca^{2+} blockers, ethosuximide and trimethadione (Shen et al. 2007). Both of these drugs significantly reduced PTS, and increased OHC survival in the hook region of the mouse cochlea, perhaps accounting for the observed protection (Shen et al. 2007). In contrast, neither mibefradil or flunarizine, two other T-type Ca^{2+} blockers, decreased noise-induced OHC death or PTS (Uemaetomari et al. 2009). The divergent outcomes with the T-type Ca^{2+} blockers may reflect drug differences; however, experimental paradigms differed significantly as well. The T-type Ca^{2+} blockers that were effective in the studies by Shen et al. (2007) were delivered for longer pre-noise periods than those used by Uemaetomari et al. (2009). In addition, the noise insult used by Shen et al. (2007) was shorter (30 min vs. 4 h) and less intense (110 dB SPL vs. 128 dB SPL) than that used by Uemaetomari et al. (2009). It is possible that these agents reduce the effects of less traumatic insult, with less protective benefit against the more traumatic insults; however, it is also possible that longer pretreatment paradigms would provide more effective protection against the more traumatic noise insults. As discussed in earlier sections of this chapter, studies that directly compare agents using identical dose paradigms and noise insults are essential for comparisons across agents. Further studies are needed to clarify these issues, such that the most effective drugs can be tested in clinical trials to determine the potential for reduction in human NIHL.

6.3 JNK Inhibitors

The c-Jun N-terminal kinase (JNK) group of mitogen-activated protein (MAP) kinases phosphorylate the transcription factor c-*Jun* (Kyriakis et al. 1994). Inactive JNK is normally located in cytoplasm. When activated by noise (or other stressors), JNK translocates either into the cell nucleus or the cell mitochondria. In the nucleus, it induces expression of various transcription factors that can promote cell survival or cell death. In the mitochondria, it activates second mitochondria-derived activator of caspase

(Smac) which then activates caspases that ultimately lead to cell death. The specific outcomes are noise dependent, and may reflect either cell survival or cell death pathways, depending on the noise trauma (Selivanova et al. 2007; Murai et al. 2008). JNK activation does not occur within the first 30 min subsequent to exposure to TTS-inducing noise, although activation at later times was detected, which may reflect activation of cell survival pathways during recovery (Meltser et al. 2009, 2010). After PTS noise insult, JNK is activated in cochlear tissue (as shown by Western blot of whole cochlea; see Meltser et al. 2010), and this activation specifically results in release of cytochrome *c* from the OHC mitochondria (Wang et al. 2007). Release of cytochrome *c* allows formation of apoptosomes, an early event in apoptotic cell death (for review, see Le Prell et al. 2007b).

Treatment with the JNK-inhibitor CEP-1,347 reduces noise-induced cell death and NIHL (Pirvola et al. 2000). Similar results were obtained with D-JNKI-1; this cell permeable peptide blocks the MAPK-JNK signal pathway and reduces noise-induced toxicities (Wang et al. 2003). After preliminary results suggested that D-JNKI-1 may be effective even when treatment is delayed several hours relative to noise insult (Guitton et al. 2004), AM-111, a proprietary formulation of D-JNK-1, was shown to reduce NIHL in chinchillas with treatments delayed 1–4 h post-noise (Coleman et al. 2007b). Earlier treatment was more effective, but PTS was reduced by approximately 10–12 dB even with the longer delay. AM-111 was maximally effective when delivered directly onto the round window using a gel formulation (~22–25 reductions in PTS; see Coleman et al. 2007b). When the gel formulation was tested in 11 human patients exposed to impulse noise (firecracker-induced acoustic trauma in Berlin and Munich on New Year's Eve in 2005/2006), an average improvement of 11 dB was measured post-trauma (Suckfuell et al. 2007). Those outcomes are challenging to interpret, however, given the small sample size, the lack of a placebo-treated control group, and multiple adverse event reports. Most of the adverse events were reports of tinnitus, which might be due to either noise exposure or the inner ear manipulation. Blood crusts were also observed on the tympanic membrane 3 days postinjection in two patients. Although these resolved, these observations highlight some of the challenges associated with any intratympanic therapy.

6.4 Src Kinase Inhibitors

One form of noise-induced mechanical stress is tearing or other stressing of the cell junctions (for detailed description, see Hu, Chap. 5). Importantly, oxidative stress can activate Src kinase, and because Src activation disrupts actin filament organization and cell junctions, oxidative stress can in fact *induce* mechanical changes (Chan et al. 2010; Chou et al. 2010). The molecular changes subsequent to ROS-mediated activation of Src kinase include (1) increased tyrosine phosphorylation of p120-catenin, (2) rapid translocation of p120-catenin with cadherin to the cell–cell adhesion sites, (3) activation of the Rho/Rho kinase pathway, (4) dissociation of the cell–cell contacts, and (5) cytoskeletal remodeling (Inumaru et al. 2009). Src inhibitors have thus been suggested as protective agents for prevention of apoptotic cell death triggered by mechanical stress (see Bielefeld et al. 2005).

Because the Src signaling pathway may be involved in the initiation of apoptosis after mechanical stress, as well as the generation of superoxide and other downstream ROS activations associated with metabolic stress, Src inhibitors have been suggested to be potentially more effective than other free radical scavengers when used as protective agents (see Bielefeld et al. 2005). The Src inhibitor KX1-004 reduced susceptibility to NIHL when applied to the round window membrane before noise (Harris et al. 2005), and also when injected systemically (Bielefeld et al. 2005). Using systemic treatments, 50 mg/kg of KX1-004 was generally equivalent to 325 mg/kg of NAC, with both agents providing approximately 12-dB reductions in NIHL at 4 kHz (the frequency at which the greatest deficits were observed; see Table 13.1). The low dose KX1-004 therapy was as effective as NAC despite the lower dose; however, the total protection provided by the two agents was identical and it is not clear if protection could be increased with higher doses of this Src inhibitor, or other novel Src inhibitors not yet tested in the inner ear (Bielefeld et al. 2005).

7 Potential to Influence Comorbidities

7.1 Tinnitus

Tinnitus, which can be transient or permanent, is experienced by persons exposed to impulse noise, such as soldiers. Tinnitus after automatic weapons fire or other traumatic military noise exposure has been characterized in soldiers from multiple countries. Although many other countries could be listed, we note here the compelling data from personnel in Finland (Ylikoski and Ylikoski 1994; Mrena et al. 2002, 2004), Israel (Melinek et al. 1976; Nageris et al. 2008, 2010), and most recently, the United States (Cave et al. 2007; Lew et al. 2007; Saunders and Griest 2009). Similar tinnitus issues have been reported for children using cap guns or firecrackers (Segal et al. 2003). Less impulsive noise exposure is also problematic; tinnitus is common in employees (Gunderson et al. 1997; Lee 1999; Bray et al. 2004) and patrons (Tin and Lim 2000; Emmerich et al. 2002) at nightclubs/discotheques, as well as in music festival attendees (Mercier et al. 2003). Finally, tinnitus is common in individuals experiencing hearing loss due to occupational noise exposure. In some cases, tinnitus has been specifically studied in workers exposed to impulsive noises, such as forge hammering, drop forges, and steel foundry work, with tinnitus reported at rates up to 70% (Kamal et al. 1989; Griest and Bishop 1998; Sulkowski et al. 1999). When workers are selected solely on the basis of compensation claims for occupational NIHL, tinnitus is reported at rates ranging from 23% to 58%, with tinnitus being bilateral even when hearing loss is unilateral, and with tinnitus most likely to be perceived as a higher frequency sensation (Alberti 1987; van Dijk et al. 1987; Barrs et al. 1994). In one study, tinnitus was reported to be severe enough to interfere with daily life by 30% of the population with tinnitus (Phoon et al. 1993). Tinnitus accompanied higher frequency hearing loss in airport workers, with those who had low-frequency hearing loss having a significantly lower incidence of tinnitus (Hong et al. 1998). In a group of miners exposed not only to noise, but also to vibrations, dust, and other toxins, tinnitus was accompanied by vertigo, headaches, and disturbed balance (Tzaneva et al. 2000). Taken together, tinnitus is a major comorbity accompanying NIHL.

Given the prevalence of noise-induced tinnitus in human patient populations, tinnitus has been well studied in laboratory settings. For example, in a laboratory study with human subjects, a 5-min exposure to a 110-dB SPL white noise produced transient tinnitus (Chermak and Dengerink 1987). In other investigations, animals trained to perform a behavioral response indicating the perception of sound (or silence) performed, after intense sound exposure, responses indicating that they heard sound even though the acoustic test conditions were quiet (Bauer and Brozoski 2001; Heffner and Harrington 2002). Studies with trained animal subjects are challenging, as it is difficult for the investigator to know if the animal is making an incorrect "guess" or a "correct" response reporting tinnitus (for review, see Moody 2004). One of the most elegant paradigms for measuring tinnitus in animal subjects has only recently emerged; this paradigm is a modified prepulse inhibition and gap detection paradigm used by Turner and colleagues (Turner et al. 2006; Turner 2007; Bauer et al. 2008; Turner and Parrish 2008). Presenting a brief loud sound without warning induces a reflexive "startle" response; preceding that startle stimulus with a suprathreshold sound reduces the subjects' response to the startle stimulus. If the startle sound is presented against a quiet noise background, then, a silent gap in the noise background can also serve as a warning signal that reduces the subsequent response to the startle stimulus. A subject that has tinnitus cannot detect the silent gap and thus the startle is not attenuated in the subject that has tinnitus. This paradigm provided a major advance in that investigators have an empirical measure of tinnitus sensations in animal subjects, without trying to maintain operant responses that depend on a priori knowledge of whether the subject does in fact have tinnitus or not.

The specific mechanism through which noise exposure produces tinnitus is not known; multiple possibilities are reviewed by Kaltenbach (Chap. 8). One possibility is that changes in OHC function might "trigger" changes in activity in the dorsal cochlear nucleus and hence tinnitus (Kaltenbach et al. 2002). If OHC disruptions are in fact one trigger for tinnitus, then protection of OHCs from trauma associated with noise exposure may prevent tinnitus. Consistent with the hypothesis that interventions that protect against hair cell death may reduce tinnitus, some people have taken minerals such as magnesium or zinc, herbal preparations such as Ginkgo biloba, homeopathic remedies, or B vitamins for their tinnitus and found them to be helpful (Schneider et al. 2000). The reduction in tinnitus with Ginkgo biloba may be a consequence of its antioxidant properties (for review, see Diamond et al. 2000). Alternatively, potential changes in tinnitus associated with Ginko biloba may be related to the increase in cochlear blood flow associated with long-term (4–6-week) oral administration of this substance (Didier et al. 1996). If the antioxidant or vasodilating effects of Ginkgo biloba are shown to reduce tinnitus in controlled scientific studies, it may be reasonable to predict that other free radical scavengers and vasodilators might also have the potential to reduce tinnitus in humans.

7.2 Vestibular Deficits

Data from the 2001–2004 National Health and Nutrition Examination Surveys reveal that 35% of U.S. adults ages 40 years and older (69 million Americans) had vestibular dysfunction (Agrawal et al. 2009). Some vestibular dysfunction may be related to prior noise insult. Unilateral deafness can cause vertigo (Kemink and Graham 1985), and noise exposure resulting in NIHL can induce vestibular effects. Exposure to impulse noise, for example, can increase body sway (Juntunen et al. 1987; Ylikoski et al. 1988; Kilburn et al. 1992), reduce the vestibular-ocular reflex (VOR) (Shupak et al. 1994), or induce a nystagmus (Oosterveld et al. 1982). Vestibular deficits in these investigations have been subclinical; that is, patients do not present with complaints of vestibular dysfunction. Instead, it is only through careful quantification of vestibular reflexes and postural control that deficits are revealed. Thus, there are suggestions that all patients presenting with significant NIHL should be evaluated for vestibular deficits (van der Laan 2001a, b). One potential test of interest is the vestibular evoked myogenic potential (VEMP). As average pure tone hearing thresholds increased in patients with NIHL, VEMP latencies were prolonged and peak-to-peak amplitude was reduced; VEMP was abnormal or absent in 67% of all NIHL subjects tested (Kumar et al. 2010).

There is some evidence suggesting an intensity-dependent effect of noise on vestibular function. Miners with hearing loss and tinnitus are highly prone to vertigo and disturbed balance (Tzaneva et al. 2000), and survivors of blast trauma, which involves intense acoustic overstimulation, also report specific vestibular deficits (Shupak et al. 1993; Van Campen et al. 1999). Vestibular deficits, which may be a function of peripheral vestibular system lesions or central changes, are a significant issue for veterans exposed to blast trauma and traumatic brain injuries (Fausti et al. 2009). Animal investigations support the proposed relationship between intensity of noise exposure and extent of vestibular disruption (Sohmer et al. 1999; Perez et al. 2002; Hsu et al. 2008). Studies of human patients similarly reveal greater vestibular-evoked myogenic potential (VEMP) test deficits with more intense noise exposures (Wang et al. 2006; Wang and Young 2007).

Published data clearly support the potential for a shared mechanism of noise-induced damage to cochlear and vestibular structures. After noise exposure, capillary constriction is observed in both the cochlear and vestibular tissue of rats (especially the cristae, with some effect also in the saccular tissues, see Lipscomb and Roettger 1973). In addition, after noise exposure, inducible nitric oxide synthase (iNOS) expression is evident not only in the sensory epithelium of the cochlea (i.e., hair cells, supporting cells), but also in the vestibular ganglion (Watanabe et al. 2004), suggesting free radical involvement in both auditory and vestibular disruption. The recent demonstration of 4-hydroxy-2-nonenal labeling of both the cochlea and the vestibule, indicating ROS formation post-noise in noise-exposed guinea pigs, confirms oxidative stress in the vestibular system after noise insult (Fetoni et al. 2009a). Antioxidant therapy can reduce vestibular deficits induced by ototoxic aminoglycoside treatment (Usami et al. 1996; Song et al. 1998; Schacht 1999; Sergi et al. 2004) or cisplatin injections (Cheng et al. 2006). Prevention of noise-induced changes in vestibular function has not been reported, however.

8 Translation from Bench to Bedside

A number of antioxidant agents described in the preceding sections have been awarded patent protection as novel therapeutics that protect or preserve hearing; there are others for which patent applications are pending. For example, *N*-acetylcysteine is the active agent in The Hearing Pill; this proprietary supplement is already being sold online by American BioHealth Group (Kopke et al. 2003). Ebselen is the active agent in a formulation identified as SPI-1,005, a proprietary oral form of a glutathione peroxidase mimic, under development by Sound Pharmaceuticals (Kil and Lynch 2010). A proprietary oral formulation of D-methionine (MRX-1,024) developed by Molecular Therapeutics is being used in human clinical trials (Campbell 2001, 2008). Finally, a proprietary formulation of β-carotene, vitamins C and E, and magnesium (Auraquell®) was developed by OtoMedicine Inc., and Hearing Health Sciences Inc. is now developing a mint-like product named “Soundbites”; that compound was recently awarded a notice of allowance of claims from the USPTO (Miller et al. 2010).[1] Other agents that act to reduce NIHL or noise-induced tinnitus are also being evaluated for potential efficacy in human subjects, and some of these are also under development as commercial agents. For example, AM-111 is a patented JNK-inhibitor being developed by Auris Medical for use in the inner ear in human patients (Bonny 2009). Patents are also pending for succinimide and its derivates (Bao 2007),[2] and patents have been awarded for glutamate receptor antagonists (Puel et al. 2009) and pancaspase inhibitors (Staecker 2008), all for the purpose of preventing NIHL and noise-induced tinnitus.

8.1 Ethical Challenges

The ethical challenges in designing and conducting human trials, without putting research subjects at risk of permanent damage to their own hearing, are clear. One strategy for meeting this challenge has been to draw upon military populations that

[1]Colleen Le Prell is a coinventor on U.S. Patent Number 7,951,845. She previously worked as a paid consultant to OtoMedicine, Inc., and she now serves as the Lead Scientific Advisor for Hearing Health Sciences.

[2]Jianxin Bao is the sole inventor on U.S. Patent Trademark Office Application No. 20,070,078,177 filed September 29, 2006.

undergo weapons training that exceeds the limits of conventional hearing protection. Thus, all subjects are guaranteed protection using the same, traditional mechanical devices (ear plugs, ear muffs) that anyone NOT participating in the study will use; however, subjects in the research study have the potential for added protection via novel therapeutic treatments if they are assigned to the treatment condition, and if the treatment is in fact effective. This was the strategy used in studies with Marine recruits (see Lynch and Kil 2005; Kopke et al. 2007), and it is the strategy that was adopted for an ongoing human clinical trial evaluating protection with the β-carotene, vitamins C and E, and magnesium formulation in Swedish military personnel undergoing weapons training. Access to military populations during and after weapons training is difficult to negotiate, however, and, there are some suggestions that hearing loss may sometimes occur in a smaller than expected subpopulation of control subjects. For example, only 28% of the placebo-treated troops had threshold shifts in the Marine population evaluated by Kopke and colleagues (as presented by Kopke 2005). Potential explanations include hearing loss that develops over a longer time frame than captured within the duration of the study, or perhaps, improved compliance with hearing protection devices (ear plugs, ear muffs) during the course of the study as a result of increased attention to hearing conservation issues during study participation.

Another strategy has therefore been to evaluate potential reductions in TTS in populations that choose not to use hearing protection, such as concert-goers and nightclub attendees (see, e.g., Kramer et al. 2006). An alternative approach that offers additional control of the exposure and the audiometric test conditions is to study TTS in users of personal music players in a controlled laboratory setting. Significant progress to develop a laboratory model of TTS using music exposures with real-world relevance has been made, and TTS can be reliably induced in controlled laboratory settings with complete recovery in all subjects tested to date (Le Prell et al. 2010a).

8.2 Test Metrics for Use in Human Subject Populations

8.2.1 Conventional Pure-Tone Audiometry

Access to subject populations is not the only hurdle for studies translating therapeutic agents from animals to humans. Selection of the test metrics used to define protection of the inner ear are also critical; test metrics must reveal protection against changes used to define NIHL in patient populations. Workers compensation laws exist in all 50 states, and hearing loss is a compensable injury in most states. Military personnel can be compensated for NIHL through the Veterans Administration, and federal (civilian) employees can be compensated for NIHL through the Department of Labor. All rules regarding compensation for NIHL have one thing in common: they are based on deficits measured using conventional pure-tone audiometry at some subset of the frequencies including 0.5, 1, 2, and 3 kHz, with some regulatory groups also considering thresholds at 4 or 6 kHz (for review, see Dobie and Megerson 2000). Given clinician dependence on conventional audiometric outcomes to define medical and legal hearing loss, the gold standard for hearing protection with novel therapeutics must include protection of hearing thresholds at the conventional test frequencies. There are a number of other tests that can be used clinically, however, and scientific data suggests these tests have merit for providing additional information about therapeutic benefits of novel agents under evaluation.

8.2.2 Extended High-Frequency Audiometry

One test that might prove useful is extended high-frequency (EHF) audiometery. The frequency range from 9 to 20 kHz is referred to as the EHF range. Over the last four decades, EHF testing has been shown to detect ototoxic changes before the conventional frequency range is affected (Jacobson et al. 1969; Fausti et al. 1984a, b; Rappaport et al. 1985; Kopelman et al. 1988). This has generated multiple suggestions for use of EHF tests to screen for early warnings of NIHL. However, noise-induced changes at EHF frequencies have proven to be variable, which reduces their clinical utility. For example, although evidence suggests PTS at EHF frequencies in music player users (Peng et al. 2007) and musicians (Schmuziger et al. 2006), TTS was not detected at EHF frequencies in musicians tested before and after rehearsal (Schmuziger et al. 2007). In contrast, studies in other noise-exposed populations have shown TTS at EHF frequencies (Kuronen et al. 2003; Balatsouras et al. 2005). It would be of significant interest to determine if there is any potential predictive value of EHF hearing for increased vulnerability in human ears exposed to TTS-producing noise. Current clinical and industrial practice does not include routine monitoring for NIHL at frequencies beyond 8 kHz.

8.2.3 Distortion Product Otoacoustic Emission

Another test of significant interest and potential clinical use is the DPOAE. DPOAEs provide a sensitive and objective measure of OHC function (Kujawa et al. 1994; Kemp 1997; Hall 2000). OHCs are particularly sensitive to noise insult (Dallos 1992; Hamernik and Qiu 2000), and DPOAEs have demonstrated high sensitivity to noise in animal and human studies too numerous to list (for examples, see Emmerich et al. 2000; Fraenkel et al. 2003; Lapsley Miller et al. 2004). Early deficits in DPOAE amplitude have been suggested as predictive for subsequent elevations in pure-tone detection thresholds (Mensh et al. 1993; LePage and Murray 1998; Lucertini et al. 2002; Seixas et al. 2005). Given early changes and high test–retest reliability (Hall and Lutman 1999), DPOAEs have been proposed for hearing surveillance in industry (Prasher and Sulkowski 1999; Sliwinska-Kowalska et al. 2003; Seixas et al. 2005); however, current clinical and industrial practice does not include routine DPOAE monitoring. One of the challenges in implementing the use of OAEs in clinical decision making is the lack of national and international standards for calibrating DPOAE test equipment, and the lack of national and international test standards. Normative data for large populations are also missing, making interpretation and clinical diagnosis potentially challenging.

9 Summary and Conclusions

The use of antioxidants to arrest disease processes, including neurodegenerative events, has significant clinical appeal. Mitochondrial dysfunction and ROS production have been implicated in progressive neurodegenerative diseases such as Alzheimer's disease, Parkinson's disease, Huntington's disease, Friedrich's ataxia, as well as acute neurodegenerative syndromes such as ischemic and hemorrhagic stroke and diseases in which cytosolic oxidative stress is the primary pathophysiology, including AIDS and diabetes, and age-related decreases in cell function. This similarity of free radicals in neurodegenerative processes and NIHL, and the putative efficacy of antioxidants in reducing neurodegenerative processes, provides a compelling rationale for continued investigations into the use of these and other antioxidant agents to reduce NIHL in human subjects.

It is critical that promising agents be compared using the same noise insult, and same animal model, so that efficacy across agents can be compared. Demonstrating dose–response curves for the agents selected will be critical as well, to determine if the most effective dose of each agent is being compared, or simply an effective dose of each agent. Some agents have a fair amount of dose–response data to draw upon; others have been tested using fewer dose levels. Finally, it is essential to evaluate other potentially more potent antioxidant therapies. In the review by Patten et al. (2010), three compelling candidate therapeutic strategies are described.

The first candidate therapeutic category is mitochondrial-targeted antioxidants, which can be delivered orally and accumulate in the mitochondria (Smith et al. 2003, 2008) . A mitochondrial-targeted version of coenzyme Q_{10}/ubiquinol (mitoQ) is several hundred times more potent than the untargeted homologue (Kelso et al. 2001, 2002; Jauslin et al. 2002, 2003). Targeted versions of vitamin E have also been developed, and like ubiquinone, the targeted versions are far more potent than the untargeted homologues (Smith et al. 1999). The second candidate therapeutic is a rechargeable antioxidant: plastoquinonyldecyltriphenyl phosphonium (SkQ1), which is a novel synthetic compound containing a plastoquinonc (antioxidant moiety), a penetrating cation, and a decane or pentane linker (Skulachev 2009; Skulachev et al. 2009). Antioxidant activity was greater than that of mitoQ, and in vivo protection of the visual system was robust and compelling in multiple animal models. The third candidate therapeutic is a class of mitochondrial antioxidants that localize to the mitochondria regardless of the mitochondrial membrane potential; these are Szeto-Schiller peptides. These cell-permeable, peptide antioxidants target the inner mitochondrial membrane with 1,000-fold concentration (Zhao et al. 2004, 2005). SS-31 provides dose-dependent protection against oxidative stress, providing complete protection with optimal dosing (Yang et al. 2009). Despite the robust protection shown against oxidative stress in other biological systems, there are no data on the use of any of these novel, targeted, potent, antioxidant compounds, which might provide protection to sensory cells in the inner ear.

The recent past brought significant advances in our understanding of the mechanisms of NIHL, including the contribution of metabolic trauma, free radical production, and endogenous defense systems that can be upregulated or supplemented. These advances have indeed led to the identification of a class of agents with the potential to protect the human inner ear. This is an exciting time in the field of auditory research, with the potential to intervene in and attenuate cell death events and hearing loss having been clearly shown in multiple animal models, with multiple therapeutic agents. With improved understanding of the mechanisms of action of the best agents to date, it is possible that novel combinations with greater efficacy will be identified, and there are of course multiple agents that have the potential for robust protection but that have not yet been evaluated in the inner ear. Although the basic research must continue at the current rapid pace, the process of translating exciting agents from animal models to human trials must accelerate. There is a clear and compelling need for translational data, evaluating, and hopefully confirming, the potential for protection of the human inner ear.

Acknowledgments We thank Kathleen Campbell, Josef Miller, and Kevin Ohlemiller for helpful comments on an earlier version of this chapter; we also thank Bianca Gomez and Danielle Rosier for helpful assistance with literature searches.

References

Abaamrane, L., Raffin, F., Gal, M., Avan, P., & Sendowski, I. (2009). Long-term administration of magnesium after acoustic trauma caused by gunshot noise in guinea pigs. *Hearing Research*, 247(2), 137–145.

Abi-Hachem, R. N., Zine, A., & Van De Water, T. R. (2010). The injured cochlea as a target for inflammatory processes, initiation of cell death pathways and application of related otoprotectives strategies. *Recent Patents on CNS Drug Discovery*, 5, 147–163.

Adam-Vizi, V., & Starkov, A. A. (2010). Calcium and mitochondrial reactive oxygen species generation: How to read the facts. *Journal of Alzheimer's Disease*, 20 (Supplement 2), 413–426.

Age-Related Eye Disease Study Research Group. (2001). A randomized, placebo-controlled, clinical trial of high-dose supplementation with vitamins C and E, beta carotene, and zinc for age-related macular degeneration and vision loss: AREDS report no. 8. *Archives of Ophthalmology*, 119(10), 1417–1436.

Agrawal, Y., Carey, J. P., Della Santina, C. C., Schubert, M. C., & Minor, L. B. (2009). Disorders of balance and vestibular function in US adults: Data from the National Health and Nutrition Examination Survey, 2001–2004. *Archives of Internal Medicine*, 169(10), 938–944.

Ahn, J. H., Kang, H. H., Kim, Y. J., & Chung, J. W. (2005). Anti-apoptotic role of retinoic acid in the inner ear of noise-exposed mice. *Biochemical and Biophysical Research Communications*, 335(2), 485–490.

Albanes, D., Virtamo, J., Rautalahti, M., Haukka, J., Palmgren, J., Gref, C. G., & Heinonen, O. P. (1992). Serum beta-carotene before and after beta-carotene supplementation. *European Journal of Clinical Nutrition*, 46(1), 15–24.

Alberti, P. W. (1987). Tinnitus in occupational hearing loss: Nosological aspects. *Journal of Otolaryngology*, 16(1), 34–35.

Attias, J., Weisz, G., Almog, S., Shahar, A., Wiener, M., Joachims, Z., Netzer, A., Ising, H., Rebentisch, E., & Guenther, T. (1994). Oral magnesium intake reduces permanent hearing loss induced by noise exposure. *American Journal of Otolaryngology*, 15(1), 26–32.

Attias, J., Bresloff, I., Haupt, H., Scheibe, F., & Ising, H. (2003). Preventing noise induced otoacoustic emission loss by increasing magnesium (Mg^{2+}) intake in guinea-pigs. *Journal of Basic and Clinical Physiology and Pharmacology*, 14(2), 119–136.

Attias, J., Sapir, S., Bresloff, I., Reshef-Haran. I., & Ising, H. (2004). Reduction in noise-induced temporary threshold shift in humans following oral magnesium intake. *Clinical Otolaryngology*, 29(6), 635–641.

Balatsouras, D. G., Homsioglou, E., & Danielidis, V. (2005). Extended high-frequency audiometry in patients with acoustic trauma. *Clinical Otolaryngology*, 30(3), 249–254.

Bao, J. (2007). Methods and compositions for treating non age related hearing impairment in a subject. U.S. Patent No. Application number 20070078177. United States Patent Trademark Office.

Barrs, D. M., Althoff, L. K., Krueger, W. W., & Olsson, J. E. (1994). Work-related, noise-induced hearing loss: Evaluation including evoked potential audiometry. *Otolaryngology – Head and Neck Surgery*, 110(2), 177–184.

Basile, A. S., Huang, J. M., Xie, C., Webster, D., Berlin, C., & Skolnick, P. (1996). *N*-methyl-D-aspartate antagonists limit aminoglycoside antibiotic-induced hearing loss.[see comment]. *Nature Medicine*, 2(12), 1338–1343.

Bauer, C. A., & Brozoski, T. J. (2001). Assessing tinnitus and prospective tinnitus therapeutics using a psychophysical animal model. *Journal of the Association for Research in Otolaryngology*, 2(1), 54–64.

Bauer, C. A., Turner, J. G., Caspary, D. M., Myers, K. S., & Brozoski, T. J. (2008). Tinnitus and inferior colliculus activity in chinchillas related to three distinct patterns of cochlear trauma. *Journal of Neuroscience Research*, 86(11), 2564–2578.

Bendich, A., Machlin, L. J., Scandurra, O., Burton, G. W., & Wayner, D. D. M. (1986). The antioxidant role of vitamin C. *Advances in Free Radical Biology and Medicine*, 2, 419–444.

Ben-Dor, A., Steiner, M., Gheber, L., Danilenko, M., Dubi, N., Linnewiel, K., Zick, A., Sharoni, Y., & Levy, J. (2005). Carotenoids activate the antioxidant response element transcription system. *Molecular Cancer Therapeutics*, 4(1), 177–186.

Bertolaso, L., Martini, A., Bindini, D., Lanzoni, I., Parmeggiani, A., Vitali, C., Kalinec, G., Kalinec, F., Capitani, S., & Previati, M. (2001). Apoptosis in the OC-k3 immortalized cell line treated with different agents. *Audiology*, 40(6), 327–335.

Bielefeld, E. C., Hynes, S., Pryznosch, D., Liu, J., Coleman, J. K., & Henderson, D. (2005). A comparison of the protective effects of systemic administration of a pro-glutathione drug and a Src-PTK inhibitor against noise-induced hearing loss. *Noise & Health*, 7(29), 24–30.

Bielefeld, E. C., Kopke, R. D., Jackson, R. L., Coleman, J. K., Liu, J., & Henderson, D. (2007). Noise protection with *N*-acetyl-l-cysteine (NAC) using a variety of noise exposures, NAC doses, and routes of administration. *Acta Oto-Laryngologica*, 127(9), 914–919.

Biesalski, H. K., Wellner, U., & Weiser, H. (1990). Vitamin A deficiency increases noise suscepti bility in guinea pigs. *Journal of Nutrition*, 120(7), 726–737.

Birney, E. C., Jenness, R., & Ayaz, K. M. (1976). Inability of bats to synthesise L-ascorbic acid. *Nature*, 260(5552), 626–628.

Boettcher, F. A. (1996). Diltiazem does not protect the ear from noise-induced hearing loss in mongolian gerbils. *Laryngoscope*, 106(6), 772–776.

Boettcher, F. A., Caldwell, R. K., Gratton, M. A., White, D. R., & Miles, L. R. (1998). Effects of nimodipine on noise-induced hearing loss. *Hearing Research*, 121(1–2), 139–146.

Bonny, C. (2009). Cell-permeable peptide inhibitors of the JNK signal transduction pathway U.S. Patent No. 7,635,681. United States Patent Trademark Office.

Branis, M., & Burda, H. (1988). Effect of ascorbic acid on the numerical hair cell loss in noise exposed guinea pigs. *Hearing Research*, 33(2), 137–140.

Bray, A., Szymanski, M., & Mills, R. (2004). Noise induced hearing loss in dance music disc jockeys and an examination of sound levels in nightclubs. *Journal of Laryngology and Otology*, 118(2), 123–128.

Burton, G. W., Joyce, A., & Ingold, K. U. (1983). Is vitamin E the only lipid-soluble, chain-breaking antioxidant in human blood plasma and erythrocyte membranes? *Archives of Biochemistry and Biophysics*, 221(1), 281–290.

Burton, G. W., Foster, D. O., Perly, B., Slater, T. F., Smith, I. C., & Ingold, K. U. (1985). Biological antioxidants. *Philosophical Transactions of the Royal Society of London - Series B: Biological Sciences*, 311(1152), 565–578.

Campbell, K. C., & Le Prell, C. G. (2011). Potential therapeutic agents. *Seminars in Hearing, in press.*

Campbell, K. C. M. (2001). Therapeutic use of D-methionine to reduce the toxicity of ototoxic drugs, noise, and radiation. U.S. Patent No. 6,265,386. United States Patent Trademark Office.

Campbell, K. C. M. (2008). Therapeutic use of methionine-derivitives to reduce the toxicity of noise U.S. Patent No. 7,423,065. United States Patent Trademark Office.

Campbell, K. C. M., Meech, R. P., Klemens, J. J., Gerberi, M. T., Dyrstad, S. S. W., Larsen, D. L., Mitchell, D. L., El-Azizi, M., Verhulst, S. J., & Hughes, L. F. (2007). Prevention of noise- and drug-induced hearing loss with D-methionine. *Hearing Research*, 226, 92–103.

Campbell, K. C. M., Nayar, R., Borgonha, S., Hughes, L., Rehemtulla, A., Ross, B. D., & Sunkara, P. (2009). Oral D-methionine (MRX-1024) significantly protects against cisplatin-induced hearing loss: A phase II study in humans. Presented at *IX European Federation of Audiology Societies (EFAS) Congress,* Tenerife, Spain, June 21–24, 2009.

Canbay, A., Feldstein, A., Baskin-Bey, E., Bronk, S. F., & Gores, G. J. (2004). The caspase inhibitor IDN-6556 attenuates hepatic injury and fibrosis in the bile duct ligated mouse. *Journal of Pharmacology and Experimental Therapeutics*, 308(3), 1191–1196.

Cassandro, E., Sequino, L., Mondola, P., Attanasio, G., Barbara, M., & Filipo, R. (2003). Effect of superoxide dismutase and allopurinol on impulse noise-exposed guinea pigs – electrophysiological and biochemical study. *Acta Oto-Laryngologica*, 123(7), 802–807.

Cave, K. M., Cornish, E. M., & Chandler, D. W. (2007). Blast injury of the ear: Clinical update from the global war on terror. *Military Medicine*, 172(7), 726–730.

Cevette, M. J., Vormann, J., & Franz, K. (2003). Magnesium and hearing. *Journal of the American Academy of Audiology*, 14(4), 202–212.

Chan, A. C. (1993). Partners in defense, vitamin E and vitamin C. *Canadian Journal of Physiology and Pharmacology*, 71(9), 725–731.

Chan, H. L., Chou, H. C., Duran, M., Gruenewald, J., Waterfield, M. D., Ridley, A., & Timms, J. F. (2010). Major role of epidermal growth factor receptor and Src kinases in promoting oxidative stress-dependent loss of adhesion and apoptosis in epithelial cells. *Journal of Biological Chemistry*, 285(7), 4307–4318.

Chang, N. C., Ho, C. K., Wu, M. T., Yu, M. L., & Ho, K. Y. (2009). Effect of manganese-superoxide dismutase genetic polymorphisms IVS3–23 T/G on noise susceptibility in Taiwan. *American Journal of Otolaryngology*, 30(6), 396–400.

Chatterjee, I. B. (1973). Evolution and the biosynthesis of ascorbic acid. *Science*, 182(118), 1271–1272.

Chatterjee, I. B., Majumder, A. K., Nandi, B. K., & Subramanian, N. (1975). Synthesis and some major functions of vitamin C in animals. *Annals of the New York Academy of Sciences*, 258, 24–47.

Cheng, A. G., Huang, T., Stracher, A., Kim, A., Liu, W., Malgrange, B., Lefebvre, P. P., Schulman, A., & Van De Water, T. R. (1999). Calpain inhibitors protect auditory sensory cells from hypoxia and neurotrophin-withdrawal induced apoptosis. *Brain Research*, 850(1–2), 234–243.

Cheng, A. G., Cunningham, L. L., & Rubel, E. W. (2005). Mechanisms of hair cell death and protection. *Current Opinion in Otolaryngology. Head and Neck Surgery*, 13(6), 343–348.

Cheng, P. W., Liu, S. H., Young, Y. H., & Lin-Shiau, S. Y. (2006). D-Methionine attenuated cisplatin-induced vestibulotoxicity through altering ATPase activities and oxidative stress in guinea pigs. *Toxicology and Applied Pharmacology*, 215(2), 228–236.

Cheng, P. W., Liu, S. H., Young, Y. H., Hsu, C. J., & Lin-Shiau, S. Y. (2008). Protection from noise-induced temporary threshold shift by D-methionine is associated with preservation of ATPase activities. *Ear and Hearing*, 29(1), 65–75.

Chermak, G. D., & Dengerink, J. E. (1987). Characteristics of temporary noise-induced tinnitus in male and female subjects. *Scandinavian Audiology*, 16(2), 67–73.

Choi, C. H., Chen, K., Vasquez-Weldon, A., Jackson, R. L., Floyd, R. A., & Kopke, R. D. (2008). Effectiveness of 4-hydroxy phenyl *N*-tert-butylnitrone (4-OHPBN) alone and in combination with other antioxidant drugs in the treatment of acute acoustic trauma in chinchilla. *Free Radical Biology and Medicine*, 44(9), 1772–1784.

Chou, H. C., Chen, Y. W., Lee, T. R., Wu, F. S., Chan, H. T., Lyu, P. C., Timms, J. F., & Chan, H. L. (2010). Proteomics study of oxidative stress and Src kinase inhibition in H9C2 cardiomyocytes: A cell model of heart ischemia-reperfusion injury and treatment. *Free Radical Biology and Medicine*, 49(1), 96–108.

Chow, C. K. (2001). Vitamin E regulation of mitochondrial superoxide generation. *Biological Signals and Receptors*, 10(1–2), 112–124.

Chow, C. K., Ibrahim, W., Wei, Z., & Chan, A. C. (1999). Vitamin E regulates mitochondrial hydrogen peroxide generation. *Free Radical Biology and Medicine*, 27(5–6), 580–587.

Colak, A., Antar, V., Karaoglan, A., Akdemir, O., Sahan, E., Celik, O., & Sagmanligil, A. (2009). Q-VD-OPh, a pancaspase inhibitor, reduces trauma-induced apoptosis and improves the recovery of hind-limb function in rats after spinal cord injury. *Neurocirugia (Astur)*, 20(6), 533–540; discussion 540.

Coleman, J. K., Kopke, R. D., Liu, J., Ge, X., Harper, E. A., Jones, G. E., Cater, T. L., & Jackson, R. L. (2007a). Pharmacological rescue of noise induced hearing loss using *N*-acetylcysteine and acetyl-L-carnitine. *Hearing Research*, 226(1–2), 104–113.

Coleman, J. K., Littlesunday, C., Jackson, R., & Meyer, T. (2007b). AM-111 protects against permanent hearing loss from impulse noise trauma. *Hearing Research*, 226(1–2), 70–78.

Corbacella, E., Lanzoni, I., Ding, D., Previati, M., & Salvi, R. (2004). Minocycline attenuates gentamicin induced hair cell loss in neonatal cochlear cultures. *Hearing Research*, 197(1–2), 11–18.

Dallos, P. (1992). The active cochlea. *Journal of Neuroscience*, 12(12), 4575–4585.

Danial, N. N. (2007). BCL-2 family proteins: Critical checkpoints of apoptotic cell death. *Clinical Cancer Research*, 13(24), 7254–7263.

Danial, N. N. (2009). BAD: Undertaker by night, candyman by day. *Oncogene*, 27 (Supplement 1), S53–70.

Dehne, N., Lautermann, J., ten Cate, W. J., Rauen, U., & de Groot, H. (2000). In vitro effects of hydrogen peroxide on the cochlear neurosensory epithelium of the guinea pig. *Hearing Research*, 143(1–2), 162–170.

Derekoy, F. S., Koken, T., Yilmaz, D., Kahraman, A., & Altuntas, A. (2004). Effects of ascorbic acid on oxidative system and transient evoked otoacoustic emissions in rabbits exposed to noise. *Laryngoscope*, 114(10), 1775–1779.

Diamond, B. J., Shiflett, S. C., Feiwel, N., Matheis, R. J., Noskin, O., Richards, J. A., & Schoenberger, N. E. (2000). Ginkgo biloba extract: Mechanisms and clinical indications. *Archives of Physical Medicine and Rehabilitation*, 81(5), 668–678.

Didier, A., Droy-Lefaix, M. T., Aurousseau, C., & Cazals, Y. (1996). Effects of Ginkgo biloba extract (EGb 761) on cochlear vasculature in the guinea pig: Morphometric measurements and laser Doppler flowmetry. *European Archives of Oto-Rhino-Laryngology*, 253(1–2), 25–30.

Dobie, R. A., & Megerson, S. C. (2000). Workers' compensation. In E. H. Berger, L. H. Royster, J. D. Royster, D. P. Driscoll & M. Layne (Eds.), *The noise manual* (5th ed., pp. 689–710). Fairfax: American Industrial Hygiene Association.

Dolk, H. M., Nau, H., Hummler, H., & Barlow, S. M. (1999). Dietary vitamin A and teratogenic risk: European Teratology Society discussion paper. *European Journal of Obstetrics, Gynecology, and Reproductive Biology*, 83(1), 31–36.

Duan, M., Agerman, K., Ernfors, P., & Canlon, B. (2000). Complementary roles of neurotrophin 3 and a *N*-methyl-D-aspartate antagonist in the protection of noise and aminoglycoside-induced ototoxicity. *Proceedings of the National Academy of Sciences of the United States of America*, 97(13), 7597–7602.

Duan, M., Qiu, J., Laurell, G., Olofsson, A., Counter, S. A., & Borg, E. (2004). Dose and time-dependent protection of the antioxidant *N*-L-acetylcysteine against impulse noise trauma. *Hearing Research*, 192(1–2), 1–9.

Emmerich, E., Richter, F., Reinhold, U., Linss, V., & Linss, W. (2000). Effects of industrial noise exposure on distortion product otoacoustic emissions (DPOAEs) and hair cell loss of the cochlea – long term experiments in awake guinea pigs. *Hearing Research*, 148(1–2), 9–17.

Emmerich, E., Richter, F., Hagner, H., Giessler, F., Gehrlein, S., & Dieroff, H. G. (2002). Effects of discotheque music on audiometric results and central acoustic evoked neuromagnetic responses. *International Tinnitus Journal*, 8(1), 13–19.

Eshraghi, A. A., & Van De Water, T. R. (2006). Cochlear implantation trauma and noise-induced hearing loss: Apoptosis and therapeutic strategies. *Anatomical Record. Part A, Discoveries in Molecular, Cellular, & Evolutionary Biology*, 288(4), 473–481.

Evans, P., & Halliwell, B. (1999). Free radicals and hearing. Cause, consequence, and criteria. *Annals of the New York Academy of Sciences*, 884, 19–40.

Faubel, S., & Edelstein, C. L. (2005). Caspases as drug targets in ischemic organ injury. *Current Drug Targets – Immune, Endocrine & Metabolic Disorders*, 5(3), 269–287.

Fausti, S. A., Rappaport, B. Z., Schechter, M. A., Frey, R. H., Ward, T. T., & Brummettt, R. E. (1984a). Detection of aminoglycoside ototoxicity by high frequency auditory evaluation: Selected case studies. *American Journal of Otolaryngology*, 5, 177–182.

Fausti, S. A., Schechter, M. A., Rappaport, B. Z., Frey, R. H., & Mass, R. E. (1984b). Early detection cisplatin ototoxicity: Selected case reports. *Cancer*, 53, 224–231.

Fausti, S. A., Wilmington, D. J., Gallun, F. J., Myers, P. J., & Henry, J. A. (2009). Auditory and vestibular dysfunction associated with blast-related traumatic brain injury. *Journal of Rehabilitation Research and Development*, 46(6), 797–810.

Fessenden, J. D., & Schacht, J. (1998). The nitric oxide/cyclic GMP pathway: A potential major regulator of cochlear physiology. *Hearing Research*, 118(1–2), 168–176.

Fetoni, A. R., Ferraresi, A., Greca, C. L., Rizzo, D., Sergi, B., Tringali, G., Piacentini, R., & Troiani, D. (2008). Antioxidant protection against acoustic trauma by coadministration of idebenone and vitamin E. *NeuroReport*, 19(3), 277–281.

Fetoni, A. R., Ferraresi, A., Picciotti, P., Gaetani, E., Paludetti, G., & Troiani, D. (2009a). Noise induced hearing loss and vestibular dysfunction in the guinea pig. *International Journal of Audiology*, 48(11), 804–810.

Fetoni, A. R., Ralli, M., Sergi, B., Parrilla, C., Troiani, D., & Paludetti, G. (2009b). Protective effects of *N*-acetylcysteine on noise-induced hearing loss in guinea pigs. *Acta Otorhinolaryngologica Italica*, 29(2), 70–75.

Fortunato, G., Marciano, E., Zarrilli, F., Mazzaccara, C., Intrieri, M., Calcagno, G., Vitale, D. F., La Manna, P., Saulino, C., Marcelli, V., & Sacchetti, L. (2004). Paraoxonase and superoxide dismutase gene polymorphisms and noise-induced hearing loss. *Clinical Chemistry*, 50(11), 2012–2018.

Fraenkel, R., Freeman, S., & Sohmer, H. (2003). Use of ABR threshold and OAEs in detection of noise induced hearing loss. *Journal of Basic and Clinical Physiology and Pharmacology*, 14(2), 95–118.

Fridberger, A., & Ulfendahl, M. (1996). Acute mechanical overstimulation of isolated outer hair cells causes changes in intracellular calcium levels without shape changes. *Acta Oto-Laryngologica*, 116(1), 17–24.

Fridberger, A., Flock, A., Ulfendahl, M., & Flock, B. (1998). Acoustic overstimulation increases outer hair cell Ca^{2+} concentrations and causes dynamic contractions of the hearing organ. *Proceedings of the National Academy of Sciences of the USA*, 95(12), 7127–7132.

Griest, S. E., & Bishop, P. M. (1998). Tinnitus as an early indicator of permanent hearing loss. A 15 year longitudinal study of noise exposed workers. *AAOHN Journal*, 46(7), 325–329.

Guitton, M. J., Wang, J., & Puel, J. L. (2004). New pharmacological strategies to restore hearing and treat tinnitus. *Acta Oto-Laryngologica*, 124(4), 411–415.

Gunderson, E., Moline, J., & Catalano, P. (1997). Risks of developing noise-induced hearing loss in employees of urban music clubs. *American Journal of Industrial Medicine*, 31(1), 75–79.

Gunther, T., Ising, H., & Joachims, Z. (1989). Biochemical mechanisms affecting susceptibility to noise-induced hearing loss. *American Journal of Otology*, 10(1), 36–41.

Hall, A., & Lutman, M. (1999). Methods for early identification of noise-induced hearing loss. *Audiology*, 38(5), 277–280.

Hall, J. W. I. (2000). *Handbook of Otoacoustic Emissions*. San Diego: Singular.

Hamacher, J., Stammberger, U., Weber, E., Lucas, R., & Wendel, A. (2009). Ebselen improves ischemia-reperfusion injury after rat lung transplantation. *Lung*, 187(2), 98–103.

Hamernik, R. P., & Qiu, W. (2000). Correlations among evoked potential thresholds, distortion product otoacoustic emissions and hair cell loss following various noise exposures in the chinchilla. *Hearing Research*, 150(1–2), 245–257.

Hamernik, R. P., Ahroon, W. A., Patterson, J. H., Jr., & Qiu, W. (2002). Relations among early postexposure noise-induced threshold shifts and permanent threshold shifts in the chinchilla. *Journal of the Acoustical Society of America*, 111(1 Pt 1), 320–326.

Hamernik, R. P., Qiu, W., & Davis, B. (2008). The effectiveness of *N*-acetyl-L-cysteine (L-NAC) in the prevention of severe noise-induced hearing loss. *Hearing Research*, 239(1–2), 99–106.

Harapanhalli, R. S., Yaghmai, V., Giuliani, D., Howell, R. W., & Rao, D. V. (1996). Antioxidant effects of vitamin C in mice following X-irradiation. *Research Communications in Molecular Pathology and Pharmacology*, 94(3), 271–287.

Harris, K. C., Hu, B., Hangauer, D., & Henderson, D. (2005). Prevention of noise-induced hearing loss with Src-PTK inhibitors. *Hearing Research*, 208(1–2), 14–25.

Haupt, H., & Scheibe, F. (2002). Preventive magnesium supplement protects the inner ear against noise-induced impairment of blood flow and oxygenation in the guinea pig. *Magnesium Research*, 15(1–2), 17–25.

Haupt, H., Scheibe, F., & Mazurek, B. (2003). Therapeutic efficacy of magnesium in acoustic trauma in the guinea pig. *ORL; Journal of Oto-Rhino-Laryngology and Its Related Specialties*, 65(3), 134–139.

Heard, K. (2010). A multicenter comparison of the safety of oral versus intravenous acetylcysteine for treatment of acetaminophen overdose. *Clinical Toxicology*, 48(5), 424–430.

Heffner, H. E., & Harrington, I. A. (2002). Tinnitus in hamsters following exposure to intense sound. *Hearing Research*, 170(1–2), 83–95.

Heinrich, U. R., Maurer, J., & Mann, W. (1999). Ultrastructural evidence for protection of the outer hair cells of the inner ear during intense noise exposure by application of the organic calcium channel blocker diltiazem. *ORL; Journal of Oto-Rhino-Laryngology and Its Related Specialties*, 61(6), 321–327.

Heinrich, U. R., Selivanova, O., Feltens, R., Brieger, J., & Mann, W. (2005). Endothelial nitric oxide synthase upregulation in the guinea pig organ of Corti after acute noise trauma. *Brain Research*, 1, 85–96.

Heinrich, U. R., Fischer, I., Brieger, J., Rumelin, A., Schmidtmann, I., Li, H., Mann, W. J., & Helling, K. (2008). Ascorbic acid reduces noise-induced nitric oxide production in the guinea pig ear. *Laryngoscope*, 118(5), 837–842.

Henderson, D., Subramaniam, M., Gratton, M. A., & Saunders, S. S. (1991). Impact noise: The importance of level, duration, and repetition rate. *Journal of the Acoustical Society of America*, 89(3), 1350–1357.

Henderson, D., Bielefeld, E. C., Harris, K. C., & Hu, B. H. (2006). The role of oxidative stress in noise-induced hearing loss. *Ear and Hearing*, 27(1), 1–19.

Hong, O. S., Chen, S. P., & Conrad, K. M. (1998). Noise induced hearing loss among male airport workers in Korea. *AAOHN Journal*, 46(2), 67–75.

Hou, F., Wang, S., Zhai, S., Hu, Y., Yang, W., & He, L. (2003). Effects of alpha-tocopherol on noise-induced hearing loss in guinea pigs. *Hearing Research*, 179(1–2), 1–8.

Hsu, W. C., Wang, J. D., Lue, J. H., Day, A. S., & Young, Y. H. (2008). Physiological and morphological assessment of the saccule in Guinea pigs after noise exposure. *Archives of Otolaryngology – Head and Neck Surgery*, 134(10), 1099–1106.

Imamura, T., Bando, N., & Yamanishi, R. (2006). Beta-carotene modulates the immunological function of RAW264, a murine macrophage cell line, by enhancing the level of intracellular glutathione. *Bioscience, Biotechnology, and Biochemistry*, 70(9), 2112–2120.

Inagaki, A., Ugawa, S., Yamamura, H., Murakami, S., & Shimada, S. (2008). The CaV3.1 T-type Ca^{2+}channel contributes to voltage-dependent calcium currents in rat outer hair cells. *Brain Research*, 1201, 68–77.

Inumaru, J., Nagano, O., Takahashi, E., Ishimoto, T., Nakamura, S., Suzuki, Y., Niwa, S., Umezawa, K., Tanihara, H., & Saya, H. (2009). Molecular mechanisms regulating dissociation of cell-cell junction of epithelial cells by oxidative stress. *Genes to Cells*, 14(6), 703–716.

Ising, H., Handrock, M., Gunther, T., Fischer, R., & Dombrowski, M. (1982). Increased noise trauma in guinea pigs through magnesium deficiency. *Archives of Oto-Rhino-Laryngology*, 236(2), 139–146.

Ison, J. R., Payman, G. H., Palmer, M. J., & Walton, J. P. (1997). Nimodipine at a dose that slows ABR latencies does not protect the ear against noise. *Hearing Research*, 106(1–2), 179–183.

Jacobson, E. J., Downs, M. P., & Fletcher, J. L. (1969). Clinical findings in high frequency thresholds during known ototoxic drug usage. *Journal of Auditory Research*, 9, 379–385.

Jagetia, G. C., Rajanikant, G. K., Rao, S. K., & Baliga, M. S. (2003). Alteration in the glutathione, glutathione peroxidase, superoxide dismutase and lipid peroxidation by ascorbic acid in the skin of mice exposed to fractionated gamma radiation. *Clinica Chimica Acta*, 332(1–2), 111–121.

Jain, A., Martensson, J., Mehta, T., Krauss, A. N., Auld, P. A., & Meister, A. (1992). Ascorbic acid prevents oxidative stress in glutathione-deficient mice: Effects on lung type 2 cell lamellar bodies, lung surfactant, and skeletal muscle. *Proceedings of the National Academy of Sciences of the USA*, 89(11), 5093–5097.

Janssen, R. (1992). Glutamate neurotoxicity in the developing rat cochlea is antagonized by kynurenic acid and MK-801. *Brain Research*, 590(1–2), 201–206.

Janssen, R., Schweitzer, L., & Jensen, K. F. (1991). Glutamate neurotoxicity in the developing rat cochlea: Physiological and morphological approaches. *Brain Research*, 552(2), 255–264.

Jauslin, M. L., Wirth, T., Meier, T., & Schoumacher, F. (2002). A cellular model for Friedreich Ataxia reveals small-molecule glutathione peroxidase mimetics as novel treatment strategy. *Human Molecular Genetics*, 11(24), 3055–3063.

Jauslin, M. L., Meier, T., Smith, R. A., & Murphy, M. P. (2003). Mitochondria-targeted antioxidants protect Friedreich ataxia fibroblasts from endogenous oxidative stress more effectively than untargeted antioxidants. *FASEB Journal*, 17(13), 1972–1974.

Joachims, Z., Babisch, W., Ising, H., Gunther, T., & Handrock, M. (1983). Dependence of noise-induced hearing loss upon perilymph magnesium concentration. *Journal of the Acoustical Society of America*, 74(1), 104–108.

Joachims, Z., Netzer, A., Ising, H., Rebentisch, E., Attias, J., Weisz, G., & Gunther, T. (1993). Oral magnesium supplementation as prophylaxis for noise-induced hearing loss: Results of a double blind field study. *Schriftenreihe des Vereins fur Wasser-, Boden-, und Lufthygiene*, 88, 503–516.

Juhn, S. K., & Ward, W. D. (1979). Alteration of oxidative enzymes (LDH and MDH) in perilymph after noise exposure. *Archives of Oto-Rhino-Laryngology*, 222(2), 103–108.

Juntunen, J., Matikainen, E., Ylikoski, J., Ylikoski, M., Ojala, M., & Vaheri, E. (1987). Postural body sway and exposure to high-energy impulse noise. *Lancet*, 2(8553), 261–264.

Kalkanis, J. G., Whitworth, C., & Rybak, L. P. (2004). Vitamin E reduces cisplatin ototoxicity. *Laryngoscope*, 114(3), 538–542.

Kaltenbach, J. A., Rachel, J. D., Mathog, T. A., Zhang, J., Falzarano, P. R., & Lewandowski, M. (2002). Cisplatin-induced hyperactivity in the dorsal cochlear nucleus and its relation to outer hair cell loss: Relevance to tinnitus. *Journal of Neurophysiology*, 88(2), 699–714.

Kamal, A. A., Mikael, R. A., & Faris, R. (1989). Follow-up of hearing thresholds among forge hammering workers. *American Journal of Industrial Medicine*, 16(6), 645–658.

Kappus, H., & Diplock, A. T. (1992). Tolerance and safety of vitamin E: A toxicological position report. *Free Radical Biology and Medicine*, 13(1), 55–74.

Kelso, G. F., Porteous, C. M., Coulter, C. V., Hughes, G., Porteous, W. K., Ledgerwood, E. C., Smith, R. A., & Murphy, M. P. (2001). Selective targeting of a redox-active ubiquinone to mitochondria within cells: Antioxidant and antiapoptotic properties. *Journal of Biological Chemistry*, 276(7), 4588–4596.

Kelso, G. F., Porteous, C. M., Hughes, G., Ledgerwood, E. C., Gane, A. M., Smith, R. A., & Murphy, M. P. (2002). Prevention of mitochondrial oxidative damage using targeted antioxidants. *Annals of the New York Academy of Sciences*, 959, 263–274.

Kemink, J. L., & Graham, M. D. (1985). Hearing loss with delayed onset of vertigo. *American Journal of Otology*, 6, 344–348.

Kemp, D. (1997). Otoacoustic emissions in perspective. In M. Robinette & T. Glattke (Eds.), *Otoacoustic emissions: clinical applications* (pp. 1–21). New York: Thieme.

Kil, J., & Lynch, E. (2010). Methods for treating hearing loss. U.S. Patent No. 7,820,640. United States Patent Trademark Office.

Kil, J., Pierce, C., Tran, H., Gu, R., & Lynch, E. D. (2007). Ebselen treatment reduces noise induced hearing loss via the mimicry and induction of glutathione peroxidase. *Hearing Research*, 226, 44–51.

Kilburn, K. H., Warshaw, R. H., & Hanscom, B. (1992). Are hearing loss and balance dysfunction linked in construction iron workers? *British Journal of Industrial Medicine*, 49, 138–141.

Kisilevsky, A. E., & Zamponi, G. W. (2008). Presynaptic calcium channels: Structure, regulators, and blockers. *Handbook of Experimental Pharmacology*, 184, 45–75.

Konig, O., Winter, E., Fuchs, J., Haupt, H., Mazurek, B., Weber, N., & Gross, J. (2003). Protective effect of magnesium and MK 801 on hypoxia-induced hair cell loss in new-born rat cochlea. *Magnesium Research*, 16(2), 98–105.

Konings, A., Van Laer, L., Pawelczyk, M., Carlsson, P. I., Bondeson, M. L., Rajkowska, E., Dudarewicz, A., Vandevelde, A., Fransen, E., Huyghe, J., Borg, E., Sliwinska-Kowalska, M., & Van Camp, G. (2007). Association between variations in CAT and noise-induced hearing loss in two independent noise-exposed populations. *Human Molecular Genetics*, 16(15), 1872–1883.

Kopelman, J., Budnick, A. S., Kramer, M. B., Sessions, R. B., & Wong, G. Y. (1988). Ototoxicity of high-dose cisplatin by bolus administration in patients with advanced cancers and normal hearing. *Laryngoscope*, 98, 858–864.

Kopke, R. D. (2005, October 9–12). *NAC for noise: From the bench top to the clinic.* Paper presented at the International Symposium – Pharmacologic Strategies for Prevention and Treatment of Hearing Loss and Tinnitus, Niagra Falls, Ottawa, Canada.

Kopke, R., Bielefeld, E., Liu, J., Zheng, J., Jackson, R., Henderson, D., & Coleman, J. K. (2005). Prevention of impulse noise-induced hearing loss with antioxidants. *Acta Oto-Laryngologica*, 125(3), 235–243.

Kopke, R. D., Coleman, J. K., Huang, X., Weisskopf, P. A., Jackson, R. L., Liu, J., Hoffer, M. E., Wood, K., Kil, J., & Van De Water, T. R. (2001). Novel strategies to prevent and reverse noise-induced hearing loss. In D. Henderson, D. Prasher, R. Kopke, R. J. Salvi & R. Hamernik (Eds.), *Noise induced hearing loss: Basic mechanisms, prevention and control.* London: Noise Research Network.

Kopke, R. D., Coleman, J. K., Liu, J., Campbell, K. C., & Riffenburgh, R. H. (2002). Candidate's thesis: Enhancing intrinsic cochlear stress defenses to reduce noise-induced hearing loss. *Laryngoscope*, 112(9), 1515–1532.

Kopke, R. D., Henderson, D., & Hoffer, M. E. (2003). Prevention or reversal of sensorineural hearing loss (SNHL) through biologic mechanisms U.S. Patent No. 6,649,621. United States Patent Trademark Office.

Kopke, R. D., Jackson, R. L., Coleman, J. K. M., Liu, J., Bielefeld, E. C., & Balough, B. J. (2007). NAC for Noise: From the bench top to the clinic. *Hearing Research*, 226, 114–125.

Kopke, R. D., Weisskopf, P. A., Boone, J. L., Jackson, R. L., Wester, D. C., Hoffer, M. E., Lambert, D. C., Charon, C. C., Ding, D. L., & McBride, D. (2000). Reduction of noise-induced hearing loss using L-NAC and salicylate in the chinchilla. *Hearing Research*, 149(1–2), 138–146.

Kramer, S., Dreisbach, L., Lockwood, J., Baldwin, K., Kopke, R. D., Scranton, S., & O'Leary, M. (2006). Efficacy of the antioxidant *N*-acetylcystein (NAC) in protecting ears exposed to loud music. *Journal of the American Academy of Audiology*, 17, 265–278.

Krinsky, N. I. (1989). Antioxidant functions of carotenoids. *Free Radical Biology and Medicine*, 7(6), 617–635.

Krinsky, N. I. (1998). The antioxidant and biological properties of the carotenoids. *Annals of the New York Academy of Sciences*, 854, 443–447.

Kujawa, S. G., Glattke, T. J., Fallon, M., & Bobbin, R. P. (1994). A nicotinic-like receptor mediates suppression of distortion product otoacoustic emissions by contralateral sound. *Hearing Research*, 74(1–2), 122–134.

Kujawa, S. G., & Liberman, M. C. (2006). Acceleration of age-related hearing loss by early noise exposure: Evidence of a misspent youth. *Journal of Neuroscience*, 26(7), 2115–2123.

Kujawa, S. G., & Liberman, M. C. (2009). Adding insult to injury: Cochlear nerve degeneration after "temporary" noise-induced hearing loss. *Journal of Neuroscience*, 29(45), 14077–14085.

Kumar, K., Vivarthini, C. J., & Bhat, J. S. (2010). Vestibular evoked myogenic potential in noise-induced hearing loss. *Noise & Health*, 12(48), 191–194.

Kuronen, P., Sorri, M. J., Paakkonen, R., & Muhli, A. (2003). Temporary threshold shift in military pilots measured using conventional and extended high-frequency audiometry after one flight. *International Journal of Audiology*, 42(1), 29–33.

Kyriakis, J. M., Banerjee, P., Nikolakaki, E., Dai, T., Rubie, E. A., Ahmad, M. F., Avruch, J., & Woodgett, J. R. (1994). The stress-activated protein kinase subfamily of c-Jun kinases. *Nature Biotechnology*, 369, 156–160.

Lapsley Miller, J. A., Marshall, L., & Heller, L. M. (2004). A longitudinal study of changes in evoked otoacoustic emissions and pure-tone thresholds as measured in a hearing conservation program. *International Journal of Audiology*, 43(6), 307–322.

Le Prell, C. G., Bledsoe, S. C., Jr., Bobbin, R. P., & Puel, J. L. (2001). Neurotransmission in the inner ear: Functional and molecular analyses. In A. F. Jahn & J. Santos-Sacchi (Eds.), *Physiology of the ear* (2nd ed., pp. 575–611). San Diego: Singular.

Le Prell, C. G., Yagi, M., Kawamoto, K., Beyer, L. A., Atkin, G., Raphael, Y., Dolan, D. F., Bledsoe, S. C. J., & Moody, D. B. (2004). Chronic excitotoxicity in the guinea pig cochlea induces temporary functional deficits without disrupting otoacoustic emissions. *Journal of the Acoustical Society of America*, 116, 1044–1056.

Le Prell, C. G., Hughes, L. F., & Miller, J. M. (2007a). Free radical scavengers vitamins A, C, and E plus magnesium reduce noise trauma. *Free Radical Biology and Medicine*, 42, 1454–1463.

Le Prell, C. G., Yamashita, D., Minami, S., Yamasoba, T., & Miller, J. M. (2007b). Mechanisms of noise-induced hearing loss indicate multiple methods of prevention. *Hearing Research*, 226, 22–43.

Le Prell, C. G., Hall, J. W. I., Sakowicz, B., Campbell, K. C. M., Kujawa, S. G., Antonelli, P. A., Green, G. E., Miller, J. M., Holmes, A. E., & Guire, K. (2010a). Temporary threshold shift subsequent to music player use: Comparison with hearing screenings in populations of adolescents and young adults. Paper presented at the National Hearing Conservation Association, Orlando, FL, February, 2010.

Le Prell, C. G., Johnson, A. W., Dossat, A., & Lang, D. (2010b). Caspase-2 expression in the inner ear after noise insult. *Abstracts of the Association for Research in Otolaryngology*, 33, 232.

Lee, L. T. (1999). A study of the noise hazard to employees in local discotheques. *Singapore Medical Journal*, 40(9), 571–574.

Le Prell, C. G., Gagnon, P. M., Bennett, D. C., & Ohlemiller, K. K., (2011a). Nutrient enhanced diet reduces noise-induced damage to the inner ear and hearing loss. *Translational Research*, 158(1), 38–53.

Le Prell, C. G., Dolan, D. F., Bennett, D. C., & Boxer, P. A., (2011b). Nutrient plasma levels achieved during treatment that reduces noise-induced hearing loss. *Translational Research*, 158(1), 54–70.

LePage, E. L., & Murray, N. M. (1998). Latent cochlear damage in personal stereo users: A study based on click-evoked otoacoustic emissions. *Medical Journal of Australia*, 169(11–12), 588–592.

Lew, H. L., Jerger, J. F., Guillory, S. B., & Henry, J. A. (2007). Auditory dysfunction in traumatic brain injury. *Journal of Rehabilitation Research and Development*, 44(7), 921–928.

Lin, C. Y., Wu, J. L., Shih, T. S., Tsai, P. J., Sun, Y. M., Ma, M. C., & Guo, Y. L. (2010). N-Acetyl-cysteine against noise-induced temporary threshold shift in male workers. *Hearing Research*, 269(1–2), 42–47.

Lipscomb, D. M., & Roettger, R. L. (1973). Capillary constriction in cochlear and vestibular tissues during intense noise stimulation. *Laryngoscope*, 83(2), 259–263.

Liu, Y. M., Li, X. D., Guo, X., Liu, B., Lin, A. H., & Rao, S. Q. (2010). Association between polymorphisms in SOD1 and noise-induced hearing loss in Chinese workers. *Acta Oto-Laryngologica*, 130(4), 477–486.

Lopez-Gonzalez, M. A., Guerrero, J. M., Rojas, F., & Delgado, F. (2000). Ototoxicity caused by cisplatin is ameliorated by melatonin and other antioxidants. *Journal of Pineal Research*, 28(2), 73–80.

Lorito, G., Giordano, P., Petruccelli, J., Martini, A., & Hatzopoulos, S. (2008). Different strategies in treating noise induced hearing loss with *N*-acetylcysteine. *Medical Science Monitor*, 14(8), BR159–164.

Lucertini, M., Moleti, A., & Sisto, R. (2002). On the detection of early cochlear damage by otoacoustic emission analysis. *Journal of the Acoustical Society of America*, 111, 972–978.

Lynch, E. D., & Kil, J. (2005). Compounds for the prevention and treatment of noise-induced hearing loss. *Drug Discovery Today*, 10(19), 1291–1298.

Lynch, E. D., Gu, R., Pierce, C., & Kil, J. (2004). Ebselen-mediated protection from single and repeated noise exposure in rat. *Laryngoscope*, 114(2), 333–337.

Maruyama, J., Miller, J. M., & Ulfendahl, M. (2007). Effects of antioxidants on auditory nerve function and survival in deafened guinea pigs. *Neurobiology of Disease*, 25(1), 309–318.

Maurer, J., Heinrich, U. R., Hinni, M., & Mann, W. (1999). Alteration of the calcium content in inner hair cells of the cochlea of the guinea pig after acute noise trauma with and without application of the organic calcium channel blocker diltiazem. *ORL; Journal of Oto-Rhino-Laryngology and Its Related Specialties*, 61(6), 328–333.

Mayne, S. T., & Parker, R. S. (1986). Subcellular distribution of dietary beta-carotene in chick liver. *Lipids*, 21(2), 164–169.

McCombe, A. W., Binnington, J., Davis, A., & Spencer, H. (1995). Hearing loss and motorcyclists. *Journal of Laryngology and Otology*, 109(7), 599–604.

McFadden, S. L., Ohlemiller, K. K., Ding, D., Shero, M., & Salvi, R. J. (2001). The influence of superoxide dismutase and glutathione peroxidase deficiencies on noise-induced hearing loss in mice. *Noise & Health*, 3(11), 49–64.

McFadden, S. L., Woo, J. M., Michalak, N., & Ding, D. (2005). Dietary vitamin C supplementation reduces noise-induced hearing loss in guinea pigs. *Hearing Research*, 202(1–2), 200–208.

Melinek, M., Naggan, L., & Altman, M. (1976). Acute acoustic trauma – a clinical investigation and prognosis in 433 symptomatic soldiers. *Israel Journal of Medical Sciences*, 12(6), 560–569.

Meltser, I., Tahera, Y., & Canlon, B. (2009). Glucocorticoid receptor and mitogen-activated protein kinase activity after restraint stress and acoustic trauma. *Journal of Neurotrauma*, 26(10), 1835–1845.

Meltser, I., Tahera, Y., & Canlon, B. (2010). Differential activation of mitogen-activated protein kinases and brain-derived neurotrophic factor after temporary or permanent damage to a sensory system. *Neuroscience*, 165(4), 1439–1446.

Mensh, B. D., Lonsbury-Martin, B. L., & Martin, G. K. (1993). Distortion-product emissions in rabbit: II. Prediction of chronic-noise effects by brief pure-tone exposures. *Hearing Research*, 70(1), 65–72.

Mercier, V., Luy, D., & Hohmann, B. W. (2003). The sound exposure of the audience at a music festival. *Noise & Health*, 5(19), 51–58.

Michikawa, T., Nishiwaki, Y., Kikuchi, Y., Hosoda, K., Mizutari, K., Saito, H., Asakura, K., Milojevic, A., Iwasawa, S., Nakano, M., & Takebayashi, T. (2009). Serum levels of retinol and other antioxidants for hearing impairment among Japanese older adults. *Journals of Gerontology. Series A, Biological Sciences and Medical Sciences*, 64(8), 910–915.

Miller, J. M., Ren, T. Y., Dengerink, H. A., & Nuttall, A. L. (1996). Cochlear blood flow changes with short sound stimulation. In A. Axelsson, H. M. Borchgrevink, R. P. Hamernik, P. A. Hellstrom, D. Henderson & R. J. Salvi (Eds.), *Scientific basis of noise-induced hearing loss* (pp. 95–109). New York: Thieme.

Miller, J. M., Le Prell, C. G., Schacht, J., & Prieskorn, D. (2011). Composition and method of treating temporary and permanent hearing loss. US Patent No. 7,951,845 United States Patent Trademark Office.

Miller, R. K., Hendrickx, A. G., Mills, J. L., Hummler, H., & Wiegand, U. W. (1998). Periconceptional vitamin A use: How much is teratogenic? *Reproductive Toxicology*, 12(1), 75–88.
Mills, J. H., Adkins, W. Y., & Gilbert, R. M. (1981). Temporary threshold shifts produced by wide-band noise. *Journal of the Acoustical Society of America*, 70(2), 390–396.
Moody, D. B. (2004). Animal models in tinnitus. In J. B. Snow (Ed.), *Tinnitus: Theory and management* (pp. 80–95). Hamilton: BC Decker.
Mrena, R., Savolainen, S., Kuokkanen, J. T., & Ylikoski, J. (2002). Characteristics of tinnitus induced by acute acoustic trauma: A long-term follow-up. *Audiology and Neuro-Otology*, 7(2), 122–130.
Mrena, R., Savolainen, S., Pirvola, U., & Ylikoski, J. (2004). Characteristics of acute acoustical trauma in the Finnish Defence Forces. *International Journal of Audiology*, 43(3), 177–181.
Murai, N., Kirkegaard, M., Jarlebark, L., Risling, M., Suneson, A., & Ulfendahl, M. (2008). Activation of JNK in the inner ear following impulse noise exposure. *Journal of Neurotrauma*, 25(1), 72–77.
Nageris, B. I., Attias, J., & Shemesh, R. (2008). Otologic and audiologic lesions due to blast injury. *Journal of Basic and Clinical Physiology and Pharmacology*, 19(3–4), 185–191.
Nageris, B. I., Attias, J., & Raveh, E. (2010). Test-retest tinnitus characteristics in patients with noise-induced hearing loss. *American Journal of Otolaryngology*, 31(3), 181–184.
Nakagawa, T., Kim, T. S., Murai, N., Endo, T., Iguchi, F., Tateya, I., Yamamoto, N., Naito, Y., & Ito, J. (2003). A novel technique for inducing local inner ear damage. *Hearing Research*, 176(1–2), 122–127.
Niki, E. (1987a). Lipid antioxidants: How they may act in biological systems. *British Journal of Cancer*, 8(Supplement), 153–157.
Niki, E. (1987b). Interaction of ascorbate and alpha-tocopherol. *Annals of the New York Academy of Sciences*, 498, 186–199.
Niki, E. (1991). Action of ascorbic acid as a scavenger of active and stable oxygen radicals. *American Journal of Clinical Nutrition*, 54(6 Supplement), 1119 S–1124 S.
Nordmann, A. S., Bohne, B. A., & Harding, G. W. (2000). Histopathological differences between temporary and permanent threshold shift. *Hearing Research*, 139(1–2), 13–30.
Ohinata, Y., Miller, J. M., & Schacht, J. (2003). Protection from noise-induced lipid peroxidation and hair cell loss in the cochlea. *Brain Research*, 966(2), 265–273.
Ohlemiller, K. K. (2008). Recent findings and emerging questions in cochlear noise injury. *Hearing Research*, 245, 5–17.
Ohlemiller, K. K., McFadden, S. L., Ding, D. L., Flood, D. G., Reaume, A. G., Hoffman, E. K., Scott, R. W., Wright, J. S., Putcha, G. V., & Salvi, R. J. (1999a). Targeted deletion of the cytosolic Cu/Zn-superoxide dismutase gene (Sod1) increases susceptibility to noise-induced hearing loss. *Audiology and Neuro-Otology*, 4(5), 237–246.
Ohlemiller, K. K., Wright, J. S., & Dugan, L. L. (1999b). Early elevation of cochlear reactive oxygen species following noise exposure. *Audiology and Neuro-Otology*, 4(5), 229–236.
Ohlemiller, K. K., McFadden, S. L., Ding, D. L., Lear, P. M., & Ho, Y. S. (2000). Targeted mutation of the gene for cellular glutathione peroxidase (Gpx1) increases noise-induced hearing loss in mice. *Journal of the Association for Research in Otolaryngology*, 1(3), 243–254.
Okuda, T., Sugahara, K., Takemoto, T., Shimogori, H., & Yamashita, H. (2005). Inhibition of caspases alleviates gentamicin-induced cochlear damage in guinea pigs. *Auris, Nasus, Larynx*, 32(1), 33–37.
Oosterveld, W. J., Polman, A. R., & Schoonheyt, J. (1982). Vestibular implications of noise-induced hearing loss. *British Journal of Audiology*, 16(4), 227–232.
Padayatty, S. J., Katz, A., Wang, Y., Eck, P., Kwon, O., Lee, J. H., Chen, S., Corpe, C., Dutta, A., Dutta, S. K., & Levine, M. (2003). Vitamin C as an antioxidant: Evaluation of its role in disease prevention. *Journal of the American College of Nutrition*, 22(1), 18–35.
Park, J. B., Park, I. C., Park, S. J., Jin, H. O., Lee, J. K., & Riew, K. D. (2006). Anti-apoptotic effects of caspase inhibitors on rat intervertebral disc cells. *Journal of Bone and Joint Surgery*, 88(4), 771–779.
Patten, D. A., Germain, M., Kelly, M. A., & Slack, R. S. (2010). Reactive oxygen species: Stuck in the middle of neurodegeneration. *Journal of Alzheimer's Disease*, 20(Supplement 2), 357–367.

Peng, J. H., Tao, Z. Z., & Huang, Z. W. (2007). Risk of damage to hearing from personal listening devices in young adults. *Journal of Otolaryngology*, 36(3), 181–185.

Perez, R., Freeman, S., Cohen, D., & Sohmer, H. (2002). Functional impairment of the vestibular end organ resulting from impulse noise exposure. *Laryngoscope*, 112(6), 1110–1114.

Phoon, W. H., Lee, H. S., & Chia, S. E. (1993). Tinnitus in noise-exposed workers. *Occupational Medicine*, 43(1), 35–38.

Pirvola, U., Xing-Qun, L., Virkkala, J., Saarma, M., Murakata, C., Camoratto, A. M., Walton, K. M., & Ylikoski, J. (2000). Rescue of hearing, auditory hair cells, and neurons by CEP-1347/KT7515, an inhibitor of c-Jun N-terminal kinase activation. *Journal of Neuroscience*, 20(1), 43–50.

Pourbakht, A., & Yamasoba, T. (2003). Ebselen attenuates cochlear damage caused by acoustic trauma. *Hearing Research*, 181(1–2), 100–108.

Prasher, D., & Sulkowski, W. (1999). The role of otoacoustic emissions in screening and evaluation of noise damage. *International Journal of Occupational Medicine and Environmental Health*, 12(2), 183–192.

Puel, J.-L., Pujol, R., & Yves, C. (2009). Delivery of modulators of glutamate-mediated neurotransmission to the inner ear. U.S. Patent No. 7,589,110. United States Patent Trademark Office.

Puel, J. L., Pujol, R., Ladrech, S., & Eybalin, M. (1991). Alpha-amino-3-hydroxy-5-methyl-4-isoxazole propionic acid electrophysiological and neurotoxic effects in the guinea-pig cochlea. *Neuroscience*, 45(1), 63–72.

Puel, J. L., Pujol, R., Tribillac, F., Ladrech, S., & Eybalin, M. (1994). Excitatory amino acid antagonists protect cochlear auditory neurons from excitotoxicity. *Journal of Comparative Neurology*, 341(2), 241–256.

Puel, J. L., Ruel, J., Gervais d'Aldin, C., & Pujol, R. (1998). Excitotoxicity and repair of cochlear synapses after noise-trauma induced hearing loss. *Neuroreport*, 9(9), 2109–2114.

Rabinowitz, P. M., Pierce Wise, J., Sr., Hur Mobo, B., Antonucci, P. G., Powell, C., & Slade, M. (2002). Antioxidant status and hearing function in noise-exposed workers. *Hearing Research*, 173(1–2), 164–171.

Rappaport, B. Z., Fausti, S. A., Schechter, M. A., Frey, R. H. N., & Hartigan, P. (1985). Detection of ototoxicity by high-frequency auditory evaluation. *Seminars in Hearing*, 6(4), 369–377.

Rezk, B. M., Haenen, G. R., Van Der Vijgh, W. J., & Bast, A. (2004). The extraordinary antioxidant activity of vitamin E phosphate. *Biochimica et Biophysica Acta*, 1683(1–3), 16–21.

Robertson, D. (1983). Functional significance of dendritic swelling after loud sounds in the guinea pig cochlea. *Hearing Research*, 9(3), 263–278.

Rothman, K. J., Moore, L. L., Singer, M. R., Nguyen, U. S., Mannino, S., & Milunsky, A. (1995). Teratogenicity of high vitamin A intake. *New England Journal of Medicine*, 333(21), 1369–1373.

Samson, J., Wiktorek-Smagur, A., Politanski, P., Rajkowska, E., Pawlaczyk-Luszczynska, M., Dudarewicz, A., Sha, S. H., Schacht, J., & Sliwinska-Kowalska, M. (2008). Noise-induced time-dependent changes in oxidative stress in the mouse cochlea and attenuation by D-methionine. *Neuroscience*, 152(1), 146–150.

Sato, K., Niki, E., & Shimasaki, H. (1990). Free radical-mediated chain oxidation of low density lipoprotein and its synergistic inhibition by vitamin E and vitamin C. *Archives of Biochemistry and Biophysics*, 279(2), 402–405.

Sauberlich, H. E. (1994). Pharmacology of vitamin C. *Annual Review of Nutrition*, 14, 371–191.

Saunders, G. H., & Griest, S. E. (2009). Hearing loss in veterans and the need for hearing loss prevention programs. *Noise & Health*, 11(42), 14–21.

Schacht, J. (1999). Antioxidant therapy attenuates aminoglycoside-induced hearing loss. *Annals of the New York Academy of Sciences*, 884, 125–130.

Schafer, F. Q., Wang, H. P., Kelley, E. E., Cueno, K. L., Martin, S. M., & Buettner, G. R. (2002). Comparing beta-carotene, vitamin E and nitric oxide as membrane antioxidants. *Biological Chemistry*, 383(3–4), 671–681.

Scheibe, F., Haupt, H., & Ising, H. (1999). Total magnesium concentrations of perilymph, cerebrospinal fluid and blood in guinea pigs fed different magnesium-containing diets. *European Archives of Oto-Rhino-Laryngology*, 256(5), 215–219.

Scheibe, F., Haupt, H., & Ising, H. (2000). Preventive effect of magnesium supplement on noise-induced hearing loss in the guinea pig. *European Archives of Oto-Rhino-Laryngology*, 257(1), 10–16.
Scheibe, F., Haupt, H., Ising, H., & Cherny, L. (2002). Therapeutic effect of parenteral magnesium on noise-induced hearing loss in the guinea pig. *Magnesium Research*, 15(1–2), 27–36.
Schmuziger, N., Patscheke, J., & Probst, R. (2006). Hearing in nonprofessional pop/rock musicians. *Ear and Hearing*, 27(4), 321–330.
Schmuziger, N., Patscheke, J., & Probst, R. (2007). An assessment of threshold shifts in nonprofessional pop/rock musicians using conventional and extended high-frequency audiometry. *Ear and Hearing*, 28(5), 643–648.
Schneider, D., Schneider, L., Shulman, A., Claussen, C. F., Just, E., Koltchev, C., Kersebaum, M., Dehler, R., Goldstein, B., & Claussen, E. (2000). Gingko biloba (Rokan) therapy in tinnitus patients and measurable interactions between tinnitus and vestibular disturbances. *International Tinnitus Journal*, 6(1), 56–62.
Segal, S., Eviatar, E., Lapinsky, J., Shlamkovitch, N., & Kessler, A. (2003). Inner ear damage in children due to noise exposure from toy cap pistols and firecrackers: A retrospective review of 53 cases. *Noise & Health*, 5(18), 13–18.
Seixas, N. S., Goldman, B., Sheppard, L., Neitzel, R., Norton, S., & Kujawa, S. G. (2005). Prospective noise induced changes to hearing among construction industry apprentices. *Occupational and Environmental Medicine*, 62(5), 309–317.
Selivanova, O., Brieger, J., Heinrich, U. R., & Mann, W. (2007). Akt and c-Jun N-terminal kinase are regulated in response to moderate noise exposure in the cochlea of guinea pigs. *ORL; Journal of Oto-Rhino-Laryngology and Its Related Specialties*, 69(5), 277–282.
Sendowski, I., Raffin, F., & Braillon-Cros, A. (2006). Therapeutic efficacy of magnesium after acoustic trauma caused by gunshot noise in guinea pigs. *Acta Oto-Laryngologica*, 126(2), 122–129.
Sergi, B., Fetoni, A. R., Ferraresi, A., Troiani, D., Azzena, G. B., Paludetti, G., & Maurizi, M. (2004). The role of antioxidants in protection from ototoxic drugs. *Acta Oto-Laryngologica*, 552(Supplement), 42–45.
Sha, S. H., Taylor, R., Forge, A., & Schacht, J. (2001). Differential vulnerability of basal and apical hair cells is based on intrinsic susceptibility to free radicals. *Hearing Research*, 155(1–2), 1–8.
Shen, H., Zhang, B., Shin, J. H., Lei, D., Du, Y., Gao, X., Wang, Q., Ohlemiller, K. K., Piccirillo, J., & Bao, J. (2007). Prophylactic and therapeutic functions of T-type calcium blockers against noise-induced hearing loss. *Hearing Research*, 226(1–2), 52–60.
Shiffman, M. L., Pockros, P., McHutchison, J. G., Schiff, E. R., Morris, M., & Burgess, G. (2010). Clinical trial: The efficacy and safety of oral PF-03491390, a pancaspase inhibitor - a randomized placebo-controlled study in patients with chronic hepatitis C. *Alimentary Pharmacology and Therapeutics*, 31(9), 969–978.
Shim, H. J., Kang, H. H., Ahn, J. H., & Chung, J. W. (2009). Retinoic acid applied after noise exposure can recover the noise-induced hearing loss in mice. *Acta Oto-Laryngologica*, 129(3), 233–238.
Shin, H. S., Cheong, E. J., Choi, S., Lee, J., & Na, H. S. (2008). T-type Ca^{2+} channels as therapeutic targets in the nervous system. *Current Opinion in Pharmacology*, 8(1), 33–41.
Shupak, A., Doweck, I., Nachtigal, D., Spitzer, O., & Gordon, C. (1993). Vestibular and audiometric conseqeunces of blast injury to the ear. *Archives of Otolaryngology – Head and Neck Surgery*, 119, 1362–1367.
Shupak, A., Bar-El, E., Podoshin, L., Spitzer, O., Gordon, C. R., & Ben-David, J. (1994). Vestibular findings associated with chronic noise induced hearing impairment. *Acta Oto-Laryngologica*, 114(6), 579–585.
Siems, W., Wiswedel, I., Salerno, C., Crifo, C., Augustin, W., Schild, L., Langhans, C. D., & Sommerburg, O. (2005). Beta-carotene breakdown products may impair mitochondrial functions – potential side effects of high-dose beta-carotene supplementation. *Journal of Nutritional Biochemistry*, 16(7), 385–397.
Skulachev, V. P. (2009). New data on biochemical mechanism of programmed senescence of organisms and antioxidant defense of mitochondria. *Biochemistry*, 74(12), 1400–1403.

Skulachev, V. P., Anisimov, V. N., Antonenko, Y. N., Bakeeva, L. E., Chernyak, B. V., Erichev, V. P., Filenko, O. F., Kalinina, N. I., Kapelko, V. I., Kolosova, N. G., Kopnin, B. P., Korshunova, G. A., Lichinitser, M. R., Obukhova, L. A., Pasyukova, E. G., Pisarenko, O. I., Roginsky, V. A., Ruuge, E. K., Senin, II, Severina, II, Skulachev, M. V., Spivak, I. M., Tashlitsky, V. N., Tkachuk, V. A., Vyssokikh, M. Y., Yaguzhinsky, L. S., & Zorov, D. B. (2009). An attempt to prevent senescence: A mitochondrial approach. *Biochimica et Biophysica Acta*, 1787(5), 437–461.

Sliwinska-Kowalska, M., Zamyslowska-Szmytke, E., Szymczak, W., Kotylo, P., Fiszer, M., Wesolowski, W., & Pawlaczyk-Luszczynska, M. (2003). Ototoxic effects of occupational exposure to styrene and co-exposure to styrene and noise. *Journal of Occupational and Environmental Medicine*, 45(1), 15–24.

Smith, R. A., Porteous, C. M., Coulter, C. V., & Murphy, M. P. (1999). Selective targeting of an antioxidant to mitochondria. *European Journal of Biochemistry*, 263(3), 709–716.

Smith, R. A., Porteous, C. M., Gane, A. M., & Murphy, M. P. (2003). Delivery of bioactive molecules to mitochondria in vivo. *Proceedings of the National Academy of Sciences of the USA*, 100(9), 5407–5412.

Smith, R. A., Adlam, V. J., Blaikie, F. H., Manas, A. R., Porteous, C. M., James, A. M., Ross, M. F., Logan, A., Cocheme, H. M., Trnka, J., Prime, T. A., Abakumova, I., Jones, B. A., Filipovska, A., & Murphy, M. P. (2008). Mitochondria-targeted antioxidants in the treatment of disease. *Annals of the New York Academy of Sciences*, 1147, 105–111.

So, H. S., Park, C., Kim, H. J., Lee, J. H., Park, S. Y., Lee, J. H., Lee, Z. W., Kim, H. M., Kalinec, F., Lim, D. J., & Park, R. (2005). Protective effect of T-type calcium channel blocker flunarizine on cisplatin-induced death of auditory cells. *Hearing Research*, 204(1–2), 127–139.

Sohmer, H., Elidan, J., & Plotnick, M. (1999). Effect of noise on the vestibular system-Vestibular evoked potential studies in rats. *Noise & Health*, 5, 41–51.

Song, B.-B., Sha, S. H., & Schacht, J. (1998). Iron chelators protect from aminoglycoside induced cochleo- and vestibulotoxicity in guinea pig. *Free Radical Biology and Medicine*, 25, 189–195.

Spankovich, C., Hood, L., Silver, H., Lambert, W., Flood, V., & Mitchell, P. (2011). Associations between dietary intakes and auditory function in an older adult population-based study. *Journal of the American Academy of Audiology*, 22(1), 1–10.

Staecker, H. (2008). Drug delivery to the inner ear and methods of using same U.S. Patent No. 7,387,614. United States Patent Trademark Office.

Suckfuell, M., Canis, M., Strieth, S., Scherer, H., & Haisch, A. (2007). Intratympanic treatment of acute acoustic trauma with a cell-permeable JNK ligand: A prospective randomized phase I/II study. *Acta Oto-Laryngologica*, 127(9), 938–942.

Sulkowski, W., Kowalska, S., Lipowczan, A., Prasher, D., & Raglan, E. (1999). Tinnitus and impulse noise-induced hearing loss in drop-forge operators. *International Journal of Occupational Medicine and Environmental Health*, 12(2), 177–182.

Suzuki, Y., Inoue, T., & Ra, C. (2010). L-type Ca^{2+} channels: A new player in the regulation of Ca^{2+} signaling, cell activation and cell survival in immune cells. *Molecular Immunology*, 47(4), 640–648.

Szydlowska, K., & Tymianski, M. (2010). Calcium, ischemia and excitotoxicity. *Cell Calcium*, 47(2), 122–129.

Tak, J. K., & Park, J. W. (2009). The use of ebselen for radioprotection in cultured cells and mice. *Free Radical Biology and Medicine*, 46(8), 1177–1185.

Takeda, S., Bando, N., & Yamanishi, R. (2008). Ingested beta-carotene enhances glutathione level and up-regulates the activity of cysteine cathepsin in murine splenocytes. *Bioscience, Biotechnology, and Biochemistry*, 72(6), 1595–1600.

Tamir, S., Adelman, C., Weinberger, J. M., & Sohmer, H. (2010). Uniform comparison of several drugs which provide protection from noise induced hearing loss. *Journal of Occupational Medicine and Toxicology*, 5, 26–32.

Teranishi, M., Nakashima, T., & Wakabayashi, T. (2001). Effects of alpha-tocopherol on cisplatin-induced ototoxicity in guinea pigs. *Hearing Research*, 151(1–2), 61–70.

Tewari, R., Sharma, V., Koul, N., Ghosh, A., Joseph, C., Hossain Sk, U., & Sen, E. (2009). Ebselen abrogates TNFalpha induced pro-inflammatory response in glioblastoma. *Molecular Oncology*, 3(1), 77–83.

Tin, L. L., & Lim, O. P. (2000). A study on the effects of discotheque noise on the hearing of young patrons. *Asia-Pacific Journal of Public Health*, 12(1), 37–40.

Toppila, E., Starck, J., Pyykko, I., & Miller, J. M. (2002). Protection against acute noise with antioxidants. Presented at *Nordic Noise: An International Symposium on Noise and Health*, in Nobel Forum, Karolinska Institutet, Stockholm, Sweden, October 25–27, 2002

Turner, J. G. (2007). Behavioral measures of tinnitus in laboratory animals. *Progress in Brain Research*, 166, 147–156.

Turner, J. G., & Parrish, J. (2008). Gap detection methods for assessing salicylate-induced tinnitus and hyperacusis in rats. *American Journal of Audiology*, 17(2), S185–192.

Turner, J. G., Brozoski, T. J., Bauer, C. A., Parrish, J. L., Myers, K., Hughes, L. F., & Caspary, D. M. (2006). Gap detection deficits in rats with tinnitus: A potential novel screening tool. *Behavioral Neuroscience*, 120(1), 188–195.

Tzaneva, L., Savov, A., & Damianova, V. (2000). Audiological problems in patients with tinnitus exposed to noise and vibrations. *Central European Journal of Public Health*, 8(4), 233–235.

Uemaetomari, I., Tabuchi, K., Nakamagoe, M., Tanaka, S., Murashita, H., & Hara, A. (2009). L-type voltage-gated calcium channel is involved in the pathogenesis of acoustic injury in the cochlea. *Tohoku Journal of Experimental Medicine*, 218(1), 41–47.

Usami, S., Hjelle, O. P., & Ottersen, O. P. (1996). Differential cellular distribution of glutathione – an endogenous antioxidant – in the guinea pig inner ear. *Brain Research*, 743(1–2), 337–340.

Van Campen, L. E., Dennis, J. M., King, S. B., Hanlin, R. C., & Velderman, A. M. (1999). One-year vestibular and balance outcomes of Oklahoma City bombing survivors. *Journal of the American Academy of Audiology*, 10(9), 467–483.

van der Laan, F. L. (2001a). Noise exposure and its effect on the labyrinth, Part I. *International Tinnitus Journal*, 7(2), 97–100.

van der Laan, F. L. (2001b). Noise exposure and its effect on the labyrinth, Part II. *International Tinnitus Journal*, 7(2), 101–104.

van Dijk, F. J., Souman, A. M., & de Vries, F. F. (1987). Non-auditory effects of noise in industry. VI. A final field study in industry. *International Archives of Occupational and Environmental Health*, 59(2), 133–145.

Vicente-Torres, M. A., & Schacht, J. (2006). A BAD link to mitochondrial cell death in the cochlea of mice with noise-induced hearing loss. *Journal of Neuroscience Research*, 83(8), 1564–1572.

Vuyyuri, S. B., Hamstra, D. A., Khanna, D., Hamilton, C. A., Markwart, S. M., Campbell, K. C., Sunkara, P., Ross, B. D., & Rehemtulla, A. (2008). Evaluation of D-methionine as a novel oral radiation protector for prevention of mucositis. *Clinical Cancer Research*, 14(7), 2161–2170.

Walden, B. E., Henselman, L. W., & Morris, E. R. (2000). The role of magnesium in the susceptibility of soldiers to noise-induced hearing loss. *Journal of the Acoustical Society of America*, 108(1), 453–456.

Wang, J., Van De Water, T. R., Bonny, C., de Ribaupierre, F., Puel, J. L., & Zine, A. (2003). A peptide inhibitor of c-Jun N-terminal kinase protects against both aminoglycoside and acoustic trauma-induced auditory hair cell death and hearing loss. *Journal of Neuroscience*, 23(24), 8596–8607.

Wang, J., Ruel, J., Ladrech, S., Bonny, C., van de Water, T. R., & Puel, J. L. (2007). Inhibition of the c-Jun N-terminal kinase-mediated mitochondrial cell death pathway restores auditory function in sound-exposed animals. *Molecular Pharmacology*, 71(3), 654–666.

Wang, Y., Hirose, K., & Liberman, M. C. (2002). Dynamics of noise-induced cellular injury and repair in the mouse cochlea. *Journal of the Association for Research in Otolaryngology*, 3(3), 248–268.

Wang, Y. P., & Young, Y. H. (2007). Vestibular-evoked myogenic potentials in chronic noise-induced hearing loss. *Otolaryngology – Head and Neck Surgery*, 137(4), 607–611.

Wang, Y. P., Hsu, W. C., & Young, Y. H. (2006). Vestibular evoked myogenic potentials in acute acoustic trauma. *Otology & Neurotology*, 27(7), 956–961.

Watanabe, K., Inai, S., Hess, A., Michel, O., & Yagi, T. (2004). Acoustic stimulation promotes the expression of inducible nitric oxide synthase in the vestibule of guinea pigs. *Acta Oto-Laryngologica*, 553(Supplement), 54–57.

Wei, X., Zhao, L., Liu, J., Dodel, R. C., Farlow, M. R., & Du, Y. (2005). Minocycline prevents gentamicin-induced ototoxicity by inhibiting p38 MAP kinase phosphorylation and caspase 3 activation. *Neuroscience*, 131(2), 513–521.

Weijl, N. I., Elsendoorn, T. J., Lentjes, E. G., Hopman, G. D., Wipkink-Bakker, A., Zwinderman, A. H., Cleton, F. J., & Osanto, S. (2004). Supplementation with antioxidant micronutrients and chemotherapy-induced toxicity in cancer patients treated with cisplatin-based chemotherapy: A randomised, double-blind, placebo-controlled study. *European Journal of Cancer*, 40(11), 1713–1723.

Wendel, A., Fausel, M., Safayhi, H., Tiegs, G., & Otter, R. (1984). A novel biologically active seleno-organic compound – II. Activity of PZ 51 in relation to glutathione peroxidase. *Biochemical Pharmacology*, 33(20), 3241–3245.

Wolf, F. I., & Trapani, V. (2008). Cell (patho)physiology of magnesium. *Clinical Science*, 114(1), 27–35.

Wolf, F. I., Maier, J. A., Nasulewicz, A., Feillet-Coudray, C., Simonacci, M., Mazur, A., & Cittadini, A. (2007). Magnesium and neoplasia: From carcinogenesis to tumor growth and progression or treatment. *Archives of Biochemistry and Biophysics*, 458(1), 24–32.

Wolf, F. I., Trapani, V., Simonacci, M., Ferre, S., & Maier, J. A. (2008). Magnesium deficiency and endothelial dysfunction: Is oxidative stress involved? *Magnesium Research*, 21(1), 58–64.

Wolf, F. I., Trapani, V., Simonacci, M., Boninsegna, A., Mazur, A., & Maier, J. A. (2009). Magnesium deficiency affects mammary epithelial cell proliferation: Involvement of oxidative stress. *Nutrition and Cancer*, 61(1), 131–136.

Yamashita, D., Jiang, H., Schacht, J., & Miller, J. M. (2004). Delayed production of free radicals following noise exposure. *Brain Research*, 1019, 201–209.

Yamashita, D., Jiang, H.-Y., Le Prell, C. G., Schacht, J., & Miller, J. M. (2005). Post-exposure treatment attenuates noise-induced hearing loss. *Neuroscience*, 134, 633–642.

Yamashita, D., Minami, S. B., Kanzaki, S., Ogawa, K., & Miller, J. M. (2008). *Bcl-2* genes regulate noise-induced hearing loss. *Journal of Neuroscience Research*, 86(4), 920–928.

Yamasoba, T., Nuttall, A. L., Harris, C., Raphael, Y., & Miller, J. M. (1998). Role of glutathione in protection against noise-induced hearing loss. *Brain Research*, 784(1–2), 82–90.

Yamasoba, T., Pourbakht, A., Sakamoto, T., & Suzuki, M. (2005). Ebselen prevents noise-induced excitotoxicity and temporary threshold shift. *Neuroscience Letters*, 380, 234–238.

Yang, L., Zhao, K., Calingasan, N. Y., Luo, G., Szeto, H. H., & Beal, M. F. (2009). Mitochondria targeted peptides protect against 1-methyl-4-phenyl-1,2,3,6-tetrahydropyridine neurotoxicity. *Antioxidants & Redox Signaling*, 11(9), 2095–2104.

Yeum, K. J., Beretta, G., Krinsky, N. I., Russell, R. M., & Aldini, G. (2009). Synergistic interactions of antioxidant nutrients in a biological model system. *Nutrition*, 25(7–8), 839–846.

Yildirim, I., Kilinc, M., Okur, E., Inanc Tolun, F., Kilic, M. A., Kurutas, E. B., & Ekerbicer, H. C. (2007). The effects of noise on hearing and oxidative stress in textile workers. *Industrial Health*, 45(6), 743–749.

Ying, Y. L., & Balaban, C. D. (2009). Regional distribution of manganese superoxide dismutase 2 (Mn SOD2) expression in rodent and primate spiral ganglion cells. *Hearing Research*, 253(1–2), 116–124.

Ylikoski, M. E., & Ylikoski, J. S. (1994). Hearing loss and handicap of professional soldiers exposed to gunfire noise. *Scandinavian Journal of Work, Environment and Health*, 20(2), 93–100.

Ylikoski, J., Juntunen, J., Matikainen, E., Ylikoski, M., & Ojala, M. (1988). Subclinical vestibular pathology in patients with noise-induced hearing loss from intense impulse noise. *Acta Oto-Laryngologica*, 105(5–6), 558–563.

Zhang, M., Liu, W., Ding, D., & Salvi, R. (2003). Pifithrin-alpha suppresses p53 and protects cochlear and vestibular hair cells from cisplatin-induced apoptosis. *Neuroscience*, 120(1), 191–205.

Zhao, K., Zhao, G. M., Wu, D., Soong, Y., Birk, A. V., Schiller, P. W., & Szeto, H. H. (2004). Cell-permeable peptide antioxidants targeted to inner mitochondrial membrane inhibit mitochondrial swelling, oxidative cell death, and reperfusion injury. *Journal of Biological Chemistry*, 279(33), 34682–34690.

Zhao, K., Luo, G., Giannelli, S., & Szeto, H. H. (2005). Mitochondria-targeted peptide prevents mitochondrial depolarization and apoptosis induced by tert-butyl hydroperoxide in neuronal cell lines. *Biochemical Pharmacology*, 70(12), 1796–1806.

Chapter 14
Frontiers in the Treatment of Hearing Loss

Tatsuya Yamasoba, Josef M. Miller, Mats Ulfendahl, and Richard A. Altschuler

1 Introduction

In the last decade, a paradigm shift has occurred in our vision for the prevention and treatment of hearing impairment. No longer are the solutions restricted to hearing aids, surgery, and implants to restore hearing, control of serum levels to prevent drug-induced ototoxicity, hearing protectors to prevent noise-induced hearing loss (NIHL), and for hereditary loss: wait and hope. Obviously all but the latter practices are of vital continued value, but the promise of more varied and more effective opportunities to prevent hearing loss and to restore hearing have provided increased hope and opportunity. Our future vision is now filled with complex pharmaceutical, cellular, and molecular strategies to modulate hereditary loss, replace and regenerate tissues of the inner ear, and prevent drug-induced hearing loss and NIHL. This future holds the promise of dramatically reducing the lost educational and job opportunities, the social isolation, and the reduced quality of life that accompanies hearing impairment and deafness, and with it the enormous economic costs associated with health care and lost productivity (estimated by the World Health Organization at >2% world GNP). This future molds and reshapes the practices of audiology and otolaryngology to place far greater efforts on the prevention of hearing impairment and the use of local and systemic drug treatment to restore hearing.

A new vision for treatment is based on an increased understanding of the cellular and molecular mechanisms underlying the progression of pathology from an initiating event to hearing impairment. Figure 14.1 diagrams this progression and

J.M. Miller (✉)
Department of Otolaryngology and Department of Cell & Developmental Biology
Kresge Hearing Research Institute, 1150 West Medical Center Drive,
Ann Arbor, MI 48109-5616, USA
e-mail: josef@umich.edu

C.G. Le Prell et al. (eds.), *Noise-Induced Hearing Loss: Scientific Advances*,
Springer Handbook of Auditory Research 40, DOI 10.1007/978-1-4419-9523-0_14,

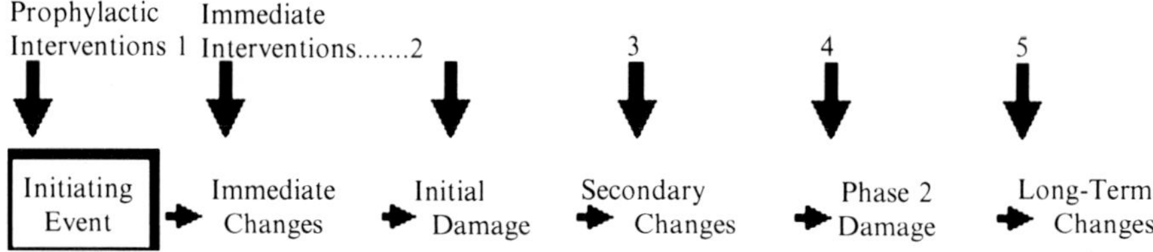

Fig. 14.1 A schematic of the events in the progression of pathology from an initiating event such as noise overstimulation to the long-term changes associated with hearing impairment. Arrows mark the opportunities for interventions for prevention, repair, and rescue, ranging from prophylactic interventions before the initiating event, and immediate interventions after the event, to treatments after damage has progressed

indicates multiple potential timings for interventions. The initiating event could be intense noise, an ototoxic drug, a viral agent, an autoimmune response, or any other traumatic event.

This initiating event leads to immediate changes, which could be common to most traumatic events or could be restricted to a specific trauma such as noise. Resulting changes could be in metabolic activity, reactive oxygen species (ROS) formation, blood flow, stress response, or excitotoxicity, all of which occur after noise (for detailed reviews, see Hu, Chap. 4; Le Prell & Bao, Chap. 13). Some of these could result in immediate initial damage, for example, excitotoxicity resulting in bursting of auditory nerve peripheral processes. Prophylactic or immediate post-trauma interventions could target these immediate changes and prevent the initial damage. In the absence of immediate intervention, there is a subsequent progression of secondary changes. These can induce and influence intracellular pathways, such as those leading toward cell death or protection, and can also set in motion cellular and molecular changes both in the cochlea and in central auditory pathways (see, e.g., Kaltenbach, Chap. 8, for detailed discussion of central auditory system plasticity post-noise). Thus, both the cochlea and the central auditory system provide targets for interventions. The progression of these secondary insult pathways can lead to apoptotic cell death and additional waves of cell damage progressing over hours, days, and weeks, with targets for interventions diminishing over time as events proceed. Once the damage is complete, the long-term changes remain, including hearing loss or hearing disorders, and the need for treatments for cure or improvement.

Many of the cellular and molecular mechanisms associated with the different pathologies and changes along this progression, as well as interventions and methods of accomplishing interventions, are considered in more detail in the other chapters. Here we relate these new insights in mechanisms to the potential clinical interventions that may treat the inner ear to prevent hearing loss or restore lost hearing. The most optimal intervention and intervention time may not always be practical when brought into a real-life situation, and alternative approaches must be considered. Translational studies based on basic research, identification of mechanisms, and potential interventions (sites and times) will have a feed-forward influence on validating (or not) our understanding and interpretation of mechanisms underlying pathologies of the inner ear. This increasing knowledge and understanding will then feed back to create new interventions and novel technologies to further enable the interventions.

2 Prevention of Hearing Loss

2.1 *Timing of Interventions for Preventing Hearing Loss*

There are many cases wherein a subject knowingly enters into a situation that provides a risk of generating hearing loss. Such an "initiating event" (Fig. 14.1) could be from noise in the working or recreational environment, or from drugs that are taken to treat diseases such as cancer or bacterial infection. It could then be possible to have interventions before the initiating event, shown as the prophylactic interventions in Fig. 14.1, during or immediately after the event (listed as Immediate Intervention in Fig. 14.1) before the immediate changes induced by the noise, drug, or other "event" occurs. In this case, our knowledge of the mechanisms leading to early pathology is critical to identify early interventions. However, many of the pathways that are induced continue to progress over hours, days, and even weeks, and prevention/intervention can still be possible well after the initiating event.

Outer hair cells are specialized sensory cells that actively expand and contract during acoustic transduction and thus contribute to the exquistive sensitivity of the auditory system. Necrotic- or apoptotic-induced hair cell death represents the primary cause of hearing impairment for most, if not all, environmental stress-induced cell death (e.g., Hu, Chap. 5). In addition, many instances of genetic stress-induced cell death appear to reflect metabolically driven mitochondrial derived oxidative stress (e.g., Gong and Lomax, Chap. 9). Noise stress can be considered a representative model of environmental stress-induced inner ear cell death. During noise stress, energy demands induce mitochondrial free radical formation, causing lipid peroxidation and the upregulation of cell death pathways, producing hair cell death by necrosis or apoptosis. Free radical formation occurs in the organ of Corti and lateral wall soft tissues, and this free radical formation is enhanced by reduced blood flow during noise and a "stroke-like" rebound reperfusion after the noise. Free radical formation continues after exposure, and increased accumulations have been linked to progressive cell death over a 10-day post-noise period (Yamashita et al. 2004). Genetic- or diet-induced upregulation of endogenous antioxidant pathways, or exogenous treatment with antioxidants and vasodilators, modulates the free radical formation, subsequent cell death, and hearing loss. Similar findings show the same mechanism, mitochondrial-derived oxidative stress, underlies aminoglycoside-induced hair cell death, may underlie age-related cell death, and has been speculated as a factor in Ménière's disease, sudden sensorineural hearing loss, and trauma of cochlear implantation. From other fields, it is clearly established that free radical formation is key to hyperoxia-, hypoxia-, reoxygenation-, radiation-, cigarette smoke-, and stroke-induced cell death (Circu and Aw 2010; Roberts et al. 2010;

for recent reviews). To the extent that mitochondrial-derived oxidative stress represents a common element to the final pathway to cell death, it represents an "upstream" target of opportunity for intervention and prevention.

Most mechanism-based therapeutic strategies take one of two approaches. One approach is to mimic or enhance endogenous "good" mechanisms, those that provide protection. Three such protective pathways are discussed herein: antioxidants, neurotrophic factors, and heat shock proteins. The other approach is to block the progression of "bad" pathways, those that lead to cell death. This could involve blocking apoptotic and excitotoxic pathways, using agents such as calcium channel blockers, calpain and calcineurin inhibitors, Bcl-2 anti-apoptotic proteins, caspase inhibitors, and JNK-inhibitors. These agents were reviewed in Le Prell et al. (2007b), and a more recent discussion is provided by Abi-Hachem et al. (2010). Recent data on calcium channel blockers and JNK inhibitors are reviewed in Le Prell and Bao (Chap. 13).

2.1.1 Antioxidants

Endogenous antioxidant systems are a major protective mechanism in the cochlea that can respond to a variety of trauma, stresses, and "initiating events" such as intense noise that generates free radicals in the cochlea for hours and days after exposure, which then induce cell death signals (for detailed reviews, see Le Prell et al. 2007b; Le Prell and Bao, Chap. 13). Administration of exogenous antioxidants has great potential for therapeutic intervention. In fact, a variety of antioxidant agents have been shown to attenuate NIHL effectively in animal studies. Such agents include glutathione monoethyl ester (GSHE; Ohinata et al. 2000; Kopke et al. 2002; Miller et al. 2003b), resveratrol (Seidman et al. 2003), allopurinol (Seidman et al. 1993; Cassandro et al. 2003), superoxide dismutase-polyethylene glycol (Seidman et al. 1993), lazaroid (a drug that inhibits lipid peroxidation and scavenges free radicals) (Quirk et al. 1994), vitamin A (Ahn et al. 2005), vitamin C or ascorbate (Derekoy et al. 2004; McFadden et al. 2005), alpha-tocopherol (Hou et al. 2003), salicylate and trolox (Yamashita et al. 2005), and (R)-phenylisopropyl-adenosine (R-PIA; Hu et al. 1997). 2-Oxothiazolidine-4-carboxylate (OTC) (Yamasoba et al. 1998), N-acetylcysteine (NAC) (Ohinata et al. 2003; Duan et al. 2004), NAC and salicylate (Kopke et al. 2000), D-methionine (Kopke et al. 2002), and ebselen (Pourbakht and Yamasoba 2003; Lynch and Kil 2005; Yamasoba et al. 2005). Other potential agents such as coenzyme Q10 (Hirose et al. 2008) and ferulic acid (Fetoni et al. 2010) continue to be added.

Dietary supplements that reduce NIHL are of particular interest given their easy over-the-counter accessibility, but therapy with any single micronutrient may need to be initiated days to weeks in advance of noise exposure to obtain clinically meaningful results. Whereas a 35-day pretreatment with vitamin C significantly reduced NIHL and cochlear hair cell death (McFadden et al. 2005), vitamin C treatment initiated 48 h before noise exposure failed to prevent noise-induced cell death (Branis and Burda 1988). Pretreatment requirements may vary across micronutrients, as vitamin E reduced NIHL with treatment initiated 3 days pre-noise (Hou et al. 2003) and vitamin A reduced NIHL with treatment initiated 2 days pre-noise (Ahn et al. 2005).

Although dietary treatments may need to be provided for some longer period of time pre-noise to be maximally effective, high-dose vitamin C did not completely prevent NIHL even with 35 days pretreatment (McFadden et al. 2005), and stable plasma and tissue levels of vitamin C are obtained in humans approximately 3 weeks after beginning dietary treatment (Levine et al. 1996). Taken together, these data suggest that dietary antioxidants may be more useful in combination than as single-agent therapeutics. The work of Le Prell et al. (2007a) demonstrating robust attenuation of NIHL with 1-h pre-exposure administration of the antioxidants beta-carotene, vitamins C and E, plus magnesium, supports this view.

With respect to the propagation of oxidative stress reactions, it is clear that iron (Halliwell and Gutteridge 1986) and other transition metals (for review, see Halliwell and Gutteridge 2007) contribute to the generation and propagation cycles of free radicals. Ferrous iron (II) is known to be oxidized by hydrogen peroxide to ferric iron (III), a hydroxyl radical and a hydroxyl anion. Iron (III) is then reduced back to iron (II), a peroxide radical and a proton by the same hydrogen peroxide. This process is known as the Fenton reaction. Because iron is involved in ROS generation, iron chelators are also potential candidates to reduce NIHL. An iron chelator, deferoxamine mesylate (DFO), alone or in combination with mannitol, a hydroxyl scavenger and weak iron chelator, attenuated NIHL in guinea pigs with little evidence for additive effects (Yamasoba et al. 1999). Because an oral iron chelator is available and used safely for humans (Oliva et al. 2010), such agents may be applied prophylactically for humans, especially for those who are scheduled to be exposed to intense noise, including those in the military such as bomber crews.

2.1.2 Neurotrophic Factors

Neurotrophic factors (NTFs) provide another endogenous protective mechanism that can be mimicked or enhanced to provide therapeutic intervention in the progression toward hearing loss. NTFs have multiple functions and, therefore provide different options. For example, NTFs will scavenge free radicals, interrupt cell death pathways, and modulate calcium homeostasis; any of which may attenuate the progression toward hearing loss. Withdrawal of NTFs leads to ROS formation and initiates a cascade of events that lead to cell death (for review, see Kirkland and Franklin 2003).

Most of the exogenous NTFs delivered into the cochlea have been reported to prevent noise-induced hair cell death, which include acidic fibroblast growth factor (FGF1) (Sugahara et al. 2001), basic FGF or FGF2 (Zhai et al. 2004), glial cell line–derived neurotrophic factor (GDNF) (Ylikoski et al. 1998; Yamasoba et al. 1999), and neurotrophic factor 3 (NT3) (Shoji et al. 2000a). Brain-derived neurotrophic factor (BDNF) (Shoji et al. 2000a) and, in some studies, FGF1 and FGF2 (Yamasoba et al. 2001) did not reduce noise-induced injury, suggesting that (1) the effect is growth factor specific, which could be a consequence of different NTF receptors on the hair cells (Ylikoski et al. 1993; Pirvola et al. 1997), and (2) the protective effects are dependent on multiple factors such as optimal drug dosage and nature or severity of injury.

In addition to preserving hair cell survival after noise, NTFs have been shown to be extremely effective at preserving neural survival in the absence of surviving hair cells. In the presence of intact hair cells, damaged auditory nerve peripheral processes may be able regrow and restore auditory sensation (Puel et al. 1991, 1995; Le Prell et al. 2004), whereas with loss of hair cell targets, auditory nerve regrowth is limited (Bohne and Harding 1992; Lawner et al. 1997; McFadden et al. 2004). It has, however, recently been shown that an acoustic overexposure that causes moderate, reversible, temporary shift of hearing threshold (TTS) may leave cochlear sensory cells intact but cause loss of afferent nerve terminal connections and delayed degeneration of the auditory nerve and cell bodies (Kujawa and Liberman 2009), suggesting that regrowth can be absent or inefficient. Although delayed auditory nerve degeneration is frequently observed as a consequence of NTF deprivation that occurs when sensory cells in the organ of Corti are damaged, the finding of loss of inner hair cell–auditory nerve connections and nerve degeneration post-noise, in the presence of intact hair cells, is a component of NIHL that should not be ignored. Indeed, much of the basic research defining protection via NTFs in the auditory system has only been in the context of neural preservation after noise- or aminoglycoside-induced cell death and deafness and not considered connections to auditory nerve.

Use of growth factor combinations, or combinations of growth factors with other non-growth factor substances, enhances efficacy over single agents both in vivo and in vitro (for review see Le Prell et al. 2007b). Importantly, a single NTF or combinations of NTFs can be highly efficacious in promoting auditory nerve survival even with temporal delay in onset of treatment relative to deafening. Nerve growth factor (NGF) delivered alone (Shah et al. 1995), or BDNF, NT3, and neurotrophin-4/5 alone (Gillespie et al. 2004), each enhanced neural survival even when administration was delayed by 2 weeks. The combination of BDNF and ciliary neurotrophic factor (CNTF) enhanced auditory nerve survival even at delays of up to 6 weeks post-deafening (Yamagata et al. 2004). Consistent with an important role for FGF1 in neurite outgrowth in the immature auditory system (Dazert et al. 1998; Hossain and Morest 2000), it has recently been demonstrated that BDNF plus FGF1 was effective in promoting systematic regrowth of the peripheral process of the auditory nerve even after a 6-week period of deafening (Miller et al. 2007; Glueckert et al. 2008). Together, these results suggest post-noise treatment with NTFs may prevent neural degeneration that occurs consequent to noise-induced sensory cell death.

2.1.3 Heat Shock Proteins

The classical stress response, involving heat shock proteins, provides another endogenous pathway that could be induced to provide protection from noise or other initiating events. Heat shock proteins provide protection by stabilizing proteins and preventing stress-induced misfolding and may further interface with the endoplasmic reticulum (ER)-related pathways and pathologies. Yoshida et al. (1999) found that providing a heat stress in mice that induced the heat shock response in the cochlea provided protection from a noise exposure that might otherwise be damaging to the cochlea and hearing. Fairfield et al. (2005) did the opposite, removing the protection by using mice with the heat shock response compromised by knockout (KO) of HSF1, the transcription factor that induces activation of the pathway. Results showed more damage and hearing loss after noise in the HSF1 KO mice compared to wild-type littermates. While heating one's ear before noise might not be practical, recently small molecules have been developed that can act at the cellular level to activate HSF1 and induce the heat shock protective response (Neef et al. 2010) providing the potential for a more applicable therapeutic intervention.

2.1.4 Blockers of Excitotoxicity

Although prevention of cell death is a major target of interventions to prevent hearing loss, there can also be excitotoxicity leading to loss of connections between inner hair cells (IHCs) and auditory nerve, contributing to hearing disorders. Although regrowth and reconnection of lost processes to surviving IHCs has been shown (Puel et al. 1998; Pujol and Puel 1999), recent studies show this reconnection is not always efficient (Kujawa and Liberman 2009), and loss of these connections could contribute to reduced speech comprehension, particulary in a noisey environment. Prevention of excitotoxicity must, therefore, also be a goal for therapeutic interventions. Excitotoxic trauma and the development of novel calcium channel blockers as potential therapeutics for prevention of NIHL are reviewed in detail in Le Prell and Bao (Chap. 13) and are not discussed further in this chapter.

2.1.5 Blood Flow Promoting Drugs

Trauma-mediated changes in cochlear blood flow influence the progression of hearing loss and interventions influencing blood flow can also be a therapeutic target. While in most other tissues increased metabolism is associated with increased blood flow to provide additional oxygen to stressed cells; in the cochlea, intense noise decreases blood flow and is followed by a subsequent rebound and overshoot in blood flow (for review see Le Prell et al. 2007b). The decreased blood flow in the cochlea is associated with noise-induced reductions in blood vessel diameter and red blood cell velocity (Quirk et al. 1992; Quirk and Seidman 1995). This appears to be caused by a byproduct of noise-induced free radical formation, particularly in tissues associated with the cochlear vasculature (lateral wall) (Miller et al. 2003b) and reducing the vasoconstriction that occurs with ROS production could contribute to the reduction of NIHL achieved by antioxidants. Agents that reduce vasoconstriction or have vasodilating effects such as hydroxyethyl starch (HES, e.g., Lamm and Arnold 2000) or magnesium (e.g., Scheibe et al. 2000), have been shown to reduce NIHL (Le Prell et al. 2007b for review). The protective effects of enhancing blood flow during noise exposure may be based on reducing the noise-induced blood flow reduction directly or by blocking the subsequent blood flow rebound and overshoot that follows the noise-induced reduction. In addition to the well-characterized effects on vasodilatation, biochemical effects of magnesium include modulation of calcium channel permeability, influx of calcium into cochlear hair cells, and glutamate release (Gunther et al. 1989; Cevette et al. 2003). Regardless of the specific mechanism, magnesium clearly attenuates NIHL and is safe for use in humans within the recommended dose range.

2.1.6 Post-trauma Interventions

The question of timing for therapeutic interventions along the progression of noise-induced damages (Fig. 14.1) is a critical one. How late can interventions be applied in the process and pathway to cell death to prevent the cell from dying? Will the preserved cell be completely healthy and functioning if it is saved late in the process? This may depend on the mechanism applied for the intervention and how far along a cell is in the apoptotic pathway; however, this question remains to be carefully studied.

One exciting development is that because cell death pathways progress over a period of time, it is possible to intervene well after the initiating event and still prevent cells from progressing toward the end state of cell death. Noise-induced oxidative stress begins early and becomes substantial over time (first suggested by Ohlemiller et al. 1999), which would explain observations of hair cell death that accelerates with time after exposure for a period of up to 14 days (Bohne et al. 1999; Yamashita et al. 2004). Yamashita et al. (2004) found peak ROS and RNS production in cells of the organ of Corti was at 7–10 days after noise insult, and the final extent of damage to cochlear tissues could reflect cell death pathways initiated by late-forming free radicals in the inner ear. Therapeutic interventions after noise exposure have proven to be effective. Treatment with salicylate and vitamin E initiated 24 h after noise exposure was almost as effective as pretreatment in preventing loss of sensory elements and treatment initiated 3 days postexposure also reduced NIHL and sensory cell death relative to untreated controls (Yamashita et al. 2005). Treatment delayed 5 days relative to noise insult was not effective. D-Methionine reduced NIHL and cochlear damage when provided 1 h after noise overstimulation (Campbell et al. 2007), and all-trans retinoic acid could reduce NIHL and cochlear damage when provided up to 2 days after a noise overstimulation (Shim et al. 2009), though efficacy decreased over time. These studies suggest there is a window of opportunity of several days after noise overstimulation where therapeutic intervention can provide benefit, even if pretreatment or treatment shortly after the noise is most effective.

2.1.7 Combination Effects

Given that none of the interventions tested to date completely prevents NIHL and noise-induced sensory cell death, it would seem reasonable to seek an additive effect with a combination of factors that intervene at multiple sites in the biochemical cell death cascade. When the effect of a combination of an antioxidant (mannitol, a hydroxyl scavenger), a neurotrophic factor (GDNF), and an iron chelator (deferoxamine mesylate [DFO]), each of which individually attenuate NIHL, was evaluated, there was little evidence for additive effects; that is, treatment with a combination of agents yielded no greater protection than the most effective agent delivered alone (Yamasoba et al. 1999).

Other studies similarly failed to find evidence for additive or synergistic effects. Another study that evaluated the potential for additive effects of various combinations of antioxidants and vasodilators, including betahistine, vitamin E, and a combination of these agents, and salicylate, vitamin E, and a combination of these agents also demonstrated no evidence for additive effects (Miller et al. 2006). When the individual and combined effects of creatine, a cellular energy enhancer, and tempol were compared in guinea pigs exposed to noise, the effects of the combination treatment were similar to those treated with creatine alone (Minami et al. 2007).

Only recently, with combinations of antioxidant vitamins and magnesium, have additive effects on prevention of NIHL or otoxicity been demonstrated. Yeum et al. (2009) have shown additive effects with β-carotene and α-tocopherol, ascorbic acid and α-tocopherol, and β-carotene and ascorbic acid on antioxidant activity in reconstituted human serum. A robust additive effect on protection from NIHL was demonstrated with the combination of β-carotene, vitamins C and E, and magnesium (Le Prell et al. 2007a). The identification of specific combinations of agents that act in additive or synergistic (i.e., multiplicative) ways is a compelling goal for future research activities. Because activation of calcineurin depends on ROS production and ROS-induced deficits in calcium homeostasis (Huang et al. 2001; Gooch et al. 2004; Rivera and Maxwell 2005), one might predict that blocking early ROS production would reduce activation of the calcineurin-initiated apoptotic pathway. If so, pretreatment with antioxidant agents that are highly efficient hydroxyl radical scavengers, in combination with FK506 to directly intervene in the calcineurin pathway, might more effectively reduce NIHL and noise-induced cell death. This hypothesis has not been directly tested, and identification of the most effective combinations remains a challenge for future research efforts.

2.1.8 Novel Therapeutic Tools: Hydrogen Gas and Water

Molecular hydrogen (hydrogen gas and hydrogen-rich water) was recently established as a unique antioxidant that selectively reduces the hydroxyl radical, the most cytotoxic ROS, but that does not react with other ROS that possess beneficial physiological roles. Inhalation of hydrogen gas markedly suppresses brain injury induced by focal ischemia and reperfusion by buffering the effects of oxidative stress in rats (Ohsawa et al. 2007). Further, the inhalation of hydrogen gas suppressed hepatic injury caused by ischemia-reperfusion in mice (Fukuda et al. 2007) and limited the extent of myocardial infarction in rats (Hayashida et al. 2008). In the nervous system, hydrogen-rich water was shown to prevent superoxide formation in brain slices of vitamin C-depleted senescence marker protein 30/gluconolactonase-knockout mice (Sato et al. 2008) and to prevent stress-induced impairments in learning tasks during chronic physical restraint in mice (Nagata et al. 2009). Moreover, a clinical study showed that consuming hydrogen-rich pure water improves lipid and glucose metabolism in type 2 diabetes patients (Kajiyama et al. 2008).

Hydrogen gas is permeable to cell membranes and can target organelles, including mitochondria and nuclei. This is especially favorable for inner-ear medicine, because many therapeutic compounds are blocked by the blood–labyrinthine barrier and can not get access to the inner ear. In a recent ex vivo study, hydrogen gas markedly decreased oxidative stress by scavenging ROS and protected cochlear cells and tissues against oxidative stress (Kikkawa et al. 2009). When antimycin A was applied to organotypic explant cultures of mouse auditory epithelia, incubation with a hydrogen-saturated medium significantly reduced ROS generation and subsequent lipid peroxidation. Reduced free radical insult increased survival of the hair cells. Considering the safety and easy accessibility of hydrogen to cells in the inner ear, hydrogen gas or hydrogen-rich water seems to be a promising agent to investigate for potential prevention of NIHL in human subjects exposed to noise.

3 Treatment of Hearing Disorders

Although prevention of hearing disorders would clearly be optimal, protective treatments have not yet been shown to work in human trials, are not yet approved by the FDA for hearing protection, and even once they are more developed they may be too late or insufficient for many subjects. Therefore, treatment of hearing loss and hearing disorders remains an important and critical goal, the last intervention target in Fig. 14.1. Treatments fall into two general categories of "maintenance" and "restoration." Maintenance can involve prevention of further pathology, where it overlaps with preventions. Restoration rests upon the three Rs of "repair," "regeneration," and "replacement." Repair involves treating remaining cells in the damaged ear to return the auditory pathways to their condition before the hearing loss. Regeneration requires treatments to induce repopulation from endogenous progenitors or redifferentiation of cells remaining in the damaged ear, although replacement could involve a variety of approaches ranging from the use of exogenous cell implants to replace lost cells to cochlear prostheses to bypass lost cells. The combination of repair, regeneration, and replacement is frequently termed "tissue engineering."

3.1 Maintenance

3.1.1 Survival Factors: Neurotrophic Factors

Just as NTFs can have multiple roles in protection, they also have roles in maintenance, repair, and restoration. NTFs have an important function as survival factors, and deafferentation can result in NTF deprivation for the auditory nerve that can lead to free radical formation and the upregulation of cell death pathways (NTF hypothesis; Mattson 1998 for review). Thus, hair cell loss results in a secondary and progressive loss of auditory nerve and its spiral ganglion neurons (SGN). If exogenous NTFs such as BDNF, NT-3, and GDNF are supplied to the auditory nerve to replace lost endogenous NTFs, they will promote maintenance and survival (e.g., Ernfors et al. 1996; Staecker et al. 1996; Miller et al. 1997; Green et al. 2008). Supplying NTFs will enhance not only the survival of SGN (Green et al. 2008 for review) but also the electrical responsiveness of the neurons (Maruyama et al. 2008). Today, the cochlear prosthesis offers an important treatment option for patients with severe hair cell loss. Because efficacy of the cochlear prosthesis is dependent on the number and functionality of the remaining SGN (e.g., Nadol et al. 1989; Incesulu and Nadol 1998), it is of therapeutic interest to prevent degeneration of auditory sensory neurons, and neurotrophic treatment has been suggested for use with cochlear prostheses to protect and support the SGN.

3.1.2 Survival Factors: Electrical Stimulation

Electrical activity within the auditory nerve provides another important survival factor (Green et al. 2008), and providing electrical stimulation to the auditory nerve has been shown to increase SGN survival after the deafferentation associated with IHC loss (Green et al. 2008 for review). The combination of chronic cochlear electrical stimulation and application of NTFs has been shown to be more effective than either alone (for examples, see Kanzaki et al. 2002; Scheper et al. 2009).

3.1.3 Regrowth of Auditory Nerve Peripheral Processes

An early event on the long-term path to SGN death after loss of IHCs is the relatively rapid degeneration of the deafferented peripheral processes of the auditory nerve, first to the level of the habenula perforata and later to the soma (Webster and Webster 1981; Spoendlin 1984; Spoendlin and Schrott 1990). If hair cell replacement becomes possible, then regrowth of the peripheral process will need to be successfully induced. Moreover, it will ultimately be necessary to connect the new hair cells to the cochlear nucleus via regrown peripheral processes when hair cell restoration or replacement becomes possible (see next section). In the present, SGN peripheral process regrowth might provide benefit to cochlear prostheses. Regrowth of the SGN peripheral process to the vicinity of the electrode would provide a closer target for cochlear electrical stimulation that would allow lower thresholds for excitation, a larger dynamic range of responsiveness, and provide less current spread and better channel separation. Lower thresholds would require less energy, allowing more complex signal processing strategies and increased battery life.

Several NTFs including BDNF, NT-3, GDNF, fibroblast growth factor (FGF), and CNTF play a role in inducing, directing, and modulating connections in the cochlea during normal development (Fritzsch et al. 1997 for review) and have also been shown to induce a robust growth of neurites in cultured SGN (Staecker et al. 1995; Green et al. 2008 for reviews). Several NTFs have been shown to induce regrowth of afferent and efferent peripheral processes into the cochlea after hair cell loss in vivo when provided either intrascalar by mini-osmotic pumps (Altschuler et al. 1999; Miller et al. 2007; Glueckert et al. 2008) or more recently after gene transfer (Shibata et al. 2010).

In preclinical implant studies, treatment with these factors has been shown to enhance electrical responsiveness, increasing both threshold sensitivity and dynamic range of electrical auditory brain stem responses (ABR) (Miller et al. 2002; Yamagata et al. 2004; Maruyama et al. 2007, 2008). In these studies, it is not clear to what extent this enhanced responsiveness reflects maintenance of SGN and to what extent it reflects regrowth of peripheral afferent processes; it is likely that both factors contribute. Electrical stimulation may also induce regrowth of peripheral processes (Altschuler et al. 1999), and NTF-induced regrowth has been shown to be further enhanced by antioxidants (Maruyama et al. 2007, 2008). Immediately after implantation, it may be appropriate to infuse NTFs to initiate a burst of neurite regrowth, followed (or accompanied by) electrical stimulation with particular parameters for the first weeks, followed then by different parameters of electrical stimulation for maintenance of the connection and signal processing. Antioxidants may be used over a period before and after implantation to enhance regrowth, as well as protect from the trauma of implantation (Abi-Hachem et al. 2010), with little risk.

3.2 *Restoration*

3.2.1 Regeneration

The exciting discovery of hair cell regeneration after sensory cell death in the chick (e.g., Corwin and Cotanche 1988; Ryals and Rubel 1988) provided the great promise that key factors driving regeneration in birds could be introduced in mammals, including humans. Although this task has not yet been fully accomplished, great progress has been made. These efforts have spawned a set of strategies to identify and analyze the inducing factors, and the first steps toward creating new hair cells in the damaged mammalian ear have been taken. In species that naturally regenerate sensory cells when damaged, the source appears to be the supporting cells, and the mechanism often involves a dedifferentiation, reentry to cell cycling, and division, with one daughter cell becoming a hair cell and the second maturing to a replacement supporting cell, thus maintaining the mosaic of the sensory epithelium critical to mechanoelectric transduction (Kwan et al. 2009; Cotanche and Kaiser 2010 for recent reviews). If the factors that induce, modulate, and guide regeneration in the chick can be induced in mammals, perhaps a comparable regeneration can occur.

Important guidance has also come from an increased understanding of the transcription factors, their downstream pathways, and the molecular mechanisms that control the normal development of the mammalian cochlea and guide an eventual hair cell versus supporting cell fate decision. Atoh1 is a key transcription factor in the hair cell fate choice (Maricich et al. 2009), and forced upregulation of Atoh1 by gene transfer can induce supporting cells into a hair cell phenotype in the mature cochlea in the profoundly deafened guinea pig with nerve fiber innervation and, remarkably, the return of hearing (Izumikawa et al. 2005). These findings provide a key validation of our understanding of many of the mechanisms involved in hair cell development and repair. However, translation to human application will be technically difficult when involving gene therapy (see Sect. 3.4.1 for further discussion), and a gene product (protein) approach affecting other key events in the differentiation process is also discussed later (Sect. 3.4.2).

3.2.2 Replacement: Cellular

An alternative to gene therapy for replacement of lost sensory cells or auditory nerve is use of exogenous cell implants. This approach has been applied to the neurodegenerative disorder Parkinson's disease, with initial promising results (see Winkler et al. 2005 for a review). Although technical hurdles need to be resolved before cell therapy becomes a realistic clinical tool for the treatment of Parkinson's disease, the promise of this strategy is clear. Importantly, the same approach could be applied to the dysfunctional inner ear. One could implant exogenous hair cells or auditory neurons or implant progenitor cells that are induced to become sensory cells or neurons. However, because the cochlea has an extremely complex three-dimensional structure, every cellular element needs to be precisely placed and oriented to achieve proper function. It is therefore difficult to imagine externally applied cells reaching the appropriate location and assuming the necessary functional connections to adequately replace missing hair cells and provide a functional replacement. The more common approach to restoring sensory cells in the inner ear has, therefore, focused on repair (as previously described) rather than replacement. Because the structural organization of the spiral ganglion is much less restrictive, it is conceivable to imagine a cell therapy approach focusing on the SGN being successful (see Li et al. 2004; Ulfendahl et al. 2007; Altschuler et al. 2008; for reviews).

Several cell types have been tested for the purpose of implantation into the inner ear for nerve or hair cell replacement. These range from the most immature embryonic stem cells to well-differentiated neural tissue (Ulfendahl et al. 2007; Altschuler et al. 2008; Edge and Chen 2008; for reviews). Stem cells are characterized by their capacity for self-renewal and give rise to many different cell types. Embryonic stem cells have been a major focus of research as transplantation candidates because they are both proliferative and capable of generating all tissues of the mammalian body. The cells replicate indefinitely in vitro, which makes it possible to culture them on a large scale and could create a nearly unlimited source of transplantable cells for auditory nerve replacement. Adult stem cells are found also in several tissues of the adult organism, where they normally produce new differentiated cells necessary for restoring degenerated cells.

The challenge in the use of undifferentiated stem cells, whether embryonic or adult, is to induce them to the appropriate phenotype. This could be done before placement in the target site or after placement. The cochlear fluids can provide an avenue for infusion of agents to influence phenotype when stem cells are placed into scala tympani. Embryonic stem cells naturally differentiate into neurons and glia; however, the percentage reaching neuronal phenotype is small when no further treatment is applied. Gene transfer of the neuronal transcription factor Neurogenin2 (Ngn2) improved the percentage reaching a neuronal phenotype (Hu et al. 2005b). Mouse embryonic stem cells engineering for inducible expression of neuronal transcription factor Neurogenin1 (Ngn1) allowed for more natural transient expression. Twenty-four hours of induced Ngn1 expression was followed by infusion of GDNF and BDNF, which are the NTFs naturally received by SGN during development. This induced the majority of the implanted stem cells into a glutamatergic neuronal phenotype both in vitro and in vivo after placement into guinea pig scala tympani (Reyes et al. 2008).

An alternative to undifferentiated stem cells is to use progenitor cells; these are more specialized cells that will develop into mature, differentiated cells of a specific type that could reduce the risk of uncontrolled proliferation after transplantation. Such cells have been found in both auditory and vestibular components of the developing inner ear (Li et al. 2003a,b; Martinez-Monedero et al. 2008; Oshima et al. 2010). However, the number of progenitor cells rapidly declines after birth and only relatively small numbers remain in the sensory epithelium of the mature mammalian cochlea (Lopez et al. 2004). Interestingly, progenitor cells have been isolated from adult human modiolus removed during surgeries (Rask-Andersen et al. 2005); these progenitor cells formed neurospheres in vitro, and could be valuable for human application. Unfortunately, as in the animal studies, the populations of stem or progenitor cells in adult tissues are relatively small and do not proliferate as readily as embryonic stem cells, and thus may not be able to give rise to enough cells for cell replacement therapies.

An ideal situation would be to use tissue from the receiving subject itself, so-called autografting. An autologous graft essentially eliminates the host reaction. Naito et al. (2004) applied an autologous graft to the inner ear with promising results. The recent technique for reprogramming somatic cells into induced pluripotent stem (iPS) cells (Takahashi and Yamanaka 2006) is exciting. This method would allow iPS cells, derived from the recipient, to be transplanted back to the same individual after necessary modifications and without the risk of rejection. Nishimura et al. (2009) have recently applied the technique to the inner ear, although they did not transplant the cells back to the same individual.

If the challenge of generating replacement cells with appropriate sensory hair cell phenotype is met, there are still three remaining challenges: survival, integration into an appropriate location/niche, and finally, forming central nervous system (CNS) connections and achieving function. Survival of new neural connections may require the same or similar neurotrophic or maintenance factors as required by endogenous auditory nerve SGN (Ulfendahl 2007; Altschuler et al. 2008). Indeed, excellent in vivo survival of mouse embryonic stem cells implanted into guinea pig cochlea was found when exogenous NTFs were provided into scala tympani (Altschuler et al. 2008; Reyes et al. 2008). Cell survival was also greatly enhanced with a cografting approach in which, in addition to the embryonic stem cells, embryonic neural tissue was implanted (Hu et al. 2004b, 2005a). Because electrical activity has been shown to enhance SGN survival after deafness in vivo (Miller et al. 2003a) or in vitro (Hansen et al. 2001, 2003; Green et al. 2008), it may also be that stem cells that reach a neuronal phenotype will have improved survival if they become activated by either cochlear electrical stimulation with a cochlear prosthesis or if they connect to remaining IHCs.

Although there is a challenge for integration into appropriate location and niche, the scala tympani provides access to the entire perilymphatic fluid compartment, and implanted donor cells may be able to travel to functionally relevant locations throughout the cochlea. Although the perilymphatic compartment is anatomically separated from the spiral ganglion, the barriers are literally "full of holes." Indeed, the separating bone structures contain microscopic fenestrae, canaliculae perforantes (Küçük et al. 1991), which provide a path for the implanted cells to reach the spiral ganglion region. An alternative, and possibly less damaging route, would be to access the perilymphatic compartment via the lateral semicircular canal of the vestibular part of the inner ear, as has been demonstrated by Iguchi et al. (2004).

For cells to replace or supplement SGN they must also bridge the connection between the ganglion region and the cochlear nucleus in the brain stem. Recent experiments have shown that embryonic stem cells or dorsal root ganglion cells transplanted to the transected auditory nerve migrated along the nerve fibers in the internal auditory meatus and, in some cases, even reached close to the cochlear nucleus in the brain stem (Hu et al. 2004a). Interestingly, embryonic brain tissue transplanted to the acutely transected ventral cochlear tract resulted not only in regeneration but also functional recovery (Ito et al. 2001). However, there are many chemical factors that produce a barrier between peripheral and central nervous system and could impede the ability of central processes of replacement neurons to make a connection in the cochlear nucleus. The central connection would also need to connect to cochlear nucleus neurons in a tonotopic manner.

It has been hypothesized that if the SGN population were to be supplemented with exogenous cells, the efficiency of the cochlear prosthesis would improve. Hu et al. (2009) reported on experiments in which embryonic dorsal root ganglion cells were implanted into the inner ears of deafened animals fitted with a scala tympani electrode for monitoring hearing function using electrically evoked ABR. NGF was infused to provide trophic support for the implanted cells. Indeed, extensive neurite projections were observed to extend from the implanted cells, through the thin bony modiolus, to the host spiral ganglion. However, no significant difference was seen in the electrical thresholds or input/output functions. The negative results could be due to the low survival rate of the implanted cells, or lack of functional contacts between the implanted cells and the host nervous system.

3.3 Replacement

3.3.1 Prostheses

Although cochlear prostheses represent one of the major treatment success stories, restoring hearing to thousands of the profoundly deaf, there are still major advances remaining in the future. The patient population continues to increase as benefits are being shown from placing prostheses into patient ears with remaining hearing, and then providing a hybrid of acoustic and electrical stimulation to those patients. In patients with significant residual hearing, but low scores in speech discrimination tasks, implants can be of remarkable benefit, yielding improved abilities to understand speech (Lenarz 2009). These patients typically will demonstrate little or no hearing at 1 kHz and above; but will have significant remaining low-frequency hearing, showing losses in the 30–40 dB range below 1 kHz. To provide electrical hearing and preserve residual acoustic hearing, implants have been modified from long, scalar filling, and modiolar hugging; to short, thin, free-floating, with the recent addition of amplified acoustic stimulation of the low frequencies, in a "hybrid" device (Woodson et al. 2010 for recent review). Enhanced performance is seen in these ears with electrical stimulation, presumably because of a more physiologic auditory nerve, reflecting functioning hair cells throughout a major apical portion of the cochlea, which is further enhanced by the acoustic stimulation, the latter contributing significantly to sound localization and discrimination of speech in noisy backgrounds. There may also be a contribution from electromotile responses of surviving hair cells (e.g., Grosh et al. 2004).

One major area of challenge for current cochlear prostheses is to improve speech discrimination in noise. Many patients demonstrate remarkable speech discrimination in quiet but their scores rapidly deteriorate in noise (Munson and Nelson 2005; for general discussion of challenges resolving speech in noise, e.g., Shrivastav and Still, Chap. 7). Another long-standing challenge is to allow improved appreciation of music (Gfeller et al. 2008). There is increasing bilateral implantation of prostheses, providing a potential for improved sound localization. Increasingly, the benefits observed have offset the earlier reservations about bilateral implantation. In the past, unilateral implants were encouraged with the hope of reserving one ear for potential later technical improvements in the implant. However, the ease of replacement surgery in the vast majority of cases where required has reduced concerns related to bilateral implantation.

One solution to provide better speech discrimination in noise and allow appreciation of music and language nuances depending on tonal modulations would be an improved channel separation, allowing an increased number of stimulation sites on the prosthesis and dividing the signal into more channels. Directed regrowth of peripheral processes toward stimulation sites or using stem cells to provide a closer target for stimulation are also potential solutions. Another approach is to place prostheses directly in the auditory nerve (Middlebrooks and Snyder 2007), providing more intimate contact of electrode to neural element, or to place prostheses in central auditory system sites such as the cochlear nucleus (Colletti and Shannon 2005; Schwartz et al. 2008) or inferior colliculus (Lim et al. 2008, 2009). Implantation into the central auditory system further increases the implant patient candidate pool, as it allows prostheses for those with unimplantable cochleae or lost auditory nerve populations. The remarkable plasticity of the central auditory system (e.g., Kaltenbach, Chap. 8) suggests the potential for successful "remapping" of these tonotopically organized nuclei with the advent of electrical stimulation via a central auditory system implant.

With electrical stimulation benefits in part dependent on hair cell survival and acoustic hearing completely dependent on hair cell survival, primary concerns have focused on reducing the trauma of cochlear implantation (hence smaller implants, with much smaller fenestrae) and eliminating any negative long-term effects of the implant or stimulation. The same strategies used for protection and repair from NIHL and ototoxicity could also be used to reduce loss of residual hearing from cochlear implantation trauma. This could include use of NTFs, immunosuppressants, cell death pathway inhibitors (Bcl-2 genes, JNK inhibitors) (Van de Water et al. 2010), antioxidants (Abi-Hachem et al. 2010), and agents that may enhance cochlear blood flow. Acute delivery into the cochlea at the time of surgery in forms that allow delayed release over time may be possible, however, risk factors should be taken into consideration (Garnham et al. 2005). The antioxidants, with and without vasodilators, that are being evaluated in multiple human trials for prevention of NIHL could also be considered for trials to improve postimplant hearing preservation and have the advantage of oral delivery, low cost, and minimal or no systemic side effects when used at recommended intake levels.

These considerations lead to a final area in the future frontiers of cochlear prostheses: the use of drug delivery systems coupled with cochlear prostheses. The use of drug interventions coupled with implants to preserve residual hearing is based on the same strategies discussed to preserve and regrow the auditory nerve. Future implant frontiers will include the integration of drug delivery with implants with the ability to deliver locally and safely NTFs, proteins, and other agents, in some cases with biopolymer–nanoparticle encapsulation of drugs, in systems that will allow burst, delayed, and sustained release. In the future, biopolymer and nanoparticle systems will be used to deliver genetically designed cells fixed to implants that can release growth factors and serve as targets for nerve growth, or extend neurites that will grow into the auditory nerve and enhance connectivity to the CNS.

3.4 Methods

3.4.1 Gene Therapy

Gene therapy technology has improved in recent years, making it a promising technique for treating inner ear disorders; the inner ear holds several unique advantages as a model for gene therapy. First, the cochlea is anatomically well suited for in vivo gene therapy both accessible and with a fluid compartment (Salt and Plontke 2009 for review). The relative isolation of the cochlear compartments minimizes unwanted effects of the introduced gene into other tissues. The inner ear is fluid filled, allowing all functionally important cells to be accessed by a transfection reagent. The concentration and dosage of complexes introduced to the cochlea can easily be modulated with a single injection or longer infusion via an osmotic pump. Cochlear endolymph and perilymph volumes have been characterized in guinea pigs, rats, mice, and also humans (e.g., Thorne et al. 1999), so adverse effects of high volume and pressure can be avoided. In addition, a variety of precise physiological measures, such as otoacoustic emissions, compound action potentials, evoked potentials, and ABR, have been developed to monitor the function of specific cells, which makes reliable assessment of efficacy and safety of gene therapy practical. Finally, many genes have been recently cloned in the mouse and human cochlea. More than 100 different genes have been identified that affect inner ear development or function, as well as many loci known to be involved in deafness (see also Gong and Lomax, Chap. 9). A transgenic technique has been demonstrated in shaker-2 mice to correct deafness (Probst et al. 1998).

Gene therapy with NTFs has been the most frequent application of gene therapy in inner ear animal research. For example, inoculation of an adenoviral vector encoding human GDNF gene (Ad.GDNF) into guinea pig cochleae via the round window membrane 4 days before injection of the ototoxic aminoglycoside antibiotic kanamycin (KM) and the loop diuretic ethacrynic acid (EA) provided better hearing and less hair cell damage compared with controls (Ad.lacZ vector) (Yagi et al. 1999). Coinoculation of two vectors, one encoding human TGF-beta1 gene and the other encoding human GDNF gene, into guinea pig cochleae 4 days prior to injection of the same ototoxic agent combination (KM and EA) provided better hearing and less hair cell loss compared to inoculation of only Ad.GDNF (Kawamoto et al. 2003). Endogenous antioxidant systems can be upregulated in the same way as endogenous NTF systems, with similarly protective benefits. Adenoviral vectors for overexpression of catalase and Mn superoxide dismutase (SOD2) protected hair cells and hearing thresholds from a combination of KM and EA when given 5 days before ototoxic insult. After inoculation, there was a significant increase in catalase and a moderate elevation in SOD2 levels in tissues of the cochlea inoculated with the respective vectors (Kawamoto et al. 2004). Gene therapy to prevent NIHL has been more challenging, perhaps because of the more complex mechanisms of cell death being initiated (e.g., Henderson et al. 2006; Hu, Chap. 5, for reviews). While exogenous GDNF administered intracochlearly can protect the inner ear from NIHL (Shoji et al. 2000a, b), Kawamoto et al. (2001) reported no difference in the protection afforded by Ad.GDNF versus control Ad. lacZ vectors.

As described previously, Atoh1 overexpression after gene transfer can promote hair cell regeneration from supporting cells after hair cell destruction (Izumikawa et al. 2005). Other more preliminary data suggest overexpression of Atoh1 may also promote recovery of the stereocilia of the cochlear hair cells after noise (Yang et al., Association for Research in Otolaryngology Meeting, 2010). The hair bundle is susceptible to acoustic trauma and ototoxic drugs, and mammalian cochlear hair cells lose the capability to regenerate the stereocilia spontaneously once lost. Atoh1 inoculated within the first week after noise exposure, however, induced stereociliary regeneration and the newly regenerated stereocilia were functional, as ABR and CM measured 1 and 2 months after Atoh1 inoculation showed significant hearing threshold improvement. These findings imply that Atoh1-based gene therapy has the potential to restore hearing after noise exposure (Izumikawa et al. 2005; Husseman and Raphael 2009).

3.4.2 Protein Transduction Therapy

The objective of gene therapy is gene delivery followed by expression of gene products that either possess a therapeutic biological activity or induce an altered cellular phenotype. Gene therapy approaches to a number of genetic disorders require long-term and appropriately regulated expression of the transgene. The short-term requirement for the presence of the therapeutic gene product raises the possibility of achieving the same objective by direct delivery of the gene product itself, rather than the gene. Recent developments in protein transduction (delivery of protein into cells) suggest this is now a realistic approach (see Tilstra et al. 2007).

Protein transduction domains (PTDs), or cell-penetrating peptides, are small peptides that are able to carry much larger molecules such as oligonucleotides, peptides, full-length proteins, 40 nm iron nanoparticles, bacteriophages, and even 200-nm liposomes across cellular membranes. They have proven useful in delivering biologically active cargoes in vivo and, remarkably, have the ability to transduce nearly all tissues, including the brain, following intraperitoneal administration of fusion proteins. At least three classes of PTDs have been described, including positively charged transduction domains (cationic), protein leader sequence–derived domains (hydrophobic), and peptides identified by phage display that are able to transduce cells in a cell-type-specific manner (tissue-specific). The positively charged cationic PTDs are the most efficient and the best characterized. These cell penetrating peptides (CPPs) include a TAT (transactivator of transcription) derived from human immunodefiency virus type 1 (HIV-1) that contains numerous cationic amino acids, where positive charges interact with the negatively charged cell membrane to facilitate permeability (Patsch and Edenhofer 2007 for review).

As described previously, a significant role of Bcl-2 genes has been implicated in NIHL as well as recovery from other auditory trauma. FNK, which has been constructed from Bcl-xL by site-directed mutagenesis based on the high-resolution crystal structure of the rat Bcl-xL, has three amino acid substitutions, Tyr-22 to Phe (F), Gln-26 to Asn (N), and Arg-165 to Lys (K), in which three hydrogen bonds stabilizing the central $\alpha 5$–$\alpha 6$ helices (the putative pore-forming domain) are abolished (Asoh et al. 2002). Compared with Bcl-xL, FNK protected cultured cells more potently from cell death induced by oxidative stress (hydrogen peroxide and paraquat), a calcium ionophore, growth factor withdraw (serum and IL-3), anti-Fas, cell cycle inhibitors (TN-16, camptothecin, hydroxyurea, and trichostatin A), a protein kinase inhibitor (staurosporine, STS), and heat treatment (Asoh et al. 2000). When FNK was fused with Tat-PTD of the HIV/Tat protein and added into culture media of human neuroblastoma cells and rat neocortical neurons, it rapidly transduced into cells and localized to mitochondria within 1 h and protected against staurosporine-induced apoptosis and glutamate-induced excitotoxicity (Asoh et al. 2002). When injected intraperitoneally, TAT-FNK gained access into mouse brain neurons and prevented delayed neuronal death in the gerbil hippocampus caused by transient global ischemia (Asoh et al. 2002). Similarly, TAT-FNK was diffusely distributed in the cochlea after an intraperitoneal administration to guinea pigs; the distribution was most prominent in the hair cells and supporting cells, followed by the SGN and peaked 3 h after the injection (Kashio et al. 2007). Further, the TAT-FNK protein intraperitoneally injected for 8 h (3 h pre-insult, 5 h post-insult) significantly attenuated ABR threshold shifts and the extent of HC death induced by a combination of EA and KM, and it significantly reduced the amount of cleaved poly-(ADP-ribose) polymerase-positive HCs compared with that in the vehicle-administered controls (Kashio et al. 2007). When TAT-FNK was topically applied on the round window membrane of guinea pigs, this protein penetrated through the membrane, distributed diffusely throughout the cochlea with the greatest expression 6 h after application and continuing up to 24 h, and significantly reduced hair cell death and caspase-9 expression induced by a combination of KM and EA (Kashio et al., ARO meeting, 2010).

Recently, to increase the biological activity of transduced protein in cells, novel carriers that transduce the target protein in its active native structural form have been designed. For example, when a PEP-1 peptide carrier, which consists of three domains – a hydrophobic tryptophan-rich motif, a spacer, and a hydrophilic lysine-rich domain – was mixed with the target protein (e.g., GFP, β-gal) and then overlaid on cultured cells, the nondenatured target protein was transduced (Morris et al. 2001). PEP-1 peptide carriers fused with SOD1 have been shown to protect cells from paraquat-induced oxidative stress in vitro and dopaminergic neuronal cell death in vivo in paraquat-induced Parkinson disease mouse models (Choi et al. 2006). Considering the rapid progress in protein transduction technology, delivery of the therapeutic gene products (e.g., anti-apoptotic agents, antioxidants, and NTFs) to the inner ear for the optimal short period seems to be promising and needs to be studied more intensively with the goal of human application.

4 Summary and Conclusions

As detailed in this chapter, and other chapters in this volume, there have been many remarkable advances in our understanding of the mechanisms associated with NIHL that have illuminated paths toward its prevention and treatment. More basic research is still needed to choose the best paths and navigate their initial hurdles, to provide guidance on which of the many approaches discussed will be the most effective, and which combinations of therapies acting by different mechanisms can provide greatest benefit. Clearly the “dirty work” of translational research is now demanded. There is sufficient knowledge of mechanisms and there are interventions with sufficient safety to begin studies in humans. There is a need for the difficult-to-fund parametric dose–response measurements of efficacy and safety, in animals and then in people; and a need to move to clinical trials. The field is much further along in some paths than others. Cochlear prostheses are, of course, already a success story, with wide application and they continue to be refined and improved. Antioxidant clinical trials are already testing for protection from noise or ototoxins. Other approaches such as stem cell therapy or induced hair cell regeneration have shown great promise on the benchtop but have yet to move from it. The fact that such a large number of approaches are being considered for prevention and treatment provides both a large opportunity and challenge for the future. They must all be tested, compared, and contrasted under the different conditions of noise and the different resulting pathologies. All the tools and knowledge are available to begin and complete that task. The promise is great; once the initial translational efforts bear fruit, there will be safe and effective measures that reduce the prevalence of deafness and tinnitus resulting from noise and other stressors. In addition, with the demonstration that NIHL can be medically treated, a paradigm change in perspective will lead to prevention and treatment of many other causes of hearing impairment.

Acknowledgments The authors' research was supported by NIH/NIDCD grants U01 DC008423, R01 DC003820, R01 DC004058 and P30 DC005188 and The Ruth and Lynn Townsend Professorship for Communication Disorders, and MECSST grants (11557125, 17659527, 20390440). We also acknowledge the editorial contributions to the paper made by Diane Prieskorn and Susan DeRemer. We also thank the editors for their helpful comments and changes.

References

Abi-Hachem, R. N., Zine, A., & Van De Water, T. R. (2010). The injured cochlea as a target for inflammatory processes, initiation of cell death pathways and application of related otoprotectives strategies. *Recent Patents on CNS Drug Delivery*, 5(2), 147–163.

Ahn, J. H., Kang, H. H., Kim, Y. J., & Chung, J. W. (2005). Anti-apoptotic role of retinoic acid in the inner ear of noise-exposed mice. *Biochemical and Biophysical Research Communications*, 335(2), 485–490.

Altschuler, R. A., Cho, Y., Ylikoski, J., Pirvola, U., Magal, E., & Miller, J. M. (1999). Rescue and regrowth of sensory nerves following deafferentation by neurotrophic factors. *Annals of the New York Academy of Sciences*, 884, 305–311.

Altschuler, R. A., O'Shea, K. S., & Miller, J. M. (2008). Stem cell transplantation for auditory nerve replacement. *Hearing Research*, 242(1–2), 110–116.

Asoh, S., Ohtsu, T., & Ohta, S. (2000). The super anti-apoptotic factor Bcl-xFNK constructed by disturbing intramolecular polar interactions in rat Bcl-xL. *The Journal of Biological Chemistry*, 275(47), 37240–37245.

Asoh, S., Ohsawa, I., Mori, T., Katsura, K., Hiraide, T., Katayama, Y., & Ohta, S. (2002). Protection against ischemic brain injury by protein therapeutics. *Proceedings of the National Academy of Sciences of the USA*, 99(26), 17107–17112.

Bohne, B. A., & Harding, G. W. (1992). Neural regeneration in the noise-damaged chinchilla cochlea. *The Laryngoscope*, 102(6), 693–703.

Bohne, B. A., Harding, G. W., Nordmann, A. S., Tseng, C. J., Liang, G. E., & Bahadori, R. S. (1999). Survival-fixation of the cochlea: A technique for following time-dependent degeneration and repair in noise-exposed chinchillas. *Hearing Research*, 134(1–2), 163–178.

Branis, M., & Burda, H. (1988). Effect of ascorbic acid on the numerical hair cell loss in noise exposed guinea pigs. *Hearing Research*, 33(2), 137–140.

Campbell, K. C., Meech, R. P., Klemens, J. J., Gerberi, M. T., Dyrstad, S. S., Larsen, D. L., & Hughes, L. F. (2007). Prevention of noise – and drug-induced hearing loss with D-methionine. *Hearing Research*, 226(1–2), 92–103.

Cassandro, E., Sequino, L., Mondola, P., Attanasio, G., Barbara, M., & Filipo, R. (2003). Effect of superoxide dismutase and allopurinol on impulse noise-exposed guinea pigs – electrophysiological and biochemical study. *Acta Oto-Laryngologica*, 123(7), 802–807.

Cevette, M. J., Vormann, J., & Franz, K. (2003). Magnesium and hearing. *Journal of the American Academy of Audiology*, 14(4), 202–212.

Choi, H. S., An, J. J., Kim, S. Y., Lee, S. H., Kim, D. W., Yoo, K. Y., & Choi, S. Y. (2006). PEP-1–SOD fusion protein efficiently protects against paraquat-induced dopaminergic neuron damage in a Parkinson disease mouse model. *Free Radical Biology & Medicine*, 41(7), 1058–1068.

Circu, M. L., & Aw, T. Y. (2010). Reactive oxygen species, cellular redox systems, and apoptosis. *Free Radic Biology & Medicine*, 48(6), 749–762.

Colletti, V., & Shannon, R. V. (2005). Open set speech perception with auditory brainstem implant? *The Laryngoscope*, 115(11), 1974–1978.

Corwin, J. T., & Cotanche, D. A. (1988). Regeneration of sensory hair cells after acoustic trauma. *Science*, 240(4860), 1772–1774.

Cotanche, D. A., & Kaiser, C. L. (2010). Hair cell fate decisions in cochlear development and regeneration. *Hearing Research*, 266(1–2), 18–25.

Dazert, S., Kim, D., Luo, L., Aletsee, C., Garfunkel, S., Maciag, T., & Ryan, A. F. (1998). Focal delivery of fibroblast growth factor-1 by transfected cells induces spiral ganglion neurite targeting in vitro. *Journal of Cellular Physiology*, 177(1), 123–129.

Derekoy, F. S., Koken, T., Yilmaz, D., Kahraman, A., & Altuntas, A. (2004). Effects of ascorbic acid on oxidative system and transient evoked otoacoustic emissions in rabbits exposed to noise. *The Laryngoscope*, 114(10), 1775–1779.

Duan, M., Qiu, J., Laurell, G., Olofsson, A., Counter, S. A., & Borg, E. (2004). Dose and time-dependent protection of the antioxidant *N*-L-acetylcysteine against impulse noise trauma. *Hearing Research*, 192(1–2), 1–9.

Edge, A. S., & Chen, Z. Y. (2008). Hair cell regeneration. *Current Opinion in Neurobiology*, 18(4), 377–382.

Ernfors, P., Duan, M. L., ElShamy, W. M., & Canlon, B. (1996). Protection of auditory neurons from aminoglycoside toxicity by neurotrophin-3. *Nature Medicine*, 2(4), 463–467.

Fairfield, D. A., Lomax, M. I., Dootz, G. A., Chen, S., Galecki, A. T., Benjamin, I. J., & Altschuler, R. A. (2005). Heat shock factor 1–deficient mice exhibit decreased recovery of hearing following noise overstimulation. *Journal of Neuroscience Research*, 81(4), 589–596.

Fetoni, A. R., Mancuso, C., Eramo, S. L., Ralli, M., Piacentini, R., Barone, E., & Troiani, D. (2010). In vivo protective effect of ferulic acid against noise-induced hearing loss in the guinea-pig. *Neuroscience*, 169(4), 1575–1588.

Fritzsch, B., Silos-Santiago, I. I., Bianchi, L. M., & Farinas, I. I. (1997). Effects of neurotrophin and neurotrophin receptor disruption on the afferent inner ear innervation. *Seminars in Cell & Developmental Biology*, 8(3), 277–284.

Fukuda, K., Asoh, S., Ishikawa, M., Yamamoto, Y., Ohsawa, I., & Ohta, S. (2007). Inhalation of hydrogen gas suppresses hepatic injury caused by ischemia/reperfusion through reducing oxidative stress. *Biochemical & Biophysical Research Communications*, 361(3), 670–674.

Garnham, C., Reetz, G., Jolly, C., Miller, J., Salt, A., & Beal, F. (2005). Drug delivery to the cochlea after implantation: Consideration of the risk factors. *Cochlear Implants International*, 6 (Supplement 1), 12–14.

Gfeller, K., Oleson, J., Knutson, J. F., Breheny, P., Driscoll, V., & Olszewski, C. (2008). Multivariate predictors of music perception and appraisal by adult cochlear implant users. *Journal of the American Academy of Audiology*, 19(2), 120–134.

Gillespie, L. N., Clark, G. M., & Marzella, P. L. (2004). Delayed neurotrophin treatment supports auditory neuron survival in deaf guinea pigs. *NeuroReport*, 15(7), 1121–1125.

Glueckert, R., Bitsche, M., Miller, J. M., Zhu, Y., Prieskorn, D. M., Altschuler, R. A., & Schrott-Fischer, A. (2008). Deafferentation-associated changes in afferent and efferent processes in the guinea pig cochlea and afferent regeneration with chronic intrascalar brain-derived neurotrophic factor and acidic fibroblast growth factor. *The Journal of Comparative Neurology*, 507(4), 1602–1621.

Gooch, J. L., Gorin, Y., Zhang, B. X., & Abboud, H. E. (2004). Involvement of calcineurin in transforming growth factor-beta-mediated regulation of extracellular matrix accumulation. *The Journal of Biological Chemistry*, 279(15), 15561–15570.

Green, S. H., Altschuler, R. A., & Miller, J. M. (2008). Cell death and cochlear protection. In J. Schacht, A. N. Popper & R. R. Fay (Eds.), *Auditory Trauma, Protection and Repair* (pp. 275–319). New York: Springer.

Grosh, K., Zheng, J., Zou, Y., de Boer, E., & Nuttall, A. L. (2004). High-frequency electromotile responses in the cochlea. *The Journal of the Acoustical Society of America*, 115(5 Pt 1), 2178–2184.

Gunther, T., Ising, H., & Joachims, Z. (1989). Biochemical mechanisms affecting susceptibility to noise-induced hearing loss. *The American Journal of Otology*, 10(1), 36–41.

Halliwell, B., & Gutteridge, J. M. (1986). Oxygen free radicals and iron in relation to biology and medicine: Some problems and concepts. *Archives of Biochemistry & Biophysics*, 246(2), 501–514.

Halliwell, B., & Gutteridge, J. M. C. (2007). *Free radicals in biology and medicine* (4th ed.). New York: Oxford University Press.

Hansen, M. R., Zha, X. M., Bok, J., & Green, S. H. (2001). Multiple distinct signal pathways, including an autocrine neurotrophic mechanism, contribute to the survival-promoting effect of depolarization on spiral ganglion neurons in vitro. *The Journal of Neuroscience*, 21(7), 2256–2267.

Hansen, M. R., Bok, J., Devaiah, A. K., Zha, X. M., & Green, S. H. (2003). Ca^{2+}/calmodulin-dependent protein kinases II and IV both promote survival but differ in their effects on axon growth in spiral ganglion neurons. *Journal of Neuroscience Research*, 72(2), 169–184.

Hayashida, K., Sano, M., Ohsawa, I., Shinmura, K., Tamaki, K., Kimura, K., & Fukuda, K. (2008). Inhalation of hydrogen gas reduces infarct size in the rat model of myocardial ischemia-reperfusion injury. *Biochemical and Biophysical Research Communications*, 373(1), 30–35.

Henderson, D., Bielefeld, E. C., Harris, K. C., & Hu, B. H. (2006). The role of oxidative stress in noise-induced hearing loss. *Ear & Hearing*, 27(1), 1–19.

Hirose, Y., Sugahara, K., Mikuriya, T., Hashimoto, M., Shimogori, H., & Yamashita, H. (2008). Effect of water-soluble coenzyme Q10 on noise-induced hearing loss in guinea pigs. *Acta Oto-Laryngologica*, 128(10), 1071–1076.

Hossain, W. A., & Morest, D. K. (2000). Fibroblast growth factors (FGF-1, FGF-2) promote migration and neurite growth of mouse cochlear ganglion cells in vitro: Immunohistochemistry and antibody perturbation. *Journal of Neuroscience Research*, 62(1), 40–55.

Hou, F., Wang, S., Zhai, S., Hu, Y., Yang, W., & He, L. (2003). Effects of alpha-tocopherol on noise-induced hearing loss in guinea pigs. *Hearing Research*, 179(1–2), 1–8.

Hu, B. H., Zheng, X. Y., McFadden, S. L., Kopke, R. D., & Henderson, D. (1997). R-phenylisopropyladenosine attenuates noise-induced hearing loss in the chinchilla. *Hearing Research*, 113(1–2), 198–206.

Hu, Z., Ulfendahl, M., & Olivius, N. P. (2004a). Central migration of neuronal tissue and embryonic stem cells following transplantation along the adult auditory nerve. *Brain Research*, 1026(1), 68–73.

Hu, Z., Ulfendahl, M., & Olivius, N. P. (2004b). Survival of neuronal tissue following xenograft implantation into the adult rat inner ear. *Experimental Neurology*, 185(1), 7–14.

Hu, Z., Andang, M., Ni, D., & Ulfendahl, M. (2005a). Neural cograft stimulates the survival and differentiation of embryonic stem cells in the adult mammalian auditory system. *Brain Research*, 1051(1–2), 137–144.

Hu, Z., Wei, D., Johansson, C. B., Holmstrom, N., Duan, M., Frisen, J., & Ulfendahl, M. (2005b). Survival and neural differentiation of adult neural stem cells transplanted into the mature inner ear. *Experimental Cell Research*, 302(1), 40–47.

Hu, Z., Ulfendahl, M., Prieskorn, D. M., Olivius, P., & Miller, J. M. (2009). Functional evaluation of a cell replacement therapy in the inner ear. *Otology & Neurotology*, 30(4), 551–558.

Huang, C., Li, J., Costa, M., Zhang, Z., Leonard, S. S., Castranova, V., & Shi, X. (2001). Hydrogen peroxide mediates activation of nuclear factor of activated T cells (NFAT) by nickel subsulfide. *Cancer Research*, 61(22), 8051–8057.

Husseman, J., & Raphael, Y. (2009). Gene therapy in the inner ear using adenovirus vectors. *Advances in Oto-Rhino-Laryngology*, 66, 37–51.

Iguchi, F., Nakagawa, T., Tateya, I., Endo, T., Kim, T. S., Dong, Y., & Ito, J. (2004). Surgical techniques for cell transplantation into the mouse cochlea. *Acta Oto-Laryngologica*, (551, Supplement), 43–47.

Incesulu, A., & Nadol, J. B., Jr. (1998). Correlation of acoustic threshold measures and spiral ganglion cell survival in severe to profound sensorineural hearing loss: Implications for cochlear implantation. *The Annals of Otology, Rhinology, & Laryngology*, 107(11 Pt 1), 906–911.

Ito, J., Murata, M., & Kawaguchi, S. (2001). Regeneration and recovery of the hearing function of the central auditory pathway by transplants of embryonic brain tissue in adult rats. *Experimental Neurology*, 169(1), 30–35.

Izumikawa, M., Minoda, R., Kawamoto, K., Abrashkin, K. A., Swiderski, D. L., Dolan, D. F., & . Raphael, Y. (2005). Auditory hair cell replacement and hearing improvement by *Atoh1* gene therapy in deaf mammals. *Nature Medicine*, 11(3), 271–276.

Kajiyama, S., Hasegawa, G., Asano, M., Hosoda, H., Fukui, M., Nakamura, N., & Yoshikawa, T. (2008). Supplementation of hydrogen-rich water improves lipid and glucose metabolism in patients with type 2 diabetes or impaired glucose tolerance. *Nutrition Research*, 28(3), 137–143.

Kanzaki, S., Stover, T., Kawamoto, K., Prieskorn, D. M., Altschuler, R. A., Miller, J. M., & Raphael, Y. (2002). Glial cell line-derived neurotrophic factor and chronic electrical stimulation prevent VIII cranial nerve degeneration following denervation. *The Journal of Comparative Neurology*, 454(3), 350–360.

Kashio, A., Sakamoto, T., Suzukawa, K., Asoh, S., Ohta, S., & Yamasoba, T. (2007). A protein derived from the fusion of TAT peptide and FNK, a Bcl-x(L) derivative, prevents cochlear hair cell death from aminoglycoside ototoxicity in vivo. *Journal of Neuroscience Research*, 85(7), 1403–1412.

Kawamoto, K., Kanzaki, S., Yagi, M., Stover, T., Prieskorn, D. M., Dolan, D. F., & Raphael, Y. (2001). Gene-based therapy for inner ear disease. *Noise Health*, 3(11), 37–47.

Kawamoto, K., Yagi, M., Stover, T., Kanzaki, S., & Raphael, Y. (2003). Hearing and hair cells are protected by adenoviral gene therapy with TGF-beta1 and GDNF. *Molecular Therapy*, 7(4), 484–492.

Kawamoto, K., Sha, S. H., Minoda, R., Izumikawa, M., Kuriyama, H., Schacht, J., & Raphael, Y. (2004). Antioxidant gene therapy can protect hearing and hair cells from ototoxicity. *Molecular Therapy*, 9(2), 173–181.

Kikkawa, Y. S., Nakagawa, T., Horie, R. T., & Ito, J. (2009). Hydrogen protects auditory hair cells from free radicals. *NeuroReport*, 20(7), 689–694.

Kirkland, R. A., & Franklin, J. L. (2003). Bax, reactive oxygen, and cytochrome *c* release in neuronal apoptosis. *Antioxidants & Redox Signaling*, 5(5), 589–596.

Kopke, R. D., Weisskopf, P. A., Boone, J. L., Jackson, R. L., Wester, D. C., Hoffer, M. E., & McBride, D. (2000). Reduction of noise-induced hearing loss using L-NAC and salicylate in the chinchilla. *Hearing Research*, 149(1–2), 138–146.

Kopke, R. D., Coleman, J. K., Liu, J., Campbell, K. C., & Riffenburgh, R. H. (2002). Candidate's thesis: Enhancing intrinsic cochlear stress defenses to reduce noise-induced hearing loss. *The Laryngoscope*, 112(9), 1515–1532.

Küçük, B., Abe, K., Ushiki, T., Inuyama, Y., Fukuda, S., & Ishikawa, K. (1991). Microstructures of the bony modiolus in the human cochlea: A scanning electron microscopic study. *Journal of Electron Microscopy (Tokyo)*, 40(3), 193–197.

Kujawa, S. G., & Liberman, M. C. (2009). Adding insult to injury: Cochlear nerve degeneration after "temporary" noise-induced hearing loss. *The Journal of Neuroscience*, 29(45), 14077–14085.

Kwan, T., White, P. M., & Segil, N. (2009). Development and regeneration of the inner ear. *Annals of the New York Academy of Sciences*, 1170, 28–33.

Lamm, K., & Arnold, W. (2000). The effect of blood flow promoting drugs on cochlear blood flow, perilymphatic pO_2 and auditory function in the normal and noise-damaged hypoxic and ischemic guinea pig inner ear. *Hearing Research*, 141(1–2), 199–219.

Lawner, B. E., Harding, G. W., & Bohne, B. A. (1997). Time course of nerve-fiber regeneration in the noise-damaged mammalian cochlea. *International Journal of Developmental Neuroscience*, 15(4–5), 601–617.

Le Prell, C. G., Yagi, M., Kawamoto, K., Beyer, L. A., Atkin, G., Raphael, Y., & Moody, D. B. (2004). Chronic excitotoxicity in the guinea pig cochlea induces temporary functional deficits without disrupting otoacoustic emissions. *The Journal of the Acoustical Society of America*, 116(2), 1044–1056.

Le Prell, C. G., Hughes, L. F., & Miller, J. M. (2007a). Free radical scavengers vitamins A, C, and E plus magnesium reduce noise trauma. *Free Radical Biology & Medicine*, 42(9), 1454–1463.

Le Prell, C. G., Yamashita, D., Minami, S. B., Yamasoba, T., & Miller, J. M. (2007b). Mechanisms of noise-induced hearing loss indicate multiple methods of prevention. *Hearing Research*, 226(1–2), 22–43.

Lenarz, T. (2009). Electro-acoustic stimulation of the cochlea. Editorial. *Audiology & Neuro-otology*, 14 (Supplement 1), 1.

Levine, M., Conry-Cantilena, C., Wang, Y., Welch, R. W., Washko, P. W., Dhariwal, K. R., & Cantilena, L. R. (1996). Vitamin C pharmacokinetics in healthy volunteers: Evidence for a recommended dietary allowance. *Proceedings of the National Academy of Sciences of the USA*, 93(8), 3704–3709.

Li, H., Liu, H., & Heller, S. (2003a). Pluripotent stem cells from the adult mouse inner ear. *Nature Medicine*, 9(10), 1293–1299.

Li, H., Roblin, G., Liu, H., & Heller, S. (2003b). Generation of hair cells by stepwise differentiation of embryonic stem cells. *Proceedings of the National Academy of Sciences of the USA*, 100(23), 13495–13500.

Li, H., Corrales, C. E., Edge, A., & Heller, S. (2004). Stem cells as therapy for hearing loss. *Trends in Molecular Medicine*, 10(7), 309–315.

Lim, H. H., Lenarz, T., Anderson, D. J., & Lenarz, M. (2008). The auditory midbrain implant: Effects of electrode location. *Hearing Research*, 242(1–2), 74–85.

Lim, H. H., Lenarz, M., & Lenarz, T. (2009). Auditory midbrain implant: A review. *Trends in Amplification*, 13(3), 149–180.

Lopez, I. A., Zhao, P. M., Yamaguchi, M., de Vellis, J., & Espinosa-Jeffrey, A. (2004). Stem/progenitor cells in the postnatal inner ear of the GFP-nestin transgenic mouse. *International Journal of Developmental Neuroscience*, 22(4), 205–213.

Lynch, E. D., & Kil, J. (2005). Compounds for the prevention and treatment of noise-induced hearing loss. *Drug Discovery Today*, 10(19), 1291–1298.

Maricich, S. M., Xia, A., Mathes, E. L., Wang, V. Y., Oghalai, J. S., Fritzsch, B., & Zoghbi, H. Y. (2009). *Atoh1*-lineal neurons are required for hearing and for the survival of neurons in the spiral ganglion and brainstem accessory auditory nuclei. *The Journal of Neuroscience*, 29(36), 11123–11133.

Martinez-Monedero, R., Yi, E., Oshima, K., Glowatzki, E., & Edge, A. S. (2008). Differentiation of inner ear stem cells to functional sensory neurons. *Developmental Neurobiology*, 68(5), 669–684.

Maruyama, J., Yamagata, T., Ulfendahl, M., Bredberg, G., Altschuler, R. A., & Miller, J. M. (2007). Effects of antioxidants on auditory nerve function and survival in deafened guinea pigs. *Neurobiology of Disease*, 25(2), 309–318.

Maruyama, J., Miller, J. M., & Ulfendahl, M. (2008). Glial cell line-derived neurotrophic factor and antioxidants preserve the electrical responsiveness of the spiral ganglion neurons after experimentally induced deafness. *Neurobiology of Disease*, 29(1), 14–21.

Mattson, M. P. (1998). Neuroprotective strategies based on targeting of postreceptor signaling events. In M. P. Mattson (Ed.), *Neuroprotective signal transduction* (pp. 301–335). Totowa, NJ: Humana Press.

McFadden, S. L., Ding, D., Jiang, H., & Salvi, R. J. (2004). Time course of efferent fiber and spiral ganglion cell degeneration following complete hair cell loss in the chinchilla. *Brain Research*, 997(1), 40–51.

McFadden, S. L., Woo, J. M., Michalak, N., & Ding, D. (2005). Dietary vitamin C supplementation reduces noise-induced hearing loss in guinea pigs. *Hearing Research*, 202(1–2), 200–208.

Middlebrooks, J. C., & Snyder, R. L. (2007). Auditory prosthesis with a penetrating nerve array. *Journal of the Association for Research in Otolaryngology*, 8(2), 258–279.

Miller, J. M., Chi, D. H., O'Keeffe, L. J., Kruszka, P., Raphael, Y., & Altschuler, R. A. (1997). Neurotrophins can enhance spiral ganglion cell survival after inner hair cell loss. *International Journal of Developmental Neuroscience*, 15(4–5), 631–643.

Miller, J. M., Miller, A. L., Yamagata, T., Bredberg, G., & Altschuler, R. A. (2002). Protection and regrowth of the auditory nerve after deafness: Neurotrophins, antioxidants and depolarization are effective in vivo. *Audiology & Neuro-otology*, 7(3), 175–179.

Miller, A. L., Prieskorn, D. M., Altschuler, R. A., & Miller, J. M. (2003a). Mechanism of electrical stimulation-induced neuroprotection: Effects of verapamil on protection of primary auditory afferents. *Brain Research*, 966(2), 218–230.

Miller, J. M., Brown, J. N., & Schacht, J. (2003b). 8–iso-prostaglandin $F_2\alpha$, a product of noise exposure, reduces inner ear blood flow. *Audiology & Neuro-otology*, 8(4), 207–221.

Miller, J., Yamashita, D., Minami, S., Yamasoba, T., & Le Prell, C. (2006). Mechanisms and prevention of noise-induced hearing loss. *Otology Japan*, 16(2), 139–153.

Miller, J. M., Le Prell, C. G., Prieskorn, D. M., Wys, N. L., & Altschuler, R. A. (2007). Delayed neurotrophin treatment following deafness rescues spiral ganglion cells from death and promotes regrowth of auditory nerve peripheral processes: Effects of brain-derived neurotrophic factor and fibroblast growth factor. *Journal of Neuroscience Research*, 85(9), 1959–1969.

Minami, S. B., Yamashita, D., Ogawa, K., Schacht, J., & Miller, J. M. (2007). Creatine and tempol attenuate noise-induced hearing loss. *Brain Research*, 1148, 83–89.

Morris, M. C., Depollier, J., Mery, J., Heitz, F., & Divita, G. (2001). A peptide carrier for the delivery of biologically active proteins into mammalian cells. *Nature Biotechnology*, 19(12), 1173–1176.

Munson, B., & Nelson, P. B. (2005). Phonetic identification in quiet and in noise by listeners with cochlear implants. *The Journal of the Acoustical Society of America*, 118(4), 2607–2617.

Nadol, J. B., Jr., Young, Y. S., & Glynn, R. J. (1989). Survival of spiral ganglion cells in profound sensorineural hearing loss: Implications for cochlear implantation. *The Annals of Otology, Rhinology & Laryngology*, 98(6), 411–416.

Nagata, K., Nakashima-Kamimura, N., Mikami, T., Ohsawa, I., & Ohta, S. (2009). Consumption of molecular hydrogen prevents the stress-induced impairments in hippocampus-dependent learning tasks during chronic physical restraint in mice. *Neuropsychopharmacology*, 34(2), 501–508.

Naito, Y., Nakamura, T., Nakagawa, T., Iguchi, F., Endo, T., Fujino, K., & Ito, J. (2004). Transplantation of bone marrow stromal cells into the cochlea of chinchillas. *NeuroReport*, 15(1), 1–4.

Neef, D. W., Turski, M. L., & Thiele, D. J. (2010). Modulation of heat shock transcription factor 1 as a therapeutic target for small molecule intervention in neurodegenerative disease. *PLoS Biology*, 8(1), e1000291.

Nishimura, K., Nakagawa, T., Ono, K., Ogita, H., Sakamoto, T., Yamamoto, N., & Ito, J. (2009). Transplantation of mouse induced pluripotent stem cells into the cochlea. *NeuroReport*, 20(14), 1250–1254.

Ohinata, Y., Yamasoba, T., Schacht, J., & Miller, J. M. (2000). Glutathione limits noise-induced hearing loss. *Hearing Research*, 146(1–2), 28–34.

Ohinata, Y., Miller, J. M., & Schacht, J. (2003). Protection from noise-induced lipid peroxidation and hair cell loss in the cochlea. *Brain Research*, 966(2), 265–273.

Ohlemiller, K. K., Wright, J. S., & Dugan, L. L. (1999). Early elevation of cochlear reactive oxygen species following noise exposure. *Audiology & Neuro-otology*, 4(5), 229–236.

Ohsawa, I., Ishikawa, M., Takahashi, K., Watanabe, M., Nishimaki, K., Yamagata, K., & Ohta, S. (2007). Hydrogen acts as a therapeutic antioxidant by selectively reducing cytotoxic oxygen radicals. *Nature Medicine*, 13(6), 688–694.

Oliva, E. N., Ronco, F., Marino, A., Alati, C., Pratico, G., & Nobile, F. (2010). Iron chelation therapy associated with improvement of hematopoiesis in transfusion-dependent patients. *Transfusion*, 50(7), 1568–1570.

Oshima, K., Shin, K., Diensthuber, M., Peng, A. W., Ricci, A. J., & Heller, S. (2010). Mechanosensitive hair cell-like cells from embryonic and induced pluripotent stem cells. *Cell*, 141(4), 704–716.

Patsch, C., & Edenhofer, F. (2007). Conditional mutagenesis by cell-permeable proteins: Potential, limitations and prospects. Handbook of Experimental. *Pharmacology*, (178), 203–232.

Pirvola, U., Hallbook, F., Xing-Qun, L., Virkkala, J., Saarma, M., & Ylikoski, J. (1997). Expression of neurotrophins and Trk receptors in the developing, adult, and regenerating avian cochlea. *Journal of Neurobiology*, 33(7), 1019–1033.

Pourbakht, A., & Yamasoba, T. (2003). Ebselen attenuates cochlear damage caused by acoustic trauma. *Hearing Research*, 181(1–2), 100–108.

Probst, F. J., Fridell, R. A., Raphael, Y., Saunders, T. L., Wang, A., Liang, Y., & Camper, S. A. (1998). Correction of deafness in shaker-2 mice by an unconventional myosin in a BAC transgene. *Science*, 280(5368), 1444–1447.

Puel, J. L., Pujol, R., Ladrech, S., & Eybalin, M. (1991). Alpha-amino-3-hydroxy-5-methyl-4-isoxazole propionic acid electrophysiological and neurotoxic effects in the guinea-pig cochlea. *Neuroscience*, 45(1), 63–72.

Puel, J. L., Saffiedine, S., Gervais d'Aldin, C., Eybalin, M., & Pujol, R. (1995). Synaptic regeneration and functional recovery after excitotoxic injury in the guinea pig cochlea. *Comptes Rendus de l' Académie des Sciences Série III*, 318(1), 67–75.

Puel, J. L., Ruel, J., Gervais d'Aldin, C., & Pujol, R. (1998). Excitotoxicity and repair of cochlear synapses after noise-trauma induced hearing loss. *NeuroReport*, 9(9), 2109–2114.

Pujol, R., & Puel, J. L. (1999). Excitotoxicity, synaptic repair, and functional recovery in the mammalian cochlea: A review of recent findings. *Annals of the New York Academy of Sciences*, 884, 249–254.

Quirk, W. S., & Seidman, M. D. (1995). Cochlear vascular changes in response to loud noise. *The American Journal of Otology*, 16(3), 322–325.

Quirk, W. S., Avinash, G., Nuttall, A. L., & Miller, J. M. (1992). The influence of loud sound on red blood cell velocity and blood vessel diameter in the cochlea. *Hearing Research*, 63(1–2), 102–107.

Quirk, W. S., Shivapuja, B. G., Schwimmer, C. L., & Seidman, M. D. (1994). Lipid peroxidation inhibitor attenuates noise-induced temporary threshold shifts. *Hearing Research*, 74(1–2), 217–220.

Rask-Andersen, H., Bostrom, M., Gerdin, B., Kinnefors, A., Nyberg, G., Engstrand, T., & Lindholm, D. (2005). Regeneration of human auditory nerve. In vitro/in video demonstration of neural progenitor cells in adult human and guinea pig spiral ganglion. *Hearing Research*, 203(1–2), 180–191.

Reyes, J. H., O'Shea, K. S., Wys, N. L., Velkey, J. M., Prieskorn, D. M., Wesolowski, K., & Altschuler, R. A. (2008). Glutamatergic neuronal differentiation of mouse embryonic stem cells after transient expression of neurogenin 1 and treatment with BDNF and GDNF: In vitro and in vivo studies. *The Journal of Neuroscience*, 28(48), 12622–12631.

Rivera, A., & Maxwell, S. A. (2005). The p53-induced gene-6 (proline oxidase) mediates apoptosis through a calcineurin-dependent pathway. *The Journal of Biological Chemistry*, 280(32), 29346–29354.

Roberts, R. A., Smith, R. A., Safe, S., Szabo, C., Tjalkens, R. B., & Robertson, F. M. (2010). Toxicological and pathophysiological roles of reactive oxygen and nitrogen species. *Toxicology*, 276(2), 85–94.

Ryals, B. M., & Rubel, E. W. (1988). Hair cell regeneration after acoustic trauma in adult Coturnix quail. *Science*, 240(4860), 1774–1776.

Salt, A. N., & Plontke, S. K. (2009). Principles of local drug delivery to the inner ear. *Audiology & Neuro-otology*, 14(6), 350–360.

Sato, Y., Kajiyama, S., Amano, A., Kondo, Y., Sasaki, T., Handa, S., & Ishigami, A. (2008). Hydrogen-rich pure water prevents superoxide formation in brain slices of vitamin C-depleted SMP30/GNL knockout mice. *Biochemical & Biophysical Research Communications*, 375(3), 346–350.

Scheibe, F., Haupt, H., & Ising, H. (2000). Preventive effect of magnesium supplement on noise-induced hearing loss in the guinea pig. *European Archives of Oto-rhino-laryngology*, 257(1), 10–16.

Scheper, V., Paasche, G., Miller, J. M., Warnecke, A., Berkingali, N., Lenarz, T., & Stover, T. (2009). Effects of delayed treatment with combined GDNF and continuous electrical stimulation on spiral ganglion cell survival in deafened guinea pigs. *The Journal of Neuroscience Research*, 87(6), 1389–1399.

Schwartz, M. S., Otto, S. R., Shannon, R. V., Hitselberger, W. E., & Brackmann, D. E. (2008). Auditory brainstem implants. *Neurotherapeutics*, 5(1), 128–136.

Seidman, M. D., Shivapuja, B. G., & Quirk, W. S. (1993). The protective effects of allopurinol and superoxide dismutase on noise-induced cochlear damage. *Otolaryngology and Head and Neck Surgery*, 109(6), 1052–1056.

Seidman, M., Babu, S., Tang, W., Naem, E., & Quirk, W. S. (2003). Effects of resveratrol on acoustic trauma. *Otolaryngology and Head and Neck Surgery*, 129(5), 463–470.

Shah, S. B., Gladstone, H. B., Williams, H., Hradek, G. T., & Schindler, R. A. (1995). An extended study: Protective effects of nerve growth factor in neomycin-induced auditory neural degeneration. *The American Journal of Otology*, 16(3), 310–314.

Shibata, S. B., Cortez, S. R., Beyer, L. A., Wiler, J. A., Di Polo, A., Pfingst, B. E., & Raphael, Y. (2010). Transgenic BDNF induces nerve fiber regrowth into the auditory epithelium in deaf cochleae. *Experimental Neurology*, 223(2), 464–472.

Shim, H. J., Kang, H. H., Ahn, J. H., & Chung, J. W. (2009). Retinoic acid applied after noise exposure can recover the noise-induced hearing loss in mice. *Acta Oto-Laryngologica*, 129(3), 233–238.

Shoji, F., Miller, A. L., Mitchell, A., Yamasoba, T., Altschuler, R. A., & Miller, J. M. (2000a). Differential protective effects of neurotrophins in the attenuation of noise-induced hair cell loss. *Hearing Research*, 146(1–2), 134–142.

Shoji, F., Yamasoba, T., Magal, E., Dolan, D. F., Altschuler, R. A., & Miller, J. M. (2000b). Glial cell line-derived neurotrophic factor has a dose dependent influence on noise-induced hearing loss in the guinea pig cochlea. *Hearing Research*, 142(1–2), 41–55.

Spoendlin, H. (1984). Factors inducing retrograde degeneration of the cochlear nerve. *The Annals of Otology, Rhinology & Laryngology*, 112(Supplement), 76–82.

Spoendlin, H., & Schrott, A. (1990). Quantitative evaluation of the human cochlear nerve. *Acta Oto-Laryngologica*, 470(Supplementum), 61–69; discussion 69–70.

Staecker, H., Liu, W., Hartnick, C., Lefebvre, P., Malgrange, B., Moonen, G., & Van de Water, T. R. (1995). NT-3 combined with CNTF promotes survival of neurons in modiolus-spiral ganglion explants. *NeuroReport*, 6(11), 1533–1537.

Staecker, H., Kopke, R., Malgrange, B., Lefebvre, P., & Van de Water, T. R. (1996). NT-3 and/or BDNF therapy prevents loss of auditory neurons following loss of hair cells. *NeuroReport*, 7(4), 889–894.

Sugahara, K., Shimogori, H., & Yamashita, H. (2001). The role of acidic fibroblast growth factor in recovery of acoustic trauma. *NeuroReport*, 12(15), 3299–3302.

Takahashi, K., & Yamanaka, S. (2006). Induction of pluripotent stem cells from mouse embryonic and adult fibroblast cultures by defined factors. *Cell*, 126(4), 663–676.

Thorne, M., Salt, A. N., DeMott, J. E., Henson, M. M., Henson, O. W., Jr., & Gewalt, S. L. (1999). Cochlear fluid space dimensions for six species derived from reconstructions of three-dimensional magnetic resonance images. *The Laryngoscope*, 109(10), 1661–1668.

Tilstra, J., Rehman, K. K., Hennon, T., Plevy, S. E., Clemens, P., & Robbins, P. D. (2007). Protein transduction: Identification, characterization and optimization. *Biochemical Society Transactions*, 35(Pt 4), 811–815.

Ulfendahl, M. (2007). Tissue transplantation into the inner ear. In A. Martini, D. Stephens, & A. P. Read (Eds.), *Genes, Hearing and Deafness*. London: Martin Dunitz & Parthenon.

Ulfendahl, M., Hu, Z., Olivius, P., Duan, M., & Wei, D. (2007). A cell therapy approach to substitute neural elements in the inner ear. *Physiology & Behavior*, 92(1–2), 75–79.

Van de Water, T. R., Dinh, C. T., Vivero, R., Hoosien, G., Eshraghi, A. A., & Balkany, T. J. (2010). Mechanisms of hearing loss from trauma and inflammation: Otoprotective therapies from the laboratory to the clinic. *Acta Oto-Laryngologica*, 130(3),

Webster, M., & Webster, D. B. (1981). Spiral ganglion neuron loss following organ of Corti loss: A quantitative study. *Brain Research*, 212(1), 17–30.

Winkler, C., Kirik, D., & Bjorklund, A. (2005). Cell transplantation in Parkinson's disease: How can we make it work? *Trends in Neurosciences*, 28(2), 86–92.

Woodson, E. A., Reiss, L. A., Turner, C. W., Gfeller, K., & Gantz, B. J. (2010). The hybrid cochlear implant: A review. *Advances in Oto-Rhino-Laryngology*, 67, 125–134.

Yagi, M., Magal, E., Sheng, Z., Ang, K. A., & Raphael, Y. (1999). Hair cell protection from aminoglycoside ototoxicity by adenovirus-mediated overexpression of glial cell line-derived neurotrophic factor. *Human Gene Therapy*, 10(5), 813–823.

Yamagata, T., Miller, J. M., Ulfendahl, M., Olivius, N. P., Altschuler, R. A., Pyykko, I., & Bredberg, G. (2004). Delayed neurotrophic treatment preserves nerve survival and electrophysiological responsiveness in neomycin-deafened guinea pigs. *The Journal of Neuroscience Research*, 78(1), 75–86.

Yamashita, D., Jiang, H. Y., Schacht, J., & Miller, J. M. (2004). Delayed production of free radicals following noise exposure. *Brain Research*, 1019(1–2), 201–209.

Yamashita, D., Jiang, H. Y., Le Prell, C. G., Schacht, J., & Miller, J. M. (2005). Post-exposure treatment attenuates noise-induced hearing loss. *Neuroscience*, 134(2), 633–642.

Yamasoba, T., Nuttall, A. L., Harris, C., Raphael, Y., & Miller, J. M. (1998). Role of glutathione in protection against noise-induced hearing loss. *Brain Research*, 784(1–2), 82–90.

Yamasoba, T., Schacht, J., Shoji, F., & Miller, J. M. (1999). Attenuation of cochlear damage from noise trauma by an iron chelator, a free radical scavenger and glial cell line-derived neurotrophic factor in vivo. *Brain Research*, 815(2), 317–325.

Yamasoba, T., Altschuler, R. A., Raphael, Y., Miller, A. L., Shoji, F., & Miller, J. M. (2001). Absence of hair cell protection by exogenous FGF-1 and FGF-2 delivered to guinea pig cochlea in vivo. In D. Henderson, D. Prasher, R. Kopke, & R. Salvi (Eds.), *Noise Induced Hearing Loss: Basic mechanisms, prevention and control* (pp. 73–86). London: Noise in Network Publications.

Yamasoba, T., Pourbakht, A., Sakamoto, T., & Suzuki, M. (2005). Ebselen prevents noise-induced excitotoxicity and temporary threshold shift. *Neurosci Letters*, 380(3), 234–238.

Yeum, K. J., Beretta, G., Krinsky, N. I., Russell, R. M., & Aldini, G. (2009). Synergistic interactions of antioxidant nutrients in a biological model system. *Nutrition*, 25(7–8), 839–846.

Ylikoski, J., Pirvola, U., Moshnyakov, M., Palgi, J., Arumae, U., & Saarma, M. (1993). Expression patterns of neurotrophin and their receptor mRNAs in the rat inner ear. *Hearing Research*, 65(1–2), 69–78.

Ylikoski, J., Pirvola, U., Virkkala, J., Suvanto, P., Liang, X. Q., Magal, E., & Saarma, M. (1998). Guinea pig auditory neurons are protected by glial cell line-derived growth factor from degeneration after noise trauma. *Hearing Research*, 124(1–2), 17–26.

Yoshida, N., Kristiansen, A., & Liberman, M. C. (1999). Heat stress and protection from permanent acoustic injury in mice. *The Journal of Neuroscience*, 19(22), 10116–10124.

Zhai, S. Q., Wang, D. J., Wang, J. L., Han, D. Y., & Yang, W. Y. (2004). Basic fibroblast growth factor protects auditory neurons and hair cells from glutamate neurotoxicity and noise exposure. *Acta Oto-Laryngologica*, 124(2), 124–129.

Index

C.G. Le Prell et al. (eds.), *Noise-Induced Hearing Loss: Scientific Advances*,
Springer Handbook of Auditory Research 40, DOI 10.1007/978-1-4419-9523-0,